Organic Electronics

Edited by
Hagen Klauk

Related Titles

Müllen, K., Scherf, U. (Eds.)

Organic Light Emitting Devices

Synthesis, Properties and Applications

2006, ISBN 3-527-31218-8

Baltes, H., Brand, O., Fedder, G. K., Hierold, C., Korvink, J. G., Tabata, O. (Eds.)

Enabling Technologies for MEMS and Nanodevices

Advanced Micro and Nanosystems

2004, ISBN 3-527-30746-X

Köhler, M., Fritzsche, W.

Nanotechnology

An Introduction to Nanostructuring Techniques

2004, ISBN 3-527-30750-8

Roth, S., Carroll, D.

One-Dimensional Metals

Conjugated Polymers, Organic Crystals, Carbon Nanotubes

2004, ISBN 3-527-30749-4

Pearce, T. C., Schiffman, S. S., Nagle, H. T., Gardner, J. W. (Eds.)

Handbook of Machine Olfaction

Electronic Nose Technology

2003, ISBN 3-527-30358-8

Komiyama, M., Takeuchi, T., Mukawa, T., Asanuma, H.

Molecular Imprinting

From Fundamentals to Applications

2003, ISBN 3-527-30569-6

Organic Electronics

Materials, Manufacturing and Applications

Edited by
Hagen Klauk

WILEY-VCH

WILEY-VCH Verlag GmbH & Co. KGaA

The Editor

Dr. Hagen Klauk
Max Planck Institute for Solid State Research
Heisenbergstr. 1
70569 Stuttgart
Germany

1 st Edition 2006

Library of Congress Card No.: applied for
British Library Cataloguing-in-Publication Data
A catalogue record for this book is available from the British Library.

Bibliographic information published by the Deutsche Nationalbibliothek
The Deutsche Nationalbibliothek lists this publication in the Deutsche Nationalbibliografie; detailed bibliographic data are available in the Internet at http://dnb.d-nb.de.

Typesetting Asco Typesetter, Hong Kong

Cover Design Grafik-Design Schulz, Fußgönheim

ISBN-13 978-3-527-31264-1

Printed and bound in Great Britain by
CPI Antony Rowe, Chippenham and Eastbourne

Contents

Organic Electronics, Materials, Manufacturing and Applications. Edited by Hagen Klauk

ISBN: 3-527-31264-1

Preface

This book is devoted to the technology of organic transistors. First reported twenty years ago, organic transistors quickly became a focus of intense research and development not only in academia, but in a rapidly increasing number of industrial laboratories. In fact, the initial reports on organic transistors originated from Japanese corporate research groups, and the first integrated circuits based on organic transistors were developed by Philips. Meanwhile, an astonishing number of university groups and non-corporate research centers have engaged in organic electronics research and have made critical contributions in many important areas such as device physics, materials chemistry, process optimization, and novel applications aspects.

For this volume, 42 authors representing 18 research groups in France, the United States, Canada, Germany, Great Britain, Austria, The Netherlands, and Japan have pooled their knowledge, experience and insights to provide a fascinating look at today's organic electronics landscape and project its future. The content of the book displays a striking balance between contributions from industry and academia, which evolved naturally in the early stages of the project. The significant number of crucial contributions from industry stands as a testament both to the progress made in developing the technology in corporate environments as well as to the continued commercial interest, while the remarkable body of work from universities and non-corporate centers is critically important to illuminate some of the fundamental problems that remain to be solved and to accelerate the advancement of a broad range of novel routes that may prove vital to the ultimate success of organic electronics.

Organic Electronics, Materials, Manufacturing and Applications. Edited by Hagen Klauk

ISBN: 3-527-31264-1

As a first-time editor I am profoundly indebted to the tremendous support provided by all of the individuals involved in this project. First and foremost I want to acknowledge and thank the authors who have given so generously of their time and expertise and with their attention and responsiveness have made the editorial task both a great privilege and a smooth ride. Second I would like to express my gratitude to Bettina Bems, Martin Ottmar, Nele Denzau and Claudia Nussbeck at Wiley-VCH for giving me the opportunity to organize this book, for all the invaluable help in the editorial process, and for the wonderful experience of working with them. Finally, I want to thank all the readers for their interest in this book and hope that it is of good scientific value to them.

Stuttgart, Spring 2006 *Hagen Klauk*

Author List

John E. Anthony
Department of Chemistry
University of Kentucky
Lexington, KY 40506-0055
USA

Ana Claudia Arias
Electronic Materials and Devices Laboratory
Palo Alto Research Center
3333 Coyote Hill Road
Palo Alto, CA 94304
USA

Martin Bergsmann
Hueck Folien GmbH
Gewerbepark 30
4342 Baumgartenberg
Austria

Graciela Blanchet
DuPont Central Sciences & Engineering
Experimental Station E356/284
PO Box 80356
Wilmington, DE 19880
USA

Eugenio Cantatore
Philips Research Laboratories
Prof. Holstlaan 4
5656 AA Eindhoven
The Netherlands

Michael L. Chabinyc
Electronic Materials and Devices Laboratory
Palo Alto Research Center
3333 Coyote Hill Road
Palo Alto, CA 94304
USA

Jürgen H. Daniel
Electronic Materials and Devices Laboratory
Palo Alto Research Center
3333 Coyote Hill Road
Palo Alto, CA 94304
USA

Gerwin H. Gelinck
Polymer Vision
Philips Technology Incubator
High Tech Campus Eindhoven 48
5656 AE Eindhoven
The Netherlands

Marcus Halik
Department of Polymer Materials
Institute of Material Science
Friedrich-Alexander-Universität
Erlangen-Nuremberg
Martensstraße 7
91058 Erlangen
Germany

Michael Heuken
AIXTRON AG
Kackertstr. 15–17
52072 Aachen
Germany

Gilles Horowitz
ITODYS
CNRS UMR 7086
Université Denis-Diderot (Paris 7)
1 rue Guy de la Brosse
75005 Paris
France

Jia Huang
Department of Materials Science and Engineering
Johns Hopkins University
3400 North Charles Street, 102 Maryland Hall
Baltimore, MD 21218
USA

H. Edzer A. Huitema
Polymer Vision
Philips Technology Incubator
High Tech Campus Eindhoven 48
5656 AE Eindhoven
The Netherlands

Howard E. Katz
Department of Materials Science and Engineering
Johns Hopkins University
3400 North Charles Street, 102 Maryland Hall
Baltimore, MD 21218
USA

Tommie Kelley
3M Advanced Optical Solution Laboratory
260-5B-09
St. Paul, MN 55144-1000
USA

Hee Hyun Lee
DuPont Central Research & Development
Experimental Section
PO Box 80356
Wilmington, DE 19880
USA
and
Department of Materials Science and Engineering
University of Illinois
1304 West Green Street
Urbana, IL 61801
USA

Dago M. de Leeuw
Philips Research Laboratories
Prof. Holstlaan 4
5656 AA Eindhoven
The Netherlands

Yuning Li
Materials Design and Integration Laboratory
Xerox Research Centre of Canada
2660 Speakman Drive
Mississauga, OT L5K 2L1
Canada

Pieter J. G. van Lieshout
Polymer Vision
Philips Technology Incubator
High Tech Campus Eindhoven 48
5656 AE Eindhoven
The Netherlands

René Lujan
Electronic Materials and Devices Laboratory
Palo Alto Research Center
3333 Coyote Hill Road
Palo Alto, CA 94304
USA

William A. MacDonald
DuPont Teijin Films
PO Box 2002
Wilton
Middlesbrough, TS90 8JF
UK

Alwin W. Marsman
Philips IP&S
Prof. Holstlaan 6
5656 AA Eindhoven
The Netherlands

Eduard J. Meijer
Philips Research Laboratories
Prof. Holstlaan 4
5656 AA Eindhoven
The Netherlands

Nico Meyer
AIXTRON AG
Kackertstr. 15–17
52072 Aachen
Germany

Shelby F. Nelson
Kodak Research Labs
Eastman Kodak Company
1999 Lake Avenue
Rochester, NY 14650-2102
USA

Beng S. Ong
Materials Design & Integration Laboratory
Xerox Research Centre of Canada
2660 Speakman Drive
Mississauga, OT L5K 2L1
Canada

Catherine Ramsdale
Plastic Logic Ltd
34 Cambridge Science Park, Milton Road
Cambridge, CB4 OFX
UK

Steven E. Ready
Electronic Materials and Devices Laboratory
Palo Alto Research Center
3333 Coyote Hill Road
Palo Alto, CA 94304
USA

John Rogers
Department of Materials Science and Engineering
University of Illinois
1304 West Green Street
Urbana, IL 61801
USA

Takayasu Sakurai
Center for Collaborative Research
University of Tokyo
4-6-1 Komaba, Neguro-ku
Tokyo 153-8505
Japan

Alberto Salleo
Department of Materials Science and Engineering
Stanford University
Stanford, Ca 94305-2205
USA

Christoph Sele
Cavendish Laboratory
University of Cambridge
J. J. Thomson Avenue
Cambridge, CB3 OHE
UK

Henning Sirringhaus
Cavendish Laboratory
University of Cambridge
J. J. Thomson Avenue
Cambridge, CB3 OHE
UK
and
Plastic Logic Ltd
34 Cambridge Science Park, Milton Road
Cambridge, CB4 OFX
UK

Takao Someya
Quantum-Phase Electronics Center
School of Engineering
University of Tokyo
7-3-1 Hongo, Bunkyo-ku
Tokyo 113-8656
Japan

Carl J. Stonley
18 St. Leonards Drive
Wollaton
Nottingham, NG8 2BB
UK

Fred J. Touwslager
Polymer Vision
Philips Technology Incubator
High Tech Campus Eindhoven 48
5656 AE Eindhoven
The Netherlands

Roland Treutlein
Hueck Folien GmbH & Co. KG
PO Box 1758
92607 Weiden
Germany

Erik van Veenendaal
Polymer Vision
Philips Technology Incubator
High Tech Campus Eindhoven 48
5656 AE Eindhoven
The Netherlands

Timothy von Werne
Plastic Logic Ltd
34 Cambridge Science Park, Milton Road
Cambridge, CB4 OFX
UK

William Wong
Electronic Materials and Devices Laboratory
Palo Alto Research Center
3333 Coyote Hill Road
Palo Alto, CA 94304
USA

Yiliang Wu
Materials Design & Integration Laboratory
Xerox Research Centre of Canada
Mississauga, OT L5K 2L1
Canada

Lisong Zhou
Electrical Engineering Department
Pennsylvania State University
121 Electrical Engineering East
University Park, PA 16802
USA

I
Introduction

1 Organic Transistors

Gilles Horowitz

1.1 Introduction

Although the beginning of the electronics age was marked by Karl Braun's cathode ray tube (1897) and Ambrose Fleming's vacuum rectifier (1904), it was actually launched by Lee de Forest's vacuum-tube "triode" (1906); by including a "grid" between the anode and the cathode, the triode transformed the rectifier into an amplifier, thus making radio communications and long-distance telephone a reality. The vacuum triode had its limitations, however – it was fragile, rather slow, difficult to miniaturize, consumed too much energy and produced too much heat. The idea of replacing the triode with a solid-state device offering an alternative to the thermionic principle can be traced back to the mid-1920s. In October 1926, Julius Edgar Lilienfeld filled a patent describing an "apparatus for controlling the flow of an electric current between two terminals of an electronically conducting solid by establishing a third potential between said terminals" [1]. He probably never got his device to work, and his patent went into obscurity. It was not until thirty years later that this early concept could be successfully demonstrated. This was *not* with the celebrated Bardeen and Brattain's "point-contact" transistor (1947), nor with Shockley's bipolar transistor (1948) – both devices were based on different principles. Actually, nearly fifteen more years of material technology research were needed to finalize the silicon–silicon dioxide metal-oxide-semiconductor field-effect transistor (MOSFET) [2]. Today, MOSFETs dominate our environment; there are millions of them in the processors used in personal computers, cellular phones, and many other microelectronic devices. The success of MOSFETs actually rests on a continuous improvement in the handling of one semi-conducting material, silicon.

Besides their numerous technological applications, FETs have also been used as tools for studying charge transport in solid materials; this is because the device gives direct access to charge-carrier mobility. A celebrated example of such a concept is with hydrogenated amorphous silicon (a-Si:H). For this, an alternative architecture was employed, the thin-film transistor (TFT) [3], which differs from the MOSFET in that the conducting channel is induced in the accumulation regime

Organic Electronics, Materials, Manufacturing and Applications. Edited by Hagen Klauk

ISBN: 3-527-31264-1

rather than through the formation of an inversion layer. The first a-Si:H TFTs were actually designed to measure the mobility of the material, which was at that time difficult to access by other techniques [4, 5]. It was only later that the technological importance of the device was recognized in applications in which large area is required and where single crystalline silicon can no longer be used. Today, a-Si:H TFTs play a crucial role in active-matrix liquid-crystal displays (AM-LCD).

Organic semiconductors have been identified as early as the late 1940 [6]. Because they are low mobility materials, the TFT structure is well suited to these solids. Apart from a handful of isolated preliminary reports [7–9], however, work on organic thin-film transistors only emerged in the late 1980s on both polymers [10, 11] and small molecules [12, 13]. Because of the poor performance of these initial devices, interest in organic thin film transistors (OTFTs) remained limited to a small number of academic groups for nearly ten more years. During that period, much research effort was devoted to improving the charge-carrier mobility; several review papers can be consulted to learn of this quest for better materials and device structure [14–18]. It is only when the mobility of organic semiconductors approached, and even surpassed, that of amorphous silicon [19] that several industrial groups decided to embark into research programs on OTFTs.

The availability of organic semiconductor devices may open the way to completely new set-ups, fabrication processes, and applications. Thus, one can envisage processing of organic materials by printing, which enables high-volume, low-cost production. New products include radio-frequency identification (RFID) tags [20], that might replace the optical bar code found on nearly all consumer goods today, single-use electronics, low-cost sensors, and flexible displays.

The purpose of this introductory chapter is to give a general overview of the topic. Emphasis will be made on recent leading advances in terms of materials and device fabrication, together with the development of models that help understanding of what controls the operating mode of the device, thus opening ways at improving its performance. It is worth pointing out that although organic semiconductors have been identified for more than half a century, the field of organic electronics is still in its infancy. Accordingly, a general consensus has not yet been achieved on several basic points. The reader should therefore keep in mind that on many points, the opinion reported here is that of the author, and may differ from what is found elsewhere in this book.

1.2 Overview of the Organic Thin-film Transistor

1.2.1 Are Organic "Semiconductors" Real Semiconductors?

Asking this question may sound provocative in a book devoted to "organic electronics". Actually, one would first ask the question "what is a semiconductor?" In most current dictionaries, a semiconductor is still defined as "a nonmetallic solid that has electrical conductivity between that of a conductor and an insulator". According to several sources, the first occurrence of the word ("Halbleiter" in German) dates back to 1911 [21], at a time when electrical conduction in solids was not fully understood. Classical physics served well at accounting for electrical conduction in metals, but was contradicted by the "anomalous" behavior of various non-metals, among which were silicon and a variety of binary compounds, for example oxides and sulfides. The "anomalous" phenomena associated with these materials, for which the word "semiconductor" was invented, included the positive variation of conductivity with temperature, photoconductivity, rectification, and photovoltaic effect. The puzzle was only resolved with the advent of quantum mechanics and the subsequent development of the band theory of solids [22]. The following definition, more in accordance with our current state of knowledge, can be found in a "modern" encyclopedia [23]: "A semiconductor is a material that is an insulator at very low temperature, but which has a sizable electrical conductivity at room temperature. The distinction between a semiconductor and an insulator is not very well-defined, but roughly, a semiconductor is an insulator with a band gap small enough that its conduction band is appreciably thermally populated at room temperature." "Semi-insulators" would certainly be a more appropriate designation of these materials, but it is too late for such a correction.

The above definition is actually that for *intrinsic* semiconductors. What make semiconductors so useful in electronics, however, is that their electronic properties can be altered in a controllable manner by adding tiny amounts of an advisedly chosen impurity. This is the well-known process of doping, which is related to the notion of *extrinsic* semiconductors.

The difference between intrinsic and extrinsic semiconductors is illustrated in Fig. 1.1, which represents the energy diagram of metals, insulators, and semi-

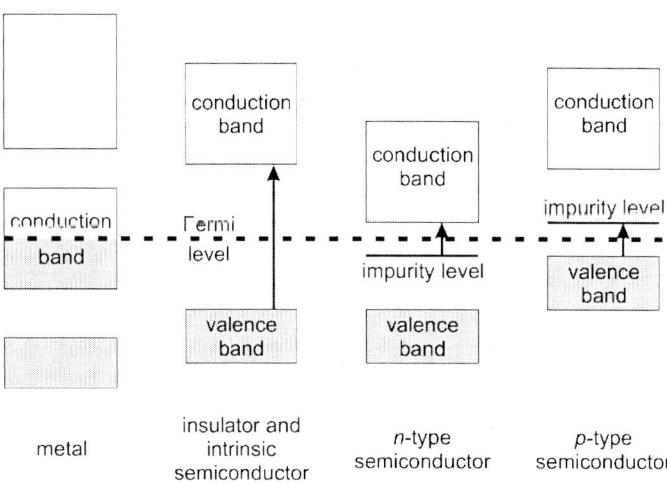

Fig. 1.1. Energy diagram of a metal, an insulator, and an extrinsic semiconductor.

conductors, as pictured in the framework of the now well-accepted band theory of solids. The theory delineates a clear distinction between a metal, which has a partially-filled conduction band, and an insulator characterized by a filled valence band and an empty conduction band. An insulator is, however, only perfectly insulating at $T = 0$ K. As soon as its temperature is elevated, electrons can be thermally excited from the valence band to the conduction band. Because electrical conduction can occur in partially filled bands, both bands contribute to conduction. Because thermal energy is low ($kT = 25$ meV at room temperature), this thermally activated conduction can only be observed in a low energy-gap insulator, also termed intrinsic semiconductors.

Doping a semiconductor results in induction of localized energy levels close to the conduction (n-type doping) or valence (p-type doping) band edge. Accordingly, the energy required to promote an electron (a hole) in the conduction (valence) band is substantially lowered, to a level comparable with thermal energy. The temperature dependence of the conductivity of a typical doped (extrinsic) semiconductor is shown in Fig. 1.2. The curve contains three separated domains. At high temperatures the intrinsic domain is characterized by thermally activated behavior. The intermediate domain is the so-called saturation (or exhaustion) regime in which the conductivity is practically temperature-independent. Finally, at low temperature the carriers are frozen. In terms of electronic devices, the only domain of interest is the exhaustion regime, where the density of charge-carriers practically equals that of the dopant.

At this stage, it is essential to note that doping requires high purity. Actually, with the notable exception of electronic grade silicon and a few other inorganic compounds used in microelectronics, most semiconductors contain approximately equal amounts of n and p-type doping impurities; such materials are termed *compensated* and behave much like intrinsic semiconductors. Almost all organic semiconductors belong to that category. In particular, the effect of intentional doping in organic semiconductors only appears for large densities of dopant (a few percent,

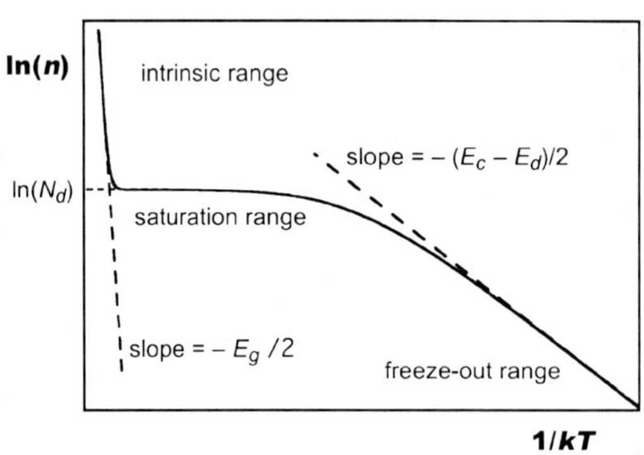

Fig. 1.2. Arrhenius plot of the temperature-dependent conductivity of an extrinsic semiconductor.

compared with ppm, or even less, in silicon). At this level, "doped" organic semiconductors are conductors rather than semiconductors, and are useless for making electronic devices.

In conclusion, we can state that organic semiconductors are closer to insulators than to semiconductors. Despite several claims, no definitive evidence has been obtained for the possibility of "microelectronic grade" doping of organic semiconductors, mainly because the purity of these materials is still too low. Their potential in constituting the basis of electronic devices is, however, now well established. This was made possible by using an alternative means of inducing charge-carriers, namely injection from electrodes. As we shall see in the following discussion, this new concept implies that the very notion of n and p-type must be redefined.

1.2.2
Thin-film Transistor Architecture

The structures of OTFTs have several variants. Roughly speaking, a TFT is made of three parts – an insulator, a thin semiconducting layer, and three electrodes. Two of the electrodes, the source and the drain, are in direct contact with the semiconductor; the third, the gate, is isolated from the semiconductor by the insulator. The structure of the device is dictated not only by its operating mode, but also by issues arising from its fabrication. The basic fabrication scheme consists of piling up thin films of the different elements. Because most organic semiconductors are fragile materials, the deposition of organic semiconductors on the insulator is much easier than the converse. So the large majority of current OTFTs are built according to the bottom-gate architecture, which in turn declines in two alternatives depicted in Fig. 1.3, top contact (TC) and bottom contact (BC). Each of these structures has its advantages and drawbacks. In the BC structure, contacts are deposited on the insulator; if the latter is an inorganic oxide for example silicon oxide, the electrodes can be patterned by means of microlithographic techniques. This is not possible with TC architecture; with this architecture contacts are deposited through shadow masks, with substantial loss of resolution. On the other hand, contact resistance has been reported to be lower in TC than BC.

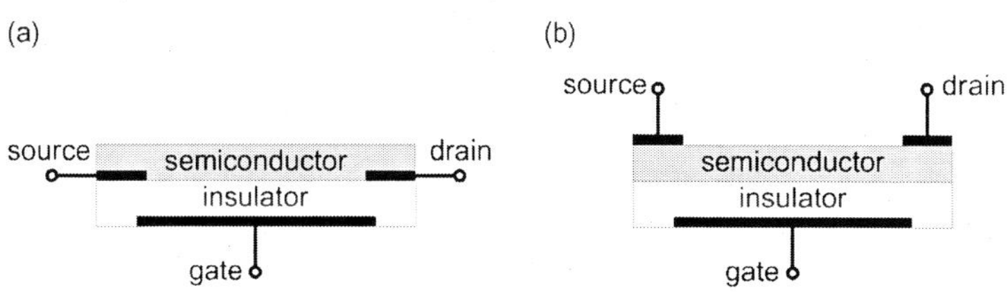

Fig. 1.3. Schematic view of the structure of organic thin film transistors. Both structures are top-gated. (a): Bottom contact (BC); (b): Top contact (TC).

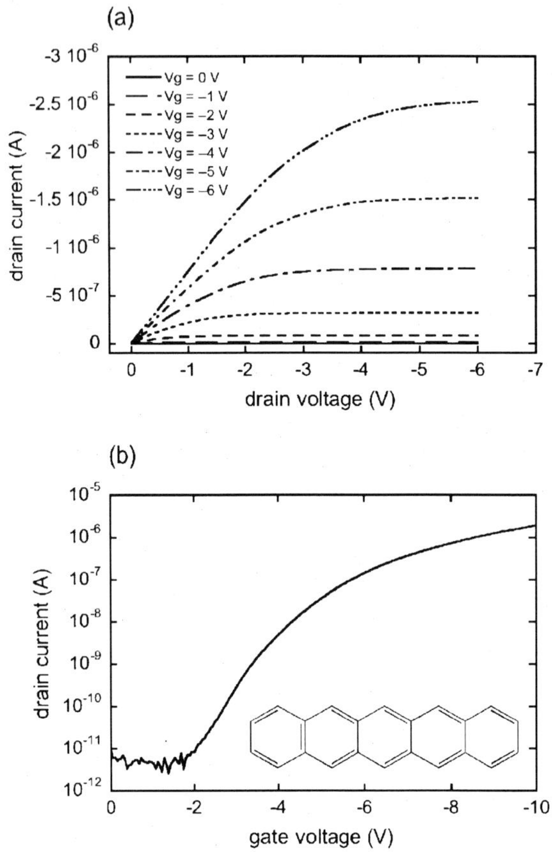

Fig. 1.4. Output (a) and transfer (b) characteristics of a typical OTFT. The inset shows the molecular structure of pentacene, which serves as semiconductor in the device.

1.2.3
Operating Mode

To demonstrate the operating mode of the OTFT, a typical set of current–voltage characteristics are shown in Fig. 1.4. These curves were measured on a device made of pentacene, the chemical structure of which appears in the inset, with gold source and drain electrodes. The energy scheme in Fig. 1.5 indicates the respective positions of the Fermi level of gold and the frontier orbitals (highest occupied molecular orbital, HOMO, and lowest unoccupied molecular orbital, LUMO) of pentacene. Data for pentacene were taken from Ref. [24]. Throughout the following discussion the source serves as the reference (grounded) electrode.

When a positive voltage is applied to the gate, negative charges are induced at

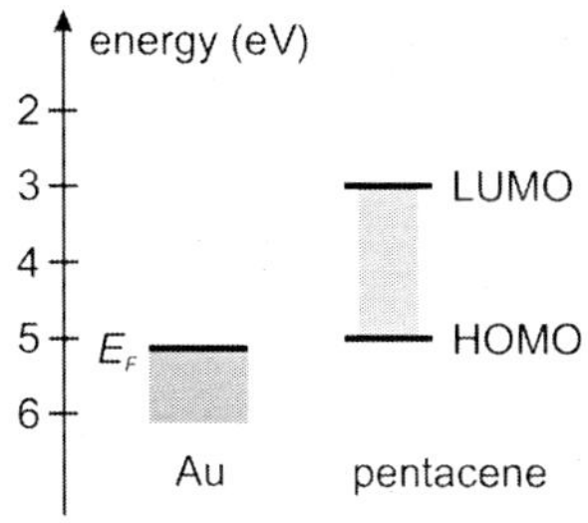

Fig. 1.5. Energy scheme of the gold-pentacene interface [24].

the source. As can be seen in Fig. 1.5, the LUMO level of pentacene is quite far away from the Fermi level of gold, so there is a substantial energy barrier for electrons and electron injection is very unlikely. Accordingly, no current passes through the pentacene layer and the small current observed in Fig. 1.4(a) essentially comes from leaks through the insulator. In contrast, when the gate voltage is reversed to negative, holes are easily injected because the Fermi level is close to the HOMO level and the barrier height is low. A conducting channel forms at the insulator–semiconductor interface and charge-carriers can be driven from source to drain by applying a second, independent, bias to the drain. Because holes are more easily injected than electrons, pentacene is said to be p-type. Note that this concept differs from that of doping in conventional semiconductors. Symmetrically, an organic semiconductor is said to be n-type when electron injection is easier than hole injection, which occurs when the LUMO is closer to the Fermi level than the HOMO. Note that in the terminology of organic light-emitting diodes, these two classes of materials are often designated "hole transport" (or hole-injecting) and "electron transport" (or electron-injecting) materials.

Basically, the thin-film transistor operates like a capacitor; when a voltage is applied to the gate an equal (but of opposite sign) charge is induced at both sides of the insulator. On the semiconductor side this charge forms a conducting channel, if the charge-carriers can be injected into the semiconducting material; because the conductance of the channel is proportional to the charge, it is also proportional to the gate voltage. At low drain voltage, the current follows Ohm's law; it is therefore proportional to both the gate and drain voltages. As the drain voltage increases, the voltage drop at drain decreases to a point at which it falls to zero. At this point, which occurs when the drain voltage approaches the gate voltage, there is a pinch off of the channel, and the channel current becomes independent of the drain bias. This regime is called the saturation regime. The gradual transition from the linear to the saturation regime is clearly shown in Fig. 1.4(a). The curve in Fig. 1.4(b) is the transfer characteristic; it shows than the device has a high on–off ratio, which makes it very useful in logic circuits.

The linear and saturation current can be quantified by use of Eqs. (1) and (2), which are derived on the basis of two assumptions:

1. the electric field along the channel is much lower than that across it. This is the so-called gradual channel approximation, which is valid when the distance between source and drain is much larger than the thickness of the insulator; and
2. the mobility, μ, is constant.

$$I_{\mathrm{Dlin}} = \frac{W}{L}\mu C_i (V_{\mathrm{G}} - V_{\mathrm{T}}) V_{\mathrm{D}} \tag{1}$$

$$I_{\mathrm{Dsat}} = \frac{W}{2L}\mu C_i (V_{\mathrm{G}} - V_{\mathrm{T}})^2 \tag{2}$$

where W and L are the channel width and length, C_i is the insulator capacitance per unit area, V_{G} and V_{D} are the gate and drain voltages, and V_{T} the threshold voltage that takes into account various potential drops through the gate–insulator–semiconductor structure. The limitations of using these over-simplified equations to estimate the mobility will be detailed below.

1.2.4
Thickness of the Channel

From the very beginning, two classes of material have been used in organic field-effect transistors – conjugated polymers and small molecules. The best performance has been obtained with materials belonging to the latter category. Conjugation means that the carbon backbone has alternating single and double bonds, which confers stiffness to the molecules, so they can be regarded as rigid rods. In the solid state, these rods tend to pack parallel to each other, thus forming layers, the width of which roughly equals the length of the molecule. The crystal structure of pentacene [25, 26], shown in Fig. 1.6, perfectly illustrates this arrangement. Typical values for the thickness of a monolayer range between 1.5 and 3 nm.

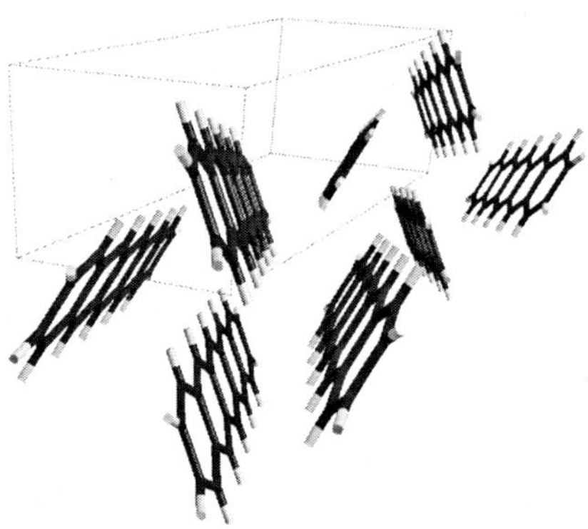

Fig. 1.6. Crystal structure of pentacene.

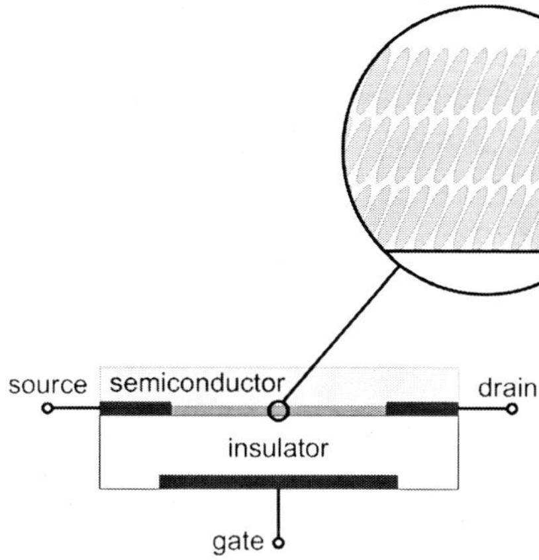

Fig. 1.7. Arrangement of the molecules in the conducting channel of an OTFT made of short conjugated molecules.

Because molecules located in the same layer are much closer to each other than those situated in different layers, charge transport is expected to be much more efficient in the direction along the layers than across them. This has largely been confirmed by X-ray diffraction measurements on sexithiophene-based devices [27, 28], which indicated that the highest performance is attained when molecules are standing upright on the insulator. Similar behavior was found for pentacene [29, 30]. This is illustrated in Fig. 1.7, in which the molecules are depicted as short rods.

Let us now turn to the thickness of the conducting channel. The concept of thickness is not that obvious, because the actual distribution of charge-carriers decreases continuously from the insulator–semiconductor interface to the semiconductor bulk, so one can more sensibly speak of an effective thickness. The distribution can be estimated by resolving Poisson's equation (Eq. 3):

$$\frac{\mathrm{d}^2 V}{\mathrm{d}x^2} = -\frac{\rho(x)}{\varepsilon_{\mathrm{C}}} \tag{3}$$

where V is the potential, x the direction perpendicular to the channel, ρ the density of charge, and ε_{S} the permittivity of the semiconductor. There is no analytical solution of Poisson's equation for the metal–insulator–semiconductor structure. An example of numerical solution can be found in Ref. [31]. A satisfactory approximate solution in the accumulation regime is given by Eq. (4):

$$n(x) = \frac{\rho(x)}{q} = \frac{(C_{\mathrm{i}} V_{\mathrm{G}})^2}{2kT\varepsilon_{\mathrm{S}}} \left(1 + \frac{x}{\sqrt{2}L_{\mathrm{D}}}\right)^{-2} \tag{4}$$

where $n(x)$ is the charge-carrier density and q the charge on the carriers. The Debye length, L_D, is given by:

$$L_D = \frac{\sqrt{2}kT\varepsilon_S}{qC_iV_G} \tag{5}$$

where k is Boltzmann's constant and T the temperature. The charge distribution was first derived by Mott and Gurney [32]. The Debye length can be viewed as a rough estimate of the thickness of the channel. Typical values of this "effective" thickness range between 0.1 and 1 nm, which is substantially lower than the thickness of a monolayer. For this reason, it is often stated that practically all the charge of the channel resides in the first monolayer next to the insulator–semiconductor interface [34].

At least two facts may temper this simple description, however. First, Eq. (4) indicates that the density of charge-carriers decreases with distance according to an inverse square law; this is a long-range law, which might imply that the density of charge remains non-negligible even far from the insulator–semiconductor interface. The second argument stems from the layer structure of the semiconducting film. Because the charge-carrier density in the conducting channel is less than one charge out of ten molecules, it is most likely that each charged molecule only bears one elemental charge (electron or hole). Quantum mechanical calculations have shown that in short conjugated molecules the additional charge almost uniformly spreads over the whole entity [33], which leads us to the conclusion that the actual distribution of charge-carriers in the film is not continuous; instead, it presents the staircase shape shown in Fig. 1.8, where the width of each step equals that of one monolayer.

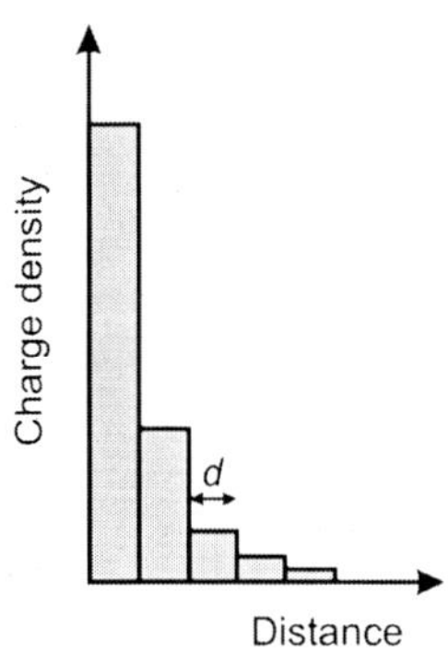

Fig. 1.8. Charge distribution across the conducting channel of an organic TFT. The width of each step corresponds to one monolayer.

Let d be the thickness of a monolayer and n the total number of layers (that is, the thickness of the film divided by d). The layers are numbered starting from the insulator–semiconductor interface. To estimate the density n_i (per unit area) of charge-carriers in the ith layer we apply Gauss's law to a cylinder of unit cross section limited by the boundaries between the ith layer and each of its neighboring layers. For a long channel device, the electric field **F** is perpendicular to the film, and we have:

$$F_{i-1} - F_i = -\frac{qn_i}{\varepsilon_S} \tag{6}$$

where F_i is the module of the electric field at the boundary between the ith and the $(i+1)$th layers. The variation of the electric field F and electrical potential V in the direction perpendicular to the layers is obtained by rewriting Poisson's equation (Eq. 3) as:

$$\frac{\mathrm{d}F}{\mathrm{d}x} = \frac{qn(x)}{\varepsilon_S} \tag{7}$$

where

$$F = -\frac{\mathrm{d}V}{\mathrm{d}x} \tag{8}$$

Because the density of charge is constant in each layer, the electric field varies linearly with distance between two boundaries, which implies, in turn, that the variation of the potential is quadratic within the same limits. Furthermore, we assume that charge transfer between adjacent layers is sufficiently efficient that the distribution of charge-carriers in the whole film is at thermodynamic equilibrium. Assuming Boltzmann's statistics holds, which is true as long as the charge density remains much lower than the density of molecules, this yields:

$$\frac{n_{i+1}}{n_i} = \exp\left[-\frac{q}{kT}(V_i - V_{i+1})\right] \tag{9}$$

After some manipulations, the following series of equations is obtained:

$$n_i = n_{i+1} \exp\left[\frac{dq^2}{kT\varepsilon_S}\left(\frac{n_{i+1}}{2} + \sum_{j=i+2}^{n} n_j\right)\right] \tag{10}$$

Although there is no analytical solution to Eqs. (10), a numerical calculation is easily performed by starting from a given density in the nth layer and cascading down to the first layer. The gate voltage is connected to the n_i values by:

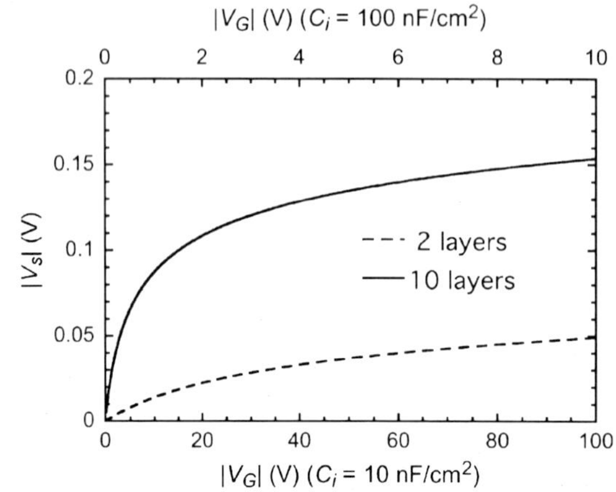

Fig. 1.9. Variation of the potential drop at the insulator–semiconductor interface as a function of gate voltage for two values of the insulator capacitance.

$$C_i(V_G - V_s) = q\sum_{1}^{n} n_i \tag{11}$$

where $V_s = V_1$, is the potential at the insulator–semiconductor interface, which is obtained by multiplying Eqs. (9) from $i = 1$ to $i = n$, with the boundary condition $V_{n+1} = 0$:

$$V_s = \frac{kT}{q}\ln\frac{n_1}{n_n} \tag{12}$$

The variation of the interface potential as a function of gate voltage is shown in Fig. 1.9. The value of the gate voltage is calculated for two values of the insulator capacitance, 10 and 100 nF cm^{-2}. In most practical cases the actual value lies between these numbers, so it can be stated that V_s can be neglected in Eq. (11).

Figure 1.10 shows the variation of the ratio of the density of charge-carriers in the first layer to the total density $n_{tot} = \sum n_i$ as a function of the total density of charge-carriers, which is connected to the gate voltage by Eq. (11). We note that the statement "all the induced charge resides in the first layer" is only true at high gate bias. It is worth remarking that at low bias the density of charge-carriers strongly depends on the thickness of the film (i.e. the number of layers). This can be understood by noting that the limit at $V_G \to 0$ is simply $n_i = n_{tot}/n$ for all values of i. We also note that, even at high gate voltages, a non-negligible part of the induced charge resides outside the first layer. This is worth mentioning because variation of the mobility by a factor of ten from sample to sample is not uncommon.

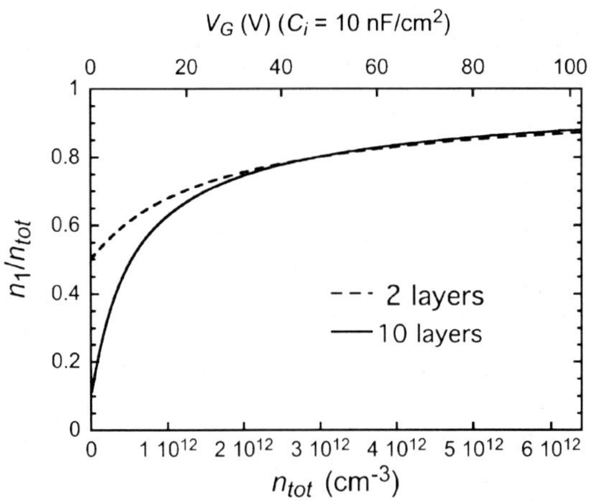

Fig. 1.10. Calculated ratio of the charge in the first layer to the total charge in the conducting channel as a function of gate voltage multiplied by insulator capacitance. The ratio is calculated for two-layer and ten-layer thin films.

1.3 Contact Resistance

Until recently, the issue of contact resistance was hardly mentioned in papers dealing with OTFTs because the performance of the devices was so low that the current flowing between source and drain was only limited by the resistance of the channel. With improvement of the charge-carrier mobility, this is no longer true; limitations by contact resistance are becoming increasingly crucial, and finding ways to reduce these limitations has become a key issue.

1.3.1 Contact Resistance Extraction

Access to contact resistance was first sought by modeling. Figure 1.11 shows equivalent circuits used for that purpose. Note that the bottom circuit includes head-to-toe diodes to account for non-linear contact resistance [35]. The model developed by Necliudov and coworkers also assumed a gate voltage-dependent mobility (this point will be discussed in more detail below).

The dependence is accounted for by a semi-empirical law depicted by Eq. (13):

$$\mu = K(V_G - V_T)^{\gamma} \tag{13}$$

where K and γ are empirical parameters. The width-normalized contact resistance (i.e. resistance multiplied by channel width) extracted from this model for both top-

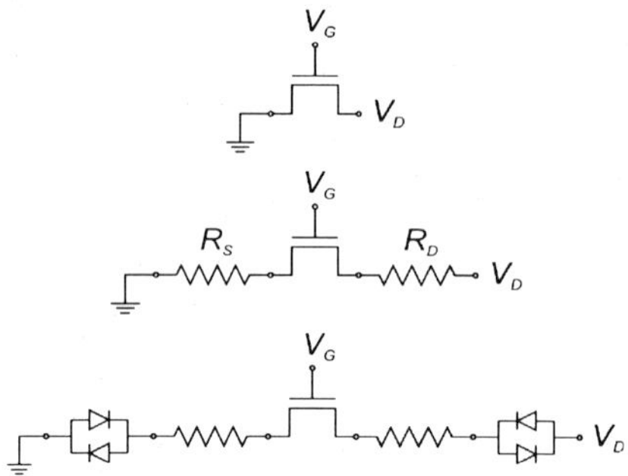

Fig. 1.11. Equivalent circuit of a TFT including contact resistance. The bottom circuit also comprises head-to-toe diodes to account for non-linearity in the contact resistance.

contact and bottom-contact architecture was of the order of $10^3\ \Omega$ cm. Data obtained for sexithiophene with TC structure [36] were approximately ten times larger. The extraction method consists of rewriting Eq. (1) by introducing an additional voltage drop $R_c I_D$ where R_c is the contact resistance. This is done by replacing V_D by $V_D - R_c I_D$ in Eq. (1). After some manipulation we obtain:

$$\frac{I_D}{V_D} = \left(\frac{1}{(W/L)C_i\mu(V_G - V_T)} + R_c \right)^{-1} \tag{14}$$

A more general approach was adopted by Street and coworkers [37]. To analyze the effect of contact resistance they add, at both ends of the channel, a small contact region of length d where there is a voltage drop V_c. The channel length is then reduced to $L - d$ and the voltage drop along the whole channel to $V_c - V_D$. In the gradual channel approximation, the drain current is given by Eq. (15) where $V(x)$ is the potential at a distance x from the source:

$$I_D = WC_i\mu[V_G - V_T - V(x)]\frac{dV}{dx} \tag{15}$$

Integrating Eq. (15) along the channel yields:

$$I_D = C_i\mu\frac{W}{L-d}\left\{(V_G - V_T)V_D - \frac{V_D^2}{2} - \left[(V_G - V_T)V_c - \frac{V_c^2}{2}\right]\right\} \tag{16}$$

For an ohmic contact resistance ($V_c = R_c I_D$) and $d \ll L$, Eq. (16) reduces to Eq. (14). While the two previous analyses [35, 36] assumed constant contact resistance and gate voltage-dependent mobility, Street and coworkers make the assumption that the mobility is constant and find a non-ohmic contact resistance that varies with gate voltage.

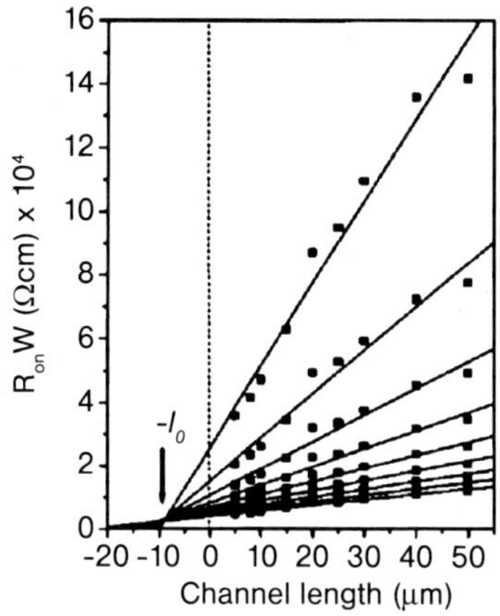

Fig. 1.12. Width-normalized resistance as a function of channel length at gate voltages ranging from −20 to −100 V [39].

At this stage, a technique that would enable independent access to the channel and contact resistances is needed. Such a feature is offered by the transfer line method (TLM) [38–41, 89], a method adapted from a classical technique use to estimate contact resistance, and first developed for the amorphous silicon thin-film transistor [42]. The method consists of measuring the channel resistance for different channel lengths. The measured resistance is actually the sum of the channel and contact resistances. As long as the measurement is performed in the linear regime (small drain voltage) the channel resistance is proportional to L (see Eq. 1) and the width-normalized ($R \times W$) total resistance is given by:

$$R \times W = \frac{L}{C_i \mu (V_G - V_T)} + R_c \times W \tag{17}$$

The contact resistance is extracted by plotting the width normalized resistance as a function of channel length. Extrapolation to zero length readily gives the contact resistance, while the slope of the curve can be used to extract device properties. The method is exemplified in Fig. 1.12 (taken from Ref. [39]). Each line corresponds to a given gate voltage. Figure 1.13 shows that the contact resistance is indeed gate voltage-dependent as assumed by Street [37]. It actually strongly decreases when the gate voltage increases. Extracting the mobility from the slope of the lines is not that easy, because this factor actually contains two parameters, mobility and threshold voltage, so a method of estimating the threshold voltage must first be found. It has, however, also been shown that the mobility is gate voltage-dependent [43, 44], even if the exact dependency cannot be undoubtedly determined from the TLM.

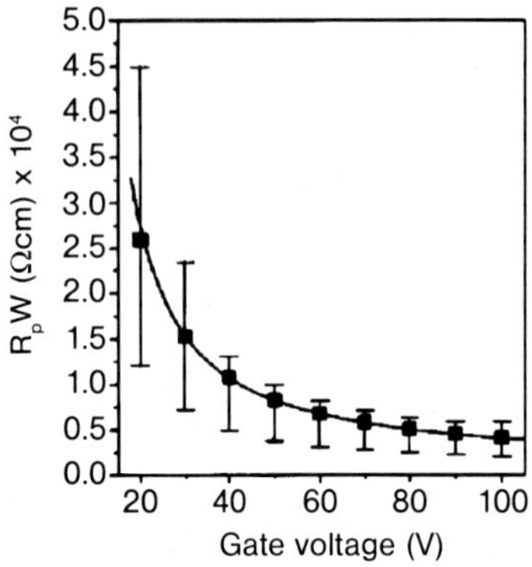

Fig. 1.13. Gate voltage-dependent contact resistance as deduced from the data in Fig. 1.11 [39].

Apart from problems in determining the mobility, the TLM has several other drawbacks. First, it requires measurements on different devices and it cannot be taken for granted that the channel and contact resistances are strictly similar for all of them, even if they are prepared during the same run. This is the reason why scattering occurs when plotting the data, as shown in Fig. 1.12. Next, the validity of Eq. (17) requires contact resistance not to depart from Ohm's law. In other words, the method cannot be used if contact resistance is non-linear. It must also be noted that, as the method requires measurements in the linear regime, that is, at low drain voltages, it is very sensitive to leaks through the insulator. Finally, the method cannot be used to make a distinction between the contact resistance at the source and that at the drain. This last point is important – theoretical modeling has shown that for ideal contacts all the ohmic drop should occur at the source electrode. As we shall see in the following discussion, this is not observed by use of other methods.

An alternative method to TLM is the four-point probe, which consists of introducing into the conducting channel two additional electrodes [45, 46]. The current remains the same all along the channel and the voltage drop between these two additional electrodes is not affected by the contact resistance, thus giving access to the true channel resistance. Moreover, as shown in Fig. 1.14, the contact resistance at each side of the channel can now be estimated independently.

An even more powerful technique makes use of an atomic force microscope (AFM) tip to probe the potential along the channel of the transistor [47, 48]. The technique and its results have been analyzed in detail by Bürgi et al. [49]. A variety of semiconductors and metals have been studied. The main features can be summarized as follows:

1. As expected, the contact resistance strongly depends on the nature of the electrode, e.g. its work function;
2. Contact resistance is gate-bias-dependent; it decreases substantially when gate

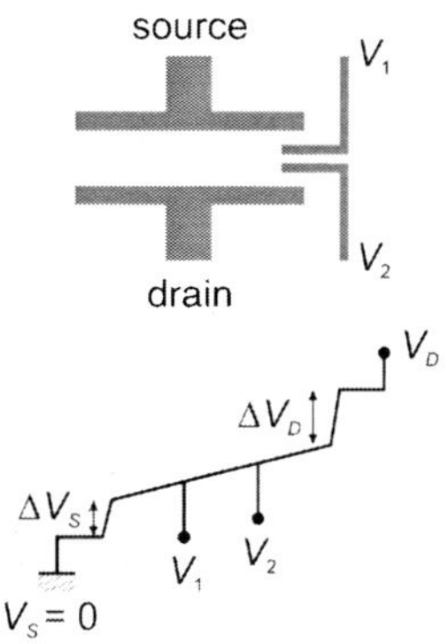

Fig. 1.14. Electrode pattern for a four-probe set up. The corresponding voltage profile along the conducting channel is shown in the bottom part of the figure.

bias is increased. This confirms what has been observed both with the TLM and by use of the four-point technique; and

3. Mobility is gate-bias-dependent.

Experiments were also conducted to study the effect of the nature of the contact on the respective part of the voltage drop at source and drain. This point will be dealt with in the next section.

1.3.2
Origin of Contact Resistance

The image most commonly used to describe source and drain contacts is that of a metal–semiconductor junction. According to the conventional Mott–Schottky (MS) model, contacts are expected to be ohmic when the work function of the metal is close to the HOMO or LUMO level of the semiconductor, depending on whether the semiconductor is p or n-type. If the reverse situation prevails, an energy barrier forms at the metal–semiconductor interface, leading to poor charge injection. From this standpoint, the Au/pentacene interface would be a good candidate as a low-resistance contact. In practice, the actual resistance is rather high. The mechanism of barrier formation at metal–organic semiconductor interfaces has been studied in great detail for organic light-emitting diodes (OLEDs), for which contact resistance is also a crucial issue. UV photoelectron spectroscopy (UPS) and inverse UPS have been used for precise determination of the energy levels at both side of the interface [50]. A typical result for the Au/pentacene interface is shown in Fig. 15, which clearly shows that the actual interface strongly deviates from the MS model. Instead, the interface has an additional "dipole" barrier, Δ, which shifts the HOMO level downward by more than 1 eV, hence increasing the barrier height by the same amount. The reason for this rather large interface dipole is that the electron density at a metal surface presents a tail that extends from the metal-free

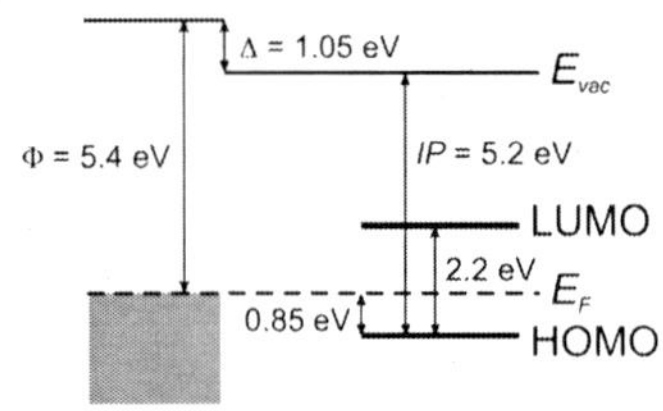

Fig. 1.15. Actual energy level diagram of the Au/pentacene interface. The position of the HOMO and the dipole barrier Δ are estimated from photoelectron spectroscopy [50].

surface into vacuum, thus forming a dipole pointing at the metal bulk. Molecules deposited on the metal tend to push back this tail, thus reducing the surface dipole and reducing the work function of the metal.

The above-described four-probe [45, 46] and scanning probe [47–49] techniques enable separate determination of the source and drain contact resistances. If contacts do behave as Schottky barriers, one would expect the voltage drop at source be substantially higher than that at drain. This is what is indeed observed with "bad" contacts. For "good" contacts, however, comparable drops are observed at both electrodes. A possible origin of this behavior has recently been suggested [51]. The model assumes the region immediately adjacent to the electrodes is made of organic material of quality different from that of the rest of the conducting channel, with very low mobility.

It is worth remarking that the contact resistance of the top contacts is usually lower than that of bottom contacts. The asymmetry of the organic-metal contact, depending on whether the organic film is deposited on the metal or the metal on the organic layer, has been studied both theoretically [52] and experimentally [53]. For example, combined UPS and XPS measurements have revealed signs of metal penetration and the formation of metal clusters after deposition of gold on top of a pentacene layer, leading to a substantial reduction of the interface barrier from 1 to 0.3 eV.

1.4 Charge Transport

Despite impressive progress in organic thin-film transistors during the past two or three decades, charge transport in organic materials remains a highly controversial topic. Basically, one can make a distinction between two families of charge-transport models. The first pertains to disordered materials, for example polymers. Models that belong to this family are based on hopping transport. An archetypal model of this group is that developed by Vissenberg and Matters [54], which assumes variable range hopping in an exponential distribution of traps. The model predicts a thermally activated mobility. It also predicts that mobility depends on the gate voltage according to a power law, a feature that is, indeed, observed in actual devices (Eq. 13), so this prediction has proved very fruitful in the analysis of current-voltage curves of OTFTs.

Dealing with well-organized molecular crystals seems much more difficult. On the basis of the inverse power law dependence of mobility on temperature (that is, mobility increases when temperature decreases) found in highly pure crystals from time-of-flight measurements [55], it is often stated that charge transport in these materials occurs via charge-carriers in delocalized states, as in conventional semiconductors. The statement does not withstand analysis, however. Except at very low temperatures, the corresponding mean free path does not exceed the intermolecular distance [56], which is not compatible with a diffusion-limited process. Although polarons have been invoked to resolve the discrepancy [56, 57], the theoretical problem is very intricate, because useful approximations used with inorganic materials, for example the one-electron limit, are no longer valid, so one must deal with huge numbers of atoms and molecules. Despite recent efforts [60–63], basic problems, for example the ultimate value of the mobility and its temperature dependence, have not been satisfactorily resolved.

An interesting aspect is the gate voltage-dependent mobility. Such dependence is predicted by the Vissenberg's model [54] and has indeed been observed in most devices made of polymers. Briefly, the dependence stems from the fact that as the gate voltage increases, injected charge-carriers tend to fill the traps, so trapping becomes less efficient and charge transport improves. Interestingly enough, similar gate-voltage dependence was reported for early devices made of small molecules. To account for that, a parent model has been developed that assumes delocalized transport limited by a distribution of traps near the band edge [64]. The model is derived from the multiple trapping and release (MTR) model developed for a-Si:H [65]. The model also predicts thermally activated mobility, which, again, was found in most early devices. It is worth pointing out that as the quality of the devices improves, both of these features (gate bias dependent and thermally activated mobility) tend to be encountered less, which tends to confirm they indeed originate from defects. This is particularly true for single-crystal devices, as will be dealt with below.

1.5 Fabrication Techniques

Organic semiconductors are so different from their inorganic analogs in terms of melting point, solubility, and mechanical properties that it is not conceivable that the fabrication techniques of conventional microelectronics would be appropriate for making organic devices. Nevertheless, these technologies are so powerful and well mastered that in the early days of OTFTs it seemed more convenient to take advantage of techniques such as thermal oxidation and photolithography, so most of the devices were actually derived from standard silicon technology. In most instances, organic transistors consisted of highly doped silicon wafers that served as the gate with thermally grown oxide as insulator and lithographically patterned source and drain. Deposition of the organic semiconductor was performed at the last step of the process, either by casting from solution or from the vapor phase.

Of course, such fabrication processes are not appropriate for taking advantage of the potential of organic materials, for example low cost, large area, and the possibility of using flexible substrates. In the following discussion we will focus on different printing techniques that have been used for the fabrication of OTFTs.

Earlier work made use of screen-printing. The major elements of the device were deposited from solutions through stainless-steel mechanical masks [58, 59]. Although this simple technique afforded devices whose performance compared well with those fabricated by "conventional" techniques, it was of much less interest for production of highly integrated circuits. Typical space resolution for screen-printing falls in the 35–100 μm range – several times larger than the critical length needed for realistic applications. To circumvent this problem, a group at Bell Laboratories developed a strategy using high resolution printing with resolution down to 2 μm [66]. Microcontact printing uses elastomeric stamps to print patterns of self-assembled monolayers (SAMs) that are then used either as resists to prevent removal of a pre-deposited material, or as initiators to promote material deposition [67]. The transistor follows a top-gate geometry, in which fabrication starts with deposition of a gold layer on the appropriately prepared substrate. A specifically designed stamp is used to produce the patterned SAM. Then, etching the gold not protected by the SAM is used to define the source and drain electrodes. The SAM is next removed with heat, ultraviolet light, or an oxygen plasma to expose the bare gold. The semiconductor (regio-regular poly(3-hexylthiophene); P3HT), insulator (poly(methyl methacrylate); PMMA), and gate electrode (carbon ink) are sequentially cast from solution. Interestingly, the technique can be extended to reel-to-reel fabrication by making use of cylindrical stamps.

Inkjet printing is an alternative printing technique that is widely used to fabricate organic light-emitting diodes [68, 69] and full-color displays [70]. The main problem with applying the technique to OTFTs is size resolution limited to 20–50 μm, because of spreading of the droplets on the substrate. A group at Cambridge University (UK) recently succeeded in overcoming the problem by confining the spreading of the water-based ink with a pattern of repelling, hydrophobic regions that define the device dimensions [71]. The pattern was fabricated by photolithography and oxygen plasma etching of a polyimide film. Again, the device was fabricated with a top-gate configuration. The elements, deposited in sequence, are: inkjet-printed source and drain (polyethylenedioxythiophene doped with polystyrene sulfonate; PEDOT–PSS), semiconductor (dioctylfluorene–bithiophene copolymer; F8T2), and insulator (polyvinylphenol; PVP).

Stamping and inkjet printing both require liquid inks, which may be a problem when dealing with conjugated polymers and small molecules. To surmount this drawback, a dry process based on thermal imaging has been developed at DuPont. The technique has been claimed to enable the patterning of organic materials at high speed and with micron size resolution [72]. It involves the transfer of a thin solid layer from a donor film on to a flexible receiver. The two flexible films are held together by vacuum. The heat is produced by a laser beam focused through the donor base at a thin metal layer. Conversion of light to heat converts the surrounding materials into gaseous products, whose expansion propels the top layer from the donor to the receiver. Not all conducting polymers can withstand the heat generated during the process, however. Good results were obtained with a polyaniline synthesized by emulsion polymerization and doped with dinonyl naphthalene sulfonic acid (DNNSA-PANI) that may constitute the source and drain electrodes. The possibility of transferring organic semiconductors has not yet been established, however. Transistors with a bottom gate configuration have been fabricated on Mylar substrates. The gate was an indium tin oxide (ITO) film on which a spin-coated glass resin served as the insulator. Heat-transferred DNNSA-PANI source and drain electrodes and vapor-deposited pentacene completed the structure. A TFT backplane containing 5000 transistors with 20 μm channel length could thus be realized on a 50×80 cm^{-2} flexible substrate.

1.6 The Materials

Organic semiconductors are traditionally classified as polymers or small molecules. The former have the advantage of being amenable to specific deposition techniques that have been developed for long for conventional polymers. Their performance is still orders of magnitude lower than that of small molecules, however. Encouraging performance has been reported with the latter, although high performance requires high ordering, particularly in the vicinity of the insulator–semiconductor interface. The importance of high ordering has been recently confirmed by measurements on single crystal devices.

1.6.1 Polymers

Two polymers are used for most work dealing with polymer-based OTFTs – polyfluorene [37, 73] and poly(3-alkylthiophene) (P3AT). We will only deal with the latter, which afford the highest mobility.

After pioneering work by Sirringhaus et al. [74] on spin coated P3AT, it is now well established that the performance of polymer OTFTs critically depends on the chemical and structural ordering of the chains at the insulator–polymer interface. High order first relies on the regio-regularity of the polymer chains; that is, the percentage of regio-regular head-to-tail attachment of the alkyl side chains to the beta position of the thiophene rings. High regio-regularity is not sufficient, however. Two orientations are observed, one with the thiophene rings flat on the surface and the other with the chain edge-on (Fig. 16). High mobility, up to 0.1 cm^2 V^{-1} s^{-1} was only found with the latter arrangement.

More recently it has been shown that the mobility can be slightly increased if the film is applied by dip-coating instead of spin-coating. Under these conditions the

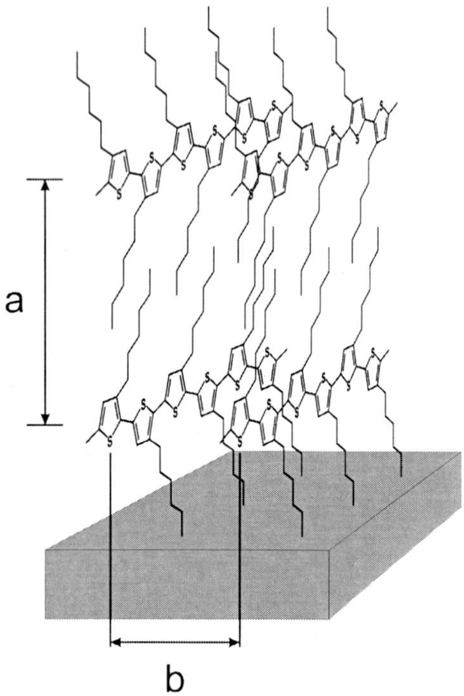

Fig. 1.16. Regio-regular poly-3-hexylthiophene with edge-on orientation of the polymer chains. Redrawn from Ref. [74].

thickness of the film could be reduced to 2–4 nm [75]. It must be remarked that the factors affecting orientation of the polymer chains on the substrate are not fully understood.

1.6.2 Small Molecules

Pentacene is the material most used for preparation of p-type OTFTs based on small molecules, with oligothiophenes and their derivatives being the nest most important. The highest reported mobility is up to 6 $cm^2\ V^{-1}\ s^{-1}$ for the former [76] and 1 $cm^2\ V^{-1}\ s^{-1}$ for the latter [77].

Major improvements have been achieved by modification of the insulator–semiconductor interface. Most devices are grown by vapor deposition on silicon oxide. Because of the different physical and chemical nature of both materials, their association may lead to highly disordered interfaces, thus leading to poor performance. Heating the substrate [27, 78, 79] and depositing at a low rate [30, 78] leads to better organization. A better alternative, however, consists in covering the surface of the oxide with an organic monomolecular layer before vapor deposition. Thus, octadecyltrichlorosilane (OTS) [80] gave good results with pentacene on SiO_2. Even better results have been claimed after use of an ultrathin film of polystyrene [76]. An alternative route is to use a polymer dielectric, which resulted in high performance with sexithiophene derivatives [77].

1.6.3
n-Type Semiconductors

As stated above, the notions of n and p-type do not have the same meaning as for inorganic semiconductors. At the current state of the art, an organic n-type material is one in which electrons are more easily injected than holes. This is therefore more a matter of HOMO and LUMO energy level rather than possibility of doping. In other words, an n-type organic semiconductor is characterized by high electron affinity.

The compounds with the highest electron mobility are currently fullerene C_{60} [81, 82] and *N*,*N*4-dialkyl-3,4,9,10-perylene tetracarboxylic diimide derivatives [83, 84], with values up to 0.5 $cm^2\ V^{-1}\ s^{-1}$. A major problem with these compounds is their high sensitivity to ambient conditions, especially oxygen and moisture. A small number of air-stable n-type compounds have been reported [85–88]. All are perfluorinated and their mobility does not exceed a few tenths of 1 $cm^2\ V^{-1}\ s^{-1}$.

A prominent aspect that promotes the search for air-stable, high-mobility n-type organic semiconductors is the possibility of access to complementary circuitry. It must be recalled that CMOS architecture is a corner stone of microelectronics. Making circuits that combine n-channel and p-channel transistors has many advantages – high robustness, low power consumption, and low noise. A new concept has recently emerged in the field of organic thin-film transistors, that of ambipolar materials, which can be defined as materials that change type depending on the nature of the contact used to inject charges. Ideally, an ambipolar semiconductor would have a low ionization potential and high electron affinity. An elegant way of realizing an ambipolar compound has been discovered by a group at Eindhoven [89]. It consists in making an interpenetrating network of two compounds, one n-type and one p-type. The former was a derivative of C_{60} (6,6-phenyl C_{61}-butyric acid methyl ester, PCBM) and the latter was either regio regular P3HT or a derivative of poly-*p*-phenylenevinylene (PPV). The group also showed that pristine pentacene could also have ambipolar behavior. It must, however, be stressed that this concept is highly controversial. Some have pointed out that complementary circuitry requires that one transistor be off while the other is on, and vice versa. Accordingly, an ambipolar material that will conduct holes for negative gate voltages and electrons for positive gate voltages will only turn off for a very limited voltage range, or for no voltage at all. This seems detrimental for any practical circuit. Separate use of a purely n-type and purely p-type compound seems a much wiser approach.

1.6.4
Single Crystals

Until recently, reports on single-crystal OTFTs were rather scarce [90, 91]. The difficulties in building a single crystal OTFT are numerous. Single crystals of organic materials useful for making OTFTs are small, fragile, and difficult to handle. The standard fabrication technique for OTFTs, which consists of depositing a semiconductor film on top of the insulator from either the vapor phase or a solution, cannot be extended to single crystals. Furthermore, many conventional fabrication processes may damage the surface of the crystal. For these reasons, the OTFT must be made of already grown crystals. To date, only two techniques have been successfully used.

1. Electrostatic bonding of the crystal on top of a previously prepared gate-insulator-source-drain structure [92–94]. Sometimes the source and drain contacts have been deposited afterwards, on top of the crystal [90].
2. Direct deposition of the contacts and gate insulator on to the crystal [95, 96]. In this technique the gate dielectric is the polymer parylene, which forms conformal coatings with good dielectric and mechanical properties. The polymer is deposited in a three-zone reactor, in which the deposition zone can be kept at room temperature.

These techniques have recently been reviewed [97] and will not be discussed further here. Instead, we will focus on the electrical characteristics of single-crystal OTFTs.

Because the concentration of defects is much lower in single crystals than in vapor-deposited or solution cast films, the main motivation in fabricating single crystal OTFTs was to explore the physical limitations on the performance of the device, with the hope of approaching the ultimate properties of the materials. Pentacene was somewhat disappointing – the highest reported single-crystal mobility was several times lower than for the best polycrystalline films. This is because the crystal structure in thin films (the so-called "thin-film" form) is slightly different from that in single crystals [98]. More encouraging results were obtained with tetracene (also called "naphthacene"), a parent molecule of pentacene made of four fused benzene rings instead of five, and even more with rubrene, the molecule of which consists of a tetracene core with four pendant benzene rings. The molecular formulas of tetracene and rubrene are given in Fig. 1.17.

Comparing single crystal and vapor-grown devices for these two compounds is difficult, because reports on evaporated tetracene OTFTs are rather scarce [99–101], and despite several (unpublished) attempts, fabrication of an operating thin-film device from rubrene has not yet been successfully achieved. For both compounds the problem seems to arise from an improper deposition mechanism, which, in contrast with experience with pentacene and sexithiophene, does not favor two-dimensional growth.

Prominent features of these single crystal devices are [102]:

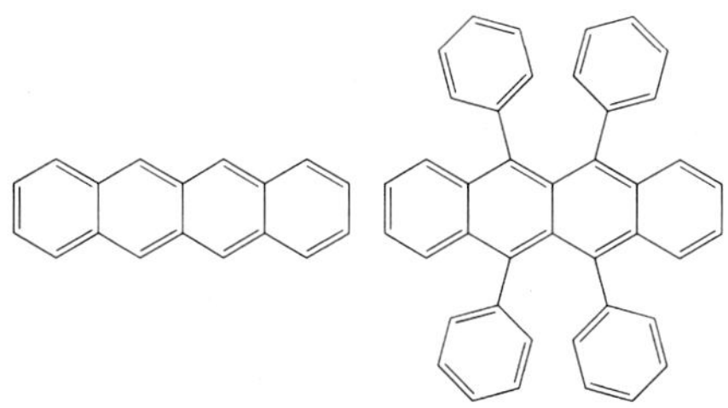

Fig. 1.17. Molecular structures of tetracene (left) and rubrene (right).

1. unexpectedly high mobility (up to 20 $cm^2 V^{-1} s^{-1}$ for rubrene at room temperature);
2. an inversion in the temperature dependence; that is, mobility tends to increase when the temperature is reduced from room temperature to ca. 280 K for tetracene and 200 K for rubrene, at which point the tendency is reversed; and
3. anisotropic mobility depending on the crystal axis.

All these features are reminiscent of results obtained from time-of-flight measurements on highly pure molecular crystals (Section 1.4). This might indicate of that the intrinsic properties of these materials are being approached.

1.6.5 Insulators

It has long been known that in an insulated gate field-effect transistor, the role of the insulator is at least as important as that of the semiconductor. For historical reasons, most of the OTFTs made so far have used inorganic dielectrics, mostly silicon oxide. We have already mentioned (Section 1.6.2) that the quality of the insulator–semiconductor interface, and hence the performance of the device, can be significantly improved by inserting an organic single layer between the insulator and the semiconductor. It seems that the performance of the OTFT is very much dependent on the physical and chemical nature of the surface of the insulator. In this respect organic insulators seem to be more flexible than their inorganic counterparts.

The nature of the insulator may intervene at different levels. First, in the bottom configuration, the insulator is capable of affecting the morphology of the semiconductor layer. For instance, its surface energy strongly affects the nucleation and growth mechanism of vapor-deposited films. According to the standard nucleation and growth models, it is expected that the nucleation rate increases, and thus the size of the grains decreases, with increasing surface energy. An archetypal example of this behavior is vapor deposition of pentacene on hydrogenated silicon (a low-energy surface) compared with silicon oxide (a high-energy surface); the nucleation rate was found to be several orders of magnitude higher on the former than on the latter [103, 104]. Similar studies on organic insulators have not yet been conducted, but would be of great interest in that respect. At this point, it is worth remarking that the effect of grain size on the performance of OTFTs is currently the subject of much debate. On the basis of conventional theories on charge transport in polycrystalline semiconductors it can be expected that mobility increases with grain size. This can be simply understood by noting that charge transport in polycrystalline media is limited by grain boundaries, so that as the grain size increases the number of grain boundaries decreases, leading to improved transport. Although this seems, indeed, to be observed for oligothiophenes, according to an early report [105], several results prove the opposite is true for pentacene [106, 107].

Besides its morphological effects, the nature of the insulator may also have chemical and electrical consequences. It has recently been claimed that the dipole field present at the surface of high dielectric constant (high-k) insulators may enhance the formation of local states that in turn induce carrier localization and reduce charge-carrier mobility. Hence, benefits have resulted from use of low-k organic dielectrics [108]. It is worth remarking that such a conclusion is at variance with previous work that established interest in using high-capacitance insulators, which can be obtained by making use of high-k dielectrics [109]. Interest in high capacitance comes from the above mentioned gate voltage (actually, charge density) dependence of the mobility. High capacitance means that high charge-carrier concentration can be reached at low gate voltage, hence mobility is expected to be higher. The respective advantages of low-k and high-k dielectrics thus seem contradictory, and more work will be necessary to clarify this point.

1.7 Concluding Remarks

The organic thin-film transistor is at a pivotal point of its history. The number of research teams that have embarked in the field has blossomed in recent years. Several important features have been uncovered or confirmed, for example the crucial role of the quality of the semiconductor at the insulator–semiconductor interface and of the resistance of the source-drain contacts. The advent of the single-crystal device has enabled charge-carrier mobility to be increased to even higher values than that reported for highly pure molecular crystal, at least at room temperature. Fundamentally, the main issue with organic thin-film transistors is the lack of reliable model for predicting the charge-transfer properties of a given molecular material. It has been largely confirmed that the performance of the device is mostly governed by the structure of the semiconductor film at the insulator–semiconductor interface. In that respect, the role of the insulator is at least as important as that of the semiconductor. This is exemplified by recalling that silicon is the universal element used in microelectronics not so much because of the intrinsic properties of the material but because of the almost perfect interface it forms with its thermally grown oxide. For this reason, the major challenge in the development of OTFTs is now to identify suitable organic insulator–semiconductor combinations rather than semiconductors by themselves.

Acknowledgements

Most of the experimental work on which this chapter is based was performed at the Laboratoire des Matériaux Moléculaires, CNRS, directed by Francis Garnier. The author gratefully acknowledges his past and present co-workers Abderrahim Yassar, Zhi-Gang Xu, Xue-Zhou Peng, Denis Fichou, Bai Xu, Didier Delabouglise, Mohamed Hmyene, Françoise Deloffre, Fayçal Kouki, Philippe Lang, Riadh Hajlaoui, Frédéric Demanze, Pratima Srivastava, Claude Noguès, Xavier Pan, Peter Spearman, Belkassem Nessakh, Ahmed El Kassmi, Mohammad Mottaghi, and Wolfgang Kalb. He is indebted to Philippe Delannoy, Jean-Luc Brédas, Jérôme Cornil, Fabio Biscarini, Antoine Kahn, Jean-Noël Chazalviel, and Libero Zuppiroli for stimulating discussions and fruitful collaborations.

References

1 J. E. Lilienfeld, US Patent 1,745,175, **1930**.
2 D. Kahng, M. M. Atalla, in *IRE Solid-State Devices Research Conference*, Pittsburgh, PA, **1960**; D. Kahng, US Patent 3,102,230, 1963.
3 P. K. Weimer, *Proc. IRE* **1962**, *50*, 1462–1469.
4 W. E. Spear, P. G. Le Comber, *J. Non-Cryst. Solids* **1972**, *8–10*, 727–738.
5 G. W. Neudeck, A. K. Malhotra, *Solid-State Electron.* **1976**, *19*, 721–729.
6 M. Pope, C. E. Swenberg, *Electronic Processes in Organic Crystals and Polymers*, Oxford University Press, New York, **1999**.
7 D. F. Barbe, C. R. Westgate, *J. Phys. Chem. Solids* **1970**, *31*, 2679–2687.
8 M. L. Petrova, L. D. Rozenshtein, *Fiz. Tverd. Tela (Soviet Phys. Solid State)* **1970**, *12*, 961–962 (756–757).
9 F. Ebisawa, T. Kurokawa, S. Nara, *J. Appl. Phys.* **1983**, *54*, 3255–3259.
10 A. Tsumura, K. Koezuka, T. Ando, *Appl. Phys. Lett.* **1986**, *49*, 1210–1212.
11 A. Tsumura, H. Koezuka, Y. Ando, *Synth. Metal* **1988**, *25*, 11–23.
12 G. Horowitz, D. Fichou, X. Z. Peng, Z. G. Xu, F. Garnier, *Solid State Commun.* **1989**, *72*, 381–384.
13 G. Horowitz, X. Z. Peng, D. Fichou, F. Garnier, *J. Appl. Phys.* **1990**, *67*, 528–532.
14 H. E. Katz, *J. Mater. Chem.* **1997**, *7*, 369–376.
15 A. R. Brown, C. P. Jarrett, D. M. de Leeuw, M. Matters, *Synth. Metal.* **1997**, *88*, 37–55.
16 F. Garnier, *Chem. Phys.* **1998**, *227*, 253–262.
17 G. Horowitz, *Adv. Mater.* **1998**, *10*, 365–377.
18 C. D. Dimitrakopoulos, P. R. L. Malenfant, *Adv. Mater.* **2002**, *14*, 99–117.
19 Y. Y. Lin, D. J. Gundlach, S. F. Nelson, T. N. Jackson, *IEEE Electron. Device Lett.* **1997**, *18*, 606–608.
20 P. F. Baude, D. A. Ender, M. A. Haase, T. W. Kelley, D. V. Muyres, S. D. Theiss, *Appl. Phys. Lett.* **2003**, *82*, 3964–3966.
21 J. Koenigsberger, J. Weiss, *Ann. Phys.* **1911**, *35*, 1–46.
22 An enlightening history of the emergence of semiconductors can be found in: F. A. Stahl, *Am. J. Phys.* **2003**, *71*, 1170–1173.
23 http://en.wikipedia.org/wiki/Semiconductor.
24 A. Kahn, N. Koch, W. Gao, *J. Polym. Sci. B, Polym. Phys.* **2003**, *41*, 2529–2548.
25 R. B. Campbell, J. M. Robertson, J. Trotter, *Acta Cryst.* **1961**, *14*, 705–711.
26 D. Holmes, S. Kumaraswamy, A. J. Matzger, K. P. C. Vollhardt, *Chem. Eur. J.* **1999**, *5*, 3399–3412.
27 B. Servet, G. Horowitz, S. Ries, O.

Lagorsse, P. Alnot, A. Yassar, F. Deloffre, P. Srivastava, R. Hajlaoui, P. Lang, F. Garnier, *Chem. Mater.* **1994**, *6*, 1809–1815.

28 F. Garnier, A. Yassar, R. Hajlaoui, G. Horowitz, F. Deloffre, B. Servet, S. Ries, P. Alnot, *J. Am. Chem. Soc.* **1993**, *115*, 8716–8721.

29 T. Minakata, I. Nagoya, M. Ozaki, *J. Appl. Phys.* **1991**, *69*, 7354–7356.

30 C. D. Dimitrakopoulos, A. R. Brown, A. Pomp, *J. Appl. Phys.* **1996**, *80*, 2501–2508.

31 C. Tanase, E. J. Meijer, P. W. M. Blom, D. M. de Leeuw, *Org. Electron.* **2003**, *4*, 33–37.

32 N. F. Mott, R. W. Gurney, *Electronic Processes in Ionic Crystals*, Clarendon Press, Oxford, **1940**.

33 J. C. Sancho-Garcia, G. Horowitz, J. L. Bredas, J. Cornil, *J. Chem. Phys.* **2003**, *119*, 12563–12568.

34 A. Dodabalapur, L. Torsi, H. E. Katz, *Science* **1995**, *268*, 270–271.

35 P. V. Necliudov, M. S. Shur, D. J. Gundlach, T. N. Jackson, *J. Appl. Phys.* **2000**, *88*, 6594–6597.

36 G. Horowitz, M. E. Hajlaoui, R. Hajlaoui, *J. Appl. Phys.* **2000**, *87*, 4456–4463.

37 R. A. Street, A. Salleo, *Appl. Phys. Lett.* **2002**, *81*, 2887–2889.

38 P. V. Necliudov, M. S. Shur, D. J. Gundlach, T. N. Jackson, *Solid-State Electron.* **2003**, *47*, 259–262.

39 J. Zaumseil, K. W. Baldwin, J. A. Rogers, *J. Appl. Phys.* **2003**, *93*, 6117–6124.

40 H. Klauk, G. Schmid, W. Radlik, W. Weber, L. Zhou, C. D. Sheraw, J. A. Nichols, T. N. Jackson, *Solid-State Electron.* **2003**, *47*, 297–301.

41 E. J. Meijer, G. H. Gelinck, E. van Veenendaal, B. H. Huisman, D. M. de Leeuw, T. M. Klapwijk, *Appl. Phys. Lett.* **2003**, *82*, 4576–4578.

42 S. Luan, G. W. Neudeck, *J. Appl. Phys.* **1992**, *72*, 766–772.

43 G. Horowitz, P. Lang, M. Mottaghi, H. Aubin, *Adv. Funct. Mater.* **2004**, *14*, 1069–1074.

44 W. Kalb, P. Lang, M. Mottaghi, H. Aubin, G. Horowitz, M. Wuttig, *Synth. Metal.* **2004**, *146*, 279–282.

45 R. J. Chesterfield, J. C. McKeen, C. R. Newman, C. D. Frisbie, P. C. Ewbank, K. R. Mann, L. L. Miller, *J. Appl. Phys.* **2004**, *95*, 6396–6405.

46 I. Yagi, K. Tsukagoshi, Y. Aoyagi, *Appl. Phys. Lett.* **2004**, *84*, 813–815.

47 K. Seshadri, C. D. Frisbie, *Appl. Phys. Lett.* **2001**, *78*, 993–995.

48 L. Bürgi, H. Sirringhaus, R. H. Friend, *Appl. Phys. Lett.* **2002**, *80*, 2913–2915.

49 L. Bürgi, T. J. Richards, R. H. Friend, H. Sirringhaus, *J. Appl. Phys.* **2003**, *94*, 6129–6137.

50 N. Koch, A. Kahn, J. Ghijsen, J.-J. Pireaux, J. Schwartz, R. L. Johnson, A. Elschner, *Appl. Surf. Sci.* **2003**, *82*, 70–72.

51 T. Li, P. P. Ruden, I. H. Campbell, D. L. Smith, *J. Appl. Phys.* **2003**, *93*, 4017–4022.

52 N. Tessler, Y. Roichman, *Appl. Phys. Lett.* **2001**, *79*, 2987–2989.

53 N. J. Watkins, L. Yan, Y. Gao, *Appl. Phys. Lett.* **2002**, *80*, 4384–4386.

54 M. C. J. M. Vissenberg, M. Matters, *Phys. Rev. B* **1998**, *57*, 12964–12967.

55 N. Karl, J. Marktanner, R. Stehle, W. Warta, *Synth. Metal.* **1991**, *42*, 2473–2481.

56 E. A. Silinsh, V. Cápek, *Organic molecular crystals: Interaction, localization, and transport phenomena*, AIP Press, New York, **1994**.

57 V. M. Kenkre, J. D. Andersen, D. H. Dunlap, C. B. Duke, *Phys. Rev. Lett.* **1989**, *62*, 1165–1168.

58 F. Garnier, R. Hajlaoui, A. Yassar, P. Srivastava, *Science* **1994**, *265*, 1684–1686.

59 Z. N. Bao, Y. Feng, A. Dodabalapur, V. R. Raju, A. J. Lovinger, *Chem. Mater.* **1997**, *9*, 1299.

60 J. L. Brédas, D. Beljonne, J. Cornil, J. P. Calbert, Z. Shuai, R. Silbey, *Synth. Metal.* **2002**, *125*, 107–116.

61 R. C. Haddon, X. Chi, M. E. Itkis, J. E. Anthony, D. L. Eaton, T. Siegrist, C. C. Mattheus, T. T. M. Palstra, *J. Phys. Chem. B* **2002**, *106*, 8288–8292.

62 Y. C. Cheng, R. J. Silbey, D. A. Da Silva, J. P. Calbert, J. Cornil, J. L.

Brdas, *J. Chem. Phys.* **2003**, *118*, 3764–3774.

63 L. Giuggioli, J. D. Andersen, V. M. Kenkre, *Phys. Rev. B* **2003**, *67*, 045110.

64 G. Horowitz, R. Hajlaoui, P. Delannoy, *J. Phys. III France* **1995**, *5*, 355–371.

65 P. G. Le Comber, W. E. Spear, *Phys. Rev. Lett.* **1970**, *25*, 509–511.

66 J. A. Rogers, Z. N. Bao, A. Makhija, P. Braun, *Adv. Mater.* **1999**, *11*, 741–745.

67 Y. Xia, G. M. Whitesides, *Angew. Chem. Int. Ed.* **1998**, *37*, 550–575.

68 T. R. Hebner, C. C. Wu, D. Marcy, M. H. Lu, J. C. Sturm, *Appl. Phys. Lett.* **1998**, *72*, 519–521.

69 Y. Yang, S. C. Chang, J. Bharathan, J. Liu, *J. Mater. Sci. Mater. Electron.* **2000**, *11*, 89–96.

70 O. Yokoyama, *Optronics* **2003**, *254*, 119–122.

71 H. Sirringhaus, T. Kawase, R. H. Friend, T. Shimoda, M. Inbasekaran, W. Wu, E. P. Woo, *Science* **2000**, *290*, 2123–2126.

72 G. B. Blanchet, Y. L. Loo, J. A. Rogers, F. Gao, C. R. Fincher, *Appl. Phys. Lett.* **2003**, *82*, 463–465.

73 H. Sirringhaus, R. J. Wilson, R. H. Friend, M. Inbasekaran, W. Wu, E. P. Woo, M. Grell, D. D. C. Bradley, *Appl. Phys. Lett.* **2000**, *77*, 406–408.

74 H. Sirringhaus, P. J. Brown, R. H. Friend, M. M. Nielsen, K. Bechgaard, B. M. W. Langeveldvoss, A. J. H. Spiering, R. A. J. Janssen, E. W. Meijer, P. Herwig, D. M. de Leeuw, Nature **1999**, *401*, 685–688.

75 G. M. Wang, J. Swensen, D. Moses, A. J. Heeger, *J. Appl. Phys.* **2003**, *93*, 6137–6141.

76 T. W. Kelley, D. V. Muyres, P. F. Baude, T. P. Smith, T. D. Jones, *MRS Symp. Proc.* **2003**, *771*, 169–179.

77 M. Halik, H. Klauk, U. Zschieschang, G. Schmid, S. Ponomarenko, S. Kirchmeyer, W. Weber, *Adv. Mater.* **2003**, *15*, 917–922.

78 D. J. Gundlach, Y. Y. Lin, T. N. Jackson, S. F. Nelson, D. G. Schlom, *IEEE Electron. Device Lett.* **1997**, *18*, 87–89.

79 W. A. Schoonveld, R. W. Stok, J. W. Weijtmans, J. Vrijmoeth, J. Wildeman, T. M. Klapwijk, *Synth. Metal.* **1997**, *84*, 583–584.

80 Y. Y. Lin, D. J. Gundlach, T. N. Jackson, S. F. Nelson, *IEEE Trans. Electron. Dev.* **1997**, *44*, 1325–1331.

81 C. P. Jarrett, K. Pichler, R. Newbould, R. H. Friend, *Synth. Metal.* **1996**, *77*, 35–38.

82 S. Kobayashi, T. Takenobu, S. Mori, A. Fujiwara, Y. Iwasa, *Appl. Phys. Lett.* **2003**, *82*, 4581–4583.

83 H. E. Katz, A. J. Lovinger, J. Johnson, C. Kloc, T. Siegrist, W. Li, Y. Y. Lin, A. Dodabalapur, *Nature* **2000**, *404*, 478–481.

84 P. R. L. Malenfant, C. D. Dimitrakopoulos, J. D. Gelorme, L. L. Kosbar, T. O. Graham, A. Curioni, W. Andreoni, *Appl. Phys. Lett.* **2002**, *80*, 2517–2519.

85 A. Facchetti, M. Mushrush, H. E. Katz, T. J. Marks, *Adv. Mater.* **2003**, *15*, 33–38.

86 H. Tada, H. Touda, M. Takada, K. Matsushige, *J. Porphyr. Phthalocyan.* **1999**, *3*, 667–671.

87 S. Hoshino, S. Nagamatsu, M. Chikamatsu, M. Misaki, Y. Yoshida, N. Tanigaki, K. Yase, *Synth. Metal.* **2003**, *137*, 953–954.

88 A. Facchetti, Y. Deng, A. C. Wang, Y. Koide, H. Sirringhaus, T. J. Marks, R. H. Friend, *Angew. Chem. Int. Ed.* **2000**, *39*, 4547–4551.

89 E. J. Meijer, D. M. De Leeuw, S. Setayesh, E. van Veenendaal, B. H. Huisman, P. W. M. Blom, J. C. Hummelen, U. Scherf, T. M. Klapwijk, *Nature Mater.* **2003**, *2*, 678–682.

90 G. Horowitz, F. Garnier, A. Yassar, R. Hajlaoui, F. Kouki, *Adv. Mater.* **1996**, *8*, 52–54.

91 M. Ichikawa, H. Yanagi, Y. Shimizu, S. Hotta, N. Suganuma, T. Koyama, Y. Taniguchi, *Adv. Mater.* **2002**, *14*, 1272–1275.

92 R. W. I. de Boer, T. M. Klapwijk, A. F. Morpurgo, *Appl. Phys. Lett.* **2003**, *83*, 4345–4347.

93 C. R. Newman, R. J. Chesterfield, J. A. Merlo, C. D. Frisbie, *Appl. Phys. Lett.* **2004**, *85*, 422–424.

94 C. Goldmann, S. Haas, C. Krellner, K. P. Pernstich, D. J. Gundlach, B. Batlogg, *J. Appl. Phys.* **2004**, *96*, 2080–2086.

95 V. Podzorov, V. M. Pudalov, M. E. Gershenson, *Appl. Phys. Lett.* **2003**, *82*, 1739–1741.

96 V. Y. Butko, X. Chi, D. V. Lang, A. P. Ramirez, *Appl. Phys. Lett.* **2003**, *83*, 4773–4775.

97 R. W. I. de Boer, M. E. Gershenson, A. F. Morpurgo, V. Podzorov, *Phys. Stat. Sol. (a)* **2004**, *201*, 1302–1331.

98 S. E. Fritz, S. M. Martin, C. D. Frisbie, M. D. Ward, M. F. Toney, *J. Am. Chem. Soc.* **2004**, *126*, 4084–4085.

99 D. J. Gundlach, J. A. Nichols, L. Zhou, T. N. Jackson, *Appl. Phys. Lett.* **2002**, *80*, 2925–2927.

100 A. Hepp, H. Heil, W. Weise, M. Ahles, R. Schmechel, H. von Seggern, *Phys. Rev. Lett.* **2003**, *91*, 157406.

101 C. Santato, R. Capelli, M. A. Loi, M. Murgia, F. Cicoira, V. A. L. Roy, P. Stallinga, R. Zamboni, C. Rost, S. F. Karg, M. Muccini, *Synth. Metal.* **2004**, *146*, 329–334.

102 V. Podzorov, E. Menard, A. Borissov, V. Kiryukhin, J. A. Rogers, M. E. Gershenson, *Phys. Rev. Lett.* **2004**, *93*, 086602.

103 F. J. Meyer zu Heringdorf, M. C. Reuter, R. M. Tromp, *Nature* **2001**, *412*, 517–520.

104 R. Ruiz, B. Nickel, N. Koch, L. C. Feldman, R. F. Haglund, A. Kahn, G. Scoles, *Phys. Rev. B* **2003**, *67*, 125406.

105 G. Horowitz, M. E. Hajlaoui, *Synth. Metal.* **2001**, *122*, 185–189.

106 D. Knipp, R. A. Street, A. Vlkel, J. Ho, *J. Appl. Phys.* **2003**, *93*, 347–355.

107 M. Shtein, J. Mapel, J. B. Benziger, S. R. Forrest, *Appl. Phys. Lett.* **2002**, *81*, 268–270.

108 J. Veres, S. D. Ogier, S. W. Leeming, D. C. Cupertino, S. M. Khaffaf, *Adv. Funct. Mater.* **2003**, *13*, 199–204.

109 C. D. Dimitrakopoulos, I. Kymissis, S. Purushothaman, D. A. Neumayer, P. R. Duncombe, R. B. Laibowitz, *Adv. Mater.* **1999**, *11*, 1372–1375.

II
Advanced Materials for Organic Electronics

2
High-performance Pentacene Transistors

Tommie Kelley

2.1
Introduction

Organic photoconductors have been known since the early 1900s [1], and much work was done in the 1940s on the various semiconducting properties of organic materials [2], but it may well have been the appearance of organic light-emitting materials [3] and conducting polymers [4] that set off the rapid, intense, and continuing research activity on organic electronic systems. It is also likely that reports of continuing improvement in thin-film transistors (TFTs) using oligothiophenes maintained some of the interest, but it was the subsequent emergence of pentacene as an organic transistor standard – a material comparable, and in some cases, higher performing than amorphous silicon, that began to attract the attention of the traditional microelectronics communities. Of course, these are only a few of the highlights from many years of work from many different research groups, but pentacene has continued to stand out an as exemplary material, especially in terms of charge carrier mobility, and one by which to gauge all other organic semiconductors. Reports of organic electronic research results now have global range with large concentrations of activity in the US, Europe, and Asia, and large companies around the world are adopting programs which include work on organic thin-film transistors (OTFTs) including 3M, Samsung, Pioneer, Infineon, Philips, Sarnoff, DuPont, and many others.

The earliest reports of charge carrier mobilities in organic small molecule materials were focused on single crystals and relied on time-of-flight or on space charge-limited current (SCLC) measurements. The values reported were in the range 10^{-3} to 1 $cm^2\ V^{-1}\ s^{-1}$, but were much lower when thin film samples were fabricated of the same materials, probably because of inadequate purification methods and limited understanding of the structure–property relationships [2]. Many of the materials begin to thermally decompose at crystal growth temperatures, so obtaining pure samples of reasonable size for device fabrication remains a challenge. Single-crystal mobility reports still remain few and far between, but thin-film transistors have become a valuable means of extracting field effect charge carrier mobilities, utilizing a construction that is easily extended to many applications. No simple re-

Organic Electronics, Materials, Manufacturing and Applications. Edited by Hagen Klauk

ISBN: 3-527-31264-1

lationship has been found among the different methods of estimating charge carrier mobility – variations in device and deposition geometry, experimental methods and measurement procedures make this nearly impossible, nor is there a simple relationship which indicates which materials will have the highest mobility (thin film or single crystal). Many of the organic small molecule semiconductors are rather insoluble, so film formation is typically performed under vacuum by thermal evaporation, and results vary widely according to processing conditions. Thin film mobilities for organics were of the same order of magnitude as for organic electroluminescent materials ($\sim 10^{-3}$ to 10^{-5} $cm^2\ V^{-1}\ s^{-1}$) and remained low until the importance of structural perfection in these van der Waals crystals began to be understood and controlled by use of optimized process conditions and surface preparation.

For many years, it seemed as if charge carrier mobilities in thin-film organics had reached a maximum for oligomeric thiophene materials, with typical device mobilities in TFT constructions of less than 0.1 $cm^2\ V^{-1}\ s^{-1}$, and large sensitivity to oxygen and humidity. The low mobility and environmental sensitivity relegated organic devices to the realm of curiosity until the early 1990s, rather than raising them to the position of possible replacement for silicon-based circuits that they now hold.

Early work on the thiophenes hinted at the importance of controlling morphology, especially at device interfaces. In comparison with some of the acenes, chemistry in the oligomeric thiophene series is relatively simple, because the precursors are obtained in several steps and remain soluble, and are easily purified by recrystallization. These factors made it possible to obtain very pure samples of a wide range of the thiophene series (bithiophene through octithiophene) fairly early in their study, and to enable researchers to vary the functional and terminal groups on the thiophene family to study the impact on structure and performance [5]. Authors' reports vary about which thin film oligomers gave the highest mobility in the series, longer is typically better, with the work of Halik et al. [5], showing α,α'-diethylsexithiophene TFTs with mobility above 1 $cm^2\ V^{-1}\ s^{-1}$. The various results reported for other oligomers with high performance probably emphasize the importance of structural perfection – different deposition systems and conditions can enhance structural order in a variety of oligomers. The thiophenes, and other rigid-rod molecules, stand nearly vertical relative to the substrate as the oligomers get longer, but alkyl chains attached to the ends of the oligomers can also affect the structural order, often resulting in improved performance.

Some researchers showed that improved crystalline structure at device interfaces could dramatically affect device performance in the acene series also. Such efforts from Jackson's group at Penn State University led to the first reports, in 1996 [6], of organic device mobilities exceeding those of amorphous silicon, which was already becoming the dominant technology for large area, transistor-controlled display applications. Pentacene mobilities continued to improve, beating all other organics for the distinction of highest performance, but researchers working independently of Jackson's group, did not arrive at OTFT mobilities $\sim 1\ cm^2\ V^{-1}\ s^{-1}$ until 2002 and 2003 [7]. Now, many groups have reported mobilities exceeding 1

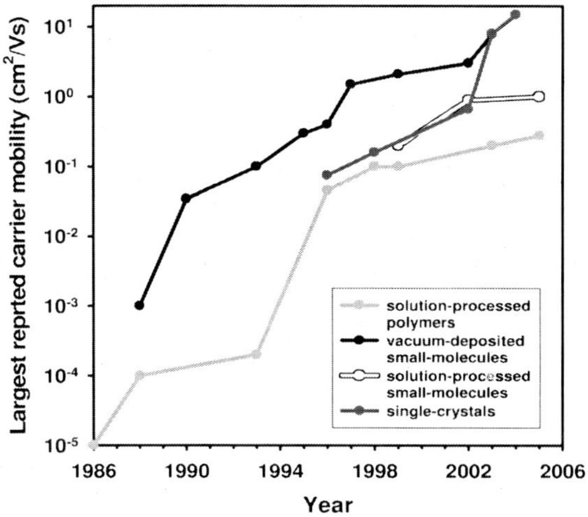

Fig. 2.1. Improvement in organic semiconductor mobility over the last two decades. Rates of mobility increase have been very similar for single crystals, polymers, and small molecules over the same time period [49].

$cm^2\ V^{-1}\ s^{-1}$ with only minimum care and effort, using a variety of process conditions and several device constructions. A further advantage of pentacene over the thiophene series is that it is commercially available from several vendors in a range of purities, making this material accessible to groups without synthetic support [8].

Most organics continued to have field effect mobilities less than 1 $cm^2\ V^{-1}\ s^{-1}$ (and usually less than 0.1 $cm^2\ V^{-1}\ s^{-1}$), even though a few had single-crystal mobilities above 1 $cm^2\ V^{-1}\ s^{-1}$ [2]. Figure 2.1 shows a compilation of reported mobilities for organic semiconductors. This figure summarizes the rapid rise of organic field-effect transistor performance starting in 1987 with early polythiophene field-effect transistor (FET) reports from Tsumura [9]. The figure includes single-crystal and polymer semiconductor results for comparison. It is interesting to note that with improved purification methods and better understanding of structure–property relationships, all material sets have improved at about the same rate. For organic thin-film devices, pentacene's consistent high mobility through simple process routes enabled it to become the favorite of the OTFT community – the first material to be tried and the material of choice to gauge the contribution of a new process to improved performance.

Optimization of pentacene TFTs for mobility has led to some very promising results, but implementation of pentacene circuits in product applications has been limited, because of poor understanding and limited data on operational and shelf life stability in different circuit architectures. A notable exception to this is the implementation of Philips' pentacene active-matrix backplanes in Polymer Vision's rollable display prototypes [10]. These organic devices have acceptable levels of performance and stability in an introductory application and are the first devices nearing commercialization to use pentacene, although it should be noted that this is not the first display prototype using OTFTs. In 2000, EInk and Lucent Technologies collaborated to demonstrate an electrophoretic flexible display using pentacene backplanes [11]. Many groups have demonstrated pentacene backplanes in combination with electrophoretic display media and simple organic light emitting diode (OLED) pixels in the years since the EInk Lucent demonstration. In addition, recent reports from Samsung Electronics [12] have shown potential utility of pentacene TFT for OLED backplanes on flexible substrates of fairly large size.

This chapter will give an overview of methods used to achieve higher mobilities in pentacene devices, will point out potential progress in understanding the nature of exceptional mobility in these devices, and will comment on possible routes to stabilize device performance.

2.2
Routes to Performance Improvement

Organic electronic performance continues to be a challenge in thin-film devices. It was not until 1998 that reports of pentacene in different devices began to appear regularly in the literature. Currently, nearly 200 publications per year are written with at least partial focus on pentacene [13]. Unfortunately, there is still a huge variation in reports of "high mobility" pentacene devices – values range from 0.5 (or less) to over 7 $cm^2\ V^{-1}\ s^{-1}$ – a good indication that there is still more to learn, and that small processing differences can have large effects on device performance.

2.2.1
Purification

The importance of material purity is still somewhat debatable, and a variety of groups have reported pentacene mobility $\sim 1\ cm^2\ V^{-1}\ s^{-1}$ using as-purchased (unpurified) material. It is generally accepted that higher purity correlates with higher performance and may also contribute to device reproducibility and stability, although careful studies are yet to be reported. Because pentacene is insoluble in most organic solvents, recrystallization cannot be used as a starting point as with many other material systems. The most common approach is a variation of thermal-gradient sublimation, often termed physical vapor transport, performed under reduced pressure and flowing inert gas [14]. This method has been widely used as a single-pass and multi-pass approach, and variations abound (atmospheric pressure and inert gas, high vacuum and inert gas, high vacuum in the absence of carrier gas, etc.). In the simplest sense, this method is effective because it separates, in the vapor phase, pure pentacene from lower-molecular-weight impurities which travel farther down the process length, and higher molecular weight impurities which remain at or near the source material. Required purity is difficult to define, but parts per million are probably the minimum purity for reliable devices, especially when considered from the standpoint of microelectronics, in which each impurity may act as a dopant.

2.2.2
Device Evolution

The first reports were largely simple trials in which silicon-based semiconductors were directly replaced with pentacene as the active layer in a TFT. It became clear, early on, that optimization of interfaces in organic devices could be even more important than in traditional silicon devices, but that the methods used for silicon were not adequate to make high quality organic devices. Researchers who brought together varied disciplines such as microfabrication, synthetic chemistry, interface engineering and surface characterization often showed the most rapid progress. Several of these research groups have succeeded in extending the capabilities of organic devices into RFID transponder tags, display backplanes, and sensors [15]. Before these groups had any success in potential applications, they spent a great deal of time and effort optimizing and understanding the performance of the individual TFTs which would eventually be combined to make circuits and devices. The simplest TFTs are fabricated on silicon wafers – a thermal oxide or high-quality device oxide is formed on one side and an ohmic gate contact is formed on the other [16]. This simplifies the processing to test the performance of organic materials and eliminates much of the uncertainty in fabricating devices from scratch, which may involve high levels of variability because of the use of a low-temperature oxides. The prepared wafer is cleaved to the appropriate sample size, and inserted into a vacuum chamber where organic materials can be thermally deposited at a vacuum level of approximately 5×10^{-6} torr or less. The sample is removed from the chamber, a shadow mask can be affixed to the sample, and the sample is then reinserted in the vacuum chamber for deposition of the metal top contacts. Alternatively, before the organic deposition, photolithography can be performed on the wafer to define source and drain contacts which will then lie under the semiconductor after it is deposited. The wafer substrate makes an easy starting place to optimize deposition conditions, or surface modification and cleaning methods before migrating to more complex device structures.

2.2.3
Structural Perfection

There is a very short list of organic semiconductors with reported thin-film field-effect mobilities greater than 1 $cm^2\ V^{-1}\ s^{-1}$. These include pentacene, sexithiophene [5a], and anthradithiophene [17]. If we extend this list to include single crystal and n-type materials, we can add perylene [18], rubrene [19], copper phthalocyanine (CuPc) [20], tetracyanoquinodimethane (TCNQ) [21], and dithiophene–tetrathiofulvalene (DT–TTF) [22] – still a short list.

What is interesting about all of these materials is the overlying similarity in their

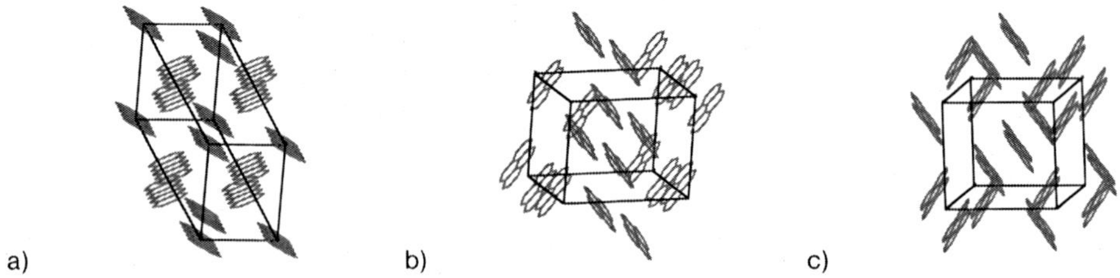

Fig. 2.2. Examples of herringbone and herringbone-like packing motifs for several rigid-backbone, small molecule, organic semiconductors: A) pentacene [54a]; B) pyrene [54b]; C) perylene [54c]

crystal structures – nearly all of these materials crystallize in a herringbone or herringbone-like structure in which molecules have consistently large overlap of their conjugated portions [23]. The molecules which comprise these lattices are typically tilted with respect to alternating rows of molecules, resulting in a zigzag of the molecular axes, as illustrated in Fig. 2.2A. In pyrene and perylene, illustrated in Fig. 2.2B and C, respectively, molecular dimers are arranged in a herringbone configuration.

There is still much controversy regarding electronic transport in these organic molecular materials. Possible explanations range from large pi–pi overlap to near band-like structure, although a true picture is still evolving. The only irrefutable conclusion at this point is that herringbone-like structure seems to be conducive to field-effect mobilities exceeding 1 $cm^2\ V^{-1}\ s^{-1}$ in organic molecular crystals. This may be because of an effective increase in bond strength because of the interlocking nature of the molecular layers. This is supported empirically by the fact that mobilities, and bond strengths, in covalent solids are usually much greater than in van der Waals crystals, whereas those in hydrogen-bonded solids (for example polymers) are almost always much less. The implication here would be that van der Waals bonded materials which adopt herringbone-like motifs could have higher crystal cohesion (higher effective bond strength) than van der Waals crystals which do not adopt the interlocking molecular motif. Recent reports on the nature of the OTFT materials in ultrathin layers at device interfaces may lead to a more detailed understanding of these effects [24]. This in no way implies that every herringbone motif van der Waals crystal will have mobility greater than 1 $cm^2\ V^{-1}\ s^{-1}$. In fact, film morphology, interface effects, and many other factors, in addition to crystal structure, all contribute to the performance of an organic semiconductor.

High structural order has also been correlated with good performance in OTFTs, through reports of very sharp, high-intensity XRD peaks, and the appearance of multiple orders of reflection in the spectra, as shown in Fig. 2.3 [25]. Also of note are rocking curve X-ray data, where several groups have observed good TFT performance corresponding to very narrow X-ray peaks – indicating high lamellar order-

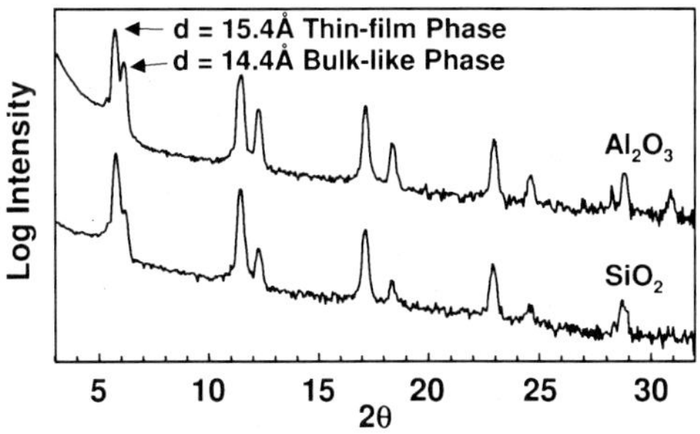

Fig. 2.3. Five orders of reflection in a thin film X-ray diffraction scan, indicating the high level of crystallinity often present in pentacene thin films deposited on different dielectrics. These results also show the presence of the two most common polymorphs of pentacene, the bulk-like phase and the thin-film phase [25].

ing in the materials. Many factors contribute to the maximum mobility obtainable in these weakly bonded organic crystals; most remain poorly understood.

2.2.4 Device Architecture

In addition to improving structural order, reliable improvements of device performance have been demonstrated through optimization of device interfaces and architectures. The architectures are perhaps more easily addressed, and are also related to improving the structural order. Several possible constructions of TFT are known from the long history of silicon-based devices. These were the first architectures adopted for OTFTs also. Top and bottom contact, indicating the location of the source and drain electrodes with regard to the semiconductor, are the most widely used.

Top-contact OTFTs typically have the highest performance. This is most probably because of reduced contact resistance at the source and drain electrodes, related to reduced structural disorder at that interface. This can be most easily understood by thinking of depositing a layered crystal structure over existing topography on a substrate (the bottom-contact construction) in which the molecular layers are forced into a less ordered structure as the material is deposited on top of the electrodes, illustrated in Fig. 2.4. The schematic diagram in Fig. 2.4A represents the disorder than can be introduced in the first few critical monolayers of a device by inadequate control or engineering of those interfaces. In addition, dramatic morphology changes (shown in Fig. 2.4B [26]) and/or dewetting of the semiconductor near metallic source and drain electrodes have been directly observed.

The performance of bottom-contact devices, fairly easy to fabricate with traditional lithographic equipment, can also be improved by control of the interface where the metal source and drain layers make contact with the semiconductor.

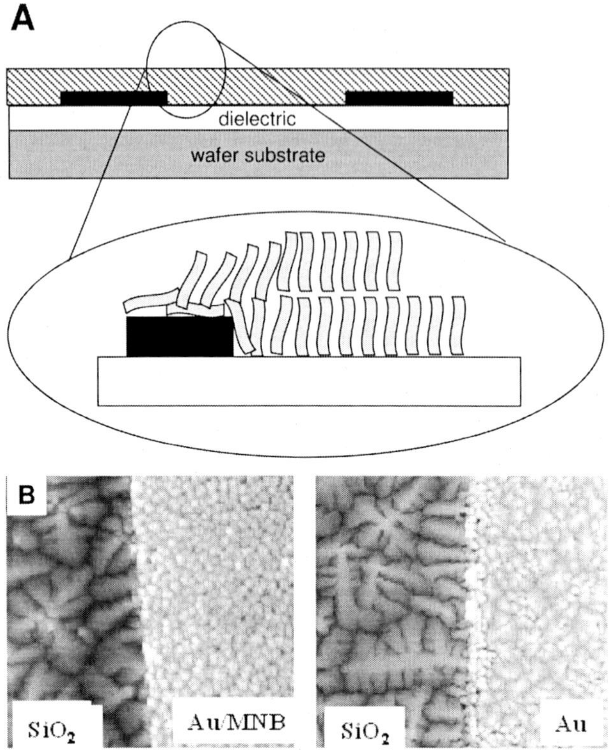

Fig. 2.4. The often dramatic effects bottom contacts can have on molecular ordering in organic semiconductors like pentacene. A. Schematic diagram of the type of disorder introduced in pentacene's lamellar structure as thin-film growth encounters a step, for example a bottom-contact electrode. B. An AFM topographical image of the change in grain size, shape, and texture as a similar edge is encountered (right image), and how that edge can be masked by use of self-assembled monolayers on the metal electrodes [26].

Early work from Jackson [26] and separately Dimitrakopoulos [27] showed that simple self-assembled monolayers (SAMs) on metal electrodes can affect molecular layer growth and enhance crystallinity at the interface where contact resistance could contribute to poor performance. Atomic-force microscopy and SEM based methods showed striking detail at the these interfaces. Organic materials were observed to nucleate and grow in dramatically different ways, or not at all, on and near the metal electrodes, as opposed to in the channel of the devices (on top of the oxide). The SAMs seem to serve as very thin layers which mask the character of the metal and serve to minimize the transition from organic to metal. The SAM can often be chosen to improve charge injection, similar to electron transport and hole injection layers in OLEDs, in which a "buffer" material is used to reduce the charge injection barrier at certain device interfaces. SAMs have the added advantage of being chemically bound to the surface of the electrode material; thereby, minimizing contributions of smaller-molecular-weight species migrating under high fields in OTFTs.

In top-contact configuration the organic material can be deposited on a perfectly planar interface, enabling the material to adopt a well-ordered morphology over the entire area. The source and drain contacts are then deposited directly on top of the organic material. Photolithography can be used, which can be tricky, because of the sensitivity of the semiconductor to process chemicals [28]. Shadow masks are very simple, but often limit resolution and complexity, and printing methods may also involve process chemicals or potentially performance-degrading heating during transfer or drying of the electrode material. A gentle method (low-temperature, low-energy, low solvent exposure) usually results in the least damage to the structural order of the semiconductor. Typical silicon-based semiconductor devices almost unilaterally use photolithography and a bottom contact configuration, but the organic electronics community has seen such dramatic improvement in device performance using top contacts, that printing and shadow masking methods are often preferred even if easy routes to scale-up and production are not yet proven. Mobility in organic devices can increase by nearly an order of magnitude on changing only the location of the source and drain electrodes. With the addition of surface treatments which better prepare the dielectric interface for a particular semiconductor, even larger improvements can be seen.

Even though shadow masking is typically regarded as less practical in a long-term production sense, it is being used as a production method for the organic layers in many OLED devices. In OLEDs also, the organic material is very sensitive to processing solvents, so shadow masks are used to pattern very fine features for red, green and blue sub pixels to approximately 150 ppi or 200 μm. Similar types of mask, with more complicated patterns, can be used to fabricate circuits in a variety of materials [29]. 3M has also developed polymeric aperture masks [30] which have proven useful in fabricating complex circuits with reduced damage to previously patterned layers, because of the soft polymeric nature of the masks, while maintaining the higher performing top-contact architecture. Figure 2.5 shows results of some of the polymer shadow masking work from 3M. Figure 2.5A shows a large, ~6-inch × 6-inch polymer shadow mask; Fig. 2.5B shows a photomicrograph of a circuit patterned using a mask as in Fig 2.5A; and Figs. 2.5C and 2.5D show device characteristics of circuits and single transistors, respectively, fabricated using the polymer masks.

Shadow masks have been made and adapted to deposition systems for many years. The simplest types of mask are often cut from metal foil or thin sheets, using heated wires, machining tools, electroplating and etching. Metal masks are inexpensive, commercially available, and easy to use. Their disadvantages are that they often have sharp/rough edges which can damage thin deposited layers, and they are prone to creasing. Silicon shadow masks are often fabricated using traditional photolithography and well-known deep-etching methods and as such are extremely thin, flat, and precise. The disadvantages of silicon masks include fragility (prone to cleave along crystal planes), limited size (based on batch lithographic processes), and high cost of manufacture. Polymer shadow masks can be made using

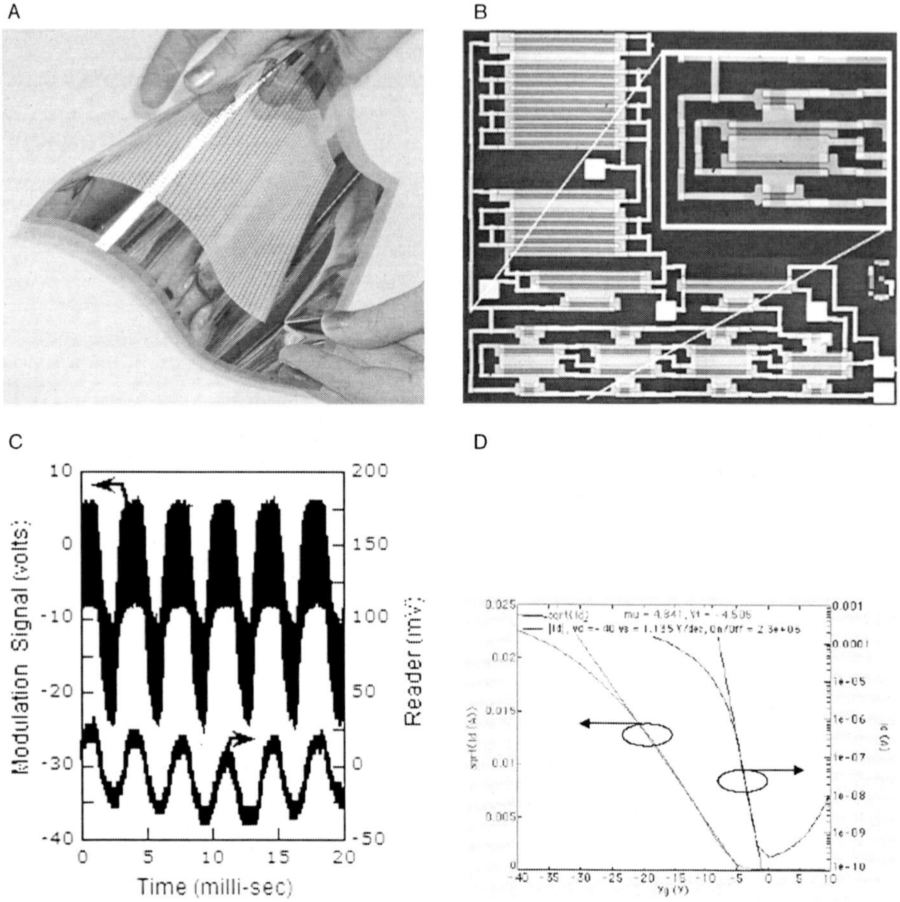

Fig. 2.5. Summary of some of the shadow masking and surface-treatment results from 3M. A. Photograph of a polymeric aperture mask [30b] used to pattern a wide range of circuits. This mask bears a ~6-inch × 6-inch deposition area, with high open aperture ratio. B. Photomicrograph taken of a one-bit RFID circuit patterned using the polymer shadow mask technology [30c]. C. Internal clock signal and the externally demodulated rf, or reader signal of the circuit in B (carrier signal was 125 kHz, and tag potential was ~30 V rms). D. Transfer characteristics for a single pentacene TFT patterned using the polymer shadow masks on a poly(alphamethyl) styrene surface treatment layer [30a]. Parameters extracted from linear fits to the data in the saturation regime are shown at the top of the figure. Specific capacitance of the alumina dielectric was 340 pF mm^{-2}. Channel dimensions were $L = 110$ μm, $W = 1000$ μm.

simple cutting tools (knives, razor blades, and die-cuts) and can be made with high precision by using excimer laser ablation through a light mask. The laser platform is very fast and accurate and can be operated using polymer roll-stock up to ~1 m wide. The polymer masks are robust, reusable, and easy to clean. They are less prone to creasing than metal masks and can be stretched to compensate for water uptake and sag in the mask over large areas. The primary disadvantage of polymer shadow masks is shared by all aperture mask methods – resolution is limited to 20–40 μm by scattering effects observed in vacuum as a "halo" of thickness non-uniformity around some features, although some researchers have gone to great lengths to show small feature sizes using shadow masks [31].

The consistent use of top-contact architecture with development of a variety of surface modifications to alter the interface where the semiconductor is deposited, detailed below, have contributed to reports of steadily increasing performance in pentacene and other organic materials.

2.2.5 Interfacial Control

Some of the earliest approaches to interfacial modification for organic semiconductors included results from the Penn State group involving vapor-deposited silanes [32], a material of choice in traditional semiconductor processing, because of ease of deposition.

Some improvement was observed with pentacene deposited on top of silane layers, but it was also observed that the silane deposition is not easy to control, and side-reactions often result in rough layers with considerable unreacted content remaining. An alternative approach relies on application of self-assembled monolayers which mimic vapor-deposited silanes [32, 33] on the dielectric interface of the organic devices. Figure 2.6 shows an overview of the different surface-treatment application methods discussed in this section.

Silane-based SAMs are typically applied from an organic solvent such as toluene at very low concentration (0.1% *w*/*w*) over a long period of time, overnight or a minimum of several hours, to form a dense, nearly full-coverage monolayer with a variety of termination chemistries possible. A simple rinse typically follows and the sample may be air dried and/or dried at moderate temperature (~100 °C) in an oven or vacuum oven. These silane-based SAMs provide molecularly well-ordered, single molecule thickness layers that assemble on silica surfaces in a well-controlled fashion and react with the surface completely leaving no residual reactive groups or unbound material at the interface with the organic semiconductor.

Phosphonic acid based SAMs [7c], which bind preferentially to alumina surfaces, can be prepared similarly, but it was also found experimentally that spin-coating a solution of 0.1% (*w*/*w*) SAM from a solvent such as toluene, followed by baking for 10–30 min on a hot plate at 100–120 °C, then a good rinse in clean toluene, and blow dry produced superior results in terms of phosphonic acid-SAM quality compared with overnight soaking in the same solution. This may be because of

Self-assembly of SAMs	Spin coating of SAMs	Spin coating of polymers
0.1-1 wt% solution Organic solvent Soak 1 hour to overnight Rinse/soak— fresh solvent Bake to dry	0.1 wt% solution Organic solvent Spin coat 500 rpm, 20 sec 2000 rpm, 40 sec Bake 100-130C ~30 min Rinse fresh solvent Blow dry (N2)	0.1 wt % polymer "good" organic solvent Spin coat 500 rpm, 20 sec 2000 rpm, 40 sec Bake 100-130C 5-30 min

Fig. 2.6. Some representative methods for effective surface-treatment preparation of both SAMs and polymers to improve the performance of pentacene TFTs.

elimination of the slow etching of the alumina dielectrics by the slightly acidic SAM solutions. Advantages of the SAMs include: wide range of terminal groups, wide range of attachment chemistries which bind molecules to the surface, and near crystalline ordering. It is likely that the spin-coating approach could save time when preparing silane samples also. Polymeric dielectrics typically require one additional step for subsequent application of SAM surface treatments. This added step is often a short exposure to oxygen plasma to generate functionality at the polymer surface to which SAM can bind [7b]. Conclusions from both types of SAMs included preferences for hydrophobic over hydrophilic-terminated SAMs, which roughly corresponds to increasing mobility as one increases contact angle observed on the SAM. Also reported is a preference for smooth and nearly defect-free layers, consistent with early reports of high structural order in the semiconductor improving performance. There has been some indication of templating effects, in which terminations which mimic periodicity in pentacene and other organic materials give somewhat higher results, but many factors complicate further conclusions. Table 2.1 summarizes some of the results reported for pentacene TFTs prepared using SAM-modified dielectrics.

Polymeric surface treatments are applied in similar fashion, namely spin-coating from a low-concentration solution of the polymer in organic solvent, e.g. 0.1% (w/w) polymer in toluene. Spin conditions can vary widely but, as reported by 3M [30a], 500 rpm for 20 s then 2000 rpm for 40 s provides $\sim$100 Å thick dry polymer films on smooth dielectrics with good surface quality and contact-angle uniformity. Spun layers are typically air-dried for several hours, or dried at an elevated temperature of 110–130 °C for 30 min in an oven or 5 min on a hot plate. These materials are typically not rinsed after they are baked, because the homopolymer version would redissolve, removing almost all of the polymer. In a recent report from Frisbie's group, in collaboration with 3M, device oxides were intentionally roughened

Tab. 2.1. Results from three groups using a wide variety of SAMs as surface treatments [7c, 33d, 52]. Several authors have shown loose correlations between improved OTFT performance and higher contact angle of the SAM. Varying results in similar surface treatments indicate that many mechanisms are at work, including contact angle, but probably also involving deposition conditions, surface roughness, and chemistry.

SAM	Mobility ($cm^2\ V^{-1}\ s^{-1}$)
Octadecyltrichlorosilane (vapor)	1.56
t-Butyldiphenylchlorosilane	1.61
Phenyltrichlorosilane	0.17
Octadecyltrichlorosilane (solution)	0.96
Butyltrichlorosilane	0.61
3-Chloropropyltrichlorosilane	0.71
3-Bromopropyltrichlorosilane	0.74
Trichloro(trifluoropropylsilane)	0.03
Perfluorooctyltrichlorosilane	0.15
Phenylethyltrichlorosilane	0.71
(Chloromethyl)phenyltrichlorosilane	0.56
(Chlorosulfonylphenyl)ethyltrichlorosilane	0.36
Alumina control	1.1
Phenylphosphonotridecane	1.8
Phenylphosphonohexadecane (exceptional)	3.3
Phenylphosphonohexadecane (grand average)	2.0
Perfluorodecane	0.4
Phosphonobenzene	0.1
Methylphosphonoundecanoate	0.4
Chlorophosphonohexadecane	0.5
Phosphonotetramethylhexadecane	1.8
Phosphonooctane	2.1

and then coated with polymeric surface treatments. TFTs fabricated with pentacene on top of these layers showed that polymeric surface treatments could mask significant roughness on the oxides, resulting in comparable or better performance than on unroughened oxides [34].

The Infineon group reported a pure polymer dielectric which has also been shown to improve pentacene performance, both as an unmodified polymer, and with a silane treatment performed on it [7b]. In addition, Samsung SDI has recently reported a proprietary polymer dielectric which enables them to achieve high mobility in pentacene TFTs [12]. Table 2.2 emphasizes some of the reports of increasing pentacene mobility.

Tab. 2.2. Reports of elevated mobility in pentacene TFTs from the past 10 years. The table shows that although very high mobility is still rather rare, mobility exceeding 1 $cm^2\ V^{-1}\ s^{-1}$ has been widely reported [53].

Year	Group	Method	Mobility ($cm^2\ V^{-1}\ s^{-1}$)
1996	Penn State	Thermal oxide	0.62
1996/7	Penn State	TC	0.7
1997	Penn State	OTS	1.3
1997	Penn State	OTS, TC, double layer	1.5
1999	Penn State	OTS, TC	2.1
2000	Penn State	OTS, BC	1.7
2000	Necliudov	OTS, TC	1.1
2002	Sheraw	OTS, BC, PVA patterned	1.2
2002	Shtein	OTS, TC	1.6
2002	Infineon	PVP + OTS	3.0
2003	3M	Polysulfonic acid SAM, TC	3.4
2003	3M	Polystyrene, TC	>5.0
2003	Knipp	OTS	1.1
2004	Choi	PVP + OTS	1.4
2004	Yoneya	Silane SAM, BC	1.1
2004	Schroeder	PVA/PVP, doped TC	2.6
2005	Jang	PVP, BC	1.8
2005	Samsung	Proprietary dielectric	7.0

TC = top contact, BC = bottom contact, OTS = octadecyltrichlorosilane, PVA = polyvinylalcohol, PVP = polyvinylphenol

2.3 Structure–Property Relationships

There has also been considerable discussion in the literature regarding the exact nature of the performance increase in organic semiconductors deposited on top of interfacial layers which modify the properties of the dielectrics. There are definitely structure–property effects, as first reported by Jackson when the "thin-film phase" [35] of pentacene was observed for higher-mobility surface-treated samples. The implication was that surface treatments were enabling the thin-film phase to maintain a metastable structure at the device interfaces of surface treated devices, as opposed to untreated devices, which typically have a larger fraction of bulk-like phase. IBM also reported direct observation of nucleation and growth differences on various surfaces imaged in LEEM [36]. 3M saw similar effects with alumina treated with SAMs and polymers [7c, 30a], supporting the conclusion that some surface treatments promote thin-film phase growth in a more energetically favorable fashion, so that a larger proportion of the film has a more perfect thin-film phase structure. The thin-film phase is generally accepted as being metastable, perhaps due to a report from Jackson's group showing thin-film phase samples which revert or restructure themselves to bulk phase on exposure to process solvents [37], but surface treatments seem to improve the structural stability of or reduce stress in the thin-film phase layers slightly, so that more of the total thickness of the organic material deposited can adopt the thin-film phase.

This has led to the possibility that materials might be designed which have high performance and only adopt a single polymorph, making that polymorph more structurally stable than a counterpart which can adopt different crystalline phases. Designing in bulky substituents on the positions most susceptible to reaction is another approach showing some potential toward structural stability, with the added benefit that some derivatives are soluble [38].

Mesoscale morphology, on the other hand, seems to be potentially misleading as a means of relating grain structure to electrical properties. Most groups reported increasing mobility with increasing grain size, but several reports indicate that the details of the relationship between grain size and mobility are more complicated. It also seems that smaller, rather than larger, grain size has since been observed to correspond to elevated mobility. This could be indicative of a different transport mechanism for high-mobility thin films. For example between 1 and 3 $cm^2 V^{-1} s^{-1}$, collected literature reports seem to indicate increasing mobility with increasing grain size, but very high mobility films, >3 $cm^2 V^{-1} s^{-1}$, seem to indicate that mobility decreases with increasing grain size (based on very limited reports of high mobilities). Atomic-force microscopy (AFM) has been widely used to characterize grain size and grain-size distributions in organic thin films. Researchers have often found that the final film structure observed by AFM in completed devices, often bears little or no resemblance to the interfacial structure which most critically determines the transport properties in a device. Imaging and observing incomplete films has been the best way for researchers to currently do to understand structure at device interfaces. AFM has been used to image various stages of growth in pentacene films, still without completely simple conclusions. "Snapshots" of film growth can be obtained by depositing many samples with varying film thickness. Unfortunately, these thin samples can rarely be made into complete functional TFTs. Figure 2.7 shows two examples of such studies which come to mixed conclusions. Figure 2.7A shows large grains growing on untreated device oxides, with smaller grains growing on SAM-treated oxides whereas Fig. 2.7B shows large grains on SAM-treated devices. These seemingly conflicting reports indicate that interfacial quality and morphology are only part of the story – deposition conditions and sample handling and testing are equally important. A second method, ultra thin-film synchrotron X-ray measurements like those from Frisbie's group, may eventually be able to distinguish very slight differences between the initial layers of organic semiconductors deposited on a variety of substrates [25]. Figure 2.8 emphasizes these results – researchers showed the subtle structural difference between device thickness films (~300 Å) of pentacene and sub-monolayer films grown on the same oxide surface, as observed by grazing incidence X-ray diffraction. Several researchers have observed that initial growth stages of many organic semiconductors are consistent with classical nucleation

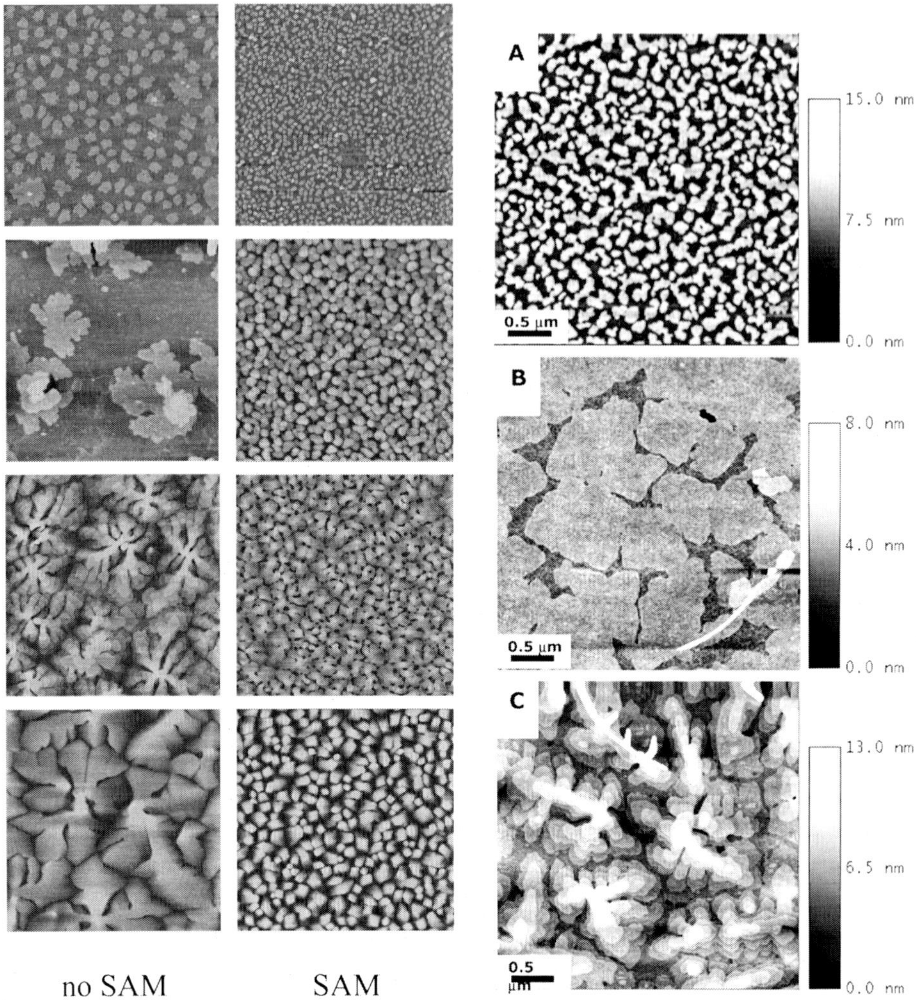

Fig. 2.7. Two sets of AFM topographs of pentacene thin films of different coverage. The left two columns, from Kalb et al. [51], compare untreated dielectrics with SAM-modified dielectrics – much greater nucleation density and smaller grain sizes are observed for the SAMs. Coverage, from the top down: 0.5, 3, 6, 30 nm; image size is 4 μm × 4 μm. Topographs in the right column, from Kelley et al. [7c], show pentacene thin films grown on SAMs. Coverage, from the top down: 1, 3.5, 7.5 nm; image size is 2.5 μm × 2.5 μm. Both groups show similar nucleation on SAM-modified dielectrics, but grain sizes in the right-most column are significantly larger than grains on the SAM-modified surfaces shown in the center column. This again emphasizes the difficulty in understanding the complex interaction of nucleation and growth in organic thin films with modification of interfaces and deposition conditions.

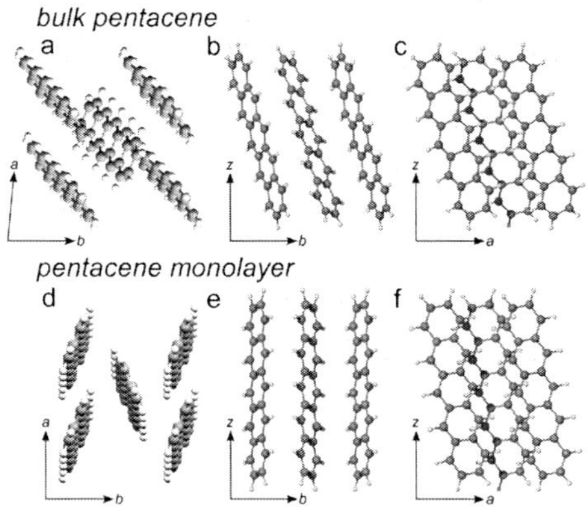

Fig. 2.8. Normal (left) and side views (center, right) of the ab planes of bulk pentacene and the proposed monolayer structures. The monolayer views were constructed from a model based on grazing-incidence X-ray diffraction data [24].

theory, but a clear understanding of the relationship between nucleation, growth, crystallinity, and device performance remains elusive [39].

2.4 Continuing Reports of High Mobilities

There have been many reports of elevated mobilities in pentacene TFTs. Table 2.2 summarizes some of these results, some of which have been discussed in the text. New reports appear frequently, continuing to push the limits for high-mobility pentacene devices.

The limited number of TFT mobility reports exceeding single crystal mobility led to some concerns, and further research in the field. Some speculated that the thin-film phase, mentioned above, could be a higher-performing phase that has yet to be isolated as a single crystal; others speculated on everything ranging from new transport mechanisms to improper measurements. Taking a more pragmatic approach, scattered reports of mobilities exceeding single crystal values have been reported for more than ten years. These scattered reports point to an inability to control process and surface conditions to the extent required to obtain these values regularly enough to understand their nature. 3M is the first to report any means of process control for high-mobility devices, with a lengthy (~6 month) study in which a large majority of freshly fabricated devices yielded high mobilities [30a]. During this study, equipment was material dedicated and conditions carefully controlled. When new materials were introduced or process conditions changed (normal operating procedure for laboratory activity), the process became less stable, and the results less predictable. Unfortunately, these initial device mobilities degrade with time to varying degrees. 3M has observed devices that remain relatively stable (mobility and threshold voltage) over the course of 6 months, with mobility dropping by <25%, but also observed devices whose mobility begin to fade within hours of fabrication and exposure to air. Initial simple experiments performed without breaking vacuum between depositions, and maintaining inert atmospheres during testing are encouraging. The major fraction of devices fabricated are typically those with limited high-mobility life times. Further work to understand the nature of the limited lifetime of the high mobilities is needed. There has been some indication there is no simple, single cause of the degradation – both structural and environmental factors may contribute. The transient nature of the high mobility makes it very difficult to isolate the causes. In addition, few reports of encapsulated devices exist, making it uncertain whether simple methods to protect devices from the environment could be sufficient to enable applications, although initial results are encouraging [10, 40]. What has been heartening though, is a new report of very high single-crystal mobility, $\sim$35 $cm^2 V^{-1} s^{-1}$ for pentacene, reported by Palstra's group, after extensive purification to remove 6,13-pentacenequinone and to inhibit the introduction of 6,13-dihydropentacene [41]. The question is no longer whether or not to believe high mobilities in organic TFTs, but how to stabilize them and harness their performance for integrated circuit applications. Since the 3M reports, other researchers are also using statistical methods to design series of experiments and interpret their results [42]. These experiments are critically important to establishing tunable fabrication methods for organic semiconductors. These statistically significant, designed experiments may be the only way to understand and compensate for differences in processing methods – each research group has differences in their deposition chambers related to geometric layout and source configuration – and these differences seem to have a large effect on each group's ability to prepare high-performance devices. It is well-known in standard microelectronics processes that statistical process control is very important; it may be found it is even more important for organic microelectronic devices.

2.5 Performance in Practice

Some progress has been made in fabricating pentacene integrated circuits with a variety of device performance ranges. The closest to a commercial product are those reported by Polymer Vision, in the Philips Technology Incubator. They are using a soluble analogue of pentacene which can be spin coated and thermally converted to pentacene in a fairly traditional lithography process to yield low-mobility devices that are sufficient to drive electrophoretic displays [10]. In addition, Polymer Vision fabricated all of their devices on flexible polymer foils laminated to rigid substrates for processing. After the devices are completed, the foils are removed from the supports, and flexible organic display backplanes result. Stability for the Polymer Vision device seems adequate for an entry product, with degradation of the order of 20% over the first three months of lifetime. Longer-term data are still being collected. Literature reports indicate device complexity is increasing; devices include a 15-inch diagonal OTFT backplane for LCD-type displays from

Samsung SDI [12] and RFID circuitry based on pentacene from 3M [30c] which has been reported to operate as high as 8 MHz.

The emergence of complex circuitry and entry products has indicated a shift in the focus of pentacene studies from maximized initial performance to designed long-term or application-specific performance. This shift has also resulted in a decrease in the number of reports highlighting details of the pentacene device fabrication process and fewer details about the design of devices using pentacene. Several groups have continued to try to understand the limitations of pentacene devices and report on those boundaries as a means of understanding the nature of transport and failure in organic devices. The PARC group is among those who have contributed most in this direction [43]. Their report later in this volume focuses on polymer TFTs, but some of their work has encouraged other groups to investigate bias stress and stability and to begin including such data in their own publications.

2.6
The Future of High-performance Organic Transistors

In short, there are several well established routes to improving performance in organic devices. The first is to construct top-contact devices using simple methods such as shadow masking. Simply putting source and drain contacts on top of the semiconductor often results in increased mobility, simply because of improved contact resistance and improved crystalline order at device interfaces. A wide range of surface treatments, placed on bottom-contact metal electrodes and/or on-device dielectrics have also been shown to be very effective in raising mobility in OTFTs. In addition, controlling and understanding deposition conditions and how they relate to structural order in the thin organic films has also contributed to improved devices.

Along with groups trying to understand and stabilize the performance of pentacene devices, and those trying to make the best of the situation by using pentacene in real devices, there are also groups trying to find alternatives, by combining the best attributes of a variety of materials. Along with optimization of architecture and interfacial control, there is also the potential to optimize the material itself. Anthony's group at the University of Kentucky follows such an approach where materials are engineered for solubility, structure, and performance [40]. His chapter will focus on some of these results, and it should be noted that his approach holds much promise for developing a feedback system, including design and testing for new materials. Understanding and designing in functional groups which would block sensitivity to impurities or atmosphere could help to stabilize device performance. Increased solubility may also aid the development of more effective purification methods. Further understanding of transport mechanisms and failure mechanisms in organics and how that relates to structure and purity could go a long way toward improving and stabilizing performance. In addition, a group at NHK has found that unmodified pentacene can be soluble in hot liquid crystal solvents, and that these solutions can be used to produce well-oriented grains of pentacene [44]. Another approach to producing aligned layers of pentacene has been shown by a group at National Cheng Kung University in Taiwan using photoaligned polyimide as substrate [45]. These approaches are interesting because they provide potential methods for more uniformly oriented and distributed grains of pentacene, but there is currently insufficient data to predict their usefulness in stabilizing device performance.

Reports from 3M [46] and from Lucent Technologies [47] have indicated that combining thiophenes and acenes may also be a promising approach. The thiophene systems are somewhat easier to synthesize and purify, because of their greater solubility, but the acenes are typically higher-performing. The combination of the two has shown some intriguing initial results.

Additionally, reports of stacked materials [48] may help to stabilize structure in the thin-film active layer. Such strategies have also worked in some OLED systems, and it is possible that these approaches will extend to the organic materials in OTFTs. Along the same lines, designed encapsulation methods which effectively bury the organic semiconductor layer, protecting it from the environment, could provide a variety of routes to achieving acceptable performance for a variety of simple applications. The initial hope was that these materials would be less sensitive to the environment than OLED materials, and that still seems to be true, but some level of encapsulation may still be required, as it is in silicon-based electronics. This is an area that has not been explored in detail, and could probably gain much from the progress made in the last 10 years in OLED encapsulation.

Lastly, continued application development should also help. Understanding the specifications for intended applications and exact knowledge of the inadequacies of a chosen set of materials often lead to breakthroughs. The goal for the last 10 years in the OTFT community has been to make the best devices possible. It seems that that goal has greatly improved device performance and has advanced our understanding of transport, architecture and crystalline order in the devices. The goal for the next 10 years should be to understand the required performance levels for specific applications and to design materials, devices and systems to accommodate those requirements – in terms of both initial performance and device lifetime.

References

1 A. Pochettino, *Acad. Lincei. Rendiconti*, **1906**, *15*, 355.

2 M. Pope, C.E. Swenberg, *Electronic Processes in Organic Crystals and Polymers*, 2nd Edition, Oxford University Press, New York, **1999**.

3 C.W. Tang, S.A. Van Slyke, *Appl. Phys. Lett.*, **1987**, *51*, 913.

4 C.K. Chiang, C.R. Fincher, Y.W. Park, A.J. Heeger, H. Shirakawa, E.J. Louis, S.C. Gau, A.G. MacDiarmid, *Phys. Rev. Lett.*, **1977**, *39*, 1098.

5 M. Halik, H. Klauk, U. Zschieschang, G. Schmid, S. Ponomarenko, S. Kirchmeyer, W. Weber, *Adv. Mater.*, **2003**, *15*, 917, and references therein.

6 (a) Y. Lin, D. Gundlach, T. Jackson, *Dig. 54th Device Research Conf.*, Univ. California, Santa Barbara, June **1996**, 80. (b) D. Gundlach, Y. Lin, T. Jackson, S. Nelson, D. Schlom, *IEEE Electr. Dev. Lett.*, **1997**, *18*, 87.

7 (a) M. Shtein, J. Mapel, J.B. Benziger, S.R. Forrest, *Appl. Phys. Lett.*, **2002**, *81*, 268. (b) H. Klauk, M. Halik, U. Zschieschang,

G. Schmid, W. Radlik, W. Weber, *J. Appl. Phys.*, **2002**, *92*, 5259. (c) T. Kelley, L. Boardman, T. Dunbar, D. Muyres, M. Pellerite, T. Smith, *J. Phys. Chem, B.*, **2003**, *107*, 5877.
8 For example, pentacene is available from Aldrich Chemical Co., H.W. Sands and TCI America (Tokyo Chemical Industry America).
9 A. Tsumura, H. Koezuka , T. Ando, *Appl. Phys. Lett.*, **1986**, *49*, 1210.
10 G. Gelinck, H. Edzer, A. Huitema, E. van Veenendaal, E. Cantatore, L. Schrijnemakers, J. van der Putten, T. Geuns, M. Beenhakkers, J. Giesbers, B.-H. Huisman, E. Meijer, E. Benito, F. Touwslager, A. Marsman, B. van Rens, D. de Leeuw, *Nat. Mater.*, **2004**, *3*, 106.
11 J. Rogers, Z. Bao, K. Baldwin, A. Dodabalapur, B. Crone, V. Raju, V. Kuck, H. Katz, K. Amundson, J. Ewing, P. Drzaic, *Proc. Natl. Acad. Sci. U.S.A.* **2001**, *98*, 4835.
12 M. Hong, B. Kim, Y. Lee, K. Song, J. Oh, J. Kim, T. Choi, M. Ryu, K. Chung, S. Lee, B. Koo, J. Shin, E. Jeong, L. Pu, *Soc. for Information Display Digest*, **2005**, 23 (paper 3.5).
13 Compendex Database keyword search returned 171 titles for "pentacene" in 2004 records.
14 Gradient sublimation has been used for many years as a means to purify materials and many variations exist and have been proven effective for organic purification. A detailed description of the vapor phase transport method applied to organic materials was given by: Ch. Kloc, P.G. Simpkins, T. Siegrist, R.A. Laudise, *J. Cryst. Growth*, **1997**, *182*, 416.
15 A complete list of accomplishments would require an inordinate amount of space – many industrial research groups including IBM, Lucent Technologies, 3M, Infineon, PARC, EInk, Samsung, Hitachi, NHK (and the list goes on....) as well as academic groups including Penn State, University of Minnesota, University of Kentucky, Chiba University, etc have shown exciting results for organic circuitry and devices. Particular examples will be called out in the text, as opposed to trying to formulate an exhaustive list.
16 See Chapter 1 in this volume for a detailed discussion of several of the TFT architectures commonly used as test vehicles.
17 M. Payne, S. Parkin, J. Anthony, C.-C. Kuo, T. Jackson, *J. Am. Chem. Soc.* **2005**, *127*, 4986.
18 S.-H. Kim, Y.-S. Yang, J.-H. Lee, J.-I. Lee, H.-Y. Chu, H. Lee, J. Oh, L.-M. Do, T. Zyung, *Optical Materials* **2003**, *21*, 439.
19 V. Sundar, J. Zaumseil, V. Podzorov, E. Menard, R. Willett, T. Someya, M. Gershenson, J. Rogers, *Science*, **2004**, *303*, 1644.
20 R. Zeis, T. Siegrist, Ch. Kloc, *Appl. Phys. Lett.* **2005**, *86*, 022103-1.
21 E. Menard, V. Podzorov, S.-H. Hur, A. Gaur, M. Gershenson, J. Rogers, *Adv. Mater.* **2004**, *16*, 2097.
22 M. Mas-Torrent, P. Hadley, X. Ribas, C. Rovira, *Synth. Met.* **2004**, *146*, 265.
23 E. Silinsh, V. Capek, *Organic Molecular Crystals*, American Institute of Physics Press, New York, **1994**.
24 S. Fritz, S. Martin, C. Frisbie, M. Ward, M. Toney, *J. Am. Chem. Soc. (Communication)*, **2004**, *126*, 4084.
25 P. Pesavento, S. Fritz, C. Frisbie, University of Minnesota, **2005**, (private communications).
26 J. Wang, D. Gundlach, C. Kuo, T. Jackson, *41st Electronic Mater. Conf. Dig.*, **1999**, 16.
27 I. Kymissis, C. Dimitrakopoulos, S. Purushothaman, *IEEE Trans. Electr. Dev.*, **2001**, *48*, 1060.
28 (a) C. Sheraw, J. Nichols, D. Gundlach, J. Huang, C. Kuo, H. Klauk, T. Jackson, M. Kane, J. Campi, F. Cuomo, B. Greening, *Tech Dig. Int. Electron Devices Meet.*, **2000**, 619. (b) C. Sheraw, L. Zhou, J. Huang, D. Gundlach, T. Jackson, M. Kane, I. Hill, M. Hammond, J. Campi, B. Greening, J. Francl, J. West, *Appl. Phys. Lett.*, **2002**, *80*, 1088. (c) I. Kymissis, Ch. Dimitrakopoulos,

S. Purushothaman, *J. Vac. Sci. Technol. B.*, **2002**, *20*, 956.

29 T. Brody, *J. Soc. for Info. Display*, **1996**, *4*, 113.

30 (a) T. Kelley, D. Muyres, P. Baude, T. Smith, T. Jones, *Mater. Res. Soc. Symp. Proc.*, **2003**, *771*, L6.5.1. (b) D. Muyres, P. Baude, S. Theiss, M. Haase, T. Kelley, P. Fleming, *J. Vac. Sci. Technol. A.*, **2004**, *22*, 1892. (c) P. Baude, D. Ender, M. Haase, T. Kelley, D. Muyres, S. Theiss, *Appl. Phys. Lett*, **2003**, *82*, 3964.

31 G. Kim, M. van den Boogaart, J. Brugger, *Microelectronic Engineering*, **2003**, *67–68*, 609.

32 (a) Y. Lin, D. Gundlach, T. Jackson, *Mater. Res. Soc. Spring Mtg. Dig.*, **1997**, 143. (b) D. Gundlach, H. Klauk, C. Sheraw, C.-C. Kuo, J. Huang, and T. Jackson, *Intl. Electron Devices Mtg Tech. Dig.*, **1999**, 111.

33 (a) L. Kosbar, Ch. Dimitrakopoulos, D. Mascaro, *Mat. Res. Soc. Symp. Proc.*, **2001**, *665*, C10.6.1. (b) M. Swiggers, G. Xia, J. Slinker, A. Gorodetsky, G. Malliaras, R. Headrick, B. Weslowski, R. Shashidhar, C. Dulcey, *Appl. Phys. Lett.*, **2001**, *79*, 1300. (c) A. Salleo, M. Chabinyc, M. Yang, R. Street, *Appl. Phys. Lett.*, **2002**, *81*, 4383. (d) K. Pernstich, C. Goldmann, C. Krellner, D. Oberhoff, D. Gundlach, B. Batlogg, *Synth. Met.*, **2004**, *146*, 325.

34 S. Fritz, T. Kelley, C. Frisbie, *J. Phys. Chem. B.*, **2005**, *109*, 10574.

35 D. Gundlach, Y. Lin, T. Jackson, S. Nelson, D. Schlom, *IEEE Electr. Dev. Lett.*, **1997**, *18*, 87. [The term "thin-film phase," was not used in this publication, but was adopted later to refer to the ~15 Å *d*-spacing observed in some, often higher mobility pentacene layers.]

36 F.-J. Meyer zu Heringdorf, M. Reuter, R. Tromp, *Nature*, **2001**, *412*, 517.

37 D. Gundlach, T. Jackson, D. Schlom, S. Nelson, *Appl. Phys. Lett.*, **1999**, *74*, 3302.

38 (a) J. Anthony, D. Eaton, S. Parkin, *Org. Lett.*, **2002**, *4*, 15. (b) J. Anthony, J. Brooks, D. Eaton, S. Parkin, *J. Am. Chem. Soc.*, **2001**, *123*, 9482.

39 (a) F. Meyer zu Heringdorf, M. Reuter, R. Tromp. *Nature*, **2001**, *412*, 517. (b) F. Meyer Zu Heringdorf, M. Reuter, R. Tromp, *Appl. Phys. A*, **2004**, *78*, 787. (c) S. Pratontep, M. Brinkmann, F. Nuesch, L. Zuppiroli, *Synth. Met.*, **2004**, *146*, 387. (d) S. Iannotta, T. Toccoli, *J. Polym. Sci. B: Polym. Phys.*, **2003**, *41*, 2501. (e) U. Haas, A. Haase, H. Maresch, B. Stadlober, G. Leising, *IEEE Internat. Conf. On Polym. Adhesives, Polytronic*, **2004**.

40 (a) T. Kelley, P. Baude, C. Gerlach, D. Ender, D. Muyres, M. Haase, D. Vogel, S. Theiss, *Chem. Mater.* **2004**, *16*, 4413. (b) Y. Inoue, S. Tokito, K. Ito, T. Suzuki, *J. Appl. Phys.*, **2004**, *95*, 5795. (c) P. Mach, S. Rodriguez, R. Nortrup, P. Wiltzius, J. Rogers, *Appl. Phys. Lett.*, **2001**, *78*, 3592. (d) Y.-L. Loo, T. Someya, K. Baldwin, Z. Bao, P. Ho, A. Dodabalapur, H. Katz, J. Rogers, *Proc. Nat. Acad. Sci.*, **2002**, *99*, 10253.

41 O. Jurchescu, J. Baas, T. Palstra, *Appl. Phys. Lett.*, **2004**, *84*, 3061.

42 W.-S. Hong, B. Koo, I. Kang, *Electrochem. Sol. St. Lett.*, **2004**, *7*, G247.

43 R. Street, D. Knipp, A. Volkel, *Appl. Phys. Lett.*, **2002**, *80*, 1658.

44 H. Fujikake, T. Suzuki, F. Isaka, F. Sato, *Jpn. J. Appl. Phys.*, **2004**, *43*, L536.

45 W. Chou, C. Kuo, H. Cheng, Y. Mai, S. Lin, C. Liao, C. Chang, F. Tang, J. Hwang, *Chem. Mater*, **2004**, *16*, 4610.

46 J. Merlo, C. Newman, C. Gerlach, T. Kelley, D. Muyres, S. Fritz, M. Toney, and C. Frisbie, *J. Am. Chem. Soc.* **2005**, *127*, 3997.

47 J. Laquindanum, H. Katz, A. Lovinger, *J. Am. Chem. Soc.*, **1998**, *120*, 664.

48 T. Jackson, Y.-Y. Lin, D. Gundlach, H. Klauk, *IEEE J. Sel. Top. Quantum Electr.*, **1998**, *4*, 100.

49 Figure 2.1 is courtesy of Hagen Klauk of the Max Planck Institute for Solid State Research in Stuttgart. References

for Fig. 2.1 include 7b, 17, 31a, b, and 30a, as well as the following: (a) V. Podzorov, S. Sysoev, E. Loginova, V. Pudalov, M. Gershenson, *Appl. Phys. Lett.*, **2003**, *83*, 27. (b) V. Sundar, J. Zaumseil, V. Podzorov, E. Menard, R. Willett, T. Someya, M. Gershenson, J. Rogers, *Science*, **2004**, *303*, 1644. (c) A. Assadi, C. Svensson, M. Willander, O. Inganas, *Appl. Phys. Lett.*, **1998**, *53*, 195. (d) H. Koezuka, A. Tsumura, H. Fuchigami, K. Kuramoto, *Appl. Phys. Lett.* **1993**, *62*, 1794. (e) Z. Bao, A. Dodabalapur, A. Lovinger, *Appl. Phys. Lett.* **1996**, *69*, 4108. (f) H. Sirringhaus, N. Tessler, R. Friend, *Science*, **1998**, *280*, 1741. (g) H. Sirringhaus, P. Brown, R. Friend, M. Nielsen, K. Bechgaard, B. Langeveld-Voss, A. Spiering, R. Janssen, E. Meijer, P. Herwig, D. de Leeuw, *Nature*, **1999**, *401*, 685. (h) G. Wang, J. Swensen, D. Moses, A. Heeger, *J. Appl. Phys.*, **2003**, *93*, 6138. (i) D. Kim, Y. Park, Y. Jang, H. Yang, Y. Kim, J. Han, D. Moon, S. Park, T. Chang, C. Chang, M. Joo, C. Ryu, K. Cho, *Adv. Funct. Mater.*, **2005**, *15*, 77. (j) G. Horowitz, F. Deloffre, F. Garnier, R. Hajlaoui, M. Hmyene, A. Yassar, *Synth. Metals*, **1993**, *54*, 435. (k) P. Herwig, K. Müllen, *Adv. Mater.* **1999**, *11*, 480. (l) A. Afzali, C. Dimitrakopoulos, T. Breen, *J. Am. Chem. Soc.* **2002**, *124*, 8812. (m) M. Ichikawa, H. Yanagi, Y. Shimizu, S. Hotta, N. Suganuma, T. Koyama, Y. Taniguchi, *Adv. Mater.*, **2002**, *14*, 1272. (n) A. Tsumura, H. Koezuka, T. Ando, *Appl. Phys. Lett.*, **1986**, *49*, 1210. (o) X. Peng, G. Horowitz, D. Fichou, F. Garnier, *Appl. Phys. Lett.*, **1990**, *57*, 2013. (p) R. Haddon, A. Perel, R. Morris, T. Palstra, A. Hebard, R. Fleming, *Appl. Phys. Lett.* **1995**, *67*, 121. (q) J. Laquindanum, H. Katz, A. Lovinger, A. Dodabalapur, *Chem. Mater.* **1996**, *8*, 2542. (r) C. Clarisse, M. Riou, M. Gauneau, M. Le Contellec, *Electronics Lett.*, **1988**, *24*, 674. (s) C. Pearson, J. Gibson, A. Moore, M. Bryce, M. Petty, *Electronics Lett.*, **1993**, *29*, 1377. (t) S. Nelson, Y.-Y. Lin, D. Gundlach, T. Jackson, *Appl. Phys. Lett.*, **1998**, *72*, 1854. (u) G. Horowitz, F. Garnier, A. Yassar, R. Hajlaoui, F. Kouki, *Adv. Mater.*, **1996**, *8*, 52. (v) G. Horowitz, R. Hajlaoui, F. Kouki, *Eur. Phys. J. AP*, **1998**, *1*, 361.

50 R. Ruiz, B. Nickel, N. Koch, L. Feldman, R. Haglund, A. Kahn, G. Scoles, *Phys. Rev. B.*, **2003**, *67*, 125406.

51 W. Kalb, P. Lang, M. Mottaghi, H. Aubin, G. Horowitz, M. Wuttig, *Synth. Met.*, **2004**, *146*, 279.

52 K. Pernstich, C. Goldmann, C. Krellner, D. Oberhoff, D. Gundlach, B. Batlogg, *Synth. Met.*, **2004**, *146*, 325.

53 References for Table 2.2 include previously cited references: 6a, b, 7a–c, 12, 28b, 30a, 31a, b, and additional references: (a) P. Necliudov, M. Shur, D. Gundlach, T. Jackson, *J. Appl. Phys.*, **2000**, *88*, 6594. (b) D. Knipp, R. Street, A. Volkel, *Appl. Phys. Lett.*, **2003**, *82*, 3907. (c) S. Kim, H. Choi, J. Jang, *Appl. Phys. Lett.*, **2004**, *85*, 4515. (d) N. Yoneya, M. Noda, N. Hirai, K. Nomoto, M. Wada, J. Kasahara, *Appl. Phys. Lett.*, **2004**, *85*, 4664. (e) R. Schroeder, L. Majewski, M. Grell, *Appl. Phys. Lett.*, **2004**, *84*, 1005. (f) J. Jang, S. Han, *SID Digest*, **2005**, *3.2*, 10.

54 Figure 2.2 is courtesy of Dr. Myles Brostrom, 3U Corporate Research Analytical Laboratory. Mercury 1.4.1 software was used to create images from structural CIF files. (The Cambridge Crystallographic Data Center, 12 Union Road, Cambridge, CB2 1EZ, UK, +44 1223336408. Copyright CCDC 2001 2005; http://www.ccdc.com.ac.uk/mercury.)

Structure data was obtained from the following references: (a) pentacene: C. Mattheus, A. Dros, J. Baas, A. Meetsma, J. de Boer, J. Palstra, *Acta Crystallog, Sect C: Cryst. Struct. Commun.* (2001), *57*, 939. (b) pyrene: C. Frampton, K. Knight, N. Shankland, K. Shankland, *J. Mol. Struct.* (2000), *520*, 29. (c) perylene: C. Nather, H. Bock, Z. Haulas, T. Hauck, *Organometallics*, (1998), *17*, 4707.

3
Engineered Pentacenes

John E. Anthony

3.1
Introduction

Pentacene (**1**) has become the premier organic semiconductor for many applications, and is the standard against which most new materials are compared [1–6]. The uniquely exploitable properties of this acene are, in part, because of the nature of the native solid-state order of pentacene and its order in crystalline films, which is dominated by strong two-dimensional edge-to-face interactions leading to a packing of molecules in a "herringbone" arrangement (Fig. 3.1) [7]. In evaporated films, pentacene deposits with its long axis roughly perpendicular to the substrate, resulting in strong two-dimensional interactions parallel to the substrate (Fig. 3.1, right) [8].

Pentacene routinely yields field-effect transistor (FET) devices with reliable hole mobility of 1 $cm^2\ V^{-1}\ s^{-1}$ [6], with mobility > 3 $cm^2\ V^{-1}\ s^{-1}$ reported for thin-film devices on polymer gate dielectrics [9]. For transistors fabricated on single crystals of pentacene, the measured mobility approaches 60 $cm^2\ V^{-1}\ s^{-1}$ [10].

Disadvantages of the use of pentacene in organic electronics include its insolubility and its susceptibility to degradation by several pathways, including endoperoxide formation (**2**) and "butterfly" dimerization (**3**, Fig. 3.2) [11]. The dominant edge-to-face arrangement of the pentacene molecules in the solid state may also limit the electronic properties of this material, because face-to-face interactions are predicted to lead to stronger electronic coupling between molecules [12, 13].

Recent efforts in the synthesis of high-performance organic materials have produced new materials such as benzodiselenophenes [14], acene–thiophene hybrid molecules [15], and functionalized oligothiophenes [16], and these compounds have been the subject of recent review articles [17]. Surprisingly few research programs have, however, successfully applied a "property engineering" approach to pentacene, even though it is already the molecule with the best device performance. The reasons behind this lack of published effort are manifold – addition of functional groups to pentacene typically alters its orbital energy levels, making an already "delicate" molecule even more susceptible to decomposition, and introduction of groups large enough to make pentacene soluble tend to do so by dis-

Organic Electronics, Materials, Manufacturing and Applications. Edited by Hagen Klauk

ISBN: 3-527-31264-1

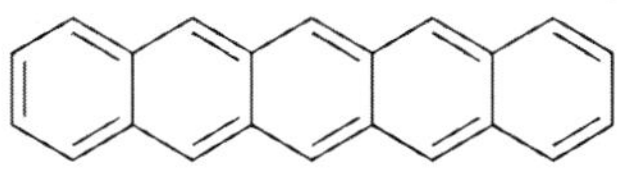

1

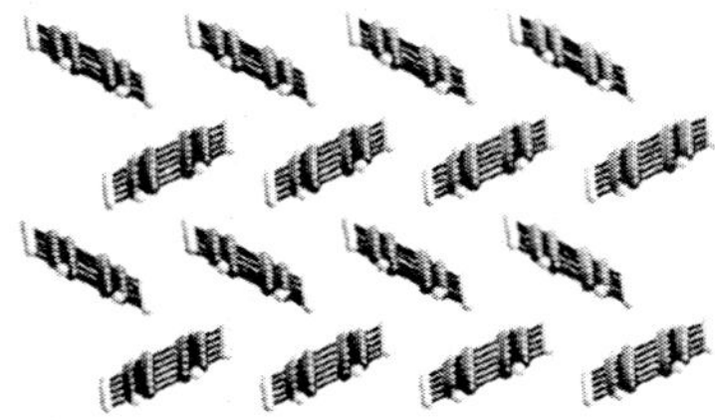

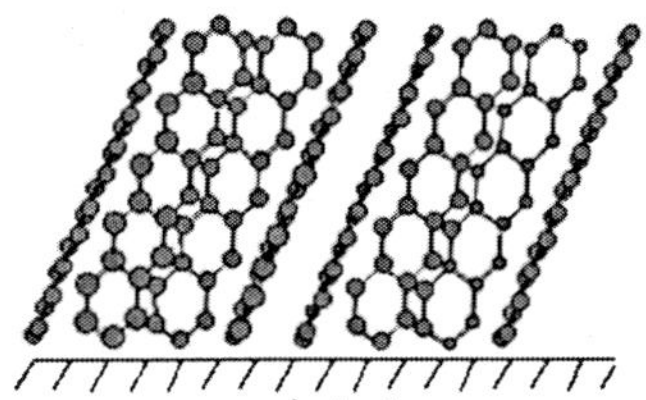

Fig. 3.1. Pentacene structure, crystalline order, and thin-film orientation.

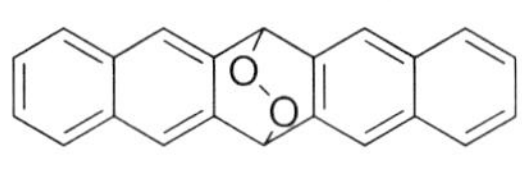

2

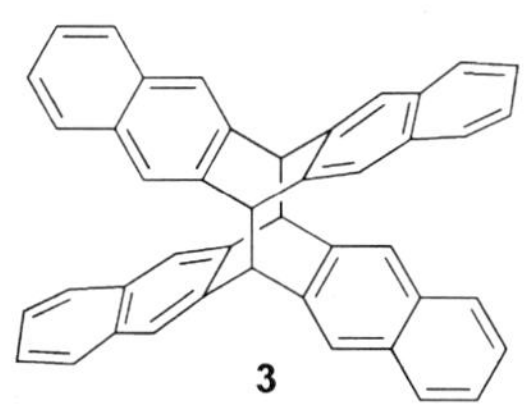

3

Fig. 3.2. Common degradation products of pentacene.

rupting the same strong intermolecular forces that are required for efficient device operation. Thus, solubilization and electronic performance often lie at different ends of the same spectrum. Despite these difficulties, there have been several reports of pentacene functionalization leading to new materials with promising properties.

3.2 Reversible Functionalization

An ingenious route to the solubilization of pentacene involves addition of substituent groups that can be removed thermally after film formation is complete. The reactive nature of the central aromatic ring in pentacene makes it a good diene for Diels–Alder type reactions, and the reversibility of this reaction makes this approach ideal for such a functionalization strategy.

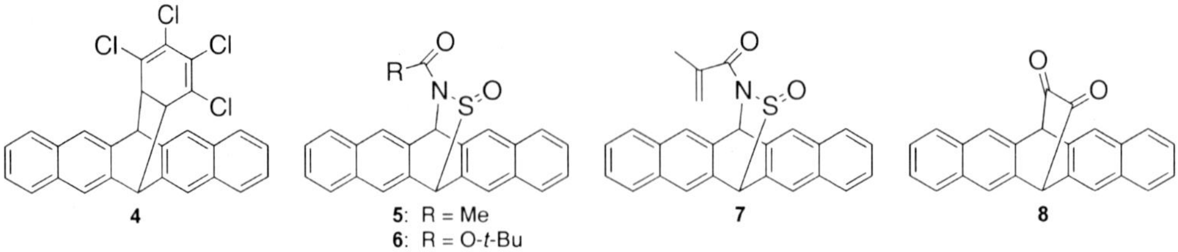

Fig. 3.3. Reversibly functionalized pentacene derivatives.

The first successful application of this method was reported by Müllen and coworkers [18]. The removal of their solubilizing group occurs thermally at temperatures as low as 180 °C, yielding pentacene and a volatile tetrahalobenzene. The tetrachloro compound **4** (Fig. 3.3) formed good-quality films from spin-cast solution, and heating the film of the soluble precursor at 200 °C for 5 s yielded the desired acene film. FET properties for devices made from solution-cast films were quite good – measured mobilities were as high as 0.2 $cm^2 V^{-1} s^{-1}$ with on/off current ratios of 10^6.

A more recent application of this method solubilizes pentacene with *N*-sulfinylamide groups, which undergo Diels–Alder reactions at the N=S double bond, reacting specifically with the central aromatic ring of pentacene. A variety of functional groups on the nitrogen atom is tolerated, yielding materials that are soluble (>50 mg mL^{-1} in chloroform) and can be converted to the pentacene by thermal treatment (>130 °C). Solution-cast films of the sulfinylamide derivative **5** yielded FETs with hole mobility as high as 0.9 $cm^2 V^{-1} s^{-1}$ after a 200 °C annealing step [19].

Use of more complicated functionality on the solubilized pentacene (**6**, **7**) yields materials that can be photopatterned to give devices with mobilities ranging from 0.02–0.2 $cm^2 V^{-1} s^{-1}$ [20, 21]. A very recent addition to the category of pentacene molecules functionalized with removable solubilizing groups is diketone **8** [22], which is relatively soluble and can be converted back to pentacene (and carbon monoxide) by irradiation with ultraviolet light.

3.3
2,3,9,10-Tetrasubstituted and 2,3-Disubstituted Pentacenes: End-substituted Derivatives

Although the first preparation of pentacene reported by Clar in 1929 proceeded by the dehydrogenation of a difficult to prepare precursor molecule (**9**, Scheme 3.1) [23], most modern syntheses of pentacene and its end-functionalized derivatives proceed via the reduction of pentacenequinones, typically by the action of aluminum amalgam in cyclohexanol (Scheme 3.2) [24].

Pentacenequinones (**11**) are in turn conveniently prepared by a fourfold aldol

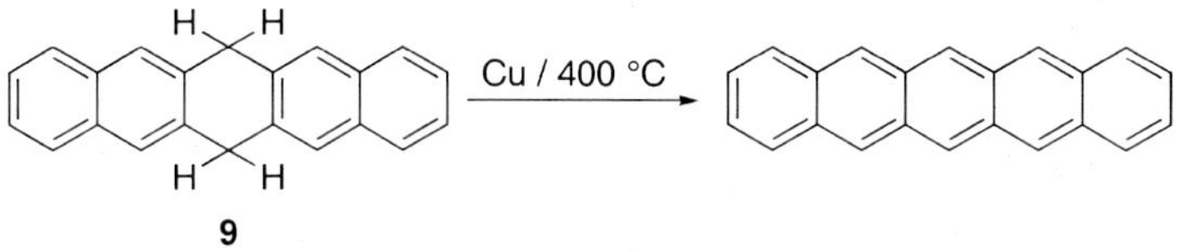

Scheme 3.1. Dehydrogenation route to pentacene.

condensation between phthalaldehydes (**10**) and 1,4-cyclohexanedione [25]. End-functionalization is the best approach to maintaining the herringbone packing in pentacene, because of the relatively weak interaction between the ends of the pentacene molecules in the crystal. Although the preparation of end-functionalized pentacene is straightforward, few of these materials have been reported in the literature, because of their generally reduced oxidative and photochemical stability. A counterpoint to this observation is tetramethyl pentacene **12**, synthesized and studied by Wudl and coworkers (Scheme 3.2) [26]. The material was reported to be as soluble as unsubstituted pentacene, but only slightly more stable. Methyl substitution did have a clear impact on the oxidation potential of the material, reducing the onset of the first oxidation to 510 mV (compared with 640 mV for unsubstituted pentacene). X-ray analysis of crystals of **12** showed that this material packs in a herringbone fashion similar to unfunctionalized pentacene. The highest mobility for this compound was obtained in a bottom-contact FET device; the hole mobility was 0.3 $cm^2\ V^{-1}\ s^{-1}$ and the on/off current ratio 6×10^3. Using a similar synthetic strategy, the Wudl group also prepared tetrachloropentacene **13** [27], but this material did not yield functioning FET devices, because of a remarkable thermal polymerization – when this compound was heated to >900 °C, it formed a carbonaceous solid with a *conductivity* of >5 S cm^{-1}.

Recently the preparation of 2,3,9,10-tetrakis(trimethylsilyl)pentacene **14** was reported [28]. Although this compound is significantly more soluble than **12** and **13**, it was reported to be quite unstable, surviving only a week in deoxygenated solution and for a few days in the crystalline state. The oxidation potential of this compound is 725 mV (vs. SCE), significantly higher than that reported for pentacene. The structure of this compound was confirmed by crystallographic analysis of one of the decomposition products – the "butterfly" dimer **15** (Fig. 3.4).

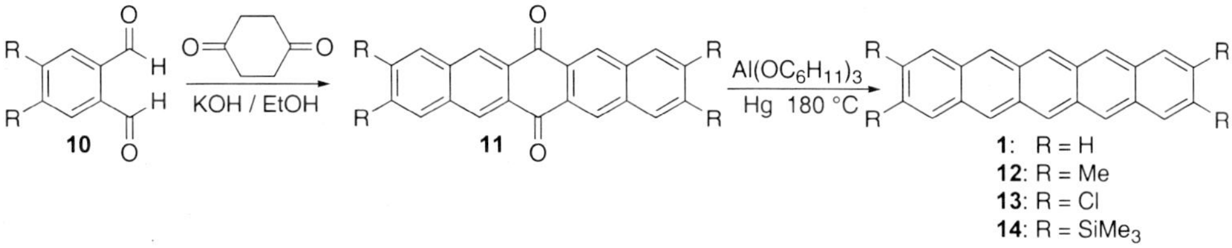

Scheme 3.2. Deoxygenation route to pentacene.

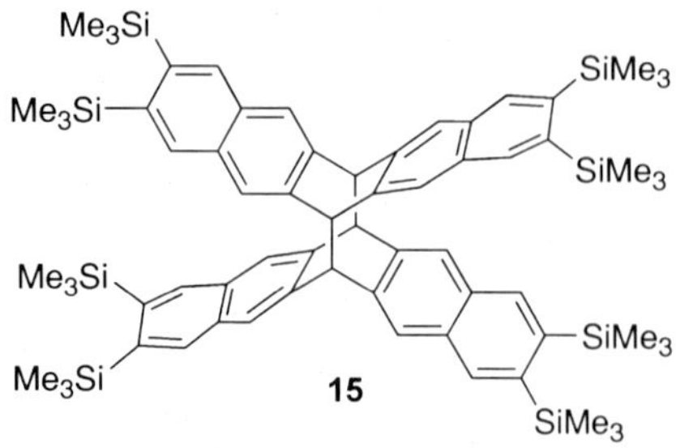

15

Fig. 3.4. Dimerized pentacene **14**.

A recent iterative approach to pentacene (Scheme 3.3) has yielded the hexacenequinone derivative **17** (plus several related compounds) [29]. The novel aspect of such compounds lies in the dipole induced by having electron-withdrawing substituents on only one end of the pentacene, and the unique self-assembly opportunities enabled by this functionalization strategy.

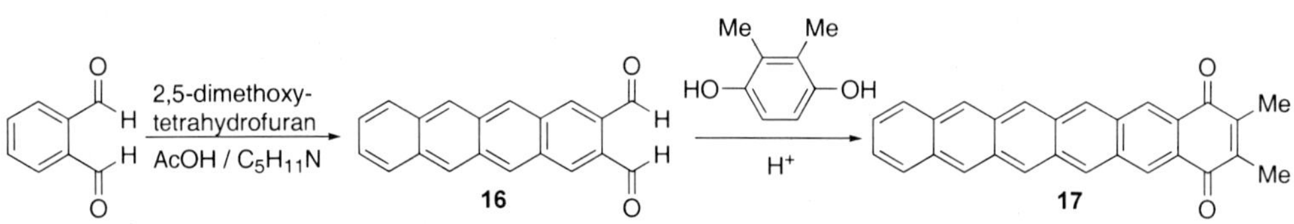

Scheme 3.3. Reiterative hexacenequinone synthesis.

Single-crystal X-ray analysis of **18** (Fig. 3.5) showed that the strongly polarized nature of this molecule led to the adoption of a cofacial arrangement in the solid state, with significant π-overlap and a separation between the π-faces of only 3.25 Å. For device characterization, films of **17** were used in a top-contact device structure to yield a hole mobility of $\sim$0.05 cm^2 V^{-1} s^{-1}, and the devices were reported to be stable over a period of several days.

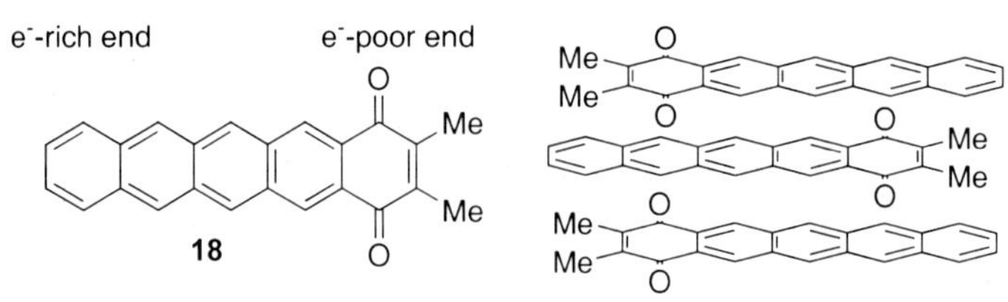

Fig. 3.5. Solid-state order of pentacenequinone **18**.

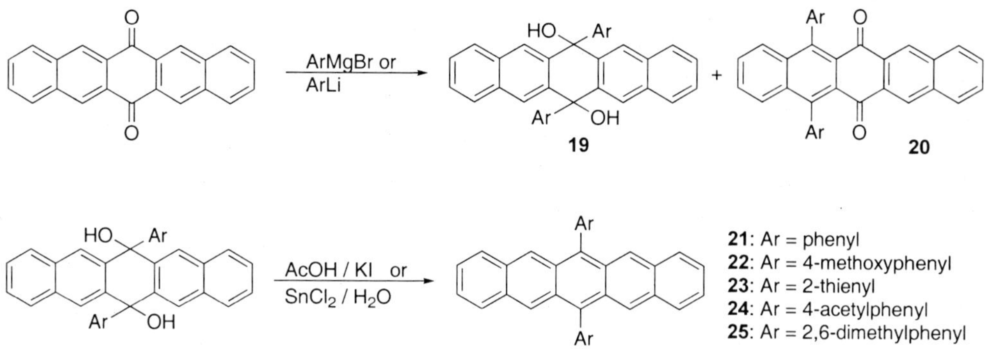

Scheme 3.4. Synthesis of diaryl pentacene.

3.4
Peri-functionalized Pentacene

An alternative to the reduction of pentacenequinone (Scheme 3.2) is addition of a carbon-based nucleophile then reductive deoxygenation to yield a *peri*-functionalized pentacene (Scheme 3.4). An early example of this procedure involves the addition of phenyl magnesium bromide, yielding diols such as **19** (often containing by-products such as conjugate addition product **20**) [30]. Removal of the alcohol moieties to yield the desired pentacene occurs under reducing conditions using reagents such as tin(II) chloride or HI.

Recent studies on the formation of diols such as **19** have shown that functional-group migration reactions are possible when very electron-rich aryl groups are used (Scheme 3.5). In the extreme case of 5-methoxythienyllithium, only the rearranged product **26** was isolated [31].

Reports of pentacene functionalized at the 6,13 positions using the aforementioned methods have appeared in the literature since the early 1940s [29]. These materials are more soluble than pentacene, and substitution at the central ring minimizes decomposition by dimerization. These materials are still sensitive to

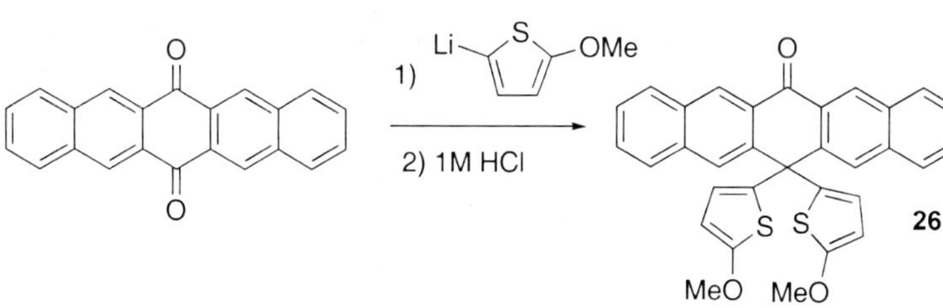

Scheme 3.5. Rearrangement product from the attempted synthesis of di(methoxythienyl)pentacene.

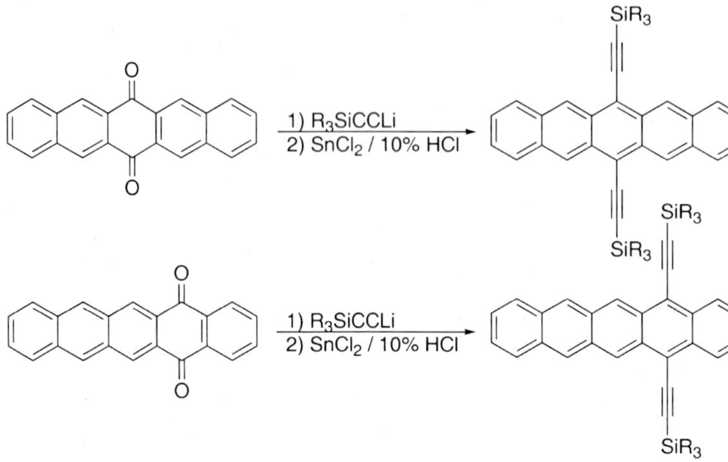

Scheme 3.6. Synthesis of trialkylsilylethynyl pentacenes.

degradation by reaction with atmospheric oxygen, and they remain potent dienes for Diels–Alder reactions – for example, 6,13-diphenylpentacene **21** will react with C_{60} to form a covalent adduct rapidly and in good yield [32]. It is also possible to substitute more than two *peri* positions using this method [33].

One of the most common uses for *peri*-substituted pentacene is as a red emitter in organic light-emitting diodes (OLEDs). Diphenylpentacene, for example, has a fluorescence quantum yield of 30% as a 0.55% dopant in Alq_3, yielding OLED devices with efficiencies near the theoretical maximum [34]. Variation of the aryl substituents improves solubility and processing and can increase fluorescence quantum yield (for example, pentacene **25** has a composite fluorescence quantum yield of 32%) [35]. There is one report of the use of diaryl pentacenes in FET devices, but the performance was generally poor (hole mobility for vapor-deposited **23** was of the order of 10^{-8} $cm^2\ V^{-1}\ s^{-1}$) [30].

Trialkylsilylethynyl pentacenes, prepared as outlined in Scheme 3.6 [36, 37], are highly soluble (>100 mg mL^{-1} in many organic solvents) and significantly more stable than unfunctionalized pentacene. A recent report claimed that trialkylsilylethynyl pentacene **29** was >50 times more stable than unfunctionalized pentacene in solution [38]. One possible explanation for the increased stability is the increased oxidation potential of the ethynylated pentacenes: Functionalized pentacene **29** has one reversible oxidation at 840 mV vs. SCE, a full 200 mV higher than that reported for unsubstituted pentacene [25]. An alternative explanation involves an electronic effect of the Si-based functional group, which has been postulated to alter the singlet–triplet energy gap of the aromatic system, making the functionalized pentacene a poor sensitizer for the formation of singlet oxygen [36].

The most dramatic difference between unfunctionalized pentacene and the silylethynylated pentacenes is the change from edge-to-face interaction to face-to-face

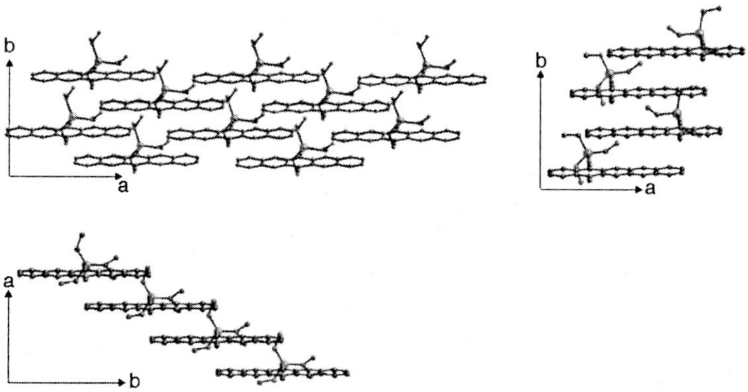

Fig. 3.6. Solid-state order of pentacenes **29** (left), **31** (center), and **27** (right), shown with their corresponding crystallographic axes.

interaction in the functionalized material. Three general solid-state arrangements are observed in the silylethynylated pentacenes – a two-dimensional π-stacked arrangement (Fig. 3.6 left), a columnar arrangement (Fig. 3.6 center), and a one-dimensional π-stacked arrangement (Fig. 3.6 right). The spacing between π-faces in all of these motifs is ~3.43 Å.

Functionalized pentacene single crystals subjected to resistivity measurement include 6,13-functionalized triisopropylsilyl (TIPS) and trimethylsilyl (TMS) derivatives **29** and **27**, 5,14-functionalized TIPS, and triethylsilyl (TES) compounds **30** and **31** [39]. Crystals used in these measurements were grown from solution, and electrodes were attached using graphite or silver paste [40]. The measurements were performed by heating the samples to 100 °C in a nitrogen-filled glove box or under vacuum and then taking resistance measurements as the sample cooled. An Arrhenius plot of the temperature dependence of resistivity provides the conductivity band gap of the materials using the relationship $\rho = \rho_0 \exp(-E_a/k_b T)$, where ρ is resistivity, E_a is activation energy, k_b is the Boltzmann constant, and T is absolute temperature.

The most conductive functionalized pentacene is derivative **29**. Along the π-stacking axis (crystallographic "b" axis), the material has a resistivity of 10^6 Ω cm, and a conductivity band gap of 0.6 eV. Measured orthogonal to the π-stacking axis along the crystallographic "a" axis, the resistivity is 10^8 Ω cm with a band gap of 0.9 eV. Finally, the resistivity measured along the crystallographic "c" axis, which is the direction incorporating the insulating silyl groups, is 10^{10} Ω cm. The change from two-dimensional to one-dimensional π-stacking motif has a dramatic impact on the electronic properties. The resistivity measured along the π-stacking axis in these systems (e.g. **27**) is typically two orders of magnitude higher (10^8 Ω cm) than in the corresponding two-dimensional π-stacked material (**29**). Because of the large amount of π-overlap, however, thermal activation of carriers leads to dramatic increases in conductivity, leading to a band gap of 0.8 eV along the π-stacking "a" axis. Because these one-dimensional π-stacks are "insulated" by the non-conductive functional groups of adjacent π-stacks, there is significant anisotropy in the resistivity of these crystals – resistivity along the crystallographic "b" and "c" axes is $>10^{10}$ Ω cm, with measured band gaps in excess of 3 eV. These measurements all support the argument that electronic properties in crystals are very sensitive to subtle changes in π-overlap [41].

The 5,14-substituted pentacenes adopt yet another solid-state arrangement in which the pentacene moieties can stack in a nearly columnar array (Fig. 3.7, left) with excellent π-overlap. In the TIPS derivative **30**, the bulk of the substituent does not enable more than a pair of acenes to stack, and this pair then arranges in a herringbone fashion (Fig. 3.7, right). Because there is no long-range π-face interaction in this material, the resistivity in the crystal is $>10^{10}$ Ω cm. By substituting a smaller substituent for the TIPS group (**31**), the 5,14-substituted pentacenes can be induced to adopt a columnar π-stacked arrangement with significant π-overlap (Fig. 3.6, center). Resistivity measured along the π-stacking "b" axis (10^7 Ω cm) lies between that of two-dimensional π-stacked **29** and one-dimensional π-stacked **27**.

Several sets of calculations of the band structure of pentacene cover the polymorphs and thin-film forms reported for this compound [42]. They show modest dispersions (0.1–0.2 eV) with a two-dimensional electronic nature. Band-structure calculations for functionalized pentacenes **27** and **29** support the measured band gap findings [43]. In the one-dimensional π-stacked compound **27**, calculations show a small gap along one crystallographic axis, with a remarkable 0.86 eV dispersion in the conduction band. In the two-dimensional π-stacked material **29**, the calculation showed small band gaps along two crystallographic axes, with dispersions of the order of 0.3 eV in the conduction band. In principle, the performance of these materials should be similar to that of unfunctionalized pentacene.

An alternative measurement of the electronic properties of single crystals relies on an all-optical pump-probe technique, which measures the mobility of photogen-

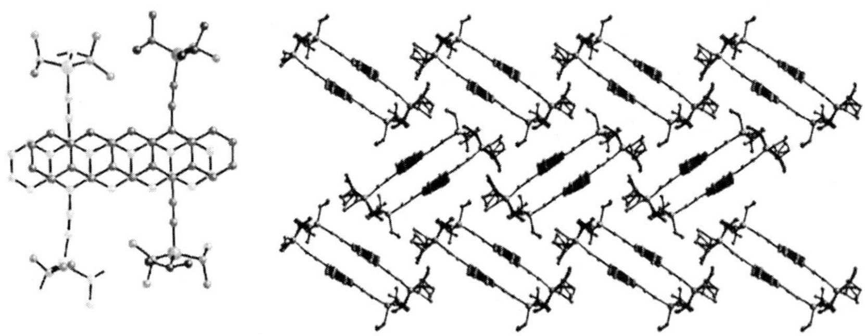

Fig. 3.7. π-overlap pattern of a pair of "offset"-substituted pentacenes **30**, and the "herringbone" arrangement of these pairs in the crystal.

erated charge carriers. Although this method does not yield an absolute charge-carrier mobility (it produces the product of mobility (μ) and efficiency of generation (η) of mobile carriers) it is an excellent technique for comparing mobility across a series of compounds [44]. To date, only one functionalized pentacene derivative has been investigated using such a method (along with unfunctionalized pentacene) [45]. Thin films of both pentacene and TIPS pentacene **29** have similar carrier mobility ($\mu\eta = 0.020$ cm^2 V^{-1} s^{-1} for pentacene, 0.025 cm^2 V^{-1} s^{-1} for **29**). It was also found that charge carrier mobility *increased* as temperature *decreased* for both pentacene and functionalized pentacene **29** [46], providing evidence for band-like transport in the thin films of these materials. A more general investigation of the photoresponse of the conductivity of functionalized pentacenes **27** and **29** revealed dramatic wavelength dependence [47], demonstrating that molecular packing is an important factor to be "engineered" in the development of materials for photovoltaics.

Peri-functionalized acenes **27–31** were also investigated for potential transistor action. Although *none* of the compounds with one-dimensional π-stacking order in the crystalline state (e.g. **27**, **28**, and **31**) yielded devices with significant hole mobility, the one derivative with two-dimensional π-stacking (**29**) yielded a device with a hole mobility of 0.4 cm^2 V^{-1} s^{-1} [48]. The high solubility of this material also enabled device fabrication by solution deposition of the active layer, yielding devices with a hole mobility of 0.2 cm^2 V^{-1} s^{-1} [49]. The similarity in FET performance between solution and vapor-deposited films of **29** is probably because of the self-assembly of the material enforced by the large solubilizing groups.

In contrast with the typical approaches to *peri*-functionalized pentacene, 1,4,8,11-tetramethoxypentacene **34** was prepared by an unusual synthetic route as outlined in Scheme 3.7 [50]. Although this deep violet material lost color if left in solution for more than an hour, it was soluble enough and stable enough for the acquisition of a proton NMR spectrum, confirming the structure. The oxidation potential of **34** was measured by cyclic voltammetry, which revealed two reversible oxidations (at 0.13 V and 0.54 V vs ferrocene/ferrocenium), demonstrating that this material is a powerful electron donor.

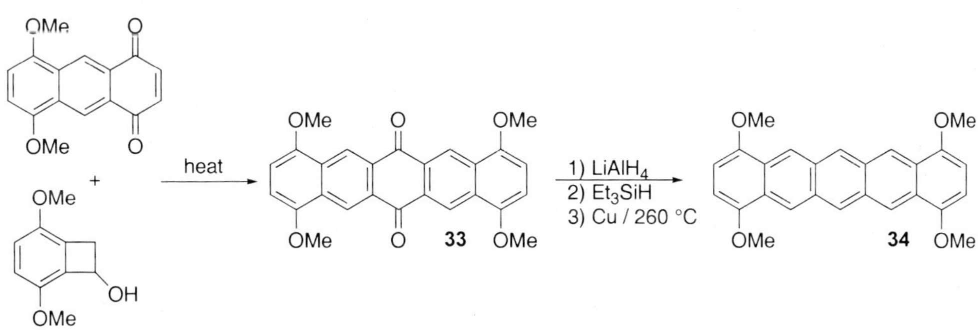

Scheme 3.7. Synthesis of tetramethoxypentacene **34**.

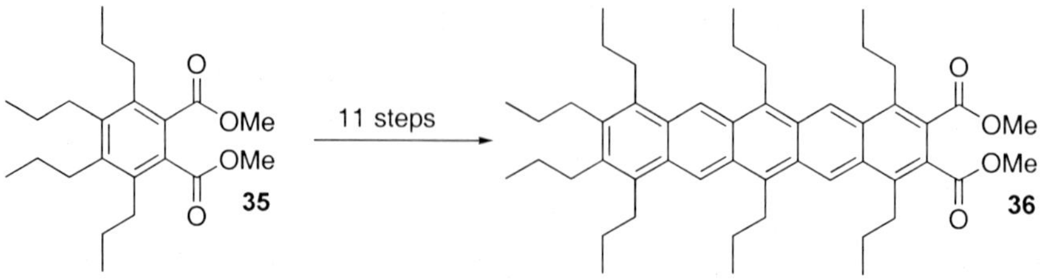

Scheme 3.8. Reiterative synthesis of pentacene **36**.

3.5
Pentacene Functionalized at Both *peri* and End Positions

A small number of stable compounds have been produced with substituents at both *peri and* end positions. The first such system was the product of repeated application of a zirconium-mediated cyclization to phthalate ester **35**, leading to the heavily substituted pentacene **36** in 11 steps (Scheme 3.8) [51]. Little discussion of the properties of this pentacene derivative was presented, but it was soluble and stable enough for complete characterization. This unusual approach to the synthesis of pentacene is one of the few to yield such high substitution.

Pentacene ethers **37–39** (Fig. 3.8) were prepared in an attempt to reduce the HOMO energy level relative to functionalized pentacene **29** [52]. These cyclic ether compounds were surprisingly stable, soluble and easily crystallized. Saturated dioxane compound **37** has an oxidation potential of 0.580 V vs SCE (similar to that of unfunctionalized pentacene), but the axial hydrogen atoms on the dioxane ring prevent π-stacking interactions in the crystal. Derivatives **38** and **39** were synthesized to overcome this structural issue. Although the change of oxidation potential

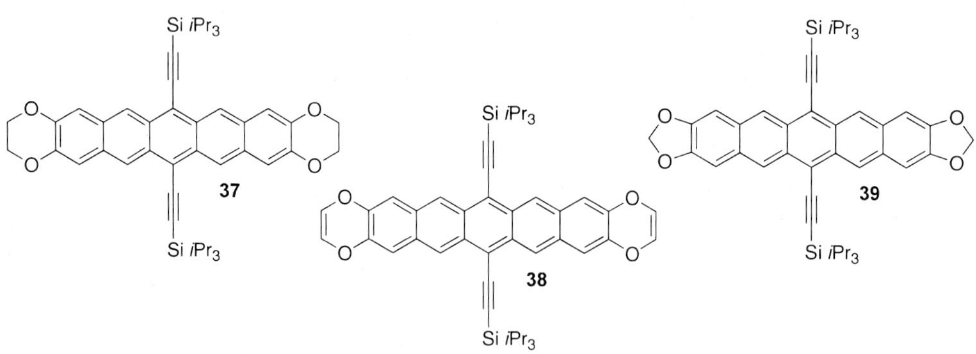

Fig. 3.8. Pentacene ethers **37–39**.

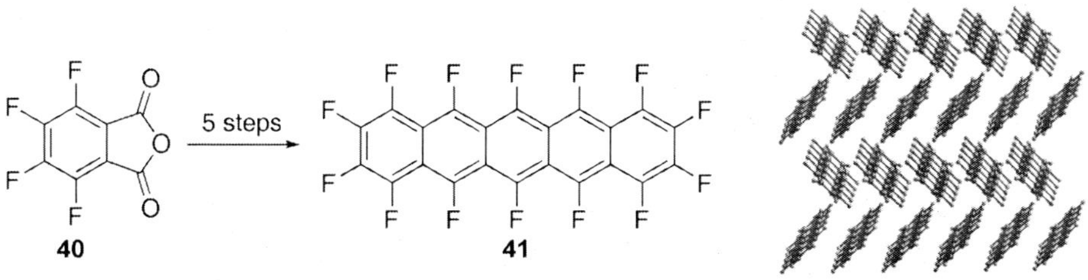

Scheme 3.9. Synthesis and solid-state order of perfluoropentacene **41**.

was not as significant for these compounds (0.776 V for **38** and 0.692 V for **39**), the materials do have strong π-stacking interactions in the solid state.

The most dramatic example of the use of functionalization to tune the properties of an organic semiconductor is found in perfluoropentacene **41** (Scheme 3.9) [53]. Perfluoropentacene has a strongly shifted LUMO, evidenced by lowering of the reduction potential by more than 700 mV. The fluorinated compound adopts a herringbone packing in the crystal that is very similar to that of the parent acene. Top-contact FET devices made from evaporated **41** had an *electron* mobility of 0.11 $cm^2\ V^{-1}\ s^{-1}$, demonstrating the change from p-type to n-type behavior.

3.6 Heteropentacenes

Heteropentacenes are formed by replacing one or more of the carbocyclic rings in pentacene with heterocyclic rings. The earliest example of the synthesis and use of heteropentacenes in electronic applications was reported by Katz and coworkers, who prepared anthradithiophenes **42–45** (Scheme 3.10) [54] from thiophene-2,3-

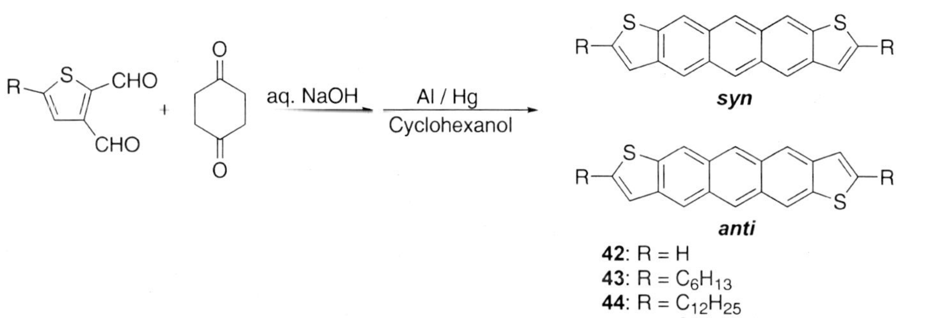

Scheme 3.10. Anthradithiophene synthesis.

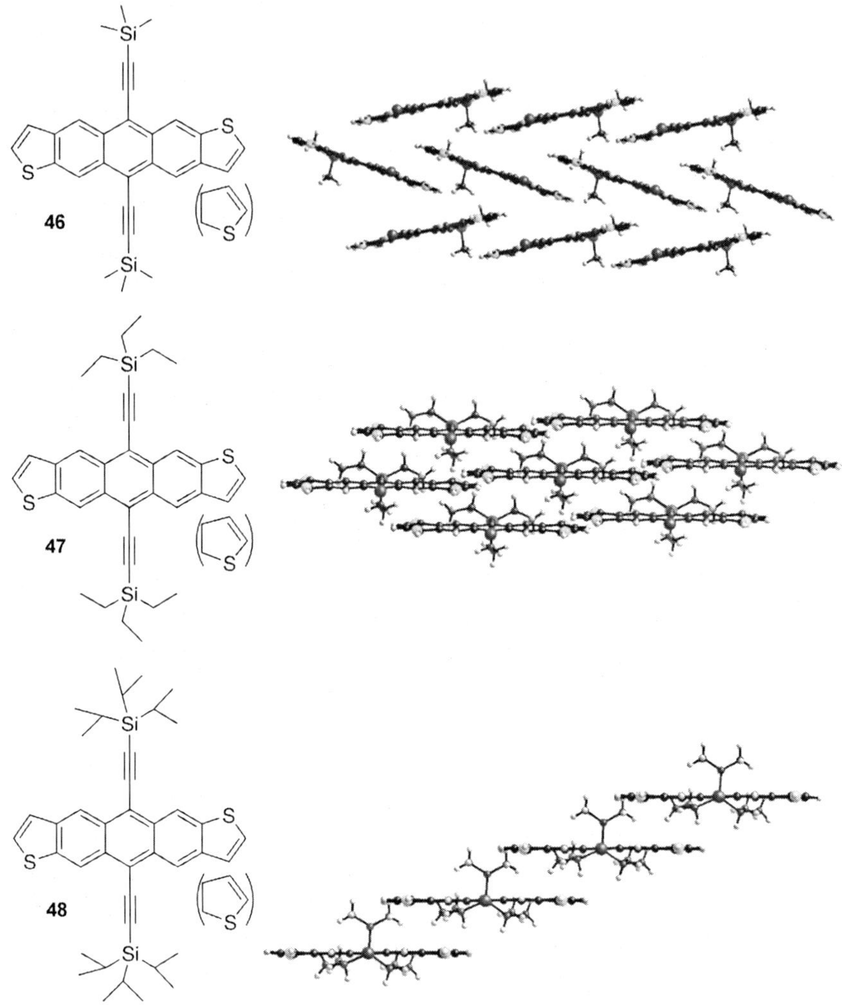

Fig. 3.9. Silylethynyl-functionalized anthradithiophenes (left) and their solid-state arrangement (right). Frontmost silyl groups omitted for clarity.

dicarboxaldehydes as an inseparable mixture of syn and anti isomers. Although all these derivatives formed highly crystalline films by evaporation, and all exhibited transistor behavior in bottom-contact devices, dihexyl derivative **43** yielded the best device properties, with hole mobility of 0.15 $cm^2\ V^{-1}\ s^{-1}$. This same dihexyl derivative could also be deposited on an FET substrate from chlorobenzene solution, yielding devices with a mobility of 0.02 $cm^2\ V^{-1}\ s^{-1}$ [55]. A similar solution-casting technique applied to pentacene was unsuccessful, because of rapid photobleaching of the acene [56] showing the increased stability of anthradithiophenes. A detailed

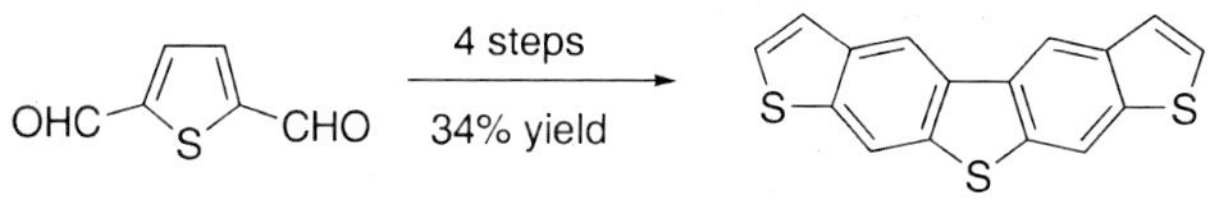

49

Scheme 3.11. Synthesis of benzothiophene derivative **46**.

computational study of the relationship between solid-state order and electronic properties of anthradithiophene showed that this class of heteroacene is a particularly promising candidate for the preparation of high-mobility FETs [57].

The addition of trialkylsilylethynyl groups to anthradithiophene leads to several solid-state arrangements, ranging from weak edge-to-face interactions (**46**) to strong two-dimensional π-overlap (**47**) (Fig. 3.9) [49]. The nature of the crystal packing has a significant effect on both film formation and device performance. TMS derivative **46** has the weakest intermolecular interactions in the crystal, and is the only one of this series that does not form continuous films from solution. Derivatives **47** and **48** both have π-face interactions in the crystal, and both form uniform films by solution casting, but TES derivative **47**, with strong π-face interactions evidenced by very close contacts between the aromatic rings in the stack (3.25 Å) yielded films with much higher crystallinity than **48**, which has only one-dimensional π-face interactions (face-to-face separation of 3.46 Å). The crystallinity is reflected in the device performance, with solution-cast films of **48** giving devices with a charge carrier mobility < 0.0001 cm^2 V^{-1} s^{-1}, and films of the more crystalline **47** yielding devices with hole mobility of 1.0 cm^2 V^{-1} s^{-1}.

A recent effort by Neckers and coworkers produced a molecule with three of pentacene's carbocyclic rings replaced with thiophene heterocycles (**49**), prepared in isomerically pure form as outlined in Scheme 3.11 [58]. The authors did not report any characterization of the electronic properties of this material, although it was reported to be sufficiently soluble in ethanol for recrystallization.

Although several tetraazapentacenes (**50–54**, Fig. 3.10) have been reported and studied in detail over the last several decades [59–62], there have been no reports

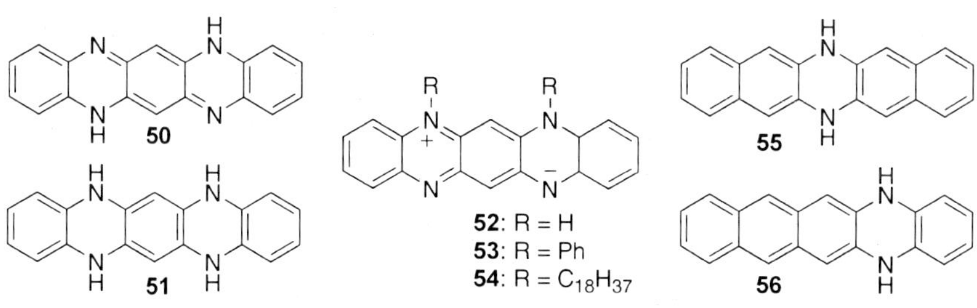

Fig. 3.10. Azapentacenes.

of their application in field-effect transistor devices. In contrast the diazadihydropentacenes **55** and **56** have been investigated for use in OFETs [63]. Although crystallographic analysis of **55** showed that this molecule arranged in a herringbone fashion similar to pentacene, the results from OFET measurements were not on par with those of this parent compound – vapor-deposited films of **55** and **56** yielded FETs with hole mobilities of only 5×10^{-5} cm^2 V^{-1} s^{-1} for **55** and 5×10^{-3} cm^2 V^{-1} s^{-1} for **56**.

3.7
Conclusion

Functionalization of pentacene with the specific aim of improving performance in devices is a recent endeavor – the first use of a functionalized pentacene in a field-effect transistor was reported only recently (2003) [26]. Functionalization of pentacene has led to the ability to engineer the solid-state arrangement, electronic, and solubility properties of this important semiconductor and to improve its stability and film-forming ability. Recent functionalized pentacene materials have yielded devices with properties comparable with those of the parent acene, have enabled the formation of devices from solution-deposited films, and have even changed the semiconductor behavior of this organic molecule from p-type to n-type. As functionalization strategies are refined, materials with *all* of the properties necessary for commercial device applications should soon be developed.

References

1 R. Ruiz, D. Choudhary, B. Nickel, T. Toccoli, K.-C. Chang, A. C. Mayer, P. Clancy, J. M. Blakely, R. L. Headrick, S. Iannotta, G. G. Malliaras, *Chem. Mater.* **2004**, *16*, 4497.

2 L. B. Roberson, J. Kowalik, L. M. Tolbert, C. Kloc, R. Zeis, X. Chi, R. Fleming, C. Wilkins, *J. Am. Chem. Soc.* **2005**, *127*, 3069.

3 (a) J. Cornil, J. Ph. Calbert, J. L. Bredas, *J. Am. Chem. Soc.* **2001**, *123*, 1250. (b) Y. C. Cheng, R. J. Silbey, D. A. da Silva Filho, J. P. Calbert, J. Cornil, J. L. Bredas, *J. Chem. Phys.* **2003**, *118*, 3764.

4 (a) G. B. Blanchet, C. R. Fincher, M. Lefenfeld, J. A. Rogers, *Appl. Phys. Lett.* **2004**, *84*, 296. (b) P. V. Necliudov, M. S. Shur, D. J. Gundlach, T. N. Jackson, *IEEE 2001 Int. Semicon. Dev. Res. Symp.* **2001**, 345. (c) P. V. Pesavento, R. J. Chesterfield, C. R. Newman, C. D. Frisbie, *J. Appl. Phys.* **2004**, *96*, 7312.

5 N. E. Gruhn, D. A. da Silva Filho, T. G. Bill, M. Malagoli, V. Coropceanu, A. Kahn, J. L. Brédas, *J. Am. Chem. Soc.* **2002**, *124*, 7918.

6 (a) Y.-Y. Lin, D. J. Gundlach, S. F. Nelson, T. N. Jackson, *IEEE Electron Device Lett.* **1997**, *18*, 606. (b) C. D. Dimitrakopoulos, P. R. L. Malenfant, *Adv. Mater.* **2002**, *14*, 99.

7 A. B. Dros, J. Baas, A. Meetsma, J. L. de Boer, T. T. M. Palstra, *Acta Cryst. C* **2001**, *57*, 939.

8 S. E. Fritz, S. M. Martin, C. D. Frisbie, M. D. Ward, M. F. Toney, *J. Am. Chem. Soc.* **2004**, *126*, 4084.

9 H. Klauk, M. Halik, U. Zschieschang, G. Schmid, W. Radlik, W. J. Weber, *J. Appl. Phys.* **2002**, *92*, 5259.

10 O. D. Jurchescu, J. Baas, T. T. M. Palstra, *Appl. Phys. Lett.* **2004**, *84*, 3061.

11 E. Clar, "Polycyclic Aromatic Hydrocarbons" Academic Press: London. 1964, Vol. 1.

12 D. E. Janzen, M. W. Burand, P. C. Ewbank, T. M. Pappenfus, H. Higuchi, D. A. da Silva Filho, V. G. Young, J. L. Brédas, K. R. Mann, *J. Am. Chem. Soc.* **2004**, *126*, 15295.

13 J. R. Sheats, *J. Mater. Res.* **2004**, *19*, 1974.

14 K. Takimiya, Y. Kunugi, Y. Konda, N. Niihara, T. Otsubo, *J. Am. Chem. Soc.* **2004**, *126*, 5084.

15 J. A. Merlo, C. R. Newman, C. P. Gerlach, T. W. Kelley, D. V. Muyres, S. E. Fritz, M. F. Toney, C. D. Frisbie, *J. Am. Chem. Soc.* **2005**, *127*, in press (ASAP web published March 1, 2005).

16 M. Halik, H. Klauk, U. Zschieschang, G. Schmid, S. Ponomarenko, S. Kirchmeyer, W. Weber, *Adv. Mater.* **2003**, *15*, 917.

17 Y. Sun, Y. Liu, D. Zhu, *J. Mater. Chem.* **2005**, *15*, 53.

18 (a) A. R. Brown, A. Pomp, D. M. de Leeuw, D. B. M. Klaassen, E. E. Havinaga, P. Herwig, K. Müllen, *J. Appl. Phys.* **1996**, *79*, 2136. (b) P. T. Herwig, K. Müllen, *Adv. Mater.* **1999**, *11*, 480.

19 A. Afzali, C. D. Dimitrakopoulos, T. L. Breen, *J. Am. Chem. Soc.* **2002**, *124*, 8812.

20 A. Afzali, C. D. Dimitrakopoulos, T. O. Graham, *Adv. Mater.* **2003**, *15*, 2066.

21 K. P. Weidkamp, A. Afzali, R. M. Tromp, R. J. Hamers, *J. Am. Chem. Soc.* **2004**, *126*, 12740.

22 H. Uno, Y. Yamashita, M. Kikuchi, H. Watanabe, H. Yamada, T. Okujima, T. Ogawa, N. Ono, *Tetrahedron Lett.* **2005**, *46*, 1981.

23 E. Clar, F. John, *Chem. Ber.* **1929**, *62*, 3021.

24 E. P. Goodings, D. A. Mitchard, G. Owen, *J. Chem. Soc. Perkin Trans. 1* **1972**, *11*, 1310.

25 W. Ried, F. Anthöfer, *Angew. Chem.* **1953**, *65*, 601.

26 H. Meng, M. Bendikov, G. Mitchell, R. Helgeson, F. Wudl, Z. Bao, T. Siegrist, C. Kloc, C.-H. Chen, *Adv. Mater.* **2003**, *15*, 1090.

27 D. F. Perepichka, M. Bendikov, H. Meng, F. Wudl, *J. Am. Chem. Soc.* **2003**, *125*, 10190.

28 S. H. Chan, H. K. Lee, Y. M. Wang, N. Y. Fu, X. M. Chen, Z. W. Cai, H. N. C. Wong, *Chem. Commun.* **2005**, 66.

29 Q. Miao, M. Lefenfeld, T.-Q. Nguyen, T. Siegrist, C. Kloc, C. Nuckolls, *Adv. Mater.* **2005**, *17*, 407.

30 C. F. H. Allen, A. Bell, *J. Am. Chem. Soc.* **1942**, *64*, 1253.

31 N. Vets, M. Smet, W. Dehaen, *Synlett* **2005**, 217.

32 G. P. Miller, J. Mack, J. Briggs, *Org. Lett.* **2000**, *2*, 3983.

33 Ref. 20 and (a) H. Hart, C. Y. Lai, G. C. Nwokogu, S. Shamouilian, *Tetrahedron* **1987**, *43*, 5203. (b) P. W. Kenny, L. L. Miller, S. F. Rak, T. H. Jozefiak, W. C. Christopfel, J. H. Kim, R. A. Uphaus, *J. Am. Chem. Soc.* **1988**, *110*, 4445.

34 L. C. Picciolo, H. Murata, Z. H. Kafafi, *Appl. Phys. Lett.* **2001**, *78*, 2378.

35 M. A. Wolak, B.-B. Jang, L. C. Palilis, Z. Kafafi, *J. Phys. Chem. B* **2004**, *108*, 5492.

36 J. E. Anthony, J. S. Brooks, D. L. Eaton, S. R. Parkin, *J. Am. Chem. Soc.* **2001**, *123*, 9482.

37 J. E. Anthony, D. L. Eaton, S. R. Parkin, *Org. Lett.* **2002**, *4*, 15.

38 A. Maliakal, K. Raghavachari, H. Katz, E. Chandross, T. Siegrist, *Chem. Mater.* **2004**, *16*, 4980.

39 J. E. Anthony, J. S. Brooks, D. L. Eaton, J. R. Matson, S. R. Parkin, *Proc. SPIE* **2003**, *5217*, 124.

40 H. C. Montgomery, *J. Appl. Phys.* **1971**, *42*, 2971.

41 (a) P. M. Kazmaier, R. Hoffmann, *J. Am. Chem. Soc.* **1994**, *116*, 9684. (b) J. L. Brédas, J. P. Calbert, D. A. da Silva Filho, J. Cornil, *PNAS* **2002**, *99*, 5804.

42 Ref. 3 and (a) A. Troisi, G. Orlandi, *J. Phys. Chem. B* **2005**, *109*, 1849. (b) T. Siegrist, C. Kloc, J. H. Schon, B.

Batlogg, R. C. Haddon, S. Berg, G. A. Thomas, *Angew. Chem. In. Ed. Engl.* **2001**, *40*, 1732.

43 R. C. Haddon, X. Chi, M. E. Itkis, J. E. Anthony, D. L. Eaton, T. Siegrist, C. C. Mattheus, T. T. M. Palstra, *J. Phys. Chem. B* **2002**, *106*, 8288.

44 (a) V. K. Thorsmolle, R. D. Averitt, X. Chi, D. J. Hilton, D. L. Smithi, A. P. Ramirez, A. J. Taylor, *Appl. Phys. Lett.* **2004**, *84*, 891. (b) C. A. Schmuttenmaer, *Chem. Rev.* **2004**, *104*, 1759.

45 F. A. Hegmann, R. R. Tykwinski, K. P. H. Lui, J. E. Bullock, J. E. Anthony, *Phys. Rev. Lett.* **2002**, *89*, 227403-(1–4).

46 O. Ostroverkhova, D. G. Cooke, S. Shcherbyna, R. F. Egerton, F. A. Hegmann, R. R. Tykwinski, J. E. Anthony, *Phys. Rev. B* **2005**, *71*, 035204.

47 T. Tokumoto, J. S. Brooks, R. Clinite, X. Wei, J. E. Anthony, D. L. Eaton, S. R. Parkin, *J. Appl. Phys.* **2002**, *92*, 5208.

48 C. D. Sheraw, T. N. Jackson, D. L. Eaton, J. E. Anthony, *Adv. Mater.* **2003**, *15*, 2009.

49 M. M. Payne, S. R. Parkin, J. E. Anthony, C.-C. Kuo, T. N. Jackson, *J. Am. Chem. Soc.* **2005**, *127*, 4986.

50 A. R. Wartini, H. A. Staab, F. A. Neugebauer, *Eur. J. Org. Chem.* **1998**, 1161.

51 T. Takahashi, M. Kitamura, B. Shen, K. Nakajima, *J. Am. Chem. Soc.* **2000**, *122*, 12876.

52 M. M. Payne, J. H. Delcamp, S. R. Parkin, J. E. Anthony, *Org. Lett.* **2004**, *6*, 3325.

53 Y. Sakamoto, T. Suzuki, M. Kobayashi, Y. Gao, Y. Fukai, Y. Inoue, F. Sato, S. Tokito, *J. Am. Chem. Soc.* **2004**, *126*, 8138.

54 J. G. Laquindanum, H. E. Katz, A. Lovinger, *J. Am. Chem. Soc.* **1998**, *120*, 664.

55 H. E. Katz, W. Li, A. J. Lovinger, J. Laquindanum, *Synth. Met.* **1999**, *102*, 897.

56 F. Garnier, *Chem. Phys.* **1998**, *10*, 457.

57 O. Kwon, V. Coropceanu, N. E. Gruhn, J. C. Durivage, J. G. Laquindanum, H. E. Katz, J. Cornil, J. L. Brédas, *J. Chem. Phys.* **2004**, *120*, 8186.

58 B. Wex, B. R. Kaafarani, D. C. Neckers, *J. Org. Chem.* **2004**, *69*, 2197.

59 E. Leete, O. Ekechukwu, P. Delvigs, *J. Org. Chem.* **1966**, *31*, 1967.

60 A. E. Riley, G. W. Mitchell, P. A. Koutentis, M. Bendikov, P. Kaszynki, F. Wudl, S. H. Tolbert, *Adv. Funct. Mater.* **2003**, *13*, 531.

61 H. Choi, X. Yang, G. W. Mitchell, C. P. Collier, F. Wudl, J. R. Heath, *J. Phys. Chem. B* **2002**, *106*, 1833.

62 F. Wudl, P. A. Koutentis, A. Weitz, B. Ma, T. Strassner, K. N. Houk, S. I. Khan, *Pure Appl. Chem.* **1999**, *71*, 295.

63 Q. Miao, T.-Q. Nguyen, T. Someya, G. B. Blanchet, C. Nuckolls, *J. Am. Chem. Soc.* **2003**, *125*, 10284.

4
Organic Semiconductors Based on Polythiophene and Indolo[3,2-*b*]carbazole

Beng S. Ong, Yiliang Wu, and Yuning Li

4.1
Introduction

The phenomenal surge in interest in organic electronics in recent years [1, 2] is primarily propelled by expectations of significantly lower cost and desirable physical features attainable with organic materials. The expected economic benefits stem from the ability to manufacture a new generation of fundamental building blocks of microelectronics, organic thin film transistors (OTFTs) at much lower cost than their prevalent silicon-based counterparts. OTFTs can be fabricated via simple liquid patterning and deposition techniques (e.g. spin coating, screen printing, flexography, inkjet printing), in contrast with the manufacture of silicon TFTs, which requires a capital-intensive facility engaging complex high temperature/vacuum photolithographic processes in a strictly controlled environment. OTFTs, being composed of organic or polymer materials, may also be built on flexible plastic substrates, thereby enabling creation of structurally compact, lightweight, and flexible electronics. OTFTs are not, however, expected to provide the performance characteristics expected of crystalline silicon TFTs with high computational powers. The capabilities of OTFTs are more in line with those of amorphous silicon TFTs, which are functionally adequate for a wide variety of electronic applications. OTFT circuits are thus particularly suitable for electronic devices which require large-area coverage (e.g. active matrix displays) and not a high transistor density or fast switching speed; they may also find applications in low-cost microelectronics (e.g. radio-frequency identification tags) where the high cost of packaging silicon circuits precludes ubiquitous usage.

One critical material for OTFTs is the channel semiconductor which dictates transistor performance. Functionally capable organic semiconductors for low-cost OTFTs are rare, because most organic semiconductors are either insoluble or very sensitive to air under ambient conditions. Material insolubility precludes the use of liquid patterning and deposition and air sensitivity requires manufacture in inert atmospheres – both of these restrictive requirements invariably lead to increased cost, thereby nullifying the fundamental economic advantage of OTFTs. Until relatively recently the challenges of designing liquid processable, high-

Organic Electronics, Materials, Manufacturing and Applications. Edited by Hagen Klauk

ISBN: 3-527-31264-1

performing organic semiconductors with sufficient air-stability to enable low-cost OTFTs seemed to be insurmountable. In recent years great strides have been made in the design of air-stable and solution-processable organic semiconductors with excellent field-effect transistor (FET) properties. These latest advances in semiconductor development are very encouraging and have strengthened the prospect of bringing the concept of low-cost OTFT circuits from laboratory to commercialization in the foreseeable future. In this chapter, we discuss the issues, challenges, and recent advances in the development of p-channel organic semiconductors, and present our design principles and structural studies for air stable, solution processable organic semiconductors for OTFTs. In particular, the design of two fundamental classes of semiconductor, one polymeric in nature (regioregular polythiophenes) and one molecular (indolocarbazoles) for high-performance OTFTs will be described. We will also describe the preparation and characterization of a novel class of organic semiconductor nanoparticles which enables inkjet patterning of OTFT channel semiconductors under ambient conditions with consistently high FET performance characteristics.

4.2 Issues and Challenges

Although FET properties in organic materials were first observed in copper phthalocyanine in the 1960s [3], little progress was made until the 1980s when several organic semiconductor compounds were demonstrated for OTFTs [4–6]. Since then significant advances in organic semiconductor development had been reported, with FET mobility being improved by as much as five orders of magnitude. The mobility of the current materials such as pentacene [7] not only rivals but supersedes those of hydrogenated amorphous silicon. Table 4.1 summarizes the performance characteristics of major classes of p-channel semiconductors for OTFTs. As can be noted, solution processability, molecular ordering and air stability of semiconductors are critical to OTFT fabrication. High mobility is mostly obtained from the vacuum deposited materials with high structural order. The FET performance of solution-processed semiconductors, particularly those fabricated in air, is generally poor. These results reveal the significance of molecular ordering on performance and reflect the great difficulties of forming structurally ordered semiconductors from solution.

The p-channel organic semiconductor compounds that have been widely studied can generally be classified into three categories:

1. acene-based small molecular compounds [7–13];
2. oligothiophenes [6, 14–19] and polythiophenes [5, 20–22]; and
3. arylene–vinylene [23] and thiophene–vinylene/arylene hybrids [24–27].

Most of these semiconductors have relatively high-lying HOMO (highest occupied molecular orbitals) and narrow band gaps. They are very susceptible to photoin-

Tab. 4.1. Typical organic semiconductors with OTFT characteristics.

Semiconductor	**Mobility ($cm^2\ V^{-1}\ s^{-1}$) (deposition conditions)**	**I_{on}/I_{off}**	**Ref.**
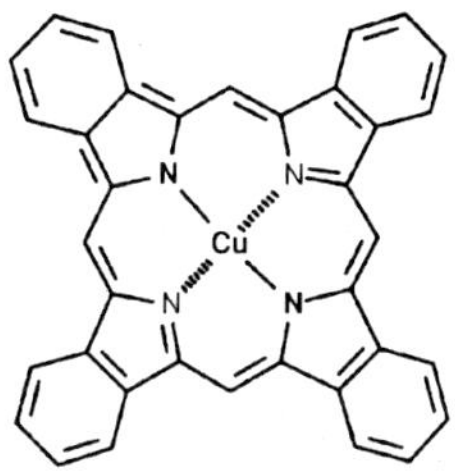 Copper phthalocyanine	Initial report: – Recent result: 0.02 (vacuum)	– 10^5	3 68
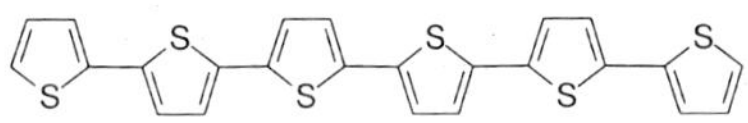 α-Sexithiophene	Initial report: 10^{-3} (vacuum) Recent result: 0.03	– $>10^6$	6 14
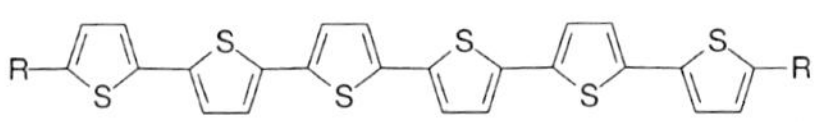 α,ω-dialkylsexithiophene	Initial report: 0.05 (vacuum) Recent result: 1.1 (vacuum) 0.01 (solution/ambient)	10^2 10^4 10^4	15 16 17, 18
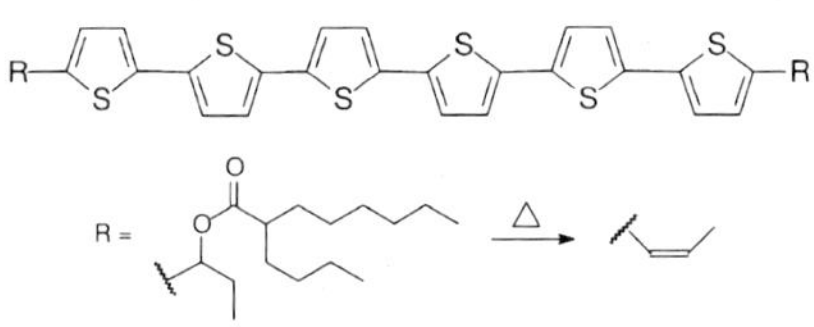 Oligothiophene derivative	Initial report: 0.05 (solution/air)[a]	10^5	19
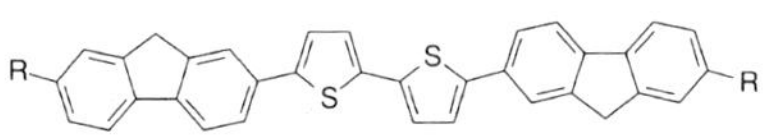 R = H or C_6H_{13}. Fluorene–thiophene oligomer	Initial report: 0.11 (vacuum)	26	
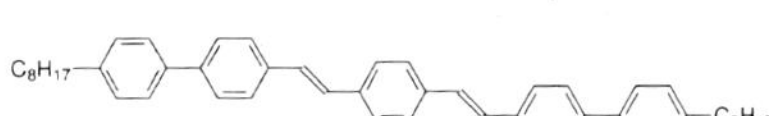 Oligo-*p*-phenylenevinylenes	Initial report: 0.12 (vacuum)	10^6	23
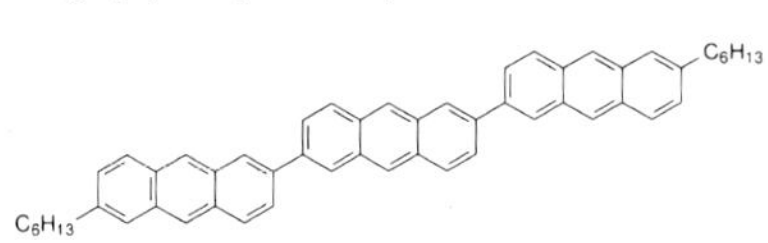 Oligo(2,6-anthrylene)s	Initial report: 0.18 (vacuum)	10^4	8
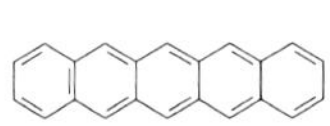 Pentacene	Initial report: 2×10^{-3} (vacuum) Recent result: 3.3 (vacuum)[b]	– 10^7	9 7

Tab. 4.1 *(continued)*

Semiconductor	**Mobility ($cm^2\ V^{-1}\ s^{-1}$) (deposition conditions)**	I_{on}/I_{off}	**Ref.**
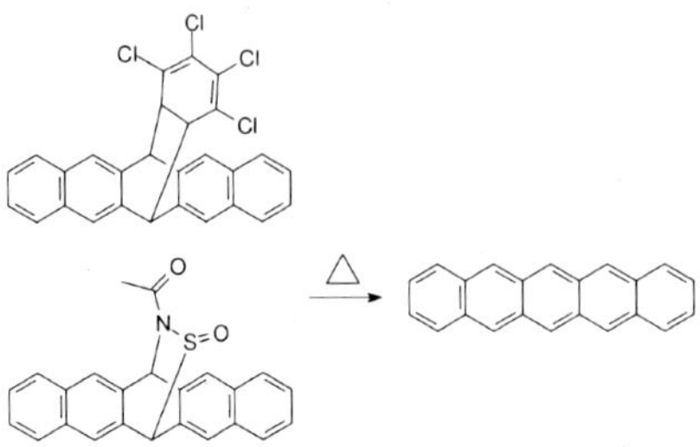 Pentacene from soluble precursors	Initial report: 0.01–0.2 (solution/ambient) Recent result: 0.42 (solution/inert)[c]	10^5 10^6	11, 12 10
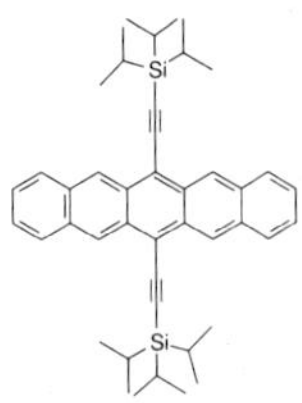 6,13-Bis(triisopropylsilylethynyl)pentacene	Initial report: 0.4 (vacuum)	10^5	13
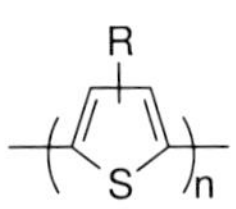 Regiorandom poly(3-alkylthiophene)	Initial report: 10^{-5} (solution/ambient)	10^3	5
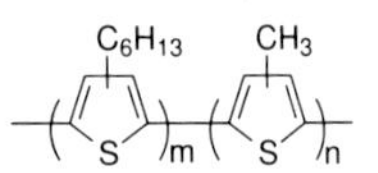 Regiorandom poly(3-methylthiophene-co-3-hexylthiophene)	Recent result: 0.046 (solution/ambient)	–	20
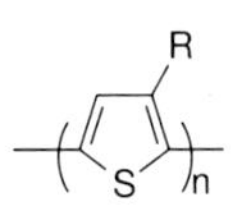 Regioregular head-to-tail poly(3-alkylthiophene)	Initial report: 0.045 (solution/ambient) Recent result: 0.1 (solution/inert)	10^2 10^6	21 22
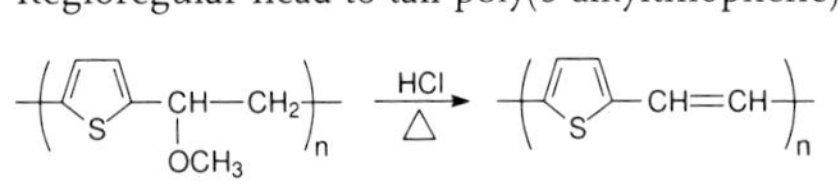 Poly(thienylene–vinylene) from soluble precursor	Initial report: 0.22 (solution)[d] Recent result: 0.0015 (solution/ambient)	10^3 10^4	24 25
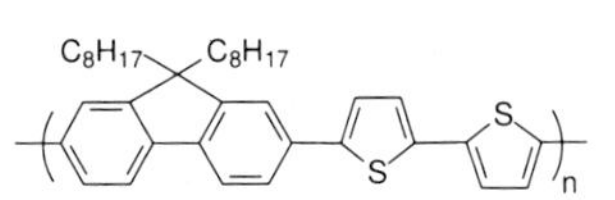 Poly(9,9′-dioctylfluorene-co-bithiophene)	Initial report: 0.02 (solution/ambient)[e]	10^6	27

[a] Annealed at 180–200 °C
[b] On alumina dielectric
[c] Annealed at 200 °C in nitrogen atmosphere
[d] Has not been reproduced
[e] Annealed at >265 °C in nitrogen atmosphere on rubbed polyimide alignment layer.

duced oxidative doping and bleaching, leading to poor semiconductor characteristics when processed under ambient conditions. These unfavorable molecular properties are particularly challenging for those electronic applications for which light emission (e.g. organic light-emitting diodes) or backlight (e.g., active matrix liquid-crystal displays) is involved.

An interesting approach to solution processed, high-mobility organic semiconductor layers proceeds via soluble Diels–Alder adducts of pentacene [10–12]. The Diels–Alder adducts are soluble in organic solvents and their solutions can be spun cast into thin films. On heating to the decomposition temperatures the pentacene adducts retro-convert to pentacene. To achieve high mobility, however, post-reaction annealing in a nitrogen atmosphere is necessary [10]. These results again reveal the challenges and complexities that confront the development of air-stable, solution-processable, and functionally capable semiconductors for OTFTs.

4.3
Structural Considerations

Charge-carrier transport in organic semiconductor systems is dominated by hopping through localized states [28], therefore proper molecular order, which is favorable to charge transport in the semiconductor, is necessary to achieve high mobility [22]. The semiconductor must also be reasonably air stable, so that the desirable FET properties can be achieved under ambient fabricating conditions before subsequent protective encapsulation is implemented for long-term operational stability. Accordingly, it is imperative that semiconductor materials for low-cost OTFTs have the following critical properties:

1. an electronic structure favorable for charge transport;
2. an ability to self-organize from solution to form proper structural order in the solid state for charge transport, and
3. a resistance to adverse environmental effects, particularly degradative interactions with atmospheric oxygen.

Most current solution-processable organic semiconductor compounds fail to satisfy all these requirements simultaneously.

Low mobility is observed for most polymer semiconductors as a result of their inability to establish higher structural orders in the solid state. Those which are capable of undergoing efficient self-organization do provide higher mobility, but the very structural regioregularity which promotes their self-assembly to a high structural order with extensive π-conjugation also renders them extremely sensitive to photoinduced oxidative doping and bleaching. These polymers require processing in an inert atmosphere or in the dark if they are to have useful FET properties. Nonetheless, polymer semiconductors are more amenable to the solution-based processing techniques. Many of these polymers form uniform, flexible thin films which are highly desirable for creating mechanically robust devices. If the high-mobility polymer semiconductors of this type, for example regioregular polythiophenes, can be structurally modified to inhibit their sensitivity to oxidative doping, then fabrication of functionally capable OTFTs would be possible.

Efficient self-organization is usually observed for small molecular semiconductors, including short-chain oligomers [6–8, 13–18], when deposited by vacuum evaporation. The crystalline semiconductor layers of this type such as those obtained from pentacene have yielded high-mobility OTFTs. High crystallinity of the semiconductor layer also has a protective shielding effect on the semiconductor molecules, thereby slowing down their degradative interactions with environmental reactants. Although vacuum deposition usually provides a continuous crystalline thin film, solution deposition often leads to "island" structures composed essentially of crystal aggregates. The noncontinuity of the island structures limits fabrication of useful channel semiconductor layers. Under well-controlled solvent evaporation conditions, solution deposition may also afford continuous semiconductor thin films. The procedural delicacy of controlled solvent evaporation is, however, impractical, if not impossible, for the high-throughput manufacturing processes needed for low-cost OTFT circuit production.

Our structural studies of organic semiconductor materials for OTFTs are aimed at resolving these material deficiencies through molecular design. We demonstrate herein that proper structural design can result in a semiconductor system with reasonable air stability and FET properties which may ultimately satisfy the performance and economic requirements for low-cost OTFTs. The application of structure–property correlation principles in our structural design and the development of an enabling semiconductor nanoparticle ink for inkjet patterning of semiconductors for high-performance OTFT circuits will be discussed herein.

4.4 Polythiophene Semiconductors

Unsubstituted polythiophenes are insoluble materials which cannot be processed by solution-based deposition techniques. Polythiophenes with substituents such as oxymethylene or long alkyl chains are soluble in common organic solvents, and have been widely studied with appropriate dopants to increase electrical conductivity for use as organic conductors [29, 30]. Undoped polythiophenes are intrinsically semiconductive in nature, and a wide variety have been explored as potential channel semiconductors for OTFTs. For charge transport in organic systems, large crystalline domains with long-range structural order that reduce grain boundary effects and aid in charge transport by hopping would lead to high mobility, whereas amorphous or disorganized materials result in poor transport. For the polythiophene system, a sufficiently long alkyl side-chain would provide the solubility needed for processing and structural regioregularity helps promote molecular self-assembly to achieve a desirable structural order for charge transport. Molecular self-organization is difficult for regio-random polythiophenes on which the substituents are not positioned regioregularly along the backbone; these are, therefore, poor semiconductors [5]. On the other hand, regioregular alkyl-substituted polythiophenes such as the head-to-tail poly(3-alkylthiophene)s with a long alkyl side-chain self-organize readily from solution when deposited on appropriate substrates [29, 31]. Specifically, when regioregular head-to-tail poly(3-hexylthiophene), or HT-P3HT (**1**), is processed as the channel semiconductor for OTFTs in an inert atmosphere, excellent FET characteristics result. The mobility and current on/off ratio are as high as 0.1 $cm^2\ V^{-1}\ s^{-1}$ and 10^6, respectively [22]. Lower mobility and significantly lower on/off ratio are achieved when the OTFTs are fabricated in air [21], however, and their FET performance rapidly degrades on continued air exposure.

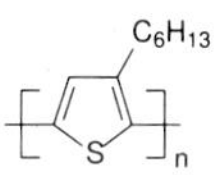

HT-P3HT

(1)

4.4.1 High-performance Polythiophene Design

The poor performance of HT-P3HT as a semiconductor in OTFTs fabricated under ambient conditions can be attributed to its sensitivity to moisture and photoinduced oxidative doping and bleaching by atmospheric oxygen. In the lamellar structural order of HT-P3HT, all the thienylene moieties along its polymer backbone are held in coplanarity by intermolecular alkyl side-chain alignment [29, 31, 32], resulting in a highly delocalized π-conjugated system. Although the lamellar structural order of this type is a constructive set-up for charge transport, it also brings about its susceptibility to photoinduced oxidative doping. Nevertheless, the fact that this regioregular polythiophene system is capable of high mobility and current on/off in OTFTs in an inert atmosphere reveals its excellent intrinsic FET properties. These results suggest that an approach to low-cost, high-performance OTFTs would be possible if appropriate structural features could be built into the regioregular polythiophene system to control its π-conjugation for air stability without compromising its FET characteristics [33].

Our studies of regioregular polythiophenes are driven by the objective of achieving functionally capable OTFTs using simple solution processes under ambient conditions. Solution processing in an environment which does not require elimination of atmospheric oxygen is considered central to achieving the low-cost benefit of OTFTs. We believe this can be accomplished by using a structurally engineered polythiophene system by proper deployment of three design principles [33–36]:

1. sufficiently long alkyl side-chains to provide solution processability;
2. structural regularity to enable molecular self-organization from solution deposition; and
3. proper control of π-conjugation to achieve a balance between air stability and FET property.

Of these design principles, the requirement for building stability against oxidative doping in a semiconductor structure without compromising its semiconductive properties is perhaps the most demanding.

Photoinduced oxidative doping of conjugated polymers under ambient conditions depends on their band gaps and ionization potentials (IPs) or their HOMO levels under vacuum, which depend on their effective π-conjugation lengths. For polythiophenes, coplanarity of the thienylene moieties along the polymer chain leads to extensive π-conjugation, thus a narrower band gap and lower IP, and a greater tendency to be oxidatively doped in air when exposed to visible light. Deviations from coplanarity result in a shorter effective π-conjugation length, a wider band gap, and higher IP, and thus a greater resistance to oxidative doping under ambient conditions. An unduly short π-conjugation length suppresses charge transport efficiency, however, and leads to low mobility. Through our structural studies we have found that simple torsional deviations from coplanarity and strategic presence of nonthienylene moieties along the polymer backbone can be effective in tuning the π-conjugation system of regioregular polythiophenes for air stability and FET properties.

4.4.2
Polydialkylterthiophenes

Regioregular polydialkylterthiophenes constitute an interesting model system for study of oxidative doping stability, self-assembly capability, and FET properties [34, 35]. These polymers can be readily prepared by $FeCl_3$-mediated oxidative polymerization of the corresponding dialkylterthiophene [37, 38] or dehalogenative coupling of dihalodialkylterthiophene (Scheme 4.1) [31].

The poly(3′,4′-alkylterthiophene), PTT (**2**) used in our studies is prepared by $FeCl_3$-mediated oxidative coupling polymerization [39]. PTT with a long alkyl side-chain ($R \geq C_6$) for example PTT-10 (**2**, $R = n\text{-}C_{10}H_{21}$) has an ability to self-organize in the solid state as reflected by a bathochromic shift in its UV–visible absorption spectra from solution to thin film (Fig. 4.1a). The solution spectrum of PTT-10 also has a progressive bathochromic shift with concomitant appearance of a longer-

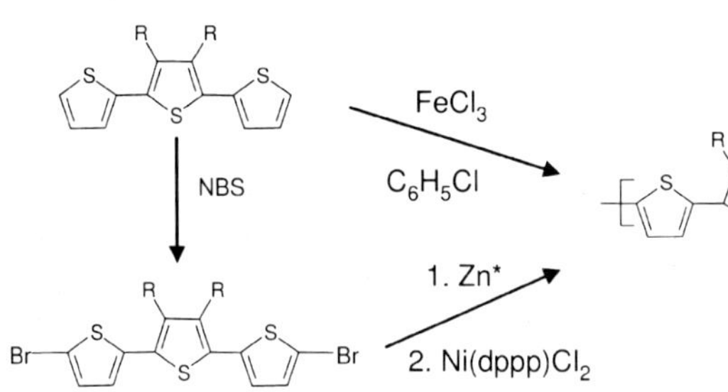

PTT
(**2**)

Scheme 4.1. Synthetic processes for PTT.

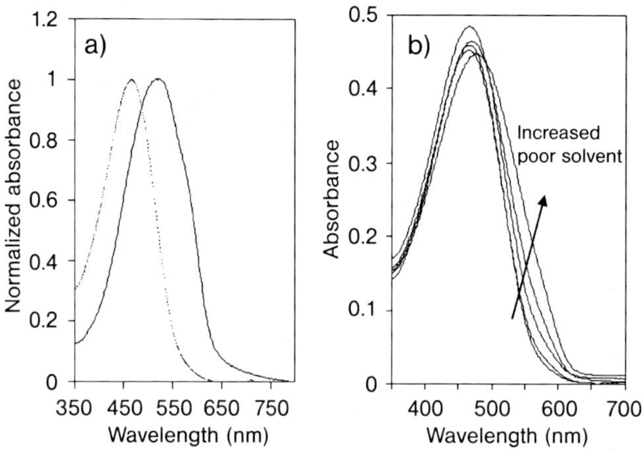

Fig. 4.1. (a) solution (dashed line) and thin film (solid line) absorption spectra of PTT-10; and (b) solution absorption spectra of PTT-10 with bathochromic shifts and appearance of long-wavelength shoulders on addition of methanol [40].

wavelength shoulder with increased addition of methanol, a poor solvent for PTT-10 (Fig. 4.1b) [40]. These spectral properties seem to suggest the formation of a cofacial π–π stacking order of PTT-10 in the solid state. No lamellar structural orders similar to that of HT-P3HT could be detected by X-ray diffraction (XRD) studies.

Bottom-gate, top-contact (Fig. 4.2a) and a bottom-gate, bottom-contact (Fig. 4.2b) TFT configurations are used to evaluate the FET performance of our semiconductors. The devices are built on an n-doped silicon wafer (gate electrode) with a 100-nm thermal silicon oxide (SiO_2) dielectric layer which is modified with a self-assembled monolayer of octyltrichlorosilane (OTS-8) to promote molecular ordering in the semiconductor layer. For the top-contact device the semiconductor layer (~20–50 nm) is deposited on the OTS-8-modified SiO_2 surface by spin coating. A

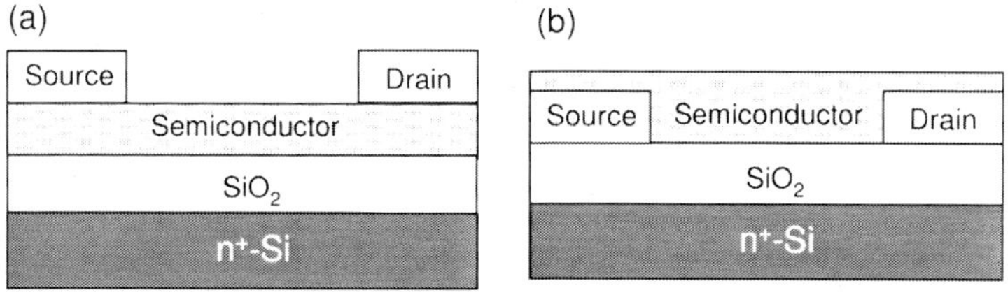

Fig. 4.2. Schematic diagrams of bottom-gate, top-contact (a) and bottom-gate, bottom-contact (b) thin film transistor test configurations.

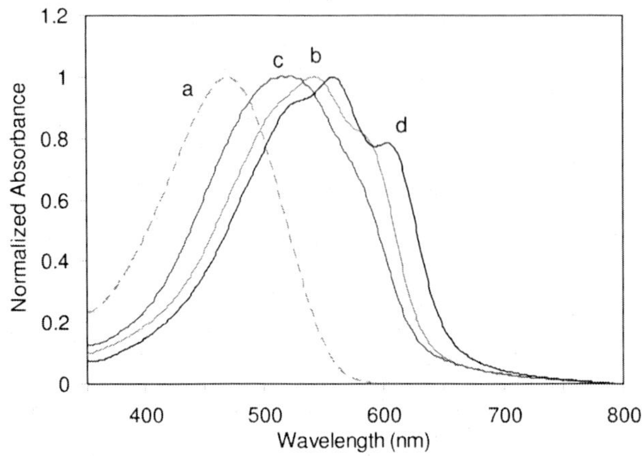

Fig. 4.3. UV–visible spectral properties: (a) dilute solution of PTT′-8 in dichlorobenzene; (b) thin film of PTT′-8, with vibronic splitting; (c) thin film of PTT-10; and (d) thin film of HT-P3HT, with vibronic splitting [35].

series of gold source-drain electrode pairs are subsequently deposited on top of the semiconductor layer by vacuum evaporation through a shadow mask, thus forming an array of OTFT with different channel length and width dimensions. For the bottom-contact device, PTT-10 is spin-coated after gold electrode deposition. OTFT fabricated with PTT-10 under ambient conditions reveal favorable FET characteristics [34]. The extracted mobility from the saturated regime is $\sim$0.01 $cm^2\ V^{-1}\ s^{-1}$ with a current on/off ratio of $\sim 10^5$. Although these FETs results are quite similar to those of HT-P3HT devices fabricated under ambient conditions, the air stability of PTT-10 OTFT is substantially better. Nonetheless, the performance of PTT is dampened by its inability to form higher lamellar π-stacking order similar to that of HT-P3HT [29, 31]. A lamellar structural order of this type in the semiconductor layer is thought to be more efficient for charge transport.

Poly(3,3″-dialkylterthiophene), PTT′ (**3**; Scheme 4.2), is a regioisomer of PTT, differing only in the regiochemistry of alkyl substitution. This “spaced-out” substitution of alkyl side-chains along the polymer backbone has conferred on PTT′ greater self-assembly ability, however. Although PTT-10 has only π–π stacking order in the solid state, PTT′-8 (**3**, R = n-C_8H_{17}) readily self-organizes into lamellar π-stacking structural order in the solid state when cast from solution, a consequence of the extensive intermolecular side-chain interactions facilitated by the “spaced-out” alkyl substitution. In our studies, PTT′-8 was prepared from the corresponding monomer by $FeCl_3$-mediated polymerization in chlorobenzene [35] instead of chloroform [41] to achieve well-defined, high-molecular weight properties. The UV–visible absorption spectrum of PTT′-8 in solution shows broad absorption at $\lambda_{max} \approx 470$ nm (Fig. 4.3a), typical of a twisted polythiophene conformation. The absorption of the thin film of PTT′-8 is, however, sigificantly red shifted together

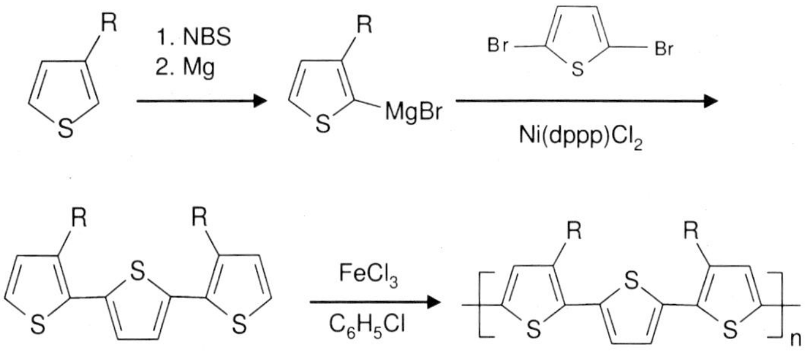

PTT'

(3)

Scheme 4.2. Synthesis of poly(3,3''-alkylterthiophene), PTT' (**3**).

with the appearance of vibronic splitting at $\lambda_{max} \approx 510$ (shoulder), 540, and 583 (shoulder) nm, indicative of higher structural order than PTT-10 in the solid state. The thin-film spectral properties of PTT'-8 are quite similar to those of HT-P3HT (Fig. 4.3d) which has a lamellar π-stacking structural order, in sharp contrast with those of PTT-10 (Fig. 4.3c) which has indiscernible vibronic features because of its lack of lamellar ordering.

The XRD pattern (Fig. 4.4) of a thin film of PTT'-8 has diffraction peaks at $2\theta = 5.8°$ (100), 11.8° (200), and 17.8° (300), corresponding to interlayer spacing of 15.1 Å. The transmission electron diffraction pattern of PTT'-8 thin film corresponds to a π–π stacking distance of ~4.0 Å. These results clearly reveal the lamellar structural order of PTT'-8 in the solid state as schematically illustrated by Fig. 4.4c. The wider than the usual π–π stacking distance of ~3.5 Å was a reflection of the loosely packed lamellar π–π stacking structure. This is probably because of steric interference between the two octyl side-chains on the same repeating units of PTT'-8, twisting the substituted thienylene moieties slightly out of coplanarity and thus sterically hindering the close packing of lamellar layers.

As expected, OTFTs prepared using the PTT'-8 semiconductor have much better FET characteristics, with the average mobility being ~0.015–0.022 $cm^2\ V^{-1}\ s^{-1}$ and a current on/off ratio of 10^5–10^6 [35], with similar air stability. Although these FET properties are among the best for OTFT solution processed at room temperature under ambient conditions without thermally assisted structural ordering, they still fell short of expectation. We envisage that a lamellar π-stacking structural order of this type with tighter packing should enable more efficient charge-carrier transport. Such a lamellar π-stacking structure can be achieved with a regioregular polythiophene system as schematically depicted in Fig. 4.5, In this structural design the side-chains and their spacing, the intervening conjugating spacers, the molecular weight properties, etc. synergistically dictate the intermolecular ordering, the intermolecular interaction, and thus the lamellar π-stacking structural formation

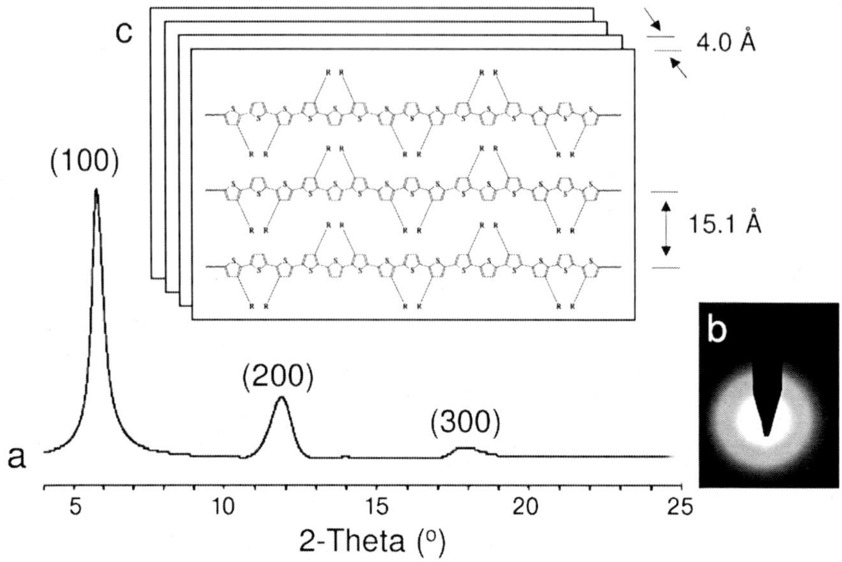

Fig. 4.4. (a) XRD of thin film of PTT′-8 with diffraction peaks at $2\theta = 5.8°$ (100), 11.8° (200), and 17.8° (300); (b) TEM of a free-standing film of PTT′-8 with diffraction equivalent to a π–π stacking distance of 4.0 Å; (c) schematic depiction of lamellar π-stacking structure of PTT′-8 [35].

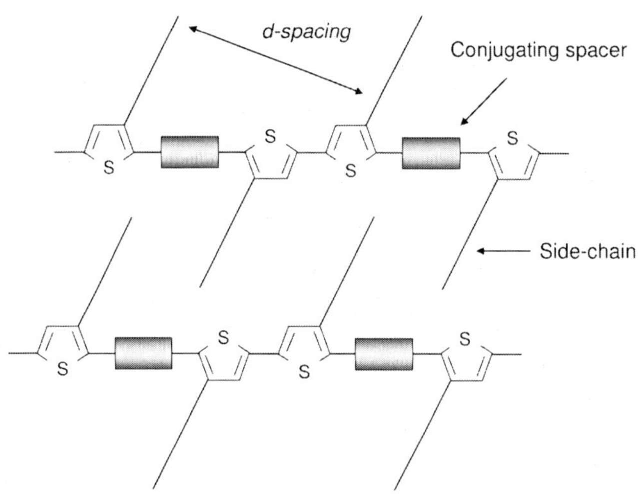

Fig. 4.5. Schematic representation of the lamellar formation of a regioregular polythiophene system via intermolecular side-chain interdigitation.

and effective π-conjugation of the system. From this perspective it is imperative that the intervening conjugating spacers maintain the relative orientations of 3-substituted thienylene moieties along the polythiophene backbone such that the side-chain ordering is optimized to enable achievement of a lamellar π-stacking order.

4.4.3
Polydialkylquaterthiophenes

Regioregular polythiophene, poly(3,3‴-dialkylquaterthiophene), PQT (**4**), is one of the regioregular polythiophene systems which satisfies our design requirements [36]. The alkyl side-chains of PQT are regioregularly positioned along the backbone such that its repeating length is approximately 15.5 Å. The d-spacing of the alkyl side-chains oriented in the same direction in the extended coplanar conformation is approximately 12 Å, because the side-chains are tilted at an angle of ~50° against the backbone. This 12-Å spacing, with a sufficiently long alkyl ($R \geq C_6$), would enable PQT to self-assemble more efficiently than PTT′ by intermolecular alternating side-chain interdigitation. This would lead to formation of a strongly held lamellar structure, thus enabling long-range lamellar π-stacking order in the solid state similar to that of PTT′, as represented schematically in Fig. 4.4c. Lamellar structures of this type have been observed in the oligomeric forms of PQT [42]. The conjugating spacer – which comprises two unsubstituted thienylene units with some rotational freedom and thus torsional deviations from coplanarity – would be expected to suppress the π-conjugation of the polythiophene to provide the air stability needed.

PQT can be prepared, in good yield, by $FeCl_3$-mediated coupling polymerization (Scheme 4.3), and purified by extraction with appropriate solvents [36]. As expected, PQT such as PQT-12 (**4**, $R = n\text{-}C_{12}H_{25}$) have an ionization potential which is ~0.1 eV higher than that of HT-P3HT and a wider band gap, indicative of its greater stability against photoinduced oxidative doping. Liquid crystalline behavior also is observed for PQT-12, as noted in its DSC thermogram, with two endo-

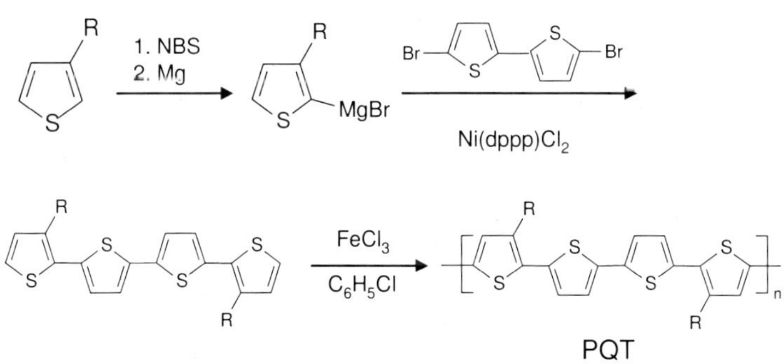

Scheme 4.3. Synthesis of poly(3,3‴-dialkylquaterthiophene), PQT (**4**).

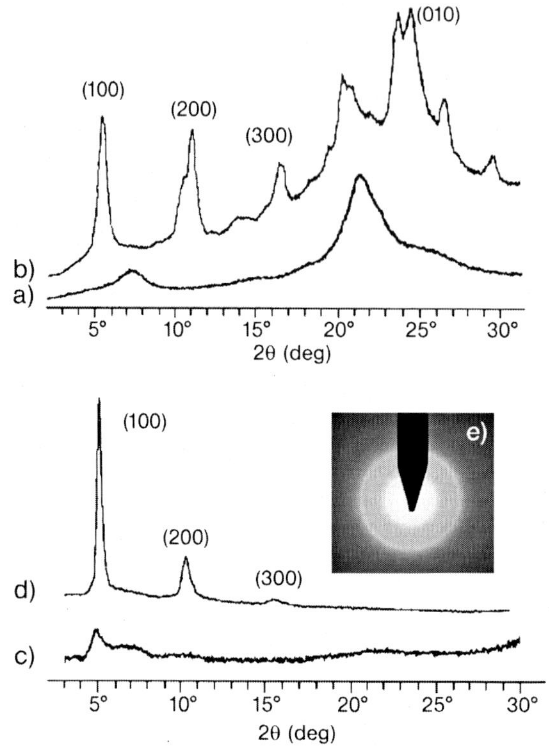

Fig. 4.6. XRD of PQT-12: (a) pressed pellet of precipitated polymer from polymerization; (b) pressed pellet annealed at 140 °C; (c) as-cast 0.2-μm thin film; (d) 0.2-μm thin film annealed at 135 °C; and (e) transmission electron diffraction pattern of PQT-12 film on carbon grid [36].

therms at ~120 °C and ~140 °C corresponding, respectively, to the crystalline-to-liquid crystalline and liquid crystalline-to-isotropic phase transitions. The XRD pattern of a powdered sample of PQT-12 has two diffraction peaks at $2\theta = 7.4$ and 21.5°, arising from side-chain (d-spacing, 12.0 Å) and π–π stacking (d-spacing, 4.1 Å) ordering, respectively (Fig. 4.6a). When the powdered sample is annealed at 120–140 °C, highly crystalline XRD patterns are observed (Fig. 4.6b), revealing a shortened interchain d-spacing of 16.4 Å and a π–π stacking distance of 3.8 Å. An annealed thin film of PQT-12 on an OTS-8-modified silicon wafer substrate has XRD diffraction peaks at $2\theta = 5.1°$ (100), 10.3° (200) and 15.4° (300) (Fig. 4.6d). These results indicate the formation of lamellar π–π stacking structures which are preferentially oriented with their lamellar axes normal to the substrate. Transmission electron diffraction analysis reveals π–π stacking of 3.7 Å (Fig. 4.6e). Thus it can be concluded that PQT-12 has excellent ability to organize into highly ordered lamellar π–π stacking structures whose orientation can be manipulated by surface-

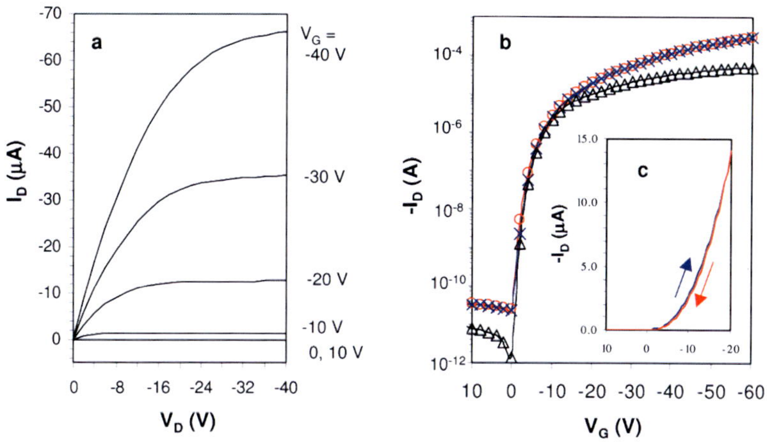

Fig. 4.7. *I–V* characteristics of a typical PQT-12 TFT device with 90 μm channel length and 5000 μm channel width: (a) output curves at different gate voltages; (b) two transfer curves in saturated regime scanned from positive to negative gate voltages (○ (red) and × (blue), $V_D = -60$ V) and transfer curve at linear regime (Δ, $V_D = -6$ V); (c) two transfer curves scanned in different directions between +10 and −20 V [36].

chemistry and alignment techniques. This is reminiscent of the self-assembly behavior of regioregular poly(3-alkylthiophene)s [29, 31].

OTFTs fabricated with the PQT-12 semiconductor by spin coating under ambient conditions have excellent FET performance on annealing at 120–145 °C. The output characteristics reveal no noticeable contact resistance, very good saturation behavior, and clear saturation currents which are quadratic to the gate bias. The devices switch on at approximately 0 V, with a sub-threshold swing of ~1.5 V decade^{-1} (Fig. 4.7). Extracted mobility of up to 0.14 cm^2 V^{-1} s^{-1} and a current on/off ratio of over 10^7 are obtained. The ability to achieve high mobility and an extremely high current on/off ratio in air attests to the significantly greater environmental stability of PQT-12. In addition, the mobility values extracted from both the linear and saturated regimes are essentially the same. The devices also have little or no hysteresis and bias stress effects at room temperature. All these data indicate that PQT-12 has ideal FET behavior.

Although spin coated OTFTs with PQT-12 semiconductors have high mobility and on/off ratio characteristics, the performance of those fabricated by inkjet patterning is more variable. This may be attributable to the tendency of PQT-12 solution to gel at room temperature, leading to nonuniformity in the jetted semiconductor channel layers. This deficiency of the PQT solution as an inkjet ink is expected to be a practical challenge in manufacture, and must therefore be addressed. By study of the solution behavior of PQTs, we have subsequently developed a nanoparticle dispersion of PQT-12, which effectively resolves this challenging problem (*vide infra*).

4.4.4
Polythiophene Nanoparticles

Molecular self-organization in solution depends critically on molecular structural features and on concentration. Molecular self-organization or aggregation in solution occurs at the critical saturation concentration when the solvency of the medium is reduced. This can be achieved by solvent evaporation, reduced temperature, addition of a nonsolvent, or a combination of all these factors. Solvatochromism and thermochromism of conjugated polymers such as regioregular polythiophenes are two illustrative examples, respectively, of solubility and temperature effects [43–45]. It should therefore be possible to use these solution phenomena to pre-establish desirable molecular organization in the semiconductor materials before deposition. Our studies of the molecular self-assembly behavior of PQT-12, which leads to the preparation of structurally ordered semiconductor nanoparticles [46], will be described. These PQT-12 nanoparticles have consistently provided excellent FET characteristics for solution-processed OTFTs, irrespective of deposition methods.

In hot, dilute dichlorobenzene solutions (>50 °C) broad absorption at $\lambda_{max} \sim 480$ nm is apparent in the UV–visible spectrum of PQT-12 (Fig. 4.8a), but the absorption is slightly red-shifted with the concomitant appearance of weak absorption at $\lambda_{max} \sim 610$ nm (Fig. 4.8b) when the solution is cooled to room temperature. This is obviously because of the migration of PQT-12 molecules from the twisted disordered conformation in hot solution to an ordered coplanar conformation at lower temperatures. Because the HOMO level of PQT-12 in the solid state, esti-

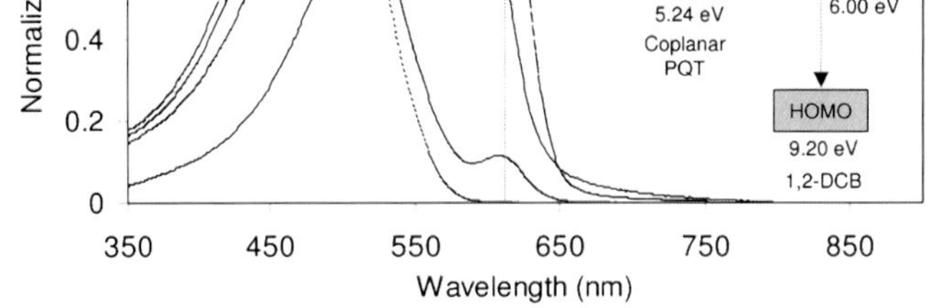

Fig. 4.8. Absorption spectra of PQT-12: (a) dilute solution in dichlorobenzene (0.002%, *w/w*) at 50 °C; (b) dilute solution in dichlorobenzene at 25 °C; (c) nanoparticle dispersion of PQT-12 (0.3%, *w/w*) in dichlorobenzene at 25 °C; and (d) thin film at 25 °C. Insert diagram shows HOMO and LUMO levels of coplanar PQT-12 and 1,2-dichlorobenzene (1,2-DCB) and the CT transition [46].

mated from its half-wave oxidation potential ($E_{1/2}$), is 5.24 eV, and that of 1,2-dichlorobenzene is 3.20 eV in a vacuum [47], the 610-nm absorption can be assigned to the charge-transfer (CT) transition (2.04 eV) from the electron-rich, highly conjugated coplanar structure of PQT-12 to the dichlorobenzene-acceptor solvent (Fig. 4.8, insert). At higher concentrations (e.g. 0.3% w/w), the solution has a tendency to gel at lower temperatures, probably because of the formation of lamellar structures and their ensuing CT interactions with dichlorobenzene. When the hot solution is subjected to ultrasonic agitation during cooling, however, nanoparticles of PQT-12 (NanoPQTs) of average diameter 5–40 nm are formed (revealed by light-scattering measurements). The UV–visible spectrum of the nanoparticle dispersion in dichlorobenzene is indicative of strong absorption with characteristic vibronic splitting ($\lambda_{max} \sim 465$ (shoulder), 510, and 558 nm) and a strong CT band at 610 nm (Fig. 4.8c). The spectrum of a thin film spun-cast from the nanoparticle dispersion on a glass slide contains well-defined vibronic splitting at λ_{max} 510 (shoulder), 545, and 585 nm (Fig. 4.8d), reflecting a highly structured thin film. Except for the absence of the 610-nm band, the thin-film spectral pattern is essentially similar to that of nanoparticle dispersion, indicating that the NanoPQT has a similarly ordered structure, and that the 610-nm band in the NanoPQT spectrum originates from the same CT transition.

The XRD pattern of a thin film spun cast from the NanoPQT dispersion on a silicon wafer contains very distinctive crystalline peaks (Fig. 4.9b), indicative of lamellar π–π stacking order (Fig. 4.9f). The reflection at $2\theta = 7.4°$ (d-spacing = 12.1 Å) is because of the ordering of dodecyl side-chains of the π–π stacked PQT-12 molecules, in excellent agreement with the theoretical estimation. The intermolecular distances of the lamellar π–π stack can be deduced from (100) and (010) reflections at $2\theta = 5.0°$ (~18.0 Å= interchain distance; 2nd-order at $2\theta = 9.9°$ and 3rd-order at $2\theta = 14.9°$), and $2\theta = 22.1°$ (~4 Å= π–π stacking), respectively. The electron diffraction micrograph reveals a reflection equivalent to a d-spacing of ~3.9 Å, typical of the π–π stacking for polythiophenes (Fig. 4.9c). These XRD diffraction peaks from the thin film cast from NanoPQT dispersion confirm the presence of lamellar π–π stacking order in NanoPQTs, because the thin film cast from a hot solution of PQT-12 in dichlorobenzene gives only one very weak diffraction peak at $2\theta = 5.0$ (18.0 Å) (Fig. 4.9a).

Upon annealing at 145 °C, the NanoPQT thin film gives an XRD diffraction pattern (Fig. 4.9d) which is similar to that from an annealed thin film from the hot solution (Fig. 4.6d). The interchain distance tightens to ~17.2 Å ($2\theta = 5.1°$), accompanied by the disappearance of $2\theta = 7.4°$ and 22.1° diffraction peaks associated with the side-chain interlayer and π–π stacking respectively. The disappearance of these diffraction peaks is indicative of the preferential orientation of PQT-12 lamellar structures with their (100) axes normal to the substrate – a result of the interaction of PQT-12 dodecyl side-chains with the octyl chains of the OTS-8 modified surface induced and facilitated by thermal annealing.

The nature of structural orders in the PQT-12 thin films can be further characterized by atomic-force microscopic (AFM) analysis. Figure 4.10 shows the AFM phase images of a 20-nm PQT-12 film on an OTS-8-modified wafer surface before

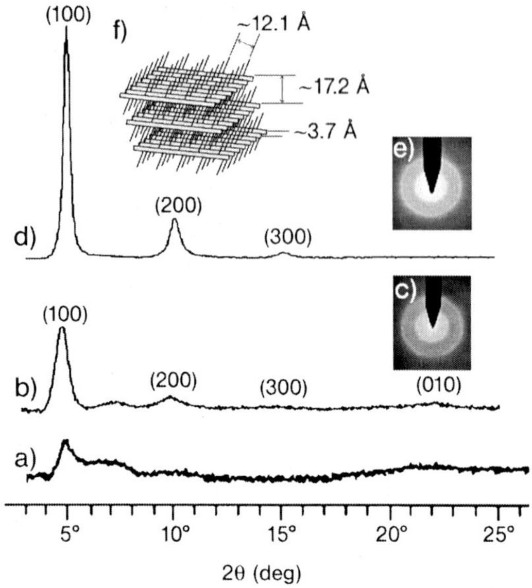

Fig. 4.9. (a) XRD of a 100-nm film cast from hot PQT-12 solution, with very weak diffraction peak at $2\theta = 5.0°$; (b) XRD of a 120-nm film cast from NanoPQT dispersion, with reflections from interchain ordering ($2\theta = 5.0°$, 9.9°, 14.9°), side-chain ordering ($2\theta = 7.2°$), and π–π stacking ($2\theta = 22.1°$); (c) electron micrograph of a free-standing film from NanoPQT dispersion, with π–π stacking reflection; (d) XRD of annealed PQT-12 film, with reflections for interchain ordering only at $2\theta = 5.1°$, 10.1°, 15.2°; (e) electron micrograph of annealed film showing π–π stack reflection; and (f) schematic representation of PQT-12 lamellar π–π stacking structure [46].

and after annealing [48]. The phase image of the as-cast film reveals the presence of an array of small ill-defined crystal domains dispersed in an amorphous polymer matrix (Fig. 4.10a). On heating at 145 °C for 10 min and then cooling to room temperature over a period of 10 min a highly crystalline film with extensive nanocrystalline domains of parallel rod-like structures is formed, and is apparent from the phase image (Fig. 4.10b). These rod-like structures are approximately 30–100 nm long and approximately 10–15 nm wide (Fig. 4.10d). For any PQT-12 sample the lengths of the rod-like structures vary with the annealing conditions (temperature and cooling rate) whereas the widths remain constant. The average widths of the structures are found to correspond to the molecular chain lengths of PQT-12, as calculated from the average molecular weights obtained by matrix-assisted laser desorption/ionization (MALDI) mass spectral analysis. On this basis and the XRD and electron diffraction data, the rod-like structure can be assigned a π–π lamellar structure composed of face-to-face stacking lamellae with the stacking direction along the length of the rod. Heating the film at 145 °C for 30 min, then slow cooling to room temperature over a period of 2 h, leads to significantly larger domains with longer rod-like structures (Fig. 4.10c), revealing the flexibility with which

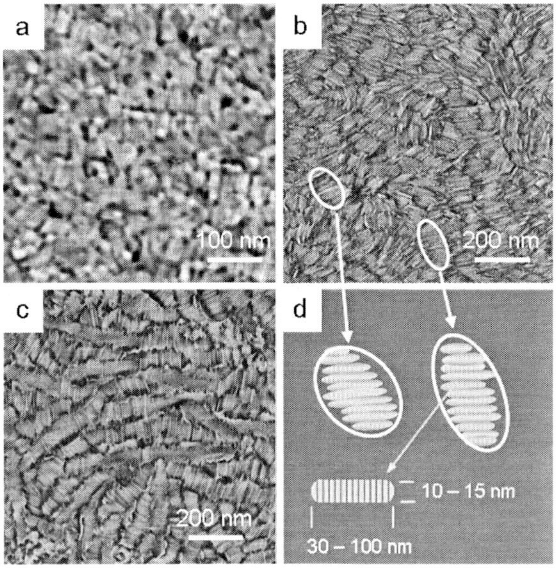

Fig. 4.10. AFM phase images of 20-nm films cast from NanoPQT dispersion on OTS-8-modified SiO_2 surfaces: (a) as cast; (b) annealed at 145 °C and cooled to 25 °C over a period of 10 min, revealing a highly crystalline film with extensive nanocrystalline domains; (c) annealed at 145 °C for 30 min and cooled to 25 °C over a period of 2 h, revealing larger domains and domain merging; (d) schematic depiction of domains of rod-like structures [46].

structural ordering of PQT-12 in the solid state can be optimized for charge transport.

Before annealing, OTFTs fabricated from the NanoPQT dispersion have significantly higher mobility than those from the PQT-12 solution. This higher mobility is because of higher structural order in the semiconductor channel layer fabricated from NanoPQT (confirmed by XRD data). The existence of pre-established structural order in NanoPQTs also renders their FET performance insensitive to surface chemistry. Specifically, OTFTs fabricated on both OTS-8-modified and unmodified substrates perform similarly. This is in sharp contrast with the performance of OTFTs from the PQT-12 hot solution, which is highly dependent on the nature of substrate surface [49]. Nevertheless, after annealing the FET performance of NanoPQT-based devices is only slightly better than that of the corresponding annealed devices from the PQT-12 solution. An extracted mobility of up to 0.2 $cm^2\ V^{-1}\ s^{-1}$ and a current on/off ratio of 10^7–10^8 are obtained, with the average mobility being in the range 0.08–0.16 $cm^2\ V^{-1}\ s^{-1}$. These OTFTs are also resistant to the effect of moisture as immersion of these devices in water for an hour results in only a very slight shift in the threshold voltages. The NanoPQT dispersion also greatly enhances the device fabrication consistency and results in reproducible FET performance characteristics. All these attributes would be particularly advantageous in the fabrication of PQT channel semiconductors by inkjet printing.

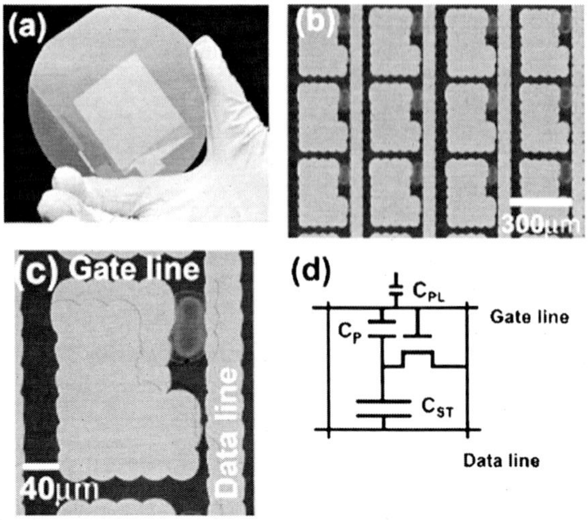

Fig. 4.11. Optical images of a printed polymer TFT array at increasing magnification, showing the whole 128 × 128 array (a), small regions of the array (b), and a single device (c). Note the printed semiconductor confined to the channel region in (b) and (c). (d) Equivalent circuit for the pixel, showing the gate and data address lines, the TFT and different capacitances. In the display pixel the storage capacitor, C_{ST}, is connected to the next gate line. C_P is the parasitic capacitance between the gate line and the pixel and C_{PL} contains all the sources of gate line capacitance, of which the main additional capacitance is to the data line and the TFT channel [50].

4.4.5
Inkjet Patterned TFT Arrays

The NanoPQT dispersion has been used in inkjet patterning of the semiconductor channels of active matrix TFT arrays (Fig. 4.11) [50]. Specifically, the NanoPQT dispersion serves as the inkjet ink for jetting on to a substrate with a pre-patterned TFT array structure. The channel lengths are approximately 30–50 µm to avoid the contact resistance observed for the OTFTs with shorter channel lengths [51]. The average drop diameter on the TFT structure was ~35 µm and the thickness of a single drop was ~25 nm in the center with an edge bead of ~100 nm. By controlling the printing speed and the overlap ratio of the droplets, a continuous layer can be formed. The printed NanoPQT ink was confined to the region between the source-drain electrodes. Ink spreading beyond the gate electrode is detrimental to FET performance because it leads to a higher leakage current from the ungated semiconductor. Also, the well defined area of jet-printed PQT-12 prevents the formation of a continuous semiconductor layer between OTFTs that would lead to significant cross-talk between pixels.

After annealing at 120–140 °C, the OTFTs have a mobility of 0.05–0.10 $cm^2\ V^{-1}\ s^{-1}$ and a current on/off ratio of ~10^6 at $V_{SD} = -40$ V. The onset voltage was close to 0 V and the subthreshold slope was 75 nF V dec^{-1} cm^{-2}. The output characteristics are indicative of good saturation with no signs of significant contact resistance. Minimum gate-bias stress effect is also observed for the OTFTs. All these properties obtained for the jet-printed TFTs are very similar to those of spin-coated devices. A high yield of working OTFTs and uniformity in both the on and off current are also observed. A typical series of measurements give an average mobility of 0.06 cm^2 V^{-1} s^{-1}, with standard deviation of 0.02 cm^2 V^{-1} s^{-1}. These results are representative of state-of-the-art polymer-based OTFTs with higher average mobility than the pentacene OTFTs from a solution processed pentacene precursor, which was recently used to drive an electrophoretic display [52].

4.5 Indocarbazole Designs

Organic semiconductors based on arylamines such as *N*,*N*,*N*′,*N*′-tetraaryl-1,1′-biphenyl-4,4′-diamines [53], and poly(carbazole) [54] have been extensively studied as hole-transport materials for optoelectronic applications (e.g. photoreceptors [55], organic light-emitting diodes [56], etc.). These amine-based semiconductors have the general attributes of not absorbing in the visible region of spectrum (larger band gaps) and low-lying HOMO; these contribute to an enhanced stability against photoinduced oxidative doping by atmospheric oxygen. Although these materials may be satisfactory for their intended applications, most lead to low FET mobility in OTFTs. This may be because of either the inability to self-assemble or the lack of a mechanism by which the generated radical cations (i.e. hole injection) could be efficiently stabilized.

Among arylamine semiconductors, indolo[3,2-*b*]carbazole (**5**) is an appealing system for studies because it has a relatively large and planar molecular structure to enable facile establishment of higher structural order for charge transport [57, 58]. Earlier, we reported the use of 5,11-bis(1-naphthyl)indolo[3,2-*b*]carbazole, a derivative of **5**, as a hole transport material in OLEDs [59]. Because of its sterically encumbered naphthyl substituents, this compound forms only an amorphous film on vacuum deposition, and mobility is low in OTFTs, as expected.

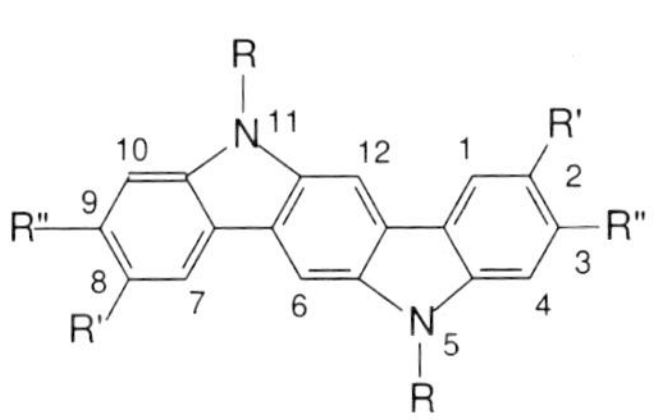

(5)

We envisage that both the ability to order and a mechanism for stabilization can be built into the indolo[3,2-*b*]carbazole system by strategic peripheral substitutions (i.e. R and R′). Indolo[3,2-*b*]carbazole (**5**) with 5,11-bis(alkylphenyl) substitution is a semiconductor of this type, because the long alkyl chains can serve to promote molecular self-organization while the phenylene moieties provide the extra resonance stabilization needed [57]. Resonance stabilization can also be achieved via electron-donor substituents at the 2,8 positions of 5,11-dialkyl-substituted indolo[3,2-*b*]carbazole (**5**) [58].

The synthesis of substituted indolo[3,2-*b*]carbazoles is presented in Scheme 4.4. Parent indolo[3,2-*b*]carbazole can be first prepared by double Fisher indolization to

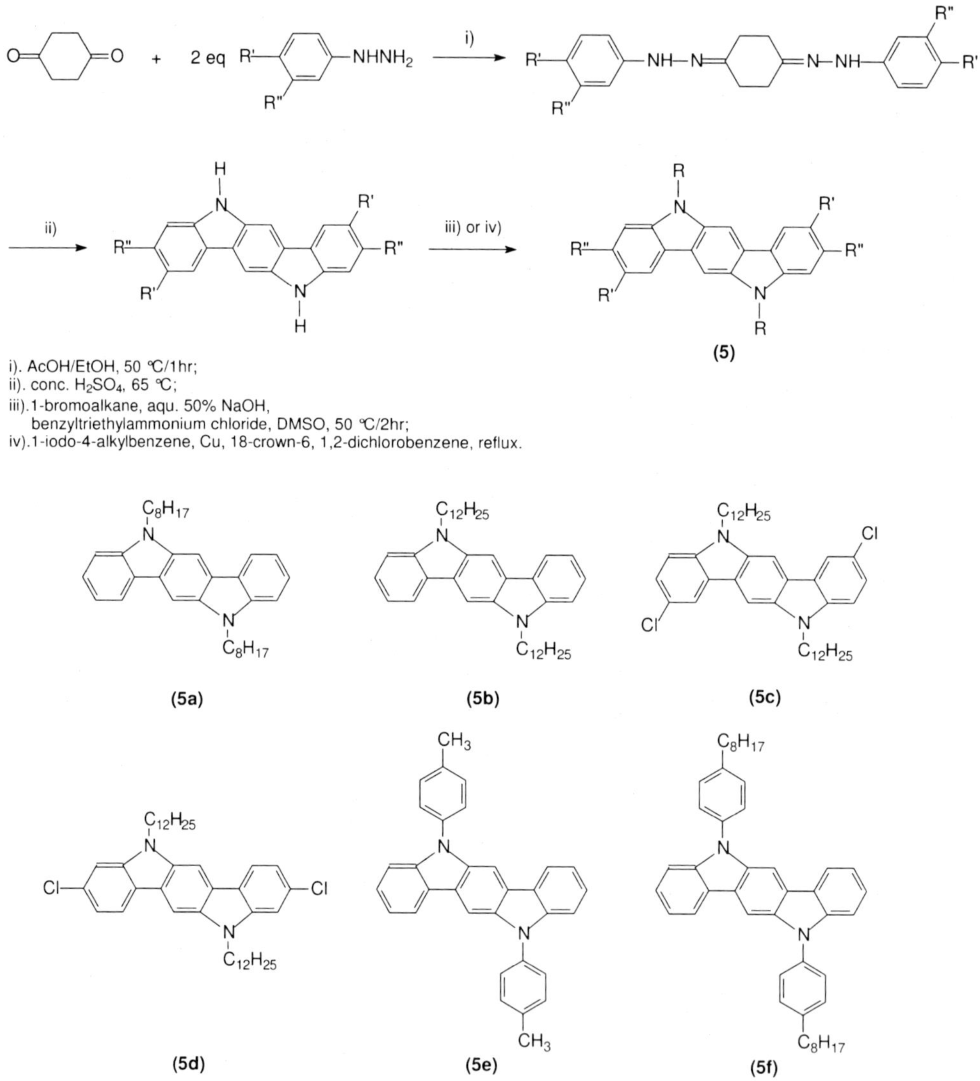

Scheme 4.4. Synthesis of substituted indolo[3,2-*b*]carbazoles.

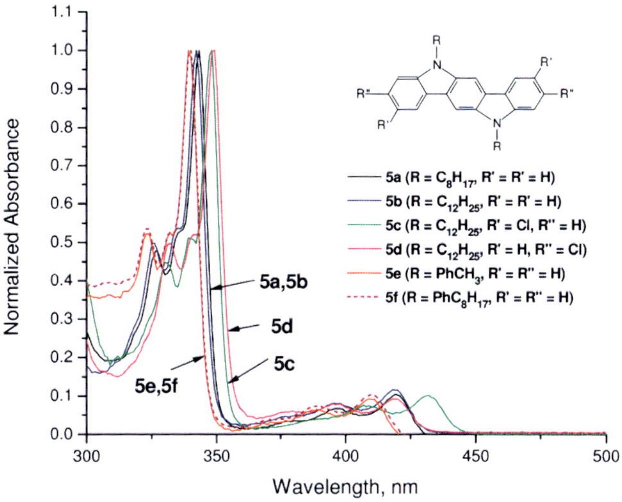

Fig. 4.12. UV–visible spectra of indolo[3,2-*b*]carbazoles (**5**) in chloroform solution.

form cyclohexane-1,4-dione bis(phenylhydrazone), followed by ring closure cyclization. 2,8-Dichloroindolo[3,2-*b*]carbazole and 3,9-dichloroindolo[3,2-*b*]carbazole are prepared in a similar manner from the corresponding chlorophenylhydrazines. Subsequent alkylation with a bromoalkane using phase-transfer reaction conditions leads to 5,11-dialkyl-substituted indolo[3,2-*b*]carbazoles. The 5,11-diaryl-substituted derivatives are, on the other hand, obtained by Ullmann condensation of parent indolo[3,2-*b*]carbazole with aryl iodide in the presence of excess copper and a catalytic amount of 18-crown-6.

UV–visible spectral measurements show that substituted indolo[3,2-*b*]carbazoles (**5**) absorb significantly below 450 nm (Fig. 4.12). Their band gaps in the solid state, estimated from the on-set UV–visible absorption, were >2.55 eV, which are significantly larger than those of most p-channel organic semiconductors for OTFTs (Table 4.1). In addition, whereas chlorine substitution at the 3,9 positions of 5,11-dialkylindolo[3,2-*b*]carbazole (i.e. **5d**) did not cause noticeable changes in the spectral properties, substitution at the 2,8 positions (i.e. **5c**) results in red-shifts in both solution and thin film absorption – a phenomenon that clearly indicates the pronounced electronic effects of substituents at positions para to the radical cation sites (5-*N* and 11-*N* positions).

All the substituted indolo[3,2-*b*]carbazoles (**5**) with long alkyl side-chains form highly crystalline thin films on vacuum deposition. The X-ray diffraction patterns of the thin films, deposited at respective optimum substrate temperatures, give sharp and intense crystalline diffraction peaks (Fig. 4.13). Table 4.2 summarizes the XRD data and extracted interlayer distances of molecular orders of indolo[3,2-*b*]carbazoles in crystalline thin films. Only **5e**, which has no long alkyl chain, has broad XRD diffraction peaks indicative of its amorphous nature. These results

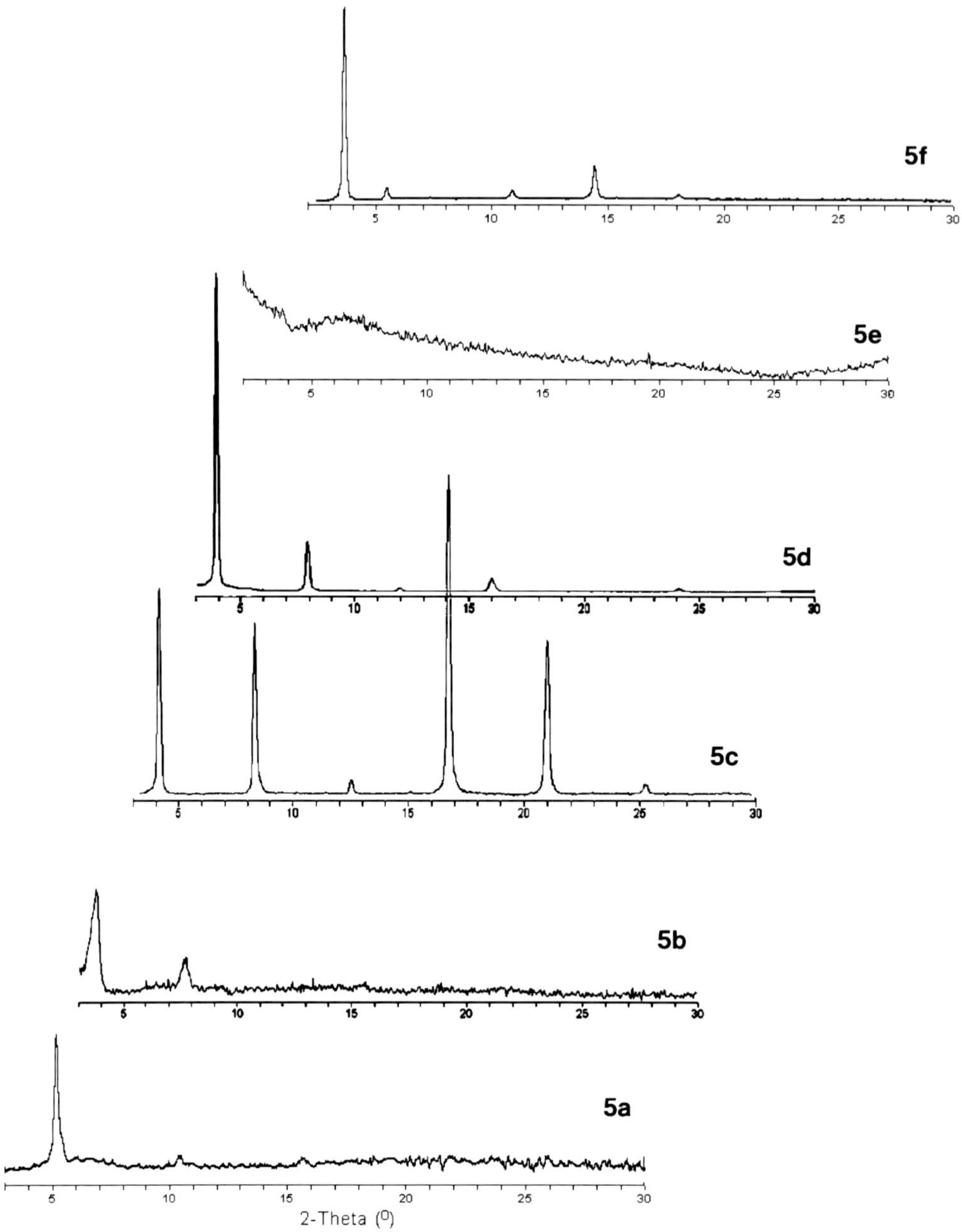

Fig. 4.13. XRD of vacuum-deposited thin films at optimum substrate temperatures: **5a**, **5b**, and **5e** at 25 °C; **5d** and **5f** at 50 °C; and **5c** at 70 °C.

Tab. 4.2. XRD data and intermolecular distances of indolo[3,2-*b*]carbazole crystalline orders in vacuum-deposited thin films.

Indocarbazole	Interlayer distance 2θ (d-spacing)	π–π stacking distance 2θ (d-spacing)
5a	5.2° (16.9 Å)	–
5b	3.8° (23.5 Å)	–
5c	4.1° (21.3 Å)	25.3° (3.5 Å)
5d	3.9° (22.5 Å)	24.1° (3.7 Å)
5f	3.6° (24.4 Å)	18.1° (4.9 Å)

clearly show the dominant effect of long alkyl side-chains on the molecular self-organization of indolo[3,2-*b*]carbazoles. In the XRD pattern of **5f**, sharp diffraction peaks up to the fifth orders of the primary diffraction peak (at $2\theta = 3.62°$) can be observed, reflecting a high crystallinity, whereas in others, for example **5c**, extra diffraction peaks indicative of polymorphism are observed. The thin film XRDs of **5** also reveal the absence of π–π stacking reflections compared with their powdered XRDs (e.g. Fig. 4.14 for **5f**), suggesting that the molecules in the thin films adopted a predominantly "edge-on" orientation relative to the substrate. This also accounts for the formation of layered structural domains as observed by atomic force microscopy (AFM, see below). On the basis of two fully extended peripheral side-chains tilted at an angle of approximately 25° to the indolo[3,2-*b*]carbazole plane [59, 60], the molecular heights are estimated to be at least 24 Å for **5a**, 30 Å for **5b**,

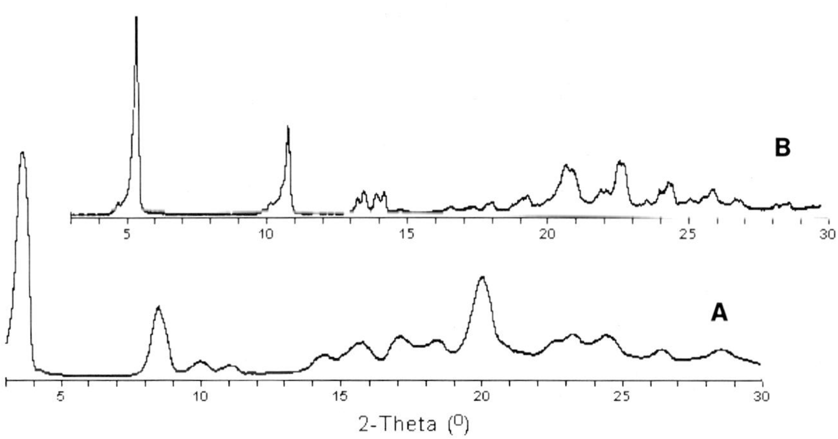

Fig. 4.14. Powdered X-ray diffraction patterns of **5f**: (A) from sublimed sample; and (B) from sample crystallized from toluene.

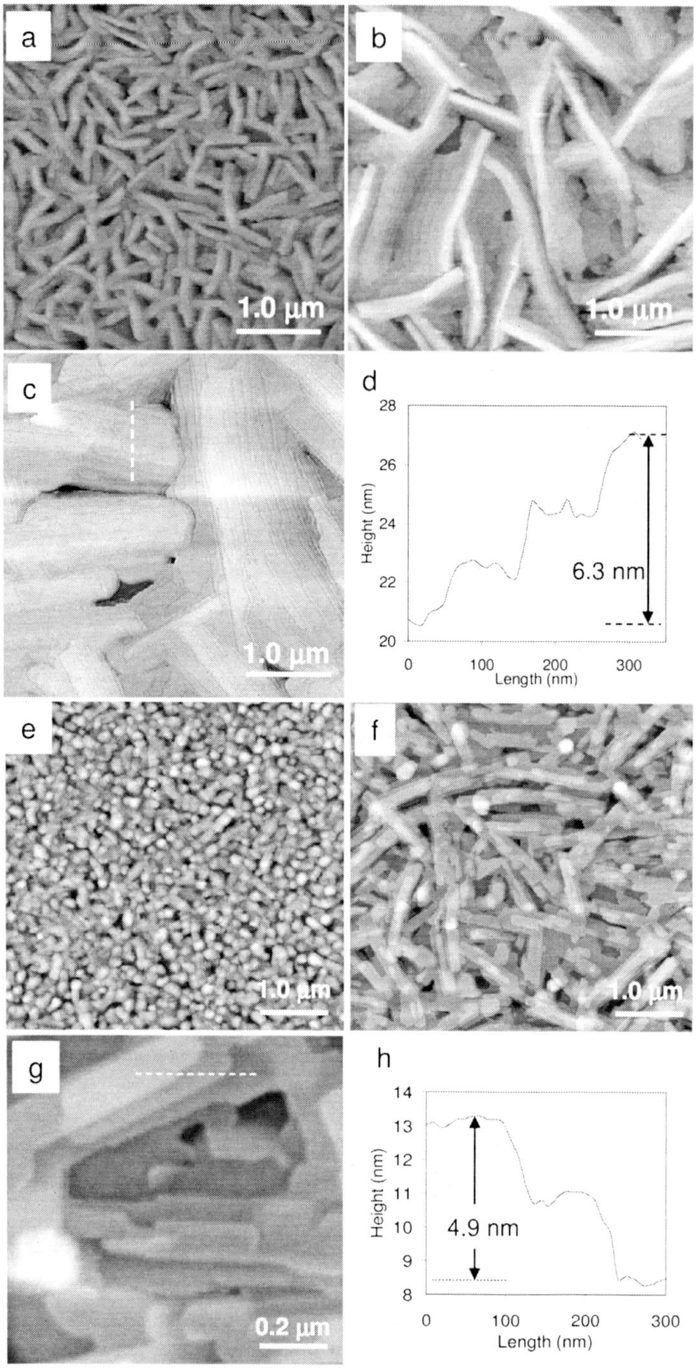

Fig. 4.15. AFM topographic images of thin films deposited at different substrate temperatures: **5c** at (a) 25 °C; (b) 50 °C; (c) 70 °C; and **5f** at (e) 25 °C; (f) 50 °C; (g) blown-up image of (f). (d) and (h) show the step heights of crystalline terrace layers for **5c** and **5f** respectively, as measured by a profiler along the dotted line direction in (c) and (h) [57, 58].

5c, and **5d**, and 34 Å for **5f**. The observed shorter interlayer distances from the XRD data (Table 4.2) suggest either that the molecules are oriented with a substantial inclination relative to the substrate or the interlayers involve some interdigitation of side-chains of the molecules in the adjacent layers.

The crystalline structures of indolo[3,2-*b*]carbazole thin films could also be directly observed by AFM. For example, the AFM topography images of thin films of **5c** and **5f** (Fig. 4.15) reveal the presence of crystalline domain structures which were highly dependent on substrate temperature. Small crystal grains form at low substrate temperatures (e.g. 25 °C) whereas elongated crystal grains as large as 5 μm in length and over 2 μm in width are observed at higher temperatures. The AFM images also indicate the presence of interconnected crystal domains with terrace-like layered structures in which the stacked terrace layers lie parallel to the substrate. Cross-section analysis of these terraced structures gives step-heights of 2.10 nm and 2.45 nm for **5c** and **5f**, respectively, identical to the interlayer distances extracted from XRD data.

Cyclic voltammetric measurements of **5c** and **5f** in 0.1 m Bu_4ClO_4/CH_2Cl_2 solution reveal reversible oxidation peaks at 0.88 V and 0.74 V, respectively, relative to the Ag/AgCl electrode, corresponding to estimated HOMO levels of, respectively, 5.26 eV and 5.12 eV from vacuum. These values are lower than those of most regioregular polythiophenes (4.9–5.0 eV) [34–36], and higher oxidative stabilities are therefore expected of these compounds.

Table 4.3 summarizes the FET mobility and current on/off ratio data for OTFTs prepared using vacuum deposited indolo[3,2-*b*]carbazoles (**5**) as the channel semi-

Tab. 4.3. FET performance of indolo[3,2-*b*]carbazoles (**5**).

Indolo[3,2-*b*]carbazole	T_d (°C)	Mobility ($cm^2\ V^{-1}\ s^{-1}$)	Current on/off ratio
5a	25	1.2×10^{-4}	10^3
5b	25	$1.3–3.0 \times 10^{-3}$	10^4
5c	25	$2.0–3.0 \times 10^{-2}$	$10^5–10^6$
	50	$6.0–8.5 \times 10^{-2}$	$10^6–10^7$
	70	$0.85–1.4 \times 10^{-1}$	10^7
	85	2.0×10^{-4}	10^3
5d	25	$4.8–5.8 \times 10^{-3}$	$10^4–10^5$
	50	$0.8–1.0 \times 10^{-2}$	10^5
	70	$2.1–3.5 \times 10^{-3}$	10^4
5e	25	1.0×10^{-5}	10^2
5f	25	$1.0–2.0 \times 10^{-2}$	$10^5–10^6$
	50	$0.7–1.2 \times 10^{-1}$	$10^6–10^7$
	70	$0.7–1.0 \times 10^{-2}$	10^5

conductors. p-Type accumulation FET behavior is observed for all the devices. Indolo[3,2-*b*]carbazole (**5e**), whose vacuum deposited film is amorphous in nature, has the lowest mobility (1.0×10^{-5} cm^2 V^{-1} s^{-1}). Indolo[3,2-*b*]carbazoles **5a** and **5b**, which have long alkyl chains and form highly crystalline semiconductor layers, have higher mobility. The mobilities of **5a** and **5b** are still relatively low, however, probably because of their inability to provide sufficient resonance stabilization to the 5,11-ammonium radical cation sites during charge injection and transport. This rationale is supported by recent independent results of low FET mobility ($\sim 10^{-3}$ cm^2 V^{-1} s^{-1}) observed for several 5,11-dialkyl-6,12-dimethylindolo[3,2-*b*]carbazoles [60]. The best FET performance of our indolo[3,2-*b*]carbazole series is that of **5c** and **5f**, for which mobility is up to 0.14 cm^2 V^{-1} s^{-1} and high current on/off ratios of 10^6–10^7 are observed. The high mobility of **5c** and **5f** can be attributed to:

1. high molecular ordering enabled by the long alkyl chain substituents; and
2. the extra resonance stabilization provided by 2,8-dichloro substituents in **5c** and *N*-phenylene moieties in **5f**.

The chloro substituent is electron withdrawing inductively and electron donating in resonance. This explains the observation that the mobility of 3,9-dichloro substituted indolo[3,2-*b*]carbazole **5d** is an order of magnitude lower than that of its regioisomeric 2,8-dichloro analog **5c**. The two chloro substituents of **5d**, which are positioned meta to 5,11-ammonium radical cation sites, exert inductive but not resonance effects.

The effect of substrate temperature (T_d) on FET performance of indolo[3,2-*b*]carbazole semiconductors is significant. Whereas FET performance of **5c** is optimum at $T_d = 70$ °C, the highest mobility of **5d** and **5f** is observed at $T_d = 50$ °C. The FET characteristics also correlate with their crystal grain sizes, because larger crystal domains of **5c** and **5f** are formed at T_d of 70 °C and 50 °C, respectively, as is apparent from their AMF images. This is reminiscent of the dependence of FET performance on crystal grain size of other vacuum-deposited semiconductors such as pentacene [7]. Although a slight contact resistance and negative turn-on voltage are observed, the overall performance of **5c** and **5f**-OTFTs are among the best reported. Figure 4.16 shows the FET characteristics of a typical OTFT using **5c** as semiconductor.

The OTFT prepared using indolo[3,2-*b*]carbazoles such as **5f** also have enhanced air and photochemical stability compared with the devices prepared using most other organic semiconductors. The transfer characteristics of **5f** devices in the dark and under white light exposure in air are essentially the same (Fig. 4.17a). The current on/off ratio was $\sim 10^7$ for both, demonstrating the excellent light and air stability of **5f** under ambient conditions. There was, on the other hand, substantial degradation of FET performance when the HT-P3HT device was exposed to white light radiation (Fig. 4.17b).

As with most organic semiconductors, semiconductor layers of **5** prepared by spin coating have significantly lower mobility, by as much as an order of magni-

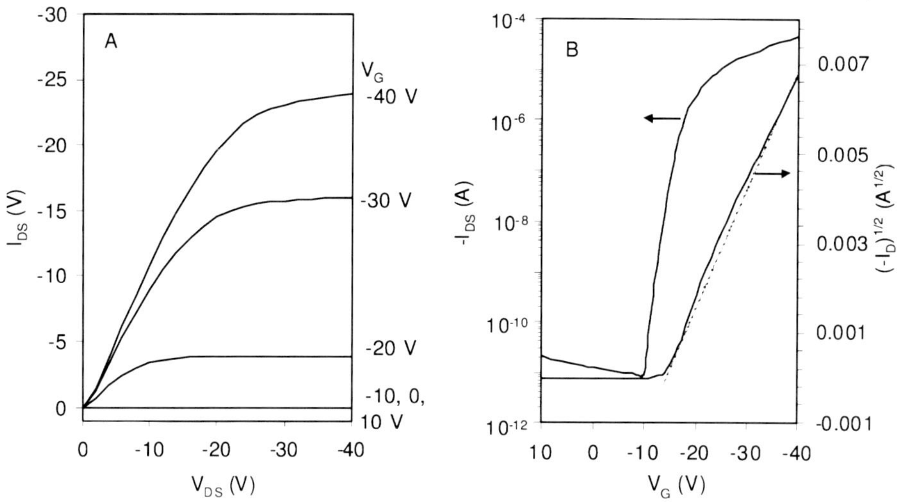

Fig. 4.16. FET characteristics of a typical OTFT prepared with vacuum-deposited **5c** (substrate at 70 °C; channel length = 90 μm. channel width = 5000 μm): (A) output curves at different gate potentials; (B) transfer curve in saturated regime at constant source-drain potential of −40 V and square root of the absolute value of current as a function of gate potential [58].

tude, than their vacuum-deposited counterparts in OTFTs. Nonetheless, we have found that drop coating of **5f** can give rise to a semiconductor with a mobility equivalent to those obtained by vacuum deposition. Under appropriate conditions drop coating of **5f** yields excellent single crystals on the substrate surface with mobility of 0.13 $cm^2\ V^{-1}\ s^{-1}$ when used as channel semiconductors in OTFTs [61]. Further process optimization of the drop-coating technique is necessary to prepare large-area semiconductor layers of this type for practical applications.

4.6 Summary and Prospects

In this chapter we have discussed issues, challenges, and recent advances in organic semiconductor development for OTFTs, and presented our structure–property correlation studies that lead to the formulation of simple design principles for air-stable, solution processable, high-performance semiconductors. Our studies of polymer semiconductors have culminated in the design of a novel class of polythiophene nanoparticles which can be used for jet printing high-mobility semiconductor channels for OTFTs. The results show that a judiciously designed polythiophene can not only have excellent self-organization properties but also enhanced air stability to counter environmental degradative effects. Both self-organization to higher

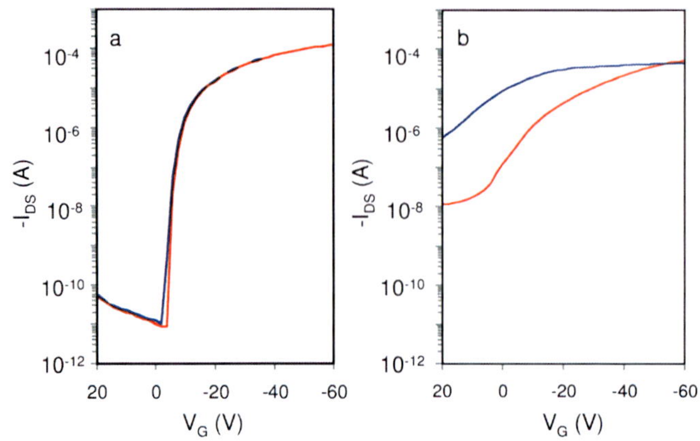

Fig. 4.17. Transfer characteristics of OTFTs (channel length = 90 μm, channel width = 5000 μm, $V_{DS} = -60$ V) in saturated regime with (a) **5f** as channel semiconductor deposited at a substrate temperature of 50 °C; and (b) HT-P3HT as channel semiconductor spin coated from 0.7% *w*/*w* solution in chlorobenzene at ambient. Transfer curves were obtained in the dark (red line) and then measured under white light (blue line) [57].

structural orders and resistance to photoinduced oxidative doping and bleaching under ambient conditions are critical to the design of organic semiconductors for low-cost OTFTs. We firmly believe that the simple polythiophene system as schematically described in Fig. 4.5, from which both PTT′ and PQT are derived, is a fundamental structural motif for self-organization and environmental stability. It should be possible to structurally tune this polythiophene skeleton, by rational choice of side-chain and conjugating spacer, to achieve desirable FET performance characteristics. Recent studies [61] on the polythiophene semiconductor systems within the structural framework of Fig. 4.5 have given strong credence to this simple set of design rules.

We have also developed a novel class of high-mobility molecular semiconductors based on indolo[3,2-*b*]carbazole, in which the dominant effects of side-chain substitution on molecular self-organization are clearly observed. Our results show that the FET performance of this class of semiconductor can be controlled by manipulation of the nature of the substituents and the substitution pattern. Appositely positioned substituents capable of providing stabilization to the indolocarbazolium radical cations (hole carriers) as a result of resonance effect give rise to high mobility. Our indolo[3,2-*b*]carbazole semiconductors have FET mobility up to 0.14 $cm^2\ V^{-1}\ s^{-1}$ and current on/off ratios up to 10^7 when fabricated by vacuum evaporation or by drop coating, although results obtained using drop coating are not as reproducible as those from vacuum deposition. These findings clearly reflect the general difficulties of establishing long-range molecular order in molecular compounds by solution deposition.

Very recently, significant advances have been made in the stabilization of organic semiconductors such as pentacene against environmentally induced degradation. Although alkyl-substituted pentacenes [62] are usually very reactive in solution, those substituted with trialkylsilylethynyl groups at the 6,13 positions have been observed to be stable under similar conditions. Of particular importance is the observation that a member of this group has a mobility of 1.0 $cm^2\ V^{-1}\ s^{-1}$ when deposited in air by blade coating [63], the highest mobility ever achieved with a solution-processed organic semiconductor in OTFTs.

The semiconductor is only one of the many critical components of OTFTs. For low-cost applications, particularly for flexible electronic devices, other solution-processable components, for example gate dielectric, electrical conductor, and protective encapsulation materials, may be required. It is encouraging to note that these topics have been receiving increasing research attention, and that significant progress has recently been made in these areas [64–67]. To propel low-cost conceptual OTFT technology from laboratory to commercialization, however, synergistic melding of materials and processing technology for roll-to-roll manufacturing must be established.

Acknowledgements

The authors are grateful to our colleagues Ping Liu and Sandra Gardner of the Xerox Research Centre of Canada, Ni Zhao and Dr Gianluigi Botton of McMaster University, and the Printed Organic Electronics group led by Dr Robert Street at Palo Alto Research Center, for contributions to the work described in this chapter. Partial financial support of our work is provided by National Institute of Standards and Technology through an Advanced Technology Program Grant (70NANB0H3033).

References

1 C. D. Dimitrakopoulos, P. R. L. Malenfant, *Adv. Mater.* **2002**, *14*, 99.
2 H. E. Katz, Z. Bao, S. L. Gilant, *Acc. Chem. Res.* **2001**, *34*, 359.
3 G. H. Heilmeier, L. A. Zanoni, *J. Phys. Chem. Solid* **1964**, *25*, 603.
4 F. Ebisawa, T. Kurokawa, S. Nara, *J. Appl. Phys.* **1983**, *54*, 3255.
5 A. Tsumura, H. Koezuka, T. Ando, *Appl. Phys. Lett.* **1986**, *49*, 1210.
6 G. Horowitz, D. Fichou, X. Peng, Z. Xu, F. Garnier, *Solid State Commun.* **1989**, *72*, 385.
7 T. W. Kelley, L. D. Boardman, T. D. Dunbar, D. V. Muyres, M. J. Pellerite, T. P. Smith, *J. Phys. Chem. B* **2003**, *107*, 5877.
8 K. Ito, T. Suzuki, Y. Sakamoto, D. Kubota, Y. Inoue, F. Sato, S. Tokito, *Angew. Chem. Int. Ed.* **2003**, *42*, 1159.
9 G. Horowitz, X. Peng, D. Fichou, F. Garnier, *Synth. Met.* **1992**, *51*, 419.
10 A. Afzali, C. D. Dimitrakopoulos, T. L. Breen, *J. Am. Chem. Soc.* **2002**, *124*, 8812.
11 A. R. Brown, A. Pomp, C. M. Hart, D. M. de Leeuw, *Science* **1995**, *270*, 972.
12 A. R. Brown, A. Pomp, D. M. de Leeuw, D. B. M. Klaassen, E. E. Havinga, P. Herwig, K. Müllen, *J. Appl. Phys.* **1996**, *79*, 2136.
13 C. D. Sheraw, T. N. Jackson, D. L.

Eaton, J. E. Anthony, *Adv. Mater.* **2003**, *15*, 2009.

14 A. Dodabalapur, L. Tousi, H. E. Katz, *Science* **1995**, *268*, 270.

15 F. Garnier, A. Yassar, R. Hajlaoui, G. Horowitz, F. Deloffre, B. Servet, S. Ries, P. Alnot, *J. Am. Chem. Soc.* **1993**, *115*, 8716.

16 M. Halik, H. Klauk, U. Zschieschang, G. Schmid, S. Ponomarenko, S. Kirchmeyer, W. Weber, *Adv. Mater.* **2003**, *15*, 917.

17 H. E. Katz, J. G. Laquindanum, A. J. Lovinger, *Chem. Mater.* **1998**, *10*, 633.

18 F. Garnier, R. Hajlaoui, A. El Kassmi, G. Horowitz, L. Laigre, W. Porzio, M. Armanini, F. Provasoli, *Chem. Mater.* **1998**, *10*, 3334.

19 A. R. Murphy, J. M. J. Frechet, P. Chang, J. Lee, V. Suramanian, *J. Am. Chem. Soc.* **2004**, *126*, 1596.

20 J. Wang, Y. Y. Lin, W. Qian, T. N. Jackson, *MRS Fall Meeting Digest* **1998**, 372.

21 Z. Bao, A. Dodabalapur, A. Lovinger, *Appl. Phys. Lett.* **1996**, *69*, 4108.

22 H. Sirringhaus, P. J. Brown, R. H. Friend, M. M. Nielen, K. Bechgaard, B. M. W. Langeveld-Voss, A. J. H. Spiering, R. A. J. Janssen, E. W. Meijer, P. Herwig, D. M. de Leeuw, *Nature* **1999**, *401*, 685.

23 T. C. Gorjanc, I. Levesque, M. D'Iorio, *Appl. Phys. Lett.* **2004**, *84*, 930.

24 H. Fuchigami, A. Tsumura, H. Koezuka, *Appl. Phys. Lett.* **1993**, *63*, 1372.

25 H. E. A. Huitema, G. H. Gelinck, J. B. P. H. van der Putter, K. E. Kujik, C. M. Hart, E. Cantatore, P. T. Herwig, A. J. J. M. van Breeman, D. M. de Leeuw, *Nature* **2001**, *414*, 599.

26 H. Meng, Z. Bao, A. J. Lovinger, B. Wang, A. M. Mujsce, *J. Am. Chem. Soc.* **2001**, *123*, 9214.

27 H. Sirringhaus, R. J. Wilson, R. H. Friend, M. Inbasekaran, W. Wu, E. P. Woo, M. Grell, D. D. C. Bradley, *Appl. Phys. Lett.* **2000**, *77*, 406.

28 G. Horowitz, *Adv. Mater.* **1998**, *10*, 365.

29 R. D. McCullough, S. Tristram-Nagle, S. P. Williams, R. D. Lowe, M. Jayaraman, *J. Am. Chem. Soc.* **1993**, *115*, 4910.

30 (a) R. D. McCullough, P. C. Ewbank, in *The Handbook of Conducting Polymers*, 2nd edition, Marcel Dekker, 1998, Chapter 9, pp. 225–258. (b) R. D. McCullough, in Handbook of Oligo- and Polythiophenes, Ed. D. Fichou, Wiley–VCH, Weinheim, 1998, Chapter 1, 1–44.

31 T. Chen, X. Wu, R. D. Rieke, *J. Am. Chem. Soc.* **1995**, *117*, 233.

32 T. Yamamoto, D. Komarudin, M. Arai, B.-L. Lee, H. Suganuma, N. Asakawa, Y. Inoue, K. Kubota, S. Sasaki, T. Fukuda, and H. Matsuda, *J. Am. Chem. Soc.* **1998**, *120*, 2047.

33 B. S. Ong, Y. Wu, P. Liu, and S. Gardner, *Polym. Prepr.* **2003**, *44*, 321.

34 B. S. Ong, Y. Wu, L. Jiang, P. Liu, and K. Murti, *Synth. Met.* **2004**, *142*, 49.

35 Y. Wu, P. Liu, S. Gardner, and B. S. Ong, *Chem. Mater.* **2005**, *17*, 221.

36 B. S. Ong, Y. Wu, P. Liu, and S. Gardner, *J. Am. Chem. Soc.* **2004**, *126*, 3378.

37 M. Pomerantz, J. J. Tseng, H. Zhu, S. J. Sproull, J. R. Reynolds, R. Uitz, H. J. Arnott, and M. I. Haider, *Synth. Met.* **1991**, *41*, 825.

38 V. M. Niemi, P. Knuuttila, J.-E. Österholm, and J. Korvola, *Polymer* **1992**, *33*, 1559.

39 C. Wang, M. E. Benz, E. LeGoff, J. L. Schindler, J. Allbritton-Thomas, C. R. Kannewurf, and M. G. Kanatzidis, *Chem. Mater.* **1994**, *6*, 401.

40 B. S. Ong, Y. Wu, P. Liu, *IEEE transaction* **2005**, *93*, 1412.

41 M. C. Gallazzi, L. Castellani, R. A. Martin, G. Zerbi, *J. Polym. Sci. Part A* **1993**, *31*, 3339.

42 P. Bäuerle, T. Fischer, B. Bidlingmeier, A. Stabel, J. P. Rabe, *Angew. Chem. Int. Ed. Engl.* **1995**, *34*, 303.

43 C. A. Sandstedt, R. D. Rieke, C. J.

ECKHARDT, *Chem. Mater.* **1995**, *7*, 1057.

44 H. TACHIBANA, N. HOSAKA, Y. TOKURA, *Macromolecules* **2001**, *34*, 1823.

45 G. DUFRESNE, J. BOUCHARD, M. BELLETETE, G. DUROCHER, M. LECLERC, *Macromolecules* **2000**, *33*, 8252.

46 B. S. ONG, Y. WU, P. LIU, S. GARDNER, *Adv. Mater.* **2005**, *17*, 1141.

47 R. KAPLAN, A. M. NAPPER, D. H. WALDECK, M. B. ZIMMT, *J. Phys. Chem. A* **2002**, *106*, 1917.

48 N. ZHAO, G. BOTTON, S. ZHU, A. DUFT, B. ONG, Y. WU, P. LIU, *Macromolecules*, **2004**, *37*, 8307.

49 Y. WU, P. LIU, B. S. ONG, T. SRIKUMAR, N. ZHAO, G. BOTTON, S. ZHU, *Appl. Phys. Lett.* **2005**, *86*, 142102.

50 A. C. ARIAS, S. E. READY, R. LUJAN, W. S. WONG, K. E. PAUL, A. SALLEO, M. L. CHABINYC, R. B. APTE, R. A. STREET, Y. WU, P. LIU, B. ONG, *Appl. Phys. Lett.*, **2004**, *85*, 3304.

51 M. L. CHABINYC, J.-P. LU, R. A. STREET, Y. WU, P. LIU, B. S. ONG, *J. Appl. Phys.* **2004**, *96*, 2063.

52 G. H. GELINCK, H. EDZER, A. HUITEMA, E. VAN VEENENDAAL, E. CANTATORE, L. SCHRIJNEMAKERS, J. B. P. H. VAN DER PUTTEN, T. C. T. GEUNS, M. BEENHAKKERS, J. B. GIESBERS, B. HUISMAN, E. J. MEIJER, E. M. BENITO, F. J. TOUWSLAGER, A. W. MARSMAN, B. J. E. VAN RENS, and D. M. DE LEEUW, *Nat. Mater.* **2004**, *3*, 106.

53 A. M. GORGAN, US patent 4047948.

54 P. M. BORSENBERGER, A. I. ATEYA, *J. Appl. Phys.* **1978**, *49*, 4035.

55 P. M. BORSENBERGER, D. S. WEISS, *Organic Photoreceptors for Xerography*; Marcel Dekker: New York, 1998.

56 G. CASALBORE-MICELI, A. DEGLI ESPOST, V. FATTORI, G. MARCONI, C. SABATINI, *Phys. Chem. Chem. Phys.* **2004**, *6*, 3092.

57 Y. WU, Y. LI, S. GARDNER, B. S. ONG, *J. Am. Chem. Soc.* **2005**, *127*, 614.

58 Y. LI, Y. WU, S. GARDNER, B. S. ONG, *Adv. Mater.* **2005**, *17*, 849.

59 N. HU, S. XIE, Z. POPOVIC, B. ONG, A. HOR, *J. Am. Chem. Soc.* **1999**, *121*, 5097.

60 S. WAKIM, J. BOUCHARD, M. SIMARD, N. DROLET, Y. TAO, M. LECLERC, *Chem. Mater.* **2004**, *16*, 4386.

61 Unpublished results.

62 H. MENG, M. BENDIKOV, G. MITCHELL, R. HELGESON, F. WUDL, Z. BAO, T. SIEGRIST, C. KLOC, C. H. CHEN, *Adv. Mater.* **2003**, *15*, 1090.

63 M. M. PAYNE, S. R. PARKIN, J. E. ANTHONY, C.-C. KUO, T. N. JACKSON, *J. Am. Chem. Soc.* **2005**, *127*, 4986.

64 J. VERES, S. OGIER, G. LLOYD, *Chem. Mater.* **2004**, *16*, 4543.

65 M. HALIK, H. KLAUK, U. ZSCHIESCHANG, G. SCHMID, C. DEHM, M. SCHÜTZ, S. MAISCH, F. EFFENBERGER, M. BRUNNBAUER, F. STELLACCI, *Nature* **2004**, *431*, 963.

66 Y. WU, Y. LI, B. S. ONG, P. LIU, S. GARDNER, B. CHIANG, *Adv. Mater.* **2005**, *17*, 184.

67 Y. LI, Y. WU, B. S. ONG, *J. Am. Chem. Soc.* **2005**, *127*, 3266.

68 Z. BAO, A. J. LOVINGER, A. DODABALPUR, *Adv. Mater.* **1997**, *9*, 42.

5
Electrical and Environmental Stability of Polymer Thin-film Transistors

Alberto Salleo and Michael L. Chabinyc

5.1
Introduction

Most recent work in polymer thin-film transistors (TFTs) has focused on improving the field-effect mobility of charge carriers in semiconducting polymers. The field-effect mobility in regio-regular polythiophene devices has now reached 0.1–0.2 $cm^2\ V^{-1}\ s^{-1}$ [1–3] bringing it close to that in amorphous silicon (mobility ≈ 1 $cm^2\ V^{-1}\ s^{-1}$), currently the most widely used material in large-area electronics. Thus, purely from the standpoint of transconductance, polymer semiconductors have the potential to replace amorphous silicon. The opportunity to adopt solution processing may also reduce fabrication costs and enable widespread introduction of electronic functionality at low cost.

The success of polymer semiconductors in the marketplace, however, depends critically on the ability of these materials to sustain their electrical performance under operation – unstable devices are of little practical use. In this chapter we will describe known issues in the stability of polymer TFTs and point out some of the challenges that must be addressed in the future development of these materials.

The electrical stability of a TFT depends both on how it is operated and how its chemical composition changes over time. Electrical instability, which is often called bias stress, manifests itself as a decay of the output current caused by the protracted operation of the transistor. Bias stress is a well known and characterized phenomenon in disordered semiconductors and is therefore not specific to polymer semiconductors. In fact, most commercial applications of amorphous silicon depend critically on the ability to predict bias stress in this material. Beyond this "intrinsic" electrical stability, if the TFT structure becomes contaminated with impurities during fabrication or during its shelf-life, compositional changes may affect its electrical stability. It is typically assumed that the three most important external factors that affect the long-term stability of polymer electronic devices are exposure to light, oxygen, and water. The details of the microscopic origin of bias stress and the effects of impurities on both the electrical performance and the lifetime of polymer TFTs are still mostly unexplored.

This chapter is organized as follows. In the first part, we will discuss the origins

Organic Electronics, Materials, Manufacturing and Applications. Edited by Hagen Klauk

ISBN: 3-527-31264-1

of bias stress in polymer TFTs. In the second part we will review existing data on electrical stress in these devices, with an emphasis on polyfluorene and polythiophene TFTs. In the third part we will summarize chemical effects on the stability of polymer TFTs.

5.2 Charge Trapping in TFTs

5.2.1 General Considerations

The vast majority of organic TFTs operate as p-channel accumulation devices. The output current I of a polymer TFTs, when the voltage V_{DS} across the device is much smaller than the gate voltage V_G, is given by Eq. (1):

$$I = \frac{W}{L}\mu C_0\left(V_G - V_T - \frac{V_{DS}}{2}\right)V_{DS} \approx \frac{W}{L}\mu C_0(V_G - V_T)V_{DS} \tag{1}$$

In Eq. (1), μ is the carrier mobility, V_T is the threshold voltage, W and L are respectively the channel width and length, and C_0 is the areal gate capacitance.

In conventional TFTs the electric field generated by the gate bias largely exceeds the longitudinal drain field. If we neglect effects at the contacts, bias stress is thus determined by the electric field across the gate dielectric or, equivalently, by the charge density in the channel $C_0 \times |V_G - V_T|$. To compare them, bias-stress measurements obtained under different experimental conditions (e.g. different device geometry and gate bias) should be normalized for the gate electric field. In fact, devices with little or no bias stress have been reported in the literature [4–6]. In all these cases, the transistors were operated at low values of $C_0 \times |V_G - V_T|$. While it is tempting to conclude that such devices are suitable for use without any appreciable electrical degradation, it is important to realize that the output current of a TFT scales with charge density. Therefore, for a fixed W/L, a device with little bias stress, because of low charge density will also have a low output current, which limits its application.

Sources of bias stress related to the presence of mobile ionic species in the dielectric are not specific to polymer semiconductor devices and will not be discussed here. With this premise, V_T and μ are the only terms in Eq. (1) for which time dependence can be introduced. Bias stress can be due to either a structural degradation of the semiconductor material – which leads to a decrease of the effective carrier mobility μ – or to trapping of mobile charge in nonconducting states – which leads to a V_T shift but no change in mobility when all the traps are filled. Trapping can occur in the semiconductor, in the gate dielectric or at the dielectric–semiconductor interface. The electrostatics of thin-film transistors dictate that all the induced charge is located within ~1 nm of the dielectric–semiconductor inter-

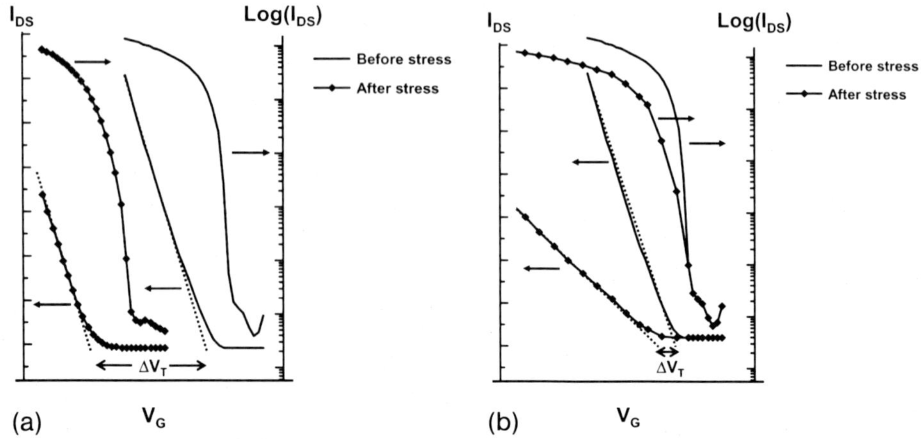

Fig. 5.1. Example of successive transfer curves obtained if bias stress is due to charge trapping in immobile states (a) and if bias stress is due to the formation of shallow donor-like states (b).

face. Therefore, the physical phenomena associated with bias stress must occur near this interface.

Different trapping processes lead to changes in the shape of transfer curves before and after electrical stressing. Carriers trapped in the dielectric, in localized states at the dielectric–semiconductor interface or in deep states in the semiconductor form an immobile charge density distribution which shields the gate voltage. When the gate voltage overcomes this electrostatic screening, the device characteristics are identical to those of a pristine device. This type of degradation gives rise to a V_T shift (ΔV_T) in the direction of the gate bias but no change in overall shape of the transfer characteristics and in carrier mobility (Fig. 5.1a). The V_T shift is equal to N_{trap}/C_0 where N_{trap} is the density of trapped charge. On the other hand, if the gate bias generates new electronic states near the top of the highest occupied molecular orbital (HOMO), the shape of the transfer characteristics of the transistor is affected. As the device is turned on, charge must fill the new states. Because these are shallow trap states, they contribute only partially to the output current through thermal excitation into conducting states. The result is a broader sub-threshold area than for an unstressed device, which also leads to a ΔV_T, and a reduced apparent mobility (Fig. 5.1b). In principle, both these stress mechanisms may occur. All experimental data yet available indicate, however, that bias stress in polymer TFTs consists of removing mobile carriers from the channel and is characterized by a translation of the transfer curves along the V_G axis in the gate-on direction, as shown in Fig. 5.1a.

If bias stress occurs on time-scales typical of device characterization (~0.1 s to a few tens of seconds), it causes the I–V curves to display hysteresis and nonideal shapes [7]. As a result, device properties often used to characterize the semiconductor material appear to depend on measurement conditions. For instance, if a TFT is measured by sweeping the gate voltage from the off state to the on state, the drain current steadily decays as the device is being held in the on state. The consequence is a pronounced sub-linear characteristic of the transfer curves and an apparent dependence of the carrier mobility on gate voltage, gate sweep rate, and dwell times. In contrast, if the TFT is measured by starting from the on state, the I versus V_G curve seems steeper as bias stress shuts off the device and leads to an overestimate of the carrier mobility. The ideal transfer curves are recovered by making use of a pulsed gate bias, which minimizes bias-stress effects during device characterization [7].

Accurate predictions of the effect of ΔV_T on TFT-based circuit operation can only be made by computer simulation. In general, a ΔV_T in the gate-on direction leads to an increase of the on-state resistance of the transistor. For example, if the TFT is a switch in a display backplane, the increased resistance determines an increase in pixel charging time. In a diode-connected inverter, a negative ΔV_T of the load transistor with respect to the V_T of the drive reduces the output swing of the inverter. On the other hand, a negative ΔV_T of the drive transistor relative to the load shifts the trip voltage of the inverter.

5.2.2
Bias Stress in Organic Transistors

The mechanisms of bias stress in organic semiconductors are still mostly unknown; many studies have been limited to observation of hysteresis in the I–V characteristics of TFTs [8, 9]. In pentacene transistors bias stress has been observed with both inorganic and organic dielectrics: positive and negative V_T shifts have been measured depending on the sign of the gate voltage [10, 11]. Little dependence of bias stress on the gate dielectric was also observed, leading to the conclusion that trapping occurs in the pentacene semiconductor. Recovery followed a power law and was not accelerated by exposure to light. *Ab-initio* calculations suggested that formation of hydrogen-induced traps may be responsible for bias stress in pentacene [12]. Experimental confirmation of the existence of such traps states was recently provided [13]. Bias stress has been observed in many polycrystalline semiconductors, for example α-sexithiophene (α6T) and other oligothiophene semiconductors [14–16]. Electrical degradation is not limited to hole conductors – Chesterfield et al. observed V_T shifts in an n-type polycrystalline organic semiconductor [17].

As with crystalline organic semiconductors, there are only few systematic and quantitative studies of bias stress in polymeric TFTs. For example, it is still unclear whether in all polymers a positive V_G leads to positive V_T shifts [7, 18]. Here we will concentrate on negative ΔV_T after application of a negative V_G, which is nearly universally observed in p-type devices with a number of semiconductor–dielectric combinations. In the next section of this chapter we will focus on studies of polyfluorene and polythiophene TFTs, because these are the polymer devices for which bias stress has been most thoroughly characterized in recent years.

5.3
Bias Stress in Polyfluorene and Polythiophene TFTs

Transistors are typically pulsed at a frequency and duty-cycle dictated by a specific application. It is, therefore, useful to understand both bias stress (i.e. device in its on state) and relaxation (i.e. device in its off state) mechanisms. A typical bias–relaxation curve for a polymeric TFT is shown in Fig. 5.2a [19]. During gate biasing the on current of the transistor decreases, because of trapping of mobile charge. Successively the gate bias is removed except for short pulses to measure device recovery. Figure 5.2a suggests the existence of two populations of trapped charge. A fraction of the trapped charge is released quickly at room temperature indicating

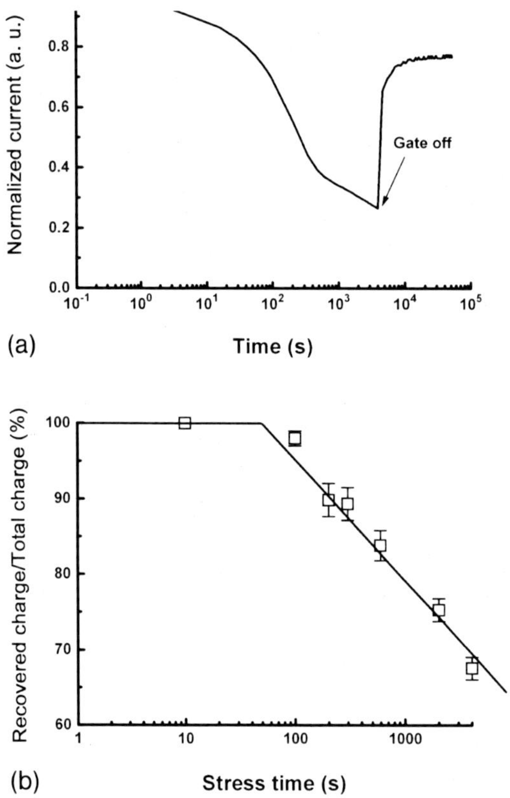

Fig. 5.2. Typical stress–relaxation curve for a polymer TFT (a). Ratio of reversible to total stress after one-day recovery for the same device (b). The device was a PQT-12 TFT on thermal SiO_2 biased at $V_G = -20$ V and $V_{DS} = -1$ V.

that these traps are shallow. The remaining fraction of the trapped charge is released extremely slowly indicating that these traps are deeper. In general, short stressing times cause only reversible trapping while trapping in deep long-lived states requires long stress times (Fig. 5.2b). The relative amount of shallow and deep traps depends on the materials, fabrication, and device operation; in principle, only one type of trap may be observed. In a TFT that is continuously switched on and off, as in normal operation, trapping – both reversible and long-lived – and relaxation all contribute to the resulting electrical degradation.

5.3.1 Reversible Bias Stress

The initial current decay in the TFT on state reflects directly the reversible trapping rate. Reversible trapping kinetic laws can therefore be determined by measuring the initial current decay rate as a function of charge density, which is controlled by the gate voltage. Fast trapping was studied in F8T2 and PQT-12 [20] (see Table 5.1 for the structure of these polymers). Transfer measurements taken immediately after stressing indicated that reversible stress in these materials consists of a V_T shift only. The fast-trapping rate was approximately proportional to the square of the charge density for both polymers. It is currently unknown whether the square law is general for all polymer semiconductors – earlier studies by Brown et al. indicated a linear dependence of trapping on V_G in PTV TFTs during longer stress times [21].

Reversible bias stress has been extensively characterized in PQT-12, providing the most complete study of electrical degradation in a polymer semiconductor [19, 20, 22]. Bias stress in PQT-12 does not depend monotonically on temperature (Fig. 5.3). Interestingly, the trapping rate in polycrystalline films of the oligothiophene α6T peaks strongly at approximately 220 K as well [14, 15]. A curve such as that shown in Fig. 5.3 is the result of both trapping kinetics and driving force. Thermal excitation of charges out of traps is more difficult at low temperature. Trapping kinetics, however, slow down as temperature is reduced. In PQT-12 these two opposing forces are such that there is a temperature range (220–240 K) in which trapping is most severe within the timeframe of the experiment (100 ms). The trapping peak may not occur in all materials as its existence depends critically on the relative values of the trap depth and the temperature-dependence of the trapping rate.

Kinetic modeling was used to estimate the depth of the fast, reversible traps in PQT-12. We assume that the square law observed during trapping is indicative of a trap state made of two correlated holes (i.e. a bipolaron, see Section 5.3.4). Making use of detailed balance we obtain:

$$\frac{dh}{dt} = -kh^2 + \frac{k}{2} N_V (h_0 - h) e^{-U/k_B T} \tag{2}$$

Tab. 5.1. Chemical structure, maximum carrier mobility, and name of the polymer semiconductors to which we refer in this chapter.

Semiconductor	Mobility ($cm^2\ V^{-1}\ s^{-1}$)	Name
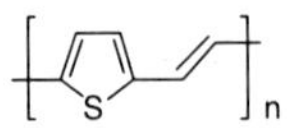 PTV	0.0015	Poly(thienylene–vinylene)
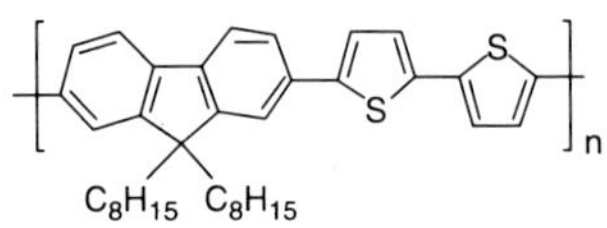 F8T2	0.008–0.02	Poly(9,9′ dioctylfluorene-co-bithiophene)
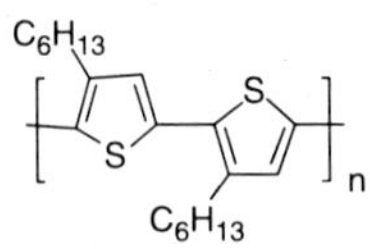 P3HT	0.01–0.1	Poly(3-hexylthiophene)
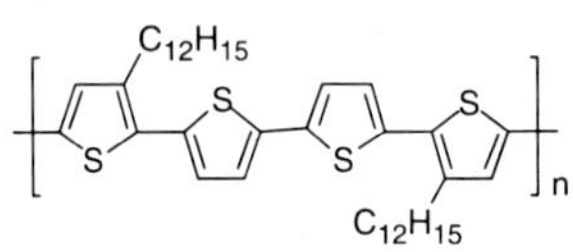 PQT-12	0.05–0.12	Poly[5,5′-bis(3-dodecyl-2-thienyl)-2,2′-bithiophene]
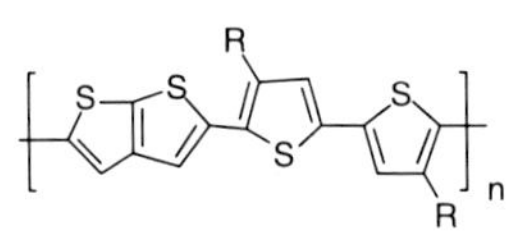	0.01–0.15	Poly[thieno[2,3-*b*]thiophene-co-bi(3-dodecyl)thiophene]

where N_V is the density of available hole states, h_0 is the initial concentration of holes, k is the trapping rate, and U is the trap depth. The second part of the right-hand term of Eq. (2) represents the thermal dissociation of the trapped species. Equation (2) was used to model the data shown in Fig. 5.3 using the experimentally determined trapping rates at different temperatures: the trap depth U compatible with the observed behavior is approximately 0.12 eV. Equation (2) is valid for the extremely simplified case in which trapping occurs in a single trap of depth U. More complex trapping behavior is obtained with a distribution of trap depths and the associated distribution of trapping rates.

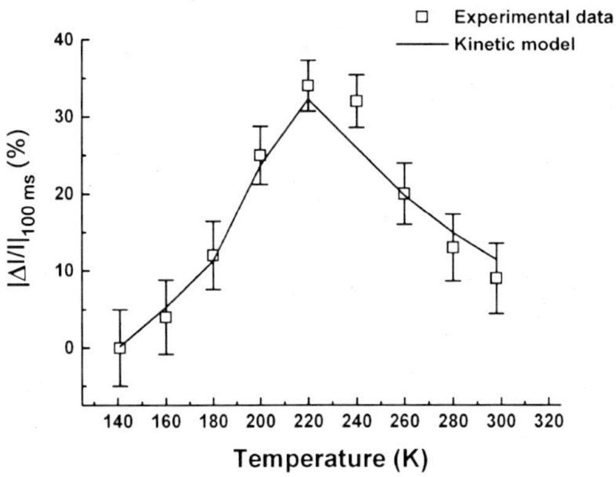

Fig. 5.3. Amount of bias stress, measured as fractional current decay after a 100 ms gate pulse ($V_G = -20$ V), as a function of temperature in PQT-12.

5.3.2
Long-lived Bias Stress

Long-lived electrical degradation in polymer semiconductors manifests itself as a stable V_T shift and no change in the shape of the transfer curves on stressing. Qualitatively similar effects, albeit with different magnitudes, have been observed in F8T2, PQT-12, P3HT, and PTV semiconductors on inorganic and organic dielectrics [7, 18, 20, 21, 23, 24]. The detrapping time is hours to several days if the TFT is left in the dark, depending on the semiconducting material. In F8T2, PQT-12, and PTV, long-lived charge-trapping was independent of the type of gate dielectric (thermal SiO_2, PECVD SiO_2, chemically modified SiO_2 or polymer). For a given stress time, the long-lived ΔV_T seemed to depend solely on the initial charge density in the channel. The V_T shift is often quickly reversed under illumination. A quantitative study of this effect on F8T2 showed that the recovery is proportional to the number of photons absorbed in the semiconductor film, irrespective of the radiation wavelength [7, 25]. This evidence suggests that the long-lived bias-stress effect is because of trapped charge located within the semiconducting polymer. Large bias-stress variations between different polymers processed on the same dielectric are consistent with trapping occurring in the semiconductor.

It is interesting to determine the depth of the traps responsible for long-lived V_T shifts. An estimate of 0.5 eV was made on the basis of the trapping and detrapping kinetics of F8T2 and assuming similar kinetics as in Eq. (2) [20]. An experimental estimate of trap depth can be obtained with thermally stimulated detrapping measurements. The dependence of the detrapping rate on temperature follows Arrhenius behavior, thus providing an upper limit of the trap depth. This limit approaches the true trap depth if the trapping activation energy is small. This method yields a maximum trap depth of 0.8–1 eV in PQT-12 (Fig. 5.4); the true

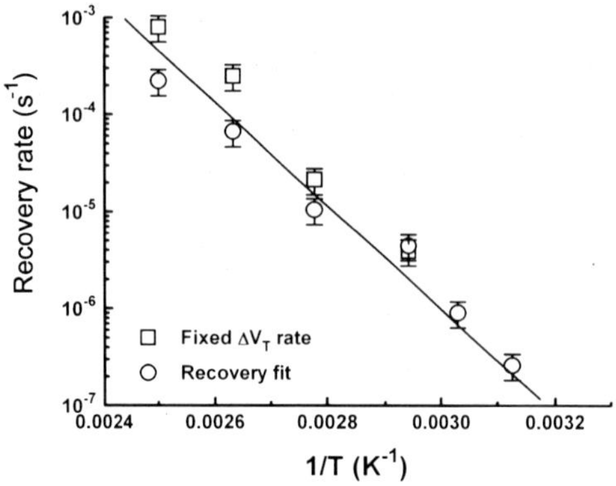

Fig. 5.4. Recovery rate of long-lived traps in PQT-12 as a function of temperature. The recovery rate was measured by fitting the slow part of the recovery curves with an exponential (recovery fit data) or by measuring the recovery rate at a fixed $\Delta V_T = -2.5$ V (fixed ΔV_T rate data). Both sets of data are thermally activated with $E_A \approx 0.8$–1 eV.

trap depth is unknown, because there are no measurements of the trapping activation energy.

5.3.3
Dependence of Bias Stress on Operating Conditions; Lifetime Predictions

In low duty-cycle operation, charge trapped in shallow traps is completely released during the rest period of the biasing cycle. Under these conditions, bias stress is purely due to long-lived traps making it virtually irreversible – when the biasing cycle is stopped, no recovery is observed (Fig. 5.5a). Thus, in low duty-cycle operation the lifetime is essentially determined by slow trapping in deep traps. Kinetics of trapping in long-lived states has been measured as a function of gate voltage in PQT-12 using low duty-cycle biasing and was found to obey a power law up to a total stress time of 1000 s. (Fig. 5.5b) [19]:

$$|\Delta V_T| = A(V_G - V_T^0)^n t^\gamma \tag{3}$$

with $n \approx 2$ and $\gamma \approx 0.37$. The fractional current drop at time t is:

$$\frac{\Delta I}{I} = A(V_G - V_T^0)^{n-1} t^\gamma \tag{4}$$

A similar power law for degradation has also been observed in a-Si:H [26] and OLEDs [27].

The recovery rate of the reversible traps determines the duty-cycle cut-off below which only long-lived traps contribute to bias stress. At higher duty cycle, the current decay is the result of the simultaneous interplay of fast and slow trapping.

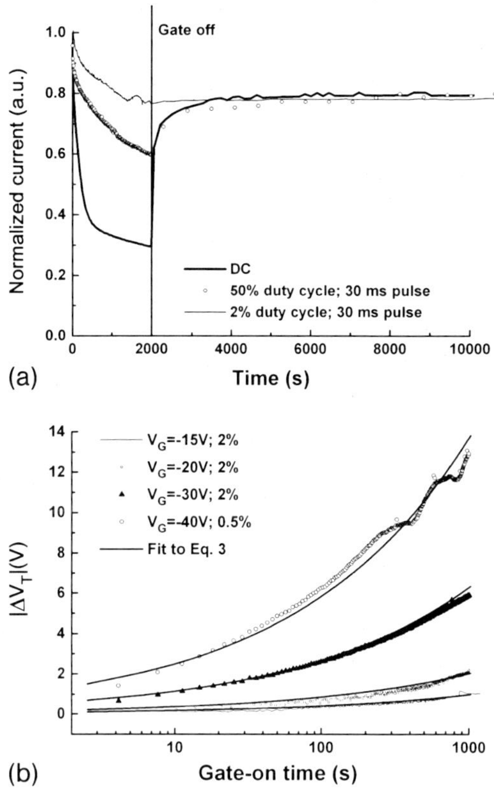

Fig. 5.5. Current decay in a PQT-12 TFT as a function of duty cycle (a). The abscissa units are defined as follows. For $t < 2000$ s, the abscissa is the total gate-on time, for $t > 2000$ s, the abscissa is the real time. Long-lived ΔV_T measured during low duty-cycle operation (b). The fits to Eq. (3) are obtained with the data: $A \approx 8 \times 10^{-4}$, $n = 2$, $\gamma = 0.37$ and $|V_T^0|$ was extracted from transfer measurements before stressing the device. $|V_T^0|$ varied between 4 V and 6 V.

Equation (4) can be used to estimate the lifetime of the TFT for a given allowed current drop at low duty cycle. Because for PQT-12 $\gamma \approx 1/3$, there is a great advantage in terms of lifetime to design a circuit that tolerates the highest possible current drop before exceeding its specification limits.

Finally, it should be noted that the stretched exponential law is independent of duty cycle only when the gate-off time is shorter than the lifetime of the long-lived traps. If the rest time between biasing pulses is long enough to enable a fraction of the long-lived traps to be released, the total amount of bias stress will depend on the duty cycle and will be lower than predicted by Eq. (4). For example, at the extremely low duty-cycle limit, detrapping during the off time may compensate completely, trapping during the off time thus leading to stable operation. The actual duty-cycle values where these different operating regimes occur are dictated by the detrapping rates of the shallow and deep traps. Finally, the effect of large positive voltages during the off part of the duty cycle must be well understood to achieve accurate lifetime predictions.

5.3.4
A Microscopic Theory of Bias Stress

The dependence of the fast-trapping rate on the square of hole concentration suggests that it is caused by the pairing of two mobile holes into a trapped state [20]. Such correlated excitations (bipolarons) are well known in conjugated polymers [28]. Bipolarons arise because the total backbone relaxation associated with pairing two holes, or polarons in polymer semiconductors, is energetically favorable and offsets the Coulomb repulsion of the two positive charges, thereby forming a negative correlation energy state. Depending on the energetics of the process and on temperature, the mechanism of formation of bipolarons can be either pure tunneling, thermally assisted tunneling, or thermal excitation over the formation barrier (Fig. 5.6). Diluted bipolarons are tightly bound states: their formation is effectively equivalent to removing mobile states from the channel.

The negative correlation energy of bipolarons is strongly affected by the localization of the positive charge: the more delocalized the single charge, the smaller the correlation energy of the bipolaron. Thus in a material with a heterogeneous microstructure, for example a semi-crystalline polymer film, we expect to find a distribution of bipolaron correlation energies: these energies are smaller in the crystallites and larger in the disordered or amorphous areas of the film.

In the framework of the bipolaron hypothesis, the trap energy U in Eq. (2) corresponds to the bipolaron correlation energy. The relatively small U estimated from the temperature dependence of the decay data in PQT-12 and the strong drop of the initial trapping rate at low temperature [22] are consistent with the assumption

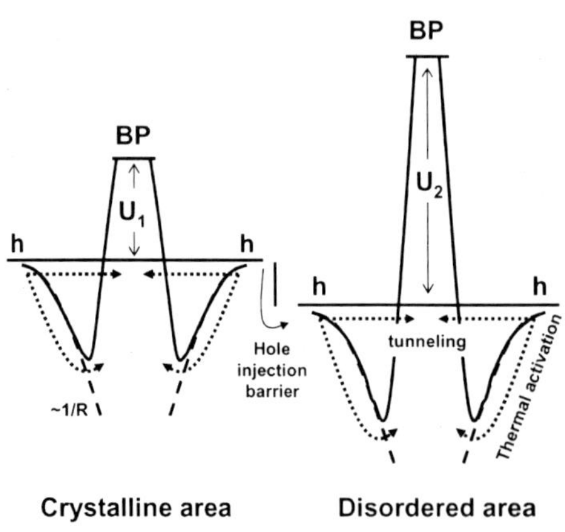

Fig. 5.6. Sketch of the energy levels of free holes and bipolarons in ordered and disordered areas of a polymer film.

that fast traps are in fact bipolarons formed in the ordered areas of the film. Indeed, in the crystallites single charge is partially delocalized – explaining the small U – and when a large fraction of charge ceases to be mobile at lower temperature, the initial trapping rate decreases abruptly.

It is more difficult to ascertain the nature of the long-lived traps although it is reasonable to hypothesize that they are deeper bipolarons, possibly associated with the disordered regions of the polymer films. The dependence of low duty-cycle decay on the square of the gate voltage ($n \approx 2$ in Eq. 3) in PQT-12 supports this hypothesis. The slower trapping kinetics are consistent with the existence of a barrier for hole injection from the ordered into the disordered areas of the film, because of the lower polymer conjugation length in the latter (Fig. 5.6). The larger trapping energy is also consistent with the larger correlation energy of bipolarons in disordered materials.

It is interesting to note that negative U states are present in different classes of disordered material (e.g. chalcogenide glasses, conjugated polymers) and profoundly affect their properties. Deep traps associated with disordered areas of the polymer film are long-lived at room temperature and are therefore most detrimental to the lifetime of electronic devices. If confirmed by direct observation of bipolarons and unequivocal correlation of their formation with electrical degradation, the bipolaron mechanism for bias stress suggests strategies for the design of electrically stable materials. For example, a successful strategy to minimize electrical degradation may consist in controlling or possibly eliminating the disordered regions of the polymer film by molecular design or film processing. Another approach is to increase the hole injection barrier between ordered and disordered regions of the film, in other words to make the disordered regions as disordered as possible to kinetically suppress the energetically favorable bipolaron formation.

5.4 Chemical Effects on Stability – Defects and Impurities

5.4.1 Introduction

Structural defects (e.g. dangling bonds) and chemical impurities (e.g. atomic substitution of hetero atoms) have been demonstrated to have a significant role in the electrical behavior of amorphous inorganic semiconductors. Semiconducting polymers also contain defects and impurities, but there are important differences between their origin and how they can be controlled compared with inorganic materials. We limit our discussion here to the effects of chemical defects and impurities on the electrical performance of TFTs. It is important, however, to be aware of work on other systems with organic materials that is beyond the scope of this chapter. Much of our understanding of chemical damage to conjugated polymers comes from studies of organic light-emitting diodes (OLEDs) [29, 30]. Although the conjugated polymers in these devices are chemically similar to those in TFTs, they are used under different carrier concentration conditions than TFTs ($\sim 10^{14}$–10^{15} cm^{-3} compared with $\sim 10^{18}$–10^{19} cm^{-3}, respectively) and are used with different sets of materials [31]. For example, OLEDs typically require at least one low-work-function metal electrode, for example calcium, that reacts with water, but the electrodes in TFTs are usually made from gold, which is oxidatively stable. Much work has also been conducted on the effects of chemical exposure on TFTs formed with molecular organic semiconductors [32, 33] and also on conductors formed from intentionally doped semiconducting polymers [34]. In many of those studies the effect of impurities was studied in the context of development of chemical sensors and was not related to the operating conditions required for applications such as displays and rf-ID tags.

5.4.2
Defects in Molecular Structure

5.4.2.1 Defects from Synthesis

Structural defects in polymers can arise as a result of chemical synthesis. Purified organic polymers consist of a distribution of molecular species with different molecular weights [35]. It is nearly impossible to obtain a "pure" sample, i.e. one with a single molecular weight, if the chain length is greater than ~10 monomer units. The distribution of molecular weights leads to an inherent microstructural disorder in thin films of semicrystalline polymers that is beyond that caused by limitations on crystallization by kinetic factors. For this reason alone it is difficult to define the nature of a "structural defect" in the microstructure of films of semiconducting polymers and they are categorized as "semicrystalline" comprising small ordered domains (~10 nm) with disordered regions. It is simpler to define defects in the molecular structure of the backbone of the polymer chain and their origins.

Imperfections in the structure of the main chain of semiconducting polymers frequently occur during chemical synthesis. For many polymers, the monomer unit is asymmetric and regioselective coupling reactions are used to control the structure of the backbone of the polymer (Fig. 5.7). These reactions are imperfect and lead to ~98% regioregular structures for materials such as poly(3-alkylthiophenes) [36]. For a sample of P3HT with a molecular weight of ~20 kDa, this regioregularity implies there should be nearly one defective coupling between repeat units in each polymer molecule. Head-to-head coupling in P3HT will prevent the near coplanarity of the thiophene units and cause difficulty in forming densely packed structures. Such problems can also be overcome by use of symmetric monomer units that enforce regioregularity during polymerization, for example the monomer for PQT-12 [3]. Even with these changes, however, one can expect that other synthetic defects, such as coupling at the 3-position on the thiophene ring, occur at levels that are difficult to detect, but may still be high enough to affect electrical transport.

The structure of organic polymers is usually described without consideration of the end groups that terminate the polymer chain. In polythiophenes, the end thio-

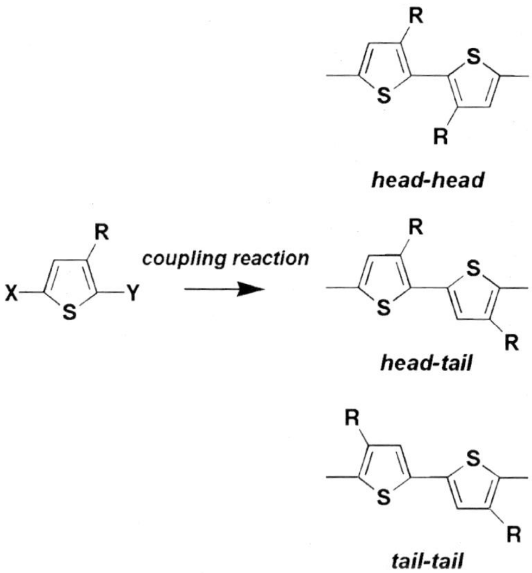

Fig. 5.7. Coupling reactions possible during the polymerization reaction to form regioregular P3HT.

phene ring may contain a bromine atom depending on the synthetic conditions [37, 38]. An estimate of the number of end-cap defects can be made. A film of P3HT that has a density of ~1 g cm^{-3} and a molecular weight of 20 kDa will have a density of end sites equal to $\sim 6 \times 10^{19}$ cm^{-3}. The number of carriers induced in an accumulation layer of a TFT is $\sim 10^{12}$ holes cm^{-2} and is approximately one tenth the number of terminal ends of the polymer molecules in a 5 nm layer. If a reaction is performed to eliminate reactive end-groups, even with a yield of 99% there are still similar numbers of end-sites and induced holes. Although there are few data exploring the effects of end-capping groups, it has been shown that for short-term measurements there is no difference in mobility in P3HT terminated with a hydrogen atom or with a bromine atom [37]. Whether the end-capping group has an effect on long-term behavior is unknown, but it is likely that chemical transformations could occur in a film over a period of time.

5.4.2.2 **Photo-induced Defects**

Although organic materials are stable under vacuum conditions, they are not equally so under ambient conditions because of to the presence of light and molecular oxygen, O_2. Many organic molecules are kinetically unreactive with O_2 in its ground state even though oxidation is a thermodynamically favorable process. The low reactivity is because the ground state of O_2 has a triplet spin multiplicity whereas most even-electron organic materials have singlet multiplicity [39]. In contrast, the singlet states of O_2 are exceptionally reactive. Singlet oxygen can be produced by photoexcitation to its excited triplet state then intersystem crossing to the singlet electronic states or by energy transfer from electronically excited organic

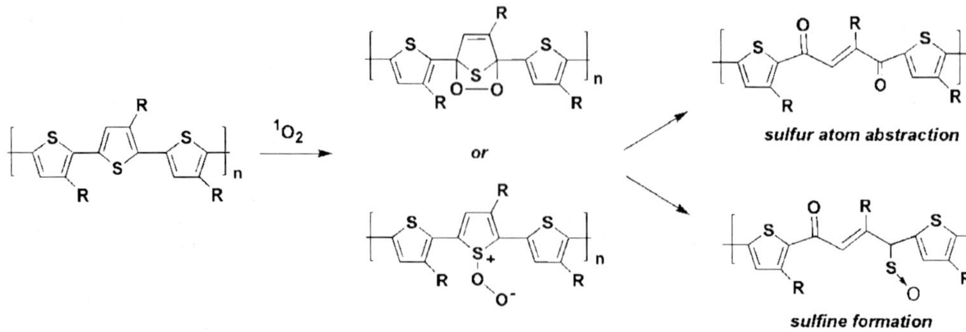

Fig. 5.8. Proposed mechanisms for the reaction of $^1O_2$ with polythiophene.

species. The energy of the first triplet state of most conjugated polymers is ~0.7 eV lower in energy than the first singlet excited state [40] and the triplet energy in O_2 is ~0.98 eV, therefore a polymer with a band gap of 1.68 eV (738 nm) or higher can potentially sensitize singlet oxygen formation.

The photochemical reactivity of thin films of conjugated polymers is typically analyzed by comparison of spectroscopic signatures with those of the products of well characterized reactions in the solution phase [29, 41]. For example, on the basis of spectroscopic analysis a mechanism for photochemical degradation of thin films of polythiophene has been proposed in which $^1O_2$ forms a transient endoperoxide with a thiophene ring followed by decomposition to either a sulfine or diketone (Fig. 5.8) [41]. Evidence for photoinduced cross-linking was also found and it was suggested, by analogy with nonconjugated polymers, that it was initiated by photolysis of residual catalyst, $FeCl_3$, in the film. The number of chemical species created during photooxidation are numerous, but the general outcome of these reactions is a loss of conjugation in the backbone of the polymer chains. A loss of conjugation in the backbone of the polymer would be expected to reduce the mobility of a TFT, simply because the number of electronically active states in the material should decrease; the chains with fewer conjugated rings have a wider HOMO–LUMO gap.

Relatively few long-term studies of the effects of light on polymeric TFTs have been reported. In a detailed study, the effects of light on P3HT-based TFTs in a top gate geometry were reported; these devices were fabricated on polyethylene terephthalate (PET) with a poly(methyl methacrylate) blend polymeric dielectric and gold gate electrode [42]. These TFTs initially had mobilities of 0.02 $cm^2\ V^{-1}\ s^{-1}$ and on-to-off ratios of ~100 (at $V_G = -30$ V relative to 0 V). Samples were irradiated on the substrate (PET) side under ambient conditions under room light and under a AM 1.5 spectrum solar simulator at 110 mW cm^{-2}. The intense light led to decrease in the on-current over a 72 h. period that followed an exponential function $I = I_0 \exp(-t/\tau)$ with a time constant of ~25 h. During irradiation significant photobleaching of the P3HT film occurred and there was infrared spectroscopic evidence of transformation of the main polymer chain into carbonyl groups, in agreement with reactivity data for polythiophenes [41]. Devices under ambient light survived for more than 1000 h. These results should be regarded as device stabilities rather than the intrinsic stability of the semiconducting material because they are essentially protected from ambient conditions by the gate dielectric layer and the gate electrode.

5.4.3 Impurities

5.4.3.1 Thermochemical Analysis

There have been several attempts to use chemical thermodynamic data to predict if a semiconducting polymer will undergo a charge-transfer process with an impurity and thereby become doped. There are extensive data on the energetics of charge-transfer reactions of organic species in aqueous or organic solvents and it would be beneficial to use these data to predict the stability of polymers. Unfortunately, because of the difficulty of accurately estimating the solvation energy of ions as they are transferred from solutions to solid films, these types of analysis are generally qualitative and cannot be used to predict absolute energetics.

Electrochemical data have been used to predict ionization energies and stability relative to doping in air for semiconducting polymers. Brédas et al. [43] derived an empirical correlation for determination of ionization energies from electrochemical oxidation and reduction potentials (Eq. 5) in which the onset of oxidation is referred to the standard calomel electrode, S.C.E.)

$$\text{I.E.} = E_{\text{onset ox}} + 6.3 \text{ eV} \tag{5}$$

The scale factor of 6.3 eV is derived from the scaling of S.C.E. to the vacuum level (4.7 eV) and a component of the polarization energy for formation of a charge in a medium (1.5 eV). It has been suggested the latter value is almost constant for a variety of organic molecules; it does, however, create a reasonable amount of uncertainty (~0.5 eV) in this relationship. DeLeeuw and coworkers have performed a simple analysis to estimate the electrochemical stability of doped semiconducting polymers in respect of reactions with water and oxygen [44]. This analysis was based on the assumption of the ability to transfer electrochemical data for oxidation and reduction reactions in aqueous environments to reactions in solid organic films. Such thermodynamic cycles only indicate the possibility of a reaction without regard to kinetic effects, for example multiple-electron processes are typically less efficient than single-electron processes [45]. From these studies a scale was derived to determine the environmental stability of n-type and p-type semiconducting polymers. These scales cannot be used to determine absolute stability; for example, using this scale polyacetylene is predicted to be stable under ambient conditions, in contrast with all experimental data that show that the material is unstable. The authors have suggested that in such circumstances reactions such as oxygen insertion into unsaturated bonds are a possible cause of the discrepancy [44]. The uncertainties in both of these analyses emphasize the difficulty of predicting thermochemical stability from tabulated electrochemical data.

5.4.3.2 Oxygen

Several experimental studies have suggested oxygen is a dopant for a variety of semiconducting polymers. Oxygen is at a relatively high concentration (20%) under ambient conditions and has been shown to interact with conducting polymers such as polyacetylene and polythiophene [46]. The effects of doping are most easily observed in TFTs in the sub-threshold region; as states begin to appear within the band gap of the material, they will affect the shape of the bandtail and thus change the turn-on characteristics of the device. Eventually, a bulk current may be observed, but if the channel current is substantially larger it will not be observed in the I–V characteristics when the channel is "on".

Several studies of P3HT-based TFTs have been conducted to determine their stability when exposed to oxygen. Interestingly, the results from these studies vary widely and are difficult to compare directly. In one of the earliest studies of the effects of exposure of polymer-TFTs to oxygen, Taylor et al. examined electrochemically deposited films of poly(3-methylthiophene) [47]. They were able to fabricate TFTs with $\mu \approx 9 \times 10^{-4}$ cm^2 V^{-1} s^{-1} and found that their conductivity increased on exposure to ambient conditions or to pure O_2, leading to the conclusion that O_2 from the atmosphere is the primary dopant for polythiophene. This conclusion is also supported by spectroscopic evidence of the presence of a charge-transfer complex (CTC) between O_2 and P3HT [48]. The solubility constant for O_2 in P3HT is ~0.2 (v/v) at standard pressure and temperature, showing that these films contain a substantial amount of O_2 when left in the atmosphere [48, 49]. The measured free energy of formation of the complex, ΔG°, is ~−1.9 kJ mol^{-1} at 298 K; these values suggest that ~30% of the solubilized oxygen molecules form CTCs or ~1% of π-conjugated segments (assumed to be ~6 repeat units). Holdcroft et al. found that the conductivity of the P3HT increased with oxygen overpressure, but found that the field-effect mobility of the devices decreased [48]. These devices had initial $\mu = 6.0 \times 10^{-4}$ cm^2 V^{-1} s^{-1}, a value that is significantly lower than that achievable currently in a TFT (~0.1 cm^2 V^{-1} s^{-1}). In higher-performance P3HT devices, Meijer and coworkers observed a large shift of the switch-on voltage and a large change in the sub-threshold slope with increasing exposure times with little change in mobility [50]. Although it was found that the doped TFTs could still be switched to an off state (~1 pA of current), in a practical device operating in a fixed voltage range this would lead to a substantial reduction in the off current of the device. These effects were found to be reversible with storage or annealing under vacuum conditions.

It is important to note that results on the stability of P3HT-based TFTs vary substantially from laboratory to laboratory. The time observed for doping of P3HT TFTs has been reported to be less than 24 h by one group [51] and several days by another [3]. Figure 5.9 shows data for P3HT and PQT-12 TFTs measured over ~3 months when stored in dry, purified air using the same device structure under identical conditions at PARC. The devices for both materials were quite stable on

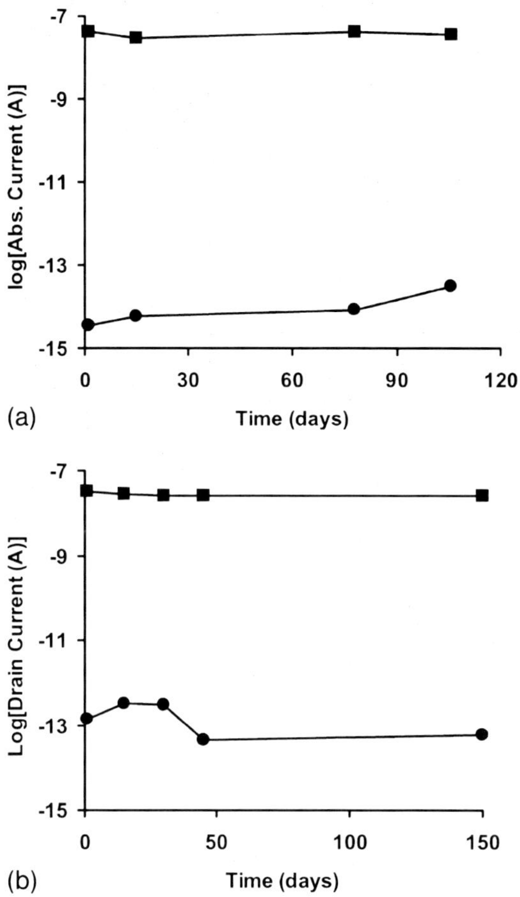

Fig. 5.9. On and off currents of TFT as a function of time in the dark in dry, purified air for (a) P3HT and (b) PQT-12. These devices were fabricated in a coplanar geometry and the semiconducting polymer was unencapsulated. The off and on currents were measured at gate voltages that were +15 V and −30 V relative to the measured V_T for each device ($C_0 = 25$ nF cm^{-2}). The currents were normalized by dividing by W/L for the device.

the basis of their mobility, on-to-off currents, and threshold voltage. These results suggest that some of the variation in the literature on device stability could be because of variations in the local ambient atmosphere and potentially because of batch-to-batch variations in the semiconducting polymer.

Direct measurements of the change in the number of states caused by doping with oxygen have only been made by a few groups. Of particular interest are electronic states that are in the band gap, but whose concentrations are too small to be observed using conventional optical transmission measurements. Photothermal deflection spectroscopy (PDS) is capable of revealing sub-gap states and has been used extensively on amorphous inorganic semiconductors [52]. PDS has been used to study the interaction of O_2 with a dialkoxy-substituted poly(phenylvinylene) and creation of states in the band gap of the material was observed on exposure to oxygen [53]. There are currently no reports of comparison of PDS spectra with current–voltage measurements for TFTs. Measurements on metal–insulator–semiconductor (MIS) diodes suggested that acceptor density in a film of P3HT is relatively unchanged by exposure to O_2 in the absence of light ($\sim 5 \times 10^{15}$ cm^{-3}), but exposure to light caused a dramatic increase in the acceptor density ($\sim 1.5 \times 10^{16}$ cm^{-3}) [49]. The efficiency of the process of creation of acceptor states depended on the energy of the incident light and was maximum near the reported absorbance of the charge-transfer complex of O_2 and P3HT.

The widely reported instability of P3HT and other conjugated polymers, for example polyacetylene, have led many groups to attempt to improve stability by chemical synthesis. Several groups have attempted to improve the stability of semiconducting polymers by decreasing the conjugation length of the thiophene rings or by incorporating other aromatic heterocycles in the backbone of the polymer (Table 5.1) [54]. For example, incorporation of thieno[2,3-*b*]thiophene into a chain produces films with ionization potentials of 5.3 eV (by UV photoelectron spectroscopy) compared with 4.8 eV for P3HT [51, 55]. In these studies, the material with higher ionization energy was shown to maintain its on-to-off ratio under ambient conditions. Interestingly, many of these materials have a field-effect mobility approximately one-tenth that of P3HT.

5.4.3.3 **Water**

Water is known to cause electrical instability in a variety of electronic devices. For example, if water is absorbed on the surface of an unencapsulated amorphous silicon TFTs it induces a charge in the semiconducting film that acts as an additional channel layer for conduction, the "back-channel effect" [56]. The situation for semiconducting polymers is more complex, because water can be absorbed by the bulk of the film and interact directly with the accumulated carrier states.

The effect of humidity on the operation of polymeric TFTs has been studied by several groups and found to exacerbate gate-bias stress. Water was found to increase off-state conduction and to cause a lack of saturation in the output characteristics of P3HT-based TFTs; the mobility of these devices was, however, relatively low, 1×10^{-4} cm^2 V^{-1} s^{-1}, and no hydrophobic layer was placed on the gate dielectric [57]. Gate-induced bias stress was found to increase in PTV TFTs under 2 mbar of water vapor compared with dry devices [18]. Interestingly, the gate-induced stress was found to be less severe if the oxide dielectric was coated with HMDS, compared with TFTs in which the dielectric was left bare. The bias stress in these TFTs was reversible and the original characteristics were recovered under positive gate bias. A detailed study of PQT-12-based TFTs has shown that the effect of absorbed water is mainly to cause carrier trapping, leading to large shifts of the threshold voltage (Fig. 5.10) [58]. The number of water molecules relative to the number of carriers in the accumulation layer was estimated. At $V_G = -30$ V with a gate capacitance of 30 nF cm^{-2}, the induced carrier concentration is $\sim 5.6 \times 10^{12}$ carriers cm^{-2}. At ~30% R.H. the film contains ~70 ng cm^{-2} water in a film

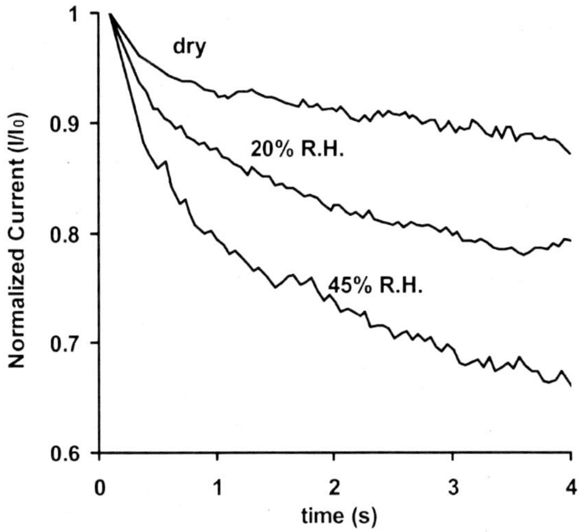

Fig. 5.10. Gate-induced current decay in the presence of water for PQT-12 TFTs at several humidities in a nitrogen atmosphere. During the transient, V_G was held at −30 V and V_{DS} at −2 V. The data were normalized to the initial current of the transient.

~30 nm thick, measured using a quartz crystal microbalance; assuming that the water is uniformly distributed in the film, these data suggest there are $\sim 8 \times 10^{13}$ water molecules cm^{-2} in the accumulation layer or ~1–10 water molecules per carrier. Despite the relatively large concentration of water, the mobility of the carriers is relatively unaffected, but deeper trap states are formed. It has been suggested the change in trapping kinetics occurs by a mechanism in which the dipolar water molecules stabilize the highly charged bipolaron states in the film and thus form deeper-lying trap states.

5.4.3.4 **Organic Solvents**

During the fabrication of most practical organic semiconductor devices they are exposed to a variety of organic solvents during processing. Torsi et al. have examined the effects of aliphatic alcohols on TFTs made with polyterthiophenes [59]. Exposure to organic alcohols caused a decrease in the saturation current of the TFTs and the effect was attributed to charge trapping at grain boundaries in the films, because of adsorption of the alcohol on the surface of the film.

5.4.3.5 **Inorganic Impurities**

Inorganic impurities may be incorporated in polymers during synthesis or as a result of contamination. Conjugated polymers are synthesized using inorganic catalysts, for example $FeCl_3$, or organometallic catalysts such as. $Ni(dppp)Cl_2$ (dppp = diphenylphosphinopropane) [36]. The concentrations of these species are reported sporadically, but can range from 1–1000 ppm [3, 54] and are high enough potentially to form electroactive traps. Indeed, polythiophenes can be intentionally doped with species such as $FeCl_3$ so it is likely that some of the behavior observed in the literature is complicated by their presence. Interestingly, if there are indeed electronic traps resulting from residual catalyst molecules (e.g. 10–100 ppm) they must be relatively shallow or they would be manifested in large onset voltages for most materials. Few studies of ionic impurities have been performed on polymeric TFTs. There is evidence that mobile ions such as Na^+ cause hysteretic effects in planar diode devices with P3HT on glass substrates [60]. The Na^+ ions in the organic layer were found to migrate toward the cathode and modify the conductivity by a hypothesized contact effect, by reduction of the concentration of holes in the material. On the basis of a reduction of the effect when electronic grade sodium-free glass was used, it was suggested the Na^+ impurities arose from sodium from the soda lime glass. The most likely consequences of drifting ionic impurities on TFTs are shifts in threshold voltage and modification of carrier injection at the contacts.

5.4.4
Studies of TFT Lifetime

It is also important to consider that in finished devices, the semiconductor in the TFT will be packaged and its exposure to the ambient environment will be minimized compared with exposure of devices used for routine tests. In most circuits the semiconductor will be shielded from light so photooxidation is critical only to the point that singlet oxygen can migrate through other layers of the film or during fabrication in ambient environments (e.g. jet-printing or spin coating). The use of sub-band gap light during fabrication is not such a stringent requirement, for example photographic film is produced in large quantities at low cost under such conditions. During the fabrication of these finished devices, the semiconducting layer in the TFT will be exposed to subsequent processing after deposition, such as addition of encapsulation layers. Thus, in addition to examining the fundamental stability of individual devices, it is important, where possible, to examine how these results compare with those of finished, packaged devices.

The lifetime of both displays and circuits using organic TFTs have been reported. The shelf life of P3HT from two sources has been studied for TFTs in a top gate geometry on polyethylene terephthalate (PET) films [5]. In these devices, the semiconducting film was covered by a polymeric insulator ~300 nm thick and a gate electrode. The organic circuits were left under ambient conditions for 1 year; the initial mobility was 0.05 $cm^2\ V^{-1}\ s^{-1}$ and degraded to 0.045 $cm^2\ V^{-1}\ s^{-1}$ and the off current increased by ~40% (value not specified) over the time period. Ring oscillator circuits with operating frequencies of 10.5 kHz were also shown to be stable during storage under ambient conditions for 1 year. Philips has reported that active-matrix backplanes using solution-processed pentacene can operate continuously for 100 h [61]. At PARC, shelf-life tests indicate that assembly displays with printed polymer TFT backplanes do not degrade significantly under ambient conditions over 1-year time-spans [62]. All of these results suggest that appropriate encapsulation methods are available and can be used to extend the lifetime of polymeric TFTs in practical applications. It is more difficult to predict their lifetime under operating conditions as there is no developed method for accelerated lifetime testing.

5.5 Conclusion

Study of the electrical stability of polymeric TFTs is essential to their adoption in practical applications. Although gate-bias stress is an important electrical characteristic for determining the lifetime of TFTs in applications such as displays, there is no fundamental understanding of how to design a semiconductor material free from bias stress. Studies of gate-bias stress will be essential for materials to determine their operating range and to optimize dielectric layers. It is doubtful that bias stress can be completely eliminated but its mitigation by materials and device design in combination with driving conditions is a necessity.

Similarly, understanding how interactions with the environment affect the operation of polymer TFTs is still in its infancy. Environmental stability seems, however, to be a solvable problem if appropriate care is taken during fabrication and if appropriate encapsulation layers are developed.

Acknowledgments

The authors gratefully acknowledge W.S. Wong, A.C. Arias, R.B. Apte, S.E. Ready, A.R. Völkel, R. Lujan, F. Endicott, J.-P. Lu and R.A. Street of PARC for assistance with the experiments and for helpful discussions. This work is partially supported by the Advanced Technology Program of the National Institute of Standards and Technology (contract 70NANB0H3033).

References

1 G. Wang, J. Swensen, D. Moses and A. J. Heeger, *J. Appl. Phys.* **2003**, *93*, 6137.

2 H. Sirringhaus, P. J. Brown, R. H. Friend, M. M. Nielsen, K. Bechgaard, B. M. W. Langeveld-Voss, A. J. H. Spiering, R. A. J. Janssen, E. W. Meijer, P. Herwig and D. M. De Leeuw, *Nature* **1999**, *401*, 685.

3 B. S. Ong, Y. Wu, P. Liu and S. Gardner, *J. Am. Chem. Soc.* **2004**, *126*, 3378.

4 H. Sirringhaus, R. J. Wilson, R. H. Friend, M. Inbasekaran, W. Wu, E. P. Woo, M. Grell and D. D. C. Bradley, *Appl. Phys. Lett.* **2000**, *77*, 406.

5 J. Ficker, A. Ullmann, W. Fix, H. Rost and W. Clemens, *J. Appl. Phys.* **2003**, *94*, 2638.

6 W. Fix, A. Ullmann, J. Ficker, H. Rost, W. Clemens, D. Brennan, D. Welsh and J. O'Brien, in *Proc. of the European Solid State Research Conference*, Editions Frontieres Paris **2002**, p. 527.

7 A. Salleo and R. A. Street, *J. Appl. Phys.* **2003**, *94*, 471.

8 A. R. Brown, A. Pomp, D. M. de

LEEUW, D. B. M. KLAASSEN, E. E. HAVINGA, P. HERWIG and K. MUELLEN, *J. Appl. Phys.* **1996**, *79*, 2136.

9 H. KLAUK, M. HALIK, U. ZSCHIESCHANG, G. SCHMID, W. RADLIK and W. WEBER, *J. Appl. Phys.* **2002**, *92*, 5259.

10 A. R. VÖLKEL, R. A. STREET and D. KNIPP, *Phys. Rev. B* **2002**, *66*, 195336.

11 D. KNIPP, R. A. STREET, A. VÖLKEL and J. HO, *J. Appl. Phys.* **2003**, *93*, 347.

12 J. E. NORTHRUP and M. L. CHABINYC, *Phys. Rev. B* **2003**, *68*, 041202.

13 D. V. LANG, X. CHI, T. SIEGRIST, A. M. SERGENT and A. P. RAMIREZ, *Phys. Rev. Lett.* **2004**, *93*, 076601.

14 P. STALLINGA, H. L. GOMES, F. BISCARINI, M. MURGIA and D. M. DE LEEUW, *J. Appl. Phys.* **2004**, *96*, 5277.

15 H. L. GOMES, P. STALLINGA, F. DINELLI, M. MURGIA, F. BISCARINI, D. M. DE LEEUW, T. MUCK, J. GEURTS, L. W. MOLENKAMP and V. WAGNER, *Appl. Phys. Lett.* **2004**, *84*, 3184.

16 W. A. SCHOONVELD, J. B. OOSTINGA, J. VRIJMOETH and T. M. KLAPWIJK, *Synth. Met.* **1999**, *101*, 608.

17 R. CHESTERFIELD, J. C. MCKEEN, C. R. NEWMAN, C. D. FRISBIE, P. C. EWBANK, K. R. MANN and L. L. MILLER, *J. Appl. Phys.* **2004**, *95*, 6396.

18 M. MATTERS, D. M. DE LEEUW, P. HERWIG and A. R. BROWN, *Synth. Met.* **1999**, *102*, 998.

19 A. SALLEO, F. ENDICOTT and R. A. STREET, *Appl. Phys. Lett.* **2005**, *86*, 263505.

20 R. A. STREET, A. SALLEO and M. L. CHABINYC, *Phys. Rev. B* **2003**, *68*, 085316.

21 A. R. BROWN, C. P. JARRETT, D. M. DE LEEUW and M. MATTERS, *Synth. Met.* **1997**, 88, 37.

22 A. SALLEO and R. A. STREET, *Phys. Rev. B* **2004**, *70*, 235324.

23 A. SALLEO, M. L. CHABINYC, M. S. YANG and R. A. STREET, *Appl. Phys. Lett.* **2002**, *81*, 4383.

24 S. SCHEINERT, T. DOLL, A. SCHERER, G. PAASCH and I. HOERSELMANN, *Appl. Phys. Lett.* **2004**, *84*, 4427.

25 M. C. HAMILTON and J. KANICKI, *IEEE J. Select. Topics Quantum Electron.* **2004**, *10*, 840.

26 T. TSUKADA, Active-Matrix Liquid-Crystal Displays in *Technology and Applications of Amorphous Silicon*, R. A. STREET (Ed), Springer Verlag, Heidelberg, **2000**.

27 W. E. HOWARD and O. F. PRACHE, *IBM J. Res. Dev.* **2001**, *45*, 1.

28 N. C. GREENHAM and R. H. FRIEND, *Solid State Physics* **1996**, *49*, 1.

29 R. D. SCURLOCK, B. WANG, P. R. OGILBY, J. R. SHEATS and R. L. CLOUGH, *J. Am. Chem. Soc.* **1995**, *117*, 10194.

30 H. AZIZ and Z. D. POPOVIC, *Chem. Mat.* **2004**, *16*, 4522.

31 R. H. FRIEND, R. W. GYMER, A. B. HOLMES, J. H. BURROUGHES, R. N. MARKS, C. TALIANI, D. D. C. BRADLEY, D. A. DOSSANTOS, J. L. BREDAS, M. LOGDLUND and W. R. SALANECK, *Nature* **1999**, *397*, 121.

32 B. CRONE, A. DODABALAPUR, A. GELPERIN, L. TORSI, H. E. KATZ, A. J. LOVINGER, R. SARPESHKAR and Z. BAO, *Appl. Phys. Lett.* **2001**, *78*, 2229.

33 L. TORSI, A. J. LOVINGER, B. CRONE, T. SOMEYA, A. DODABALAPUR, H. E. KATZ and A. GELPERIN, *J. Phys. Chem. B.* **2002**, *106*, 12563.

34 K. J. ALBERT, N. S. LEWIS, C. L. SCHAUER, G. A. SOLTZING, S. E. STITZEL, T. P. VAID and D. R. WALT, *Chem. Rev.* **2000**, *100*, 2595.

35 L. H. SPERLING, *Introduction to Physical Polymer Science*, John Wiley & Sons, New York, **1992**.

36 R. D. MCCULLOUGH, *Adv. Mater.* **1998**, *10*, 93.

37 R. J. KLINE, M. D. MCGEHEE, E. N. KADNIKOVA, J. LIU and J. M. J. FRÉCHET, *Adv. Mater.* **2003**, *15*, 1519.

38 M. JEFFRIES-EL, G. SAUVÉ and R. D. MCCULLOUGH, *Adv. Mater.* **2004**, *16*, 1017.

39 H. H. WASSERMAN and B. H. LIPSHUTZ, Reactions of Singlet Oxygen with Heterocyclic Systems in *Singlet Oxygen*, H. H. WASSERMAN and R. W. MURRAY (Eds.), Academic Press, New York, **1979**.

40 A. KOHLER, J. S. WILSON and R. H. FRIEND, *Adv. Mater.* **2002**, *14*, 701.

41 M. S. A. Abdou and S. Holdcroft, *Can. J. Chem.* **1995**, *73*, 1893.

42 J. Ficker, H. vonSeggern, H. Rost, W. Fix, W. Clemens and I. McCulloch, *Appl. Phys. Lett.* **2004**, *85*, 1377.

43 J. L. Brédas, R. Silbey, D. S. Boudreaux and R. R. Chance, *J. Am. Chem. Soc.* **1983**, *105*, 6555.

44 D. M. de Leeuw, M. M. J. Simenon, A. R. Brown and R. E. F. Einerhand, *Synth. Met.* **1997**, *87*, 53.

45 K. J. Vetter, *Electrochemical Kinetics*, Academic Press, New York, **1967**.

46 T. A. Skotheim, R. L. Elsenbaumer and J. R. Reynolds, *Handbook of Conducting Polymers*, Marcel Dekker, Inc., New York, **1998**.

47 D. M. Taylor, H. L. Gomes, A. E. Underhill, S. Edge and P. I. Clemenson, *J. Phys. D.: Appl. Phys.* **1991**, *24*, 2032.

48 M. S. A. Abdou, F. P. Orfino, Y. Son and S. Holdcroft, *J. Am. Chem. Soc.* **1997**, *119*, 4518.

49 E. J. Meijer, A. V. G. Mangnus, B.-H. Huisman, G. W. t'Hooft, D. M. deLeeuw and T. M. Kapwijk, *Synth. Met.* **2004**, *142*, 53.

50 E. J. Meijer, C. Tanase, P. W. M. Blom, E. van Veenendaal, B. H. Huisman, D. M. de Leeuw and T. M. Klapwijk, *Appl. Phys. Lett.* **2002**, *80*, 3838.

51 M. Heeney, C. Bailey, K. Genevicius, M. Shkunov, D. Sparrowe, S. Tierney and I. McCulloch, *J. Am. Chem. Soc.* **2005**, *127*, 1078.

52 W. B. Jackson and N. M. Amer, *Phys. Rev. B.* **1982**, *25*, 5559.

53 J. Manca, L. Goris, E. Kesters, L. Lutsen, T. Martens, K. Haenen, M. Nesladek, O. Sanna, D. Vanderzande, J. D'Haen and L. DeSchepper, *Mat. Res. Soc. Symp. Proc.* **2003**, *771*, L12.3.1.

54 D. J. Brennan, Y. Chen, S. Feng, J. P. Godschalx, G. E. Spilman, P. H. Townsend, S. R. Kisting, M. G. Dibbs, J. M. Shaw, D. M. Welsh, J. L. Miklovich and D. Stutts, *Mater. Res. Soc. Symp. Proc.* **2004**, *814*, 319.

55 I. McCulloch, C. Bailey, M. Giles, M. Heeney, I. Love, M. Shkunov, D. Sparrowe and S. Tierney, *Chem. Mater.* **2005**, *17*, 1381.

56 R. A. Street, *Hydrogenated amorphous silicon*, Cambridge University Press, Cambridge, **1991**.

57 S. Hoshino, M. Yoshida, S. Uemura, T. Kodzasa, N. Takada, T. Kamata and K. Yase, *J. Appl. Phys.* **2004**, *95*, 5088.

58 M. L. Chabinyc, F. Endicott, D. M. DeLongchamp, B. D. Vogt and E. K. Lin, *manuscript in preparation.*

59 L. Torsi, M. C. Tanese, N. Cioffi, M. C. Gallazzi, L. Sabbatini, P. G. Zambonin, G. Raos, S. V. Meille and M. M. Giangregorio, *J. Phys. Chem. B.* **2003**, *107*, 7589.

60 D. B. A. Rep, A. F. Morpurgo, W. G. Sloof and T. M. Kapwijk, *J. Appl. Phys.* **2003**, *93*, 2082.

61 G. H. Gelinck, H. E. A. Huitema, E. van Veenendaal, E. Cantatore, L. Schrijnemakers, J. B. P. H. van der Putten, T. C. T. Geuns, M. Beenhakkers, J. B. Giesbers, B.-H. Huisman, E. J. Meijer, E. M. Benito, F. J. Touwslager, A. W. Marsman, B. J. E. van Rens and D. M. de Leeuw, *Nature Mat.* **2004**, *3*, 106.

62 A. C. Arias and R. B. Apte, *personal communication.*

6
Gate Dielectrics

Marcus Halik

6.1
Introduction

With the focus on the potential applications of organic thin film transistors (TFTs) in low-price and/or large-area devices of low or medium complexity, the materials used in the fabrication of organic TFTs and the device design play a significant role in realizing the required electrical performance. The spectrum of applications for organic TFTs ranges from pixel driver transistors for flat panel display backplanes, which require very high uniformity of the various electrical performance characteristics and large drive currents of the TFTs, to transistors in integrated circuits for radio-frequency transponders with limited budget in supply voltage [1–6]. One of the factors determining the electrical performance of organic TFTs for specific applications is the structure of the device, which includes its layout and architecture. Although the layout dimensions channel length and channel width can, in principle, be varied from a few tens of nanometers to several centimeters, thus enabling the highest frequency of operation and the drive current to be varied over many orders of magnitude, the choices for the device architecture are quite limited [6–10]. Basically, three transistor architectures have been employed for organic TFTs – the bottom-gate, bottom-contact (inverted coplanar) structure, the bottom-gate, top-contact (inverted staggered) structure, and the top-gate configuration (noninverted staggered) structure [11–13]. The choice of transistor architecture has a significant effect on electrical performance, because of differences between charge injection [14], but often the degree of freedom is limited by restrictions in processing methods or the incompatibility of the materials used in the fabrication of organic TFTs [15]. Thus, an important set of factors in tuning device performance is choice and optimization of the materials. Early organic TFT work focused almost exclusively on organic semiconductor materials, quickly pushing the performance of organic TFTs into the range of amorphous silicon TFTs [16–18]. Later, research on organic transistors was extended to materials for the source and drain contacts and to the materials used as the substrates [2, 11, 19–24]. But perhaps the material with the greatest effect on device performance – and the organic

Organic Electronics, Materials, Manufacturing and Applications. Edited by Hagen Klauk

ISBN: 3-527-31264-1

semiconductor – is the gate dielectric. If the electrically conducting gate electrode is insulated from the organic semiconductor channel, the gate-dielectric layer plays a key role in the electronic functionality of the field-effect transistor. The capacitance (i.e. the thickness and permittivity) of the gate dielectric determines the operating voltage of the transistor. In organic TFTs that use the inverted device structure the gate-dielectric layer provides the surface on which the organic semiconductor is deposited. Thus, the gate dielectric determines the molecular growth and molecular orientation of the organic semiconductor layer, both for small-molecule semiconductors and for pre-oriented, highly ordered semiconducting polymers [20, 25], and it determines the characteristics of the interface between the gate dielectric and the semiconductor. Because the carrier channel in which the electronic charges are transported in the device is formed in close proximity to that interface, the properties of the gate dielectric surface and of the dielectric/semiconductor interface are of critical importance [26]. The gate-dielectric layer must also be sufficiently robust to withstand all subsequent steps in the device fabrication process. Finally, the processes for the deposition and patterning of the gate dielectric must be compatible with the general process flow, particularly for applications on flexible substrates associated with a restricted thermal budget. Thus, in addition to the electrical characteristics of the gate-dielectric materials, the mechanical and chemical properties and the processability must also be taken into consideration.

6.2 The Impact of Gate Dielectrics on the Electrical Functionality of Organic TFTs

The principal architecture and the electrical characteristics of an organic TFT are shown in Fig. 6.1. An organic TFT is a three-terminal device in which the electric current between two (ideally ohmic) contacts (called the source and drain contacts) is modulated by a voltage applied to the third electrode (the gate); see Fig. 6.1a. The channel for charge carrier transport is formed at the dielectric/semiconductor interface when the absolute value of the gate voltage is raised beyond the switch-on voltage. The gate as the control electrode must be electrically isolated from the semiconductor channel and from the source and drain contacts to allow an electric field to build up upon application of a gate voltage and to prevent electrical shorts. The electrical transfer characteristic of a pentacene TFT is shown in Fig. 6.1b. When a positive potential is applied to the gate relative to the source (e.g. gate – source voltage = +10 V), no carriers accumulate at the dielectric/semiconductor interface, so the drain current is very small (less than 1 pA) and the transistor is in the nonconducting ("off") state. When the gate-source voltage is negative (e.g. −20 V), a carrier channel is formed at the interface, a large drain current is observed, and the transistor is in the conducting ("on") state.

The carrier concentration in the channel and thus the drain current I_D for a given drain-source voltage V_{DS} are determined by the gate-source voltage V_{GS} and the capacitance of the gate dielectric C_i:

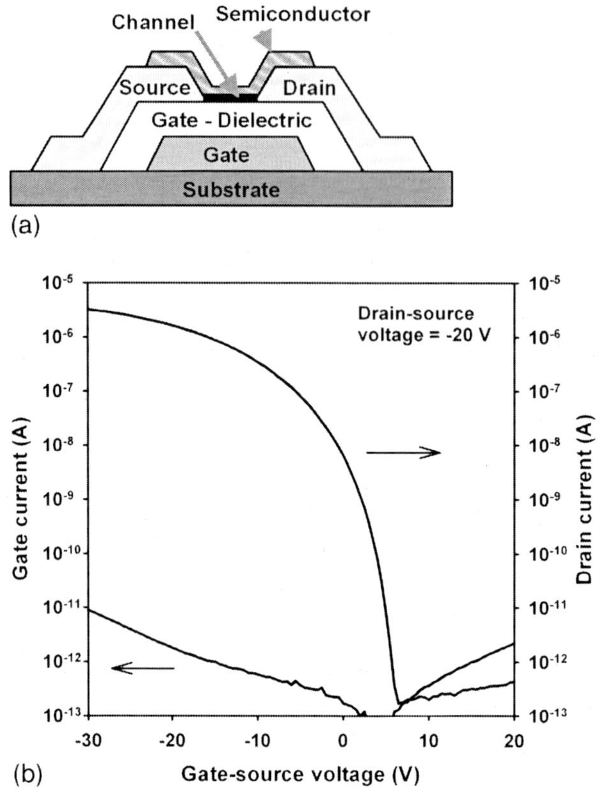

Fig. 6.1. (a) Schematic cross-section of an inverted coplanar (bottom-contact) TFT. The channel is formed at the interface between the gate dielectric and the organic semiconductor. (b) Electrical transfer characteristic of a pentacene TFT.

$$I_D = \frac{1}{2}\mu_{eff} C_i \frac{W}{L} (V_{GS} - V_T)^2 \tag{1}$$

$$C_i = \varepsilon_0 \varepsilon_i \frac{A}{d} \tag{2}$$

The drain current I_D is proportional to the charge-carrier mobility μ_{eff}, the transistor dimensions, where W and L are the channel width and length respectively, the applied voltage, where V_{GS} and V_T are the gate-source and threshold voltage, and the insulator capacitance C_i (Eq. 1). Thus, the impact of the dielectric material on transistor performance is given by the dielectric capacitance which results from a geometrical quotient of area A and distance d (distance is the film thickness of the dielectric layer) and a material factor where ε_0 and ε_r are the dielectric constant in a vacuum and of the material, respectively (Eq. 2). As consequences of these correla-

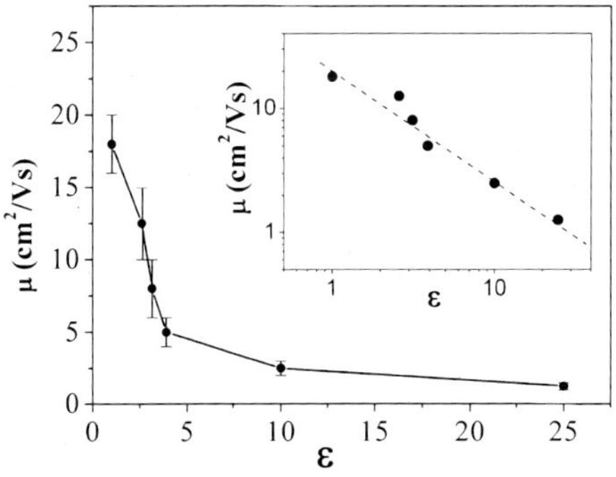

Fig. 6.2. Decrease of mobility with increasing ε, as observed in rubrene single crystal FET with different gate insulators. The bars give a measure of spread of mobility values. Inset: when plotted on a log–log scale, the available data show a linear dependence, with slope −1 (the variation in μ is proportional to ε^{-1}) [27].

tions the gate-dielectric layer has an impact on the operating voltage of a TFT; a thinner gate dielectric or one with a larger permittivity will result in a smaller operating voltage, for the same semiconductor material and the same transistor dimensions. Another effect of the dielectric material, discovered by experiments on rubrene single-crystal transistors with several dielectric materials, is related to the dependence of the charge-carrier mobility μ on the dielectric constant $\varepsilon_0\varepsilon_i$ of the insulating layer (Fig. 6.2) [27].

The advantage of using a single-crystalline semiconductor for this particular experiment is the excellent reproducibility of the transistor characteristics as a result of eliminating the random effects of molecular disorder generally present in polycrystalline thin film devices. It was found that the mobility measured for the same semiconductor (single-crystalline rubrene) increases by more than an order of magnitude when the permittivity of the gate dielectric is reduced. Furthermore, an empirical relationship between mobility and permittivity $\mu \approx 1/\varepsilon$ was proposed.

Besides the general effect on operating voltage and charge-carrier mobility in TFTs, the gate-dielectric layers have to withstand the applied electric field to prevent gate leakage. The gate current should be as small as possible, preferably no larger than a few picoamps. To make a useful device the electrical characteristics of the dielectric material must be stable over time without degradation caused by electrical or environmental stress.

6.3 Insulating Materials – An Overview

Over the past decade, a wide range of dielectric materials have been successfully employed as gate dielectrics in organic TFTs. In principal, these materials can be classified as shown in Fig. 6.3 into four groups describing materials with the pur-

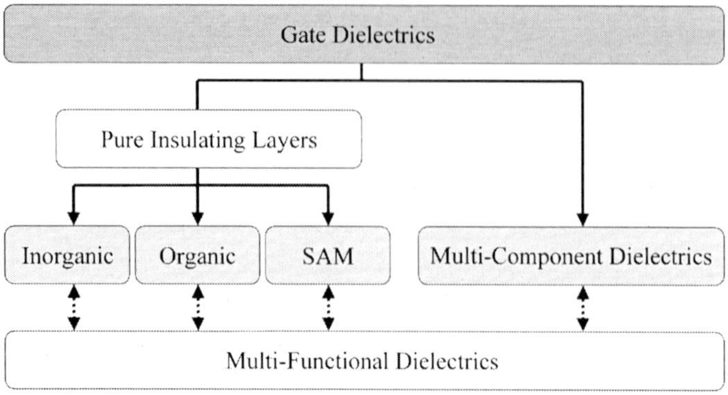

Fig. 6.3. Schematic overview and classification of gate-dielectric materials used in organic TFTs.

pose of acting as "pure" insulating layers (inorganic dielectrics, organic dielectrics, self-assembled monolayer dielectrics, and multilayer or multi-component dielectrics), distinguished by their chemical composition or their assembly as a dielectric layer. By extending the classification of dielectric layers beyond their insulating properties, a fifth group can be defined as multifunctional dielectrics, populated from all materials described above.

6.3.1
Inorganic Gate Dielectrics

Engaging in the new and exciting field of organic electronics, in particular organic transistors, early on the researchers built on their experience with state-of-the-art silicon technology. Consequently, for a long time silicon dioxide thermally grown on single-crystalline silicon has dominated as the gate-dielectric material for organic TFTs. The availability of silicon wafers and the use of heavily doped silicon as the gate electrode combined with the thermally grown silicon oxide as the gate dielectric have helped this system become an attractive test vehicle for screening new organic semiconductor materials [16, 25, 28]. However, the advantage of organic electronics, independence of expensive rigid substrates, cannot be achieved with this system. Also, the realization of devices more complex than transistors is unrealistic with thermally oxidized silicon wafers, because the gate electrode (i.e. the wafer) cannot be patterned. With the focus on substrate-independent materials, a large number of high-quality inorganic gate dielectrics has been investigated (Table 6.1). The scope of research ranges from bulk materials, for example inorganic oxides [2, 29–31] and nitrides [32–34] and their deposition methods, to multi-layers or blends consisting of inorganic dielectrics and polymer dielectrics [31, 35, 36].

The Penn State group has employed an ion-beam sputter process for silicon dioxide. This process deposits a dielectric film at low temperatures (80 °C) and is suitable for deposition on glass and flexible substrates [2, 29, 30]. With this ion-

Tab. 6.1. Overview of inorganic dielectric materials.

Material	Preparation method	Deposition temperature [°C]	Dielectric constant	Refs
SiO_2	Thermally grown	>600	3.9	16, 25, 28
SiO_2	Ion beam sputtered	80	3.9	2, 29, 30
Al_2O_3	Sputtered	not given	~5.1	31
Al_2O_3	Anodized	r.t.	~4.2	40
TiO_2	Anodized	r.t.	21	36
SiN_x	CVD	250–350	6–7	32–34
BZT	rf Sputtered	r.t.	17.3	37
BST	rf Sputtered	r.t.	16	37
Ta_2O_5	Sputtered	300	25	27

beam sputtered silicon oxide, organic TFTs for integrated circuits and for driver TFTs in backplanes for LC displays have been demonstrated. As another inorganic dielectric material, chemical vapor-deposited (CVD) silicon nitride was used in device fabrication. Groups at Lucent and Philips have demonstrated organic ICs and display backplanes on rigid substrates [32, 33]. Following the trend of reduced power consumption, materials with higher dielectric constant than silicon oxide are of interest. Inorganic oxides such as Al_2O_3, TiO_2, BZT [Ba(ZrTi)O], BST [Ba(SrTi)O], or Ta_2O_5 have been tested as pure thin-film materials, and in combination with other materials [6, 27, 31, 36–40]. While most high-ε materials suffer from expensive deposition methods (sputtering) and poor film quality (requiring relatively thick films to reduce gate leakage), the implementation of Al_2O_3 has made good progress, because of suitable deposition processes for Al_2O_3 and the direct conversion of Al by anodization [31, 40].

6.3.2
Polymer Gate Dielectrics

Because organic TFTs are targeting a wide range of potentially inexpensive applications, alternative gate dielectrics and low-cost fabrication methods are of interest, including solution-processable polymer gate dielectrics [1, 11, 20, 21, 34] that can be deposited by spin-coating, spray-coating, or printing, rather than by vacuum deposition. Because most organic polymers are insulators, this class of materials potentially leads to tailor-made materials for organic electronics applications. Several approaches for polymer dielectric films have been tested. The most common method is deposition of a polymer solution – polymer dissolved in a suitable solvent – by spin-coating, spray-coating, or printing and subsequent drying of the film by evaporating the solvent. For several materials an additional temperature or irradiation step is required (cross-link or cyclization reactions) to produce the final film. Typical examples of these polymers are polyimide (PI) [41, 42], polyvinylphenol (PVP) [19, 26], photoresists [11], poly(methyl methacrylate) (PMMA) [43], poly(vinyl alcohol) (PVA) [44], and the class of spin-on glasses (silsesquioxanes) [45, 46]. With this top-down approach, in which the properties of the resulting polymer film (thickness, stability, roughness, etc.) are determined by the polymer formulation and the corresponding process, the highest mobilities in TFTs with organic gate dielectrics have been obtained [25]. It should be noted that the advantage of solubility of polymers in the sense of inexpensive deposition methods could be a problem regarding possible intermixing and swelling in the preparation of multilayer stacks as given in "all-polymer" devices.

Another method of forming polymeric insulating films on a substrate is the "growth" of polymers from monomers directly on the gate electrode surface. An example of this "bottom-up" approach is poly-*para*-xylylene (Parylene), in which the pyrolysis of a *para*-xylylene dimer generates radicals that polymerize on a substrate held at room temperature forming an insulating film [47, 48]. Another method is the surface-initiated ring-opening metathesis polymerization (SI-ROMP) forming, for example, a polynorbornene layer, a "living" metal-catalyzed polymerization in solution. The film thickness can be tuned by altering the reaction time [49].

Besides the general insulating properties of insulating materials, chemical and thermal stability is required and excellent film-forming properties and methods for patterning the insulating layer. Therefore, the most common polymers (e.g. polyethylene, polypropylene, polyvinylchloride etc.) have not yet been used as gate-dielectric layers.

6.3.3
Self-Assembled Monolayer Gate Dielectrics

One of the most critical problems with organic TFTs are the large operating voltages, which often exceed 20 V. This is because of poor capacitive coupling through the relatively thick conventional gate dielectrics, which are often thicker than 100 nm to minimize gate leakage. One promising approach to reduce the dielectric thickness and thus the operating voltage is the implementation of self-assembled monolayers (SAMs). Given the small thickness of the molecular monolayers (approx. the length of a molecule) self-assembled on the gate electrode surface, a high quality insulating film is required to prevent leakage, i.e. no pinholes and dense packaging. By using the proper molecular design and a defined deposition process the SAMs are able to withstand electrical fields up to 16 MV cm^{-1} – comparable with or better than those of thermally grown silicon dioxide dielectrics of similar thickness [50]. Insulating layers of sufficient quality for organic TFTs have been demonstrated with bottom-up and top-down approaches. Vuillaume et al. have prepared suitable SAMs with a solution-grafted monolayer from vinyl-terminated alkoxysilanes on a silicon substrate, followed by oxidation to obtain a carboxyl-terminated monolayer [51, 52]. The Infineon group has grafted modified alkoxysilane molecules directly on a briefly activated surface to obtain a high quality SAM without chemical conversion [50]. With both approaches, functional organic TFTs with operating voltages as low as 2 V could be prepared. With direct grafted monolayers integrated circuits (ICs) on rigid and even on flexible substrates have been reported [53].

6.3.4
Multi-layer and Multi-component Gate Dielectrics

In addition to the straightforward strategy of using one single material as the gate-dielectric layer, several approaches with more complex dielectric systems were studied. The work on multi-layer or multi-component systems is basically driven by the motivation to increase TFT performance or to tune the electrical characteristics in a certain way, sometimes ignoring that one advantage of organic electronics is simplicity and most of the potential applications are low-cost products.

The best-known example of a multi-layer dielectric system is octadecyltrichlorosilane (OTS) on silicon oxide. After treating the silicon oxide with this C_{18} alkane, which forms a covalently bonded monolayer on the oxide surface with a silane anchor group, improved TFT performance (mobility, threshold voltage, subthreshold swing, and on/off ratio) and homogeneous distribution of TFT properties over a substrate were obtained for pentacene TFTs compared with devices on an untreated oxide surface [2, 28, 29]. Although the effect on the insulating properties (change in ε value) of the OTS monolayer is probably insignificant (given the large thickness of the silicon oxide), the effect on the device characteristics is significant, depending on the organic nature of the interface (insulator/semiconductor) which may provide a more efficient transistor channel, perhaps because of better matching in surface energy of these materials. Kobayashi et al. have extended this approach by changing the nature of the SAM and with this the interface between insulator and semiconductor and were able to tune the threshold voltage over a wide range from positive to negative values [54].

Similar results, supporting the hypothesis that an organic interface at the channel is favorable for high-performance organic TFTs, have also been obtained with polymer-treated inorganic dielectrics [31, 36]. Apolar polymers (e.g. PAMS – poly(α-methylstyrene)) deposited a few nanometers thick on oxide surfaces increase the mobility in pentacene TFTs by approximately a factor of three compared with devices on untreated oxide surfaces. It should be noted that the effect of increased mobility on dielectrics treated with organic materials (SAMs or polymers) or pure organic dielectrics does not correlate with the number of pentacene grains in the transistor channel as a measure of the grain size or the number of grain boundaries [31, 55]. Furthermore, this effect can be postulated as a general tendency useful in future device development.

Not only the combination of inorganic dielectrics with organic monolayers or polymers has been tested so far, but also inorganic–inorganic and organic–organic multi-layers have interesting effects as gate dielectric composites [56]. An example of an organic double-layer dielectric system is the combination of a SAM layer with a polymer dielectric film (PVP), but with architecture opposite to that typically fabricated as in the SiO_2/OTS in which the SAM covers the bulk dielectric. In this instance the SAM is grafted on an aluminum gate pattern and covered by a thin PVP layer. ICs based on pentacene TFTs containing a double layer 2.5 nm thick (18-phenoxyoctadecyl)trichlorosilane (PhO-OTS) followed by a 13 nm thick film of cross-linked PVP can operate with supply voltages from 3 to 20 V without electrical shorts. This could protect a TFT from surge currents, without losing the functionality at low voltages.

Most polymer dielectric materials have relatively low dielectric constants (2.5–4.0), depending on their molecular structure. To increase the capacitive coupling through a polymer film, addition of high-ε components to the polymer matrix is suitable, and maintains the layer-by-layer deposition in the systems described above. One approach is the dispersion of high-ε nanoparticles (TiO_2) in a traditional polymer dielectric PVP reported by Chen et al. [35]. With different concentrations of TiO_2 particles in the blend, the ε value of the dielectric film increases from 3.5 for the unmodified polymer to 5.4 for the nanocomposite dielectric containing 7% nanoparticles, corresponding to an increase of output current in the pentacene TFTs. The surface quality of the organic–inorganic nanocomposite film is poor, however, and realization of vias in these films, required for realization of ICs, will be more complicated than in pure organic or inorganic dielectric films.

6.3.5
Multifunctional Dielectrics

Besides the general function of insulating the gate electrode in a TFT and serving as the channel interface for the organic semiconductor, the dielectric layer could combine additional features extending the nature of a TFT from a transistor to a memory element. Several groups have described the use of ferroelectric polymers in capacitor elements to create nonvolatile memory cells, on the basis of the displacement in an electric field. A typical characteristic of such a cell is hysteresis in current–voltage curves depending on the direction of the voltage sweep. Hysteresis is normally a problem in organic TFTs and organic circuits, but providing a permanent and reproducible hysteresis can be used to define two digital states of the TFT. Schroeder et al. have demonstrated a so called "FerrOFET" with poly(*m*-xyleneadipamide) as dielectric layer and pentacene active layer, for which the current–voltage characteristics of the TFT are shifted by approximately 20 V, depending on whether the gate voltage is swept forward or backward. The devices have one digital state "0" with low drain current and another state "1" with higher drain current at the same gate potential. The ratio of the two states is 2.7×10^4 and the hysteresis is stable over time (Fig. 6.4). Memory retention time of hours to days was predicted for such a memory TFT [57].

6.4
Application-related Aspects of Dielectrics

Because of the complexity of the relationships between the properties of the gate dielectric and the performance of organic TFTs and ICs it may be useful to explain

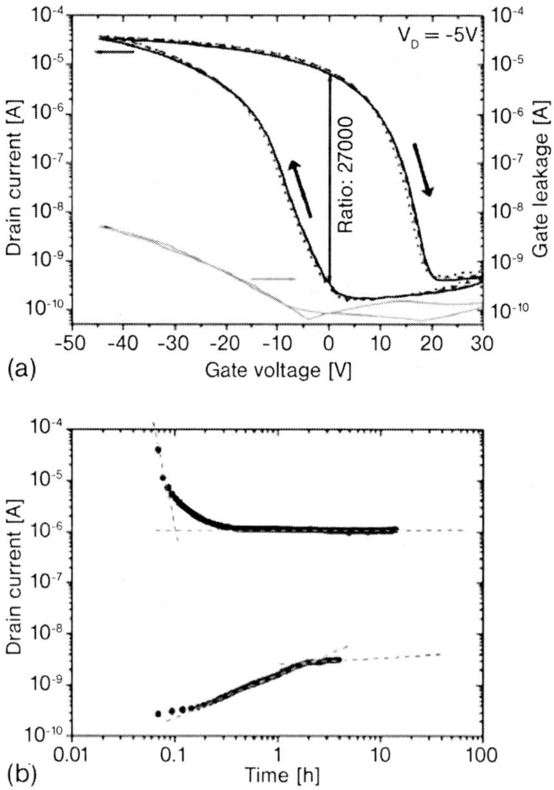

Fig. 6.4. (a) Transfer characteristics of a "FerrOFET". At zero gate voltage, the memory ratio is 2.7×10^4 (the solid curve is the first sweep, the dashed and dotted curves were recorded at later times); (b) Memory retention of the ON and OFF states [57].

these relationships in more detail with two examples. The first example is the high-quality polymer dielectric material PVP which was investigated in simple transistor devices and in more complex ICs. In a second example, self-assembled monolayer (SAM) gate dielectrics will be discussed. These have several advantages over other insulating materials. In this brief introduction application-related aspects such as hysteresis, stability of devices, variation of device properties, and their potential in low-cost production methods will be covered.

6.4.1 Poly-4-vinylphenol Dielectrics

In 1998 the commercially available polymer PVP (M_w 20,000) was introduced as a dielectric layer for organic TFTs by the Philips group [19]. It is soluble in common organic solvents (e.g. acetone, *n*-butanol, propylene glycol monomethyl ether ace-

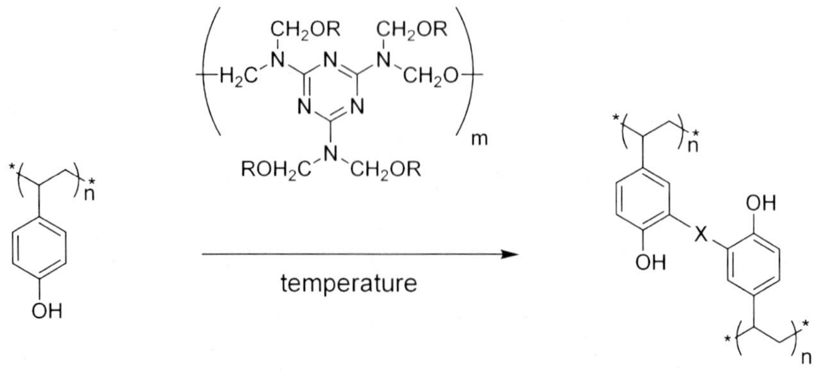

Fig. 6.5. Thermally initiated cross-linking reaction of poly-4-vinylphenol with poly(melamine-co-formaldehyde) methylated as a cross linking agent.

tate (PGMEA), *N*-methylpyrrolidone (NMP), etc.) and can thus deposited on a substrate by dip coating, spin coating, spray coating, or, potentially, by printing. Because of this excellent solubility, a wide range of formulations and resulting film thicknesses (10 nm to 1 μm) are possible. In addition to the pure PVP polymer, several copolymers (e.g. poly-4-vinylphenol-co-2-hydroxyethylmethacrylate) have also been tested as gate dielectrics [25]. Most of the PVP formulations have excellent film-forming properties, giving a homogeneous film with a smooth surface [58]. One drawback of the solubility is the sensitivity of the polymer film to solvents during subsequent process steps in the TFT fabrication. A simple route to a mechanically and chemically robust polymer film was demonstrated by cross-linking the PVP basic polymer with a cross-linking agent, forming a "Bakelite" structure. As a useful cross-linking agent (with regard to solubility, onset of the cross-link reaction, etc.) poly(melamine-co-formaldehyde) methylated (M_n 511) was identified (Fig. 6.5) [23]. The basic polymer (PVP) and the cross-linker can be mixed in one solvent and deposited in the same way as the pure PVP formulation. A typical ratio of basic polymer to cross-linking agent is 5:1. The shelf-life of these formulations is more than a year. After drying (evaporation of the solvent) a homogeneous film is formed on the substrate. In subsequent thermal treatment the cross-linking reaction starts at approximately 150 °C. An efficient reaction temperature is 200 °C. The poly(melamine-co-formaldehyde) acts as a donor of a formaldehyde moiety reacting at the activated 3-position of the phenol ring in the PVP. In a second step the cross-link is completed by reacting with another phenol ring (from the same or a neighboring polymer chain), forming a covalent-bonded bridge.

The resulting flexible films withstand treatment with chemicals such as acetone, alcohols, and neutral aqueous etch formulations, but are sensitive to very basic solutions. It should be noted that significantly larger concentrations of the cross-linking agent in the polymer formulation will result in more ridged films, and

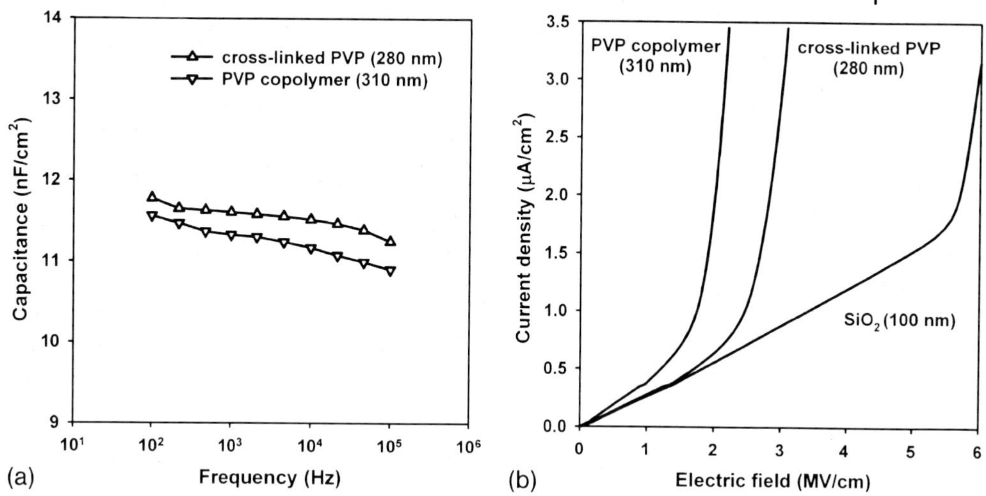

Fig. 6.6. (a) Capacitance–frequency characteristics of capacitor structures prepared with cross-linked PVP and poly(4-vinylphenol)-co-2-hydroxyethyl methacrylate (PVP copolymer); (b) Current leakage through cross-linked PVP and PVP copolymer gate dielectrics compared with thermally grown silicon oxide [25].

may lead to cracks in the cured films, whereas smaller concentrations of the cross-linking agent may increase the film's solubility.

To reduce the cross-linking temperature for applications that do not allow a temperature of 200 °C, the cross-link reaction can be performed with alternative cross-linking agents in a photo-initiated reaction. A photo acid generator (PAG) sensitive at a particular wavelength is mixed into the formulation. As an alternative cross-linking agent, 4-(hydroxymethyl)benzyl alcohol was tested successfully. In addition to the reduced process temperatures (maximum 150 °C for post-exposure bake, PEB), this system potentially opens the opportunity to produce directly patterned dielectrics, depending on the different solubility of irradiated and nonirradiated regions in the polymer film. The electrical characteristics of pentacene TFTs based on these films are comparable with those with thermally cross-linked PVP gate dielectrics [59].

To determine the permittivity of the PVP gate-dielectric materials, capacitor structures can be used in which the spin-coated polymer layer is sandwiched between radio-frequency-sputtered titanium films, using silicon wafers as substrates. Figure 6.6a shows capacitance–frequency characteristics obtained from small-signal impedance measurements on capacitors with a dielectric thickness of 280 nm (for the thermally cross-linked PVP) and 310 nm (for the PVP copolymer), for a contact area of 0.13 cm^2. From the capacitance data, a permittivity of 3.6 was determined for the cross-linked PVP gate dielectric and a permittivity of 4.0 for the PVP copolymer gate dielectric [25]. An important concern with any gate-dielectric material, but particularly with polymer gate dielectrics, is gate leakage. The measured gate leakage on capacitor structures similar to those used for the capacitance–frequency measurements is shown in Fig. 6.6b. The current density through the gate-dielectric layer is given as a function of the electric field applied across the film for a dielectric thickness of 280 nm (for the cross-linked PVP), 310 nm (for the PVP copolymer), and 100 nm (for SiO_2 thermally grown on silicon), for a contact area of 0.29 cm^2. Gate dielectric leakage for the polymer gate dielectric films is less than 50 nA cm^{-2} at a gate bias of 5 V and less than 250 nA cm^{-2} at a gate bias of 20 V, similar to the leakage currents observed for thermally grown SiO_2.

The PVP dielectric system, and in particular the thermally cross-linked PVP, has been tested by several groups in organic TFT devices. Simple test devices and photolithographically defined TFTs have been investigated. In test devices a heavily doped silicon wafer serves as the substrate and gate electrode. The PVP dielectric layer was added by spin coating and cross-linked at 200 °C, followed by the deposition of the organic semiconductor. Gate, gate dielectric and semiconductor remain unpatterned. The source and drain contacts are deposited by thermal evaporation through a shadow mask. In this set-up, a record value for charge-carrier mobility of 3.0 $cm^2 V^{-1} s^{-1}$ (measured in a pentacene TFT) was obtained with this dielectric material (Fig. 6.7); this is still the highest mobility for organic TFTs with a polymer gate-dielectric layer.

To take full advantage of organic electronics, independence from single-crystalline and expensive substrates, the silicon substrate must be replaced by a less expensive carrier substrate. A wide range of conducting materials, including metals and conducting polymers, have been used as gate electrode materials. Typically, non-noble metals such as Al or Ti will be used, because of their inexpensive processing. For applications in display backplanes or ICs each transistor must operate with an individual gate, requiring a patterned deposition or a patterning step after deposition of the gate electrode material. The processing of the PVP dielectric layer follows the same procedure as on silicon substrates. It should be noted that a cross-linked PVP layer reproduces the topography of the underlying substrate, even the surface roughness of the gate electrode. On an atomically smooth substrate the PVP layer has a smooth surface with an average surface roughness of approximately 7 Å, as demonstrated on single-crystalline silicon wafers. A "smoothing" effect of the PVP seems negligible, suggesting that the surface quality of the gate electrode material and the roughness of the substrate will have an effect on device characteristics [23]. To create access to the gate electrode layer the PVP layer is patterned by photolithography and oxygen plasma etching. To define the source and the drain contacts gold is deposited by evaporation and patterned by photolithography and wet etching. As an alternative, the conducting polymer polyethylenedioxythiophene doped with polystyrene sulfonic acid (PEDOT:PSS; Baytron P formulation) was used for the source/drain contacts. Finally, pentacene is deposited by thermal evaporation to create the organic active TFT layer. Also, different organic semiconducting materials were demonstrated on PVP gate dielectrics, including polymers and n-type materials [58, 60].

Irrespective of the process flow, the PVP withstands treatment with the chemi-

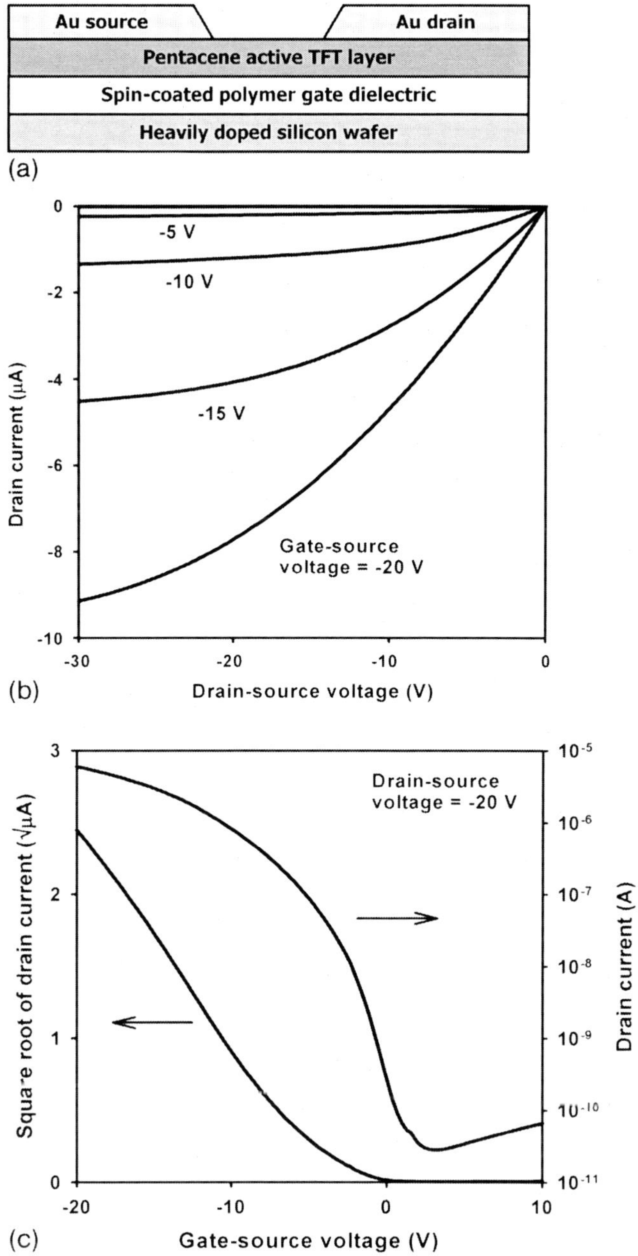

Fig. 6.7. (a) Schematic cross-section of pentacene TFT with top-contact configuration. The gate-dielectric layer is a 260 nm thick thermally cross-linked PVP; (b) electrical output characteristics; (c) electrical transfer characteristics ($\mu = 3.0$ cm^2 V^{-1} s^{-1}) [25].

cals involved after deposition and cross-link of the dielectric layer. Reference measurements of the film thickness, surface roughness, and static contact angle indicate that after each step in the processing the dielectric layer remains unchanged.

In the fully patterned bottom-contact TFTs using Ti gates on glass substrates with a thermally cross-linked PVP dielectric layer 120 nm thick, gold source and drain contacts, and pentacene as active material, a charge-carrier mobility of 0.7 $cm^2\ V^{-1}\ s^{-1}$ was obtained (Fig. 6.8) [61]. The slightly reduced mobility compared with the top-contact TFT is related to the large contact resistance in the bottom-contact TFT and is not notably affected by the PVP dielectric layer (e.g. by following process steps or because of increased surface roughness). Cross-experiments with bottom-contact TFTs using a single-crystalline silicon gate electrode instead of Ti, which serves as a smooth surface, prove that the contact resistance clearly dominates the electrical performance of the TFT and the "image effect" on the surface roughness of the gate electrode is negligible, at least for smooth glass substrates with metal electrodes.

The dielectric material not only affects the charge-carrier mobility in organic TFTs – the values of the switch-on voltage and the subthreshold swing also depend on the nature of the channel interface and so, in part, on the dielectric material. With the focus on potential products homogeneous distribution of these electrical characteristics is rather more important than their absolute values. Small or no variations of all characteristics are required for the realization of functional combinational and sequential logic circuits. Fairly tight distributions of these characteristics can be observed for TFTs fabricated on the same substrate, as has been shown by the Philips group for TFTs with a solution-processed pentacene active layer [62]. The distribution for 49 TFTs on the same substrate with thermally evaporated pentacene layer and thermally cross-linked PVP, prepared in the same way as described above and all with the same dimensions, is shown in Fig. 6.9. A mobility of 0.16 $cm^2\ V^{-1}\ s^{-1}$ (with a standard deviation of 0.005 $cm^2\ V^{-1}\ s^{-1}$, or 3%), a switch-on voltage of 1.82 V (with a standard deviation of 0.05 V, or 2.8%), and a subthreshold swing of 0.81 V $decade^{-1}$ (with a standard deviation of 0.019 V $decade^{-1}$, or 2.3%) were extracted, indicating excellent uniformity for the cross-linked PVP also.

In addition to uniformity of electrical characteristics on the same substrate, substrate-to-substrate variation is a benchmark of the stability of the process and with this an indicator for the sensitivity of materials (including dielectrics) during processing. The characteristics (mobility, switch-on voltage, and subthreshold swing) often vary significantly from substrate to substrate, which is not surprising considering that each substrate is processed individually, manually and without particular emphasis on minimizing process variations. Even under laboratory conditions and with individual processing, however, sufficiently small variations could be obtained. In a lot of 32 substrates the variation in mobility was 25% and the variation of the subthreshold swing was 22%, measured for one representative TFT on each of the 32 substrates (Fig. 6.10). Further improvement of variation should be possible by using automated substrate handling and processing.

Another important quality criterion for dielectric materials in organic TFTs is their hysteresis. Hysteresis is the shift in threshold voltage depending on the direc-

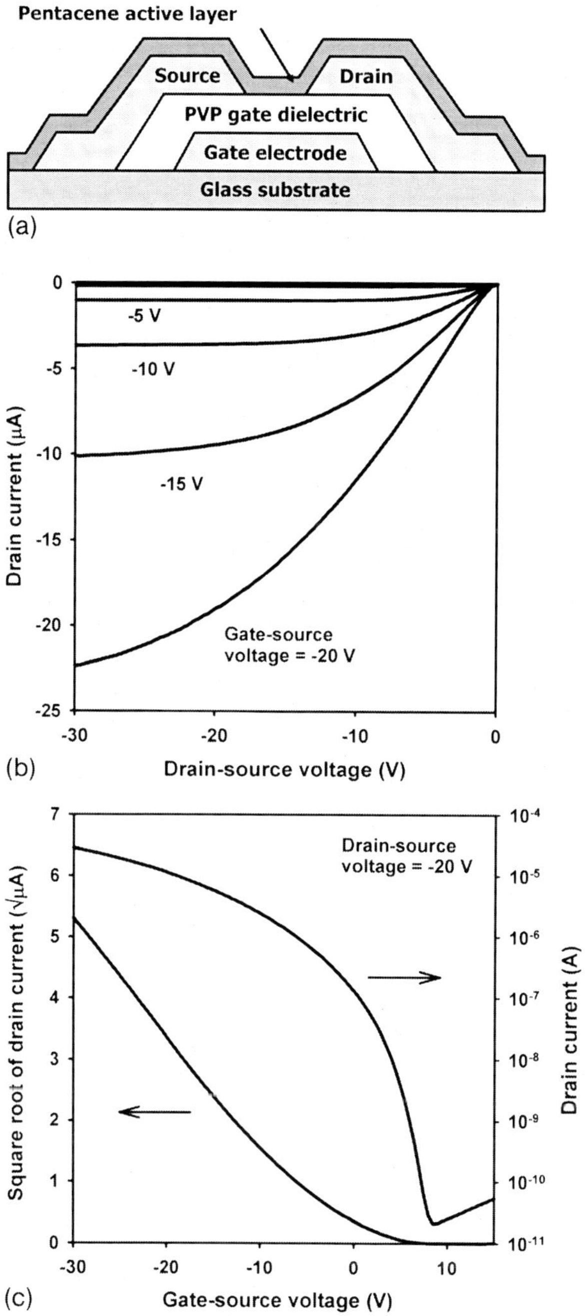

Fig. 6.8. (a) Schematic cross-section of pentacene TFT with bottom-contact configuration. The dielectric layer is 120 nm thick thermally cross-linked PVP. (b) Electrical output characteristics. (c) Electrical transfer characteristics ($\mu = 0.7\ cm^2\ V^{-1}\ s^{-1}$) [58].

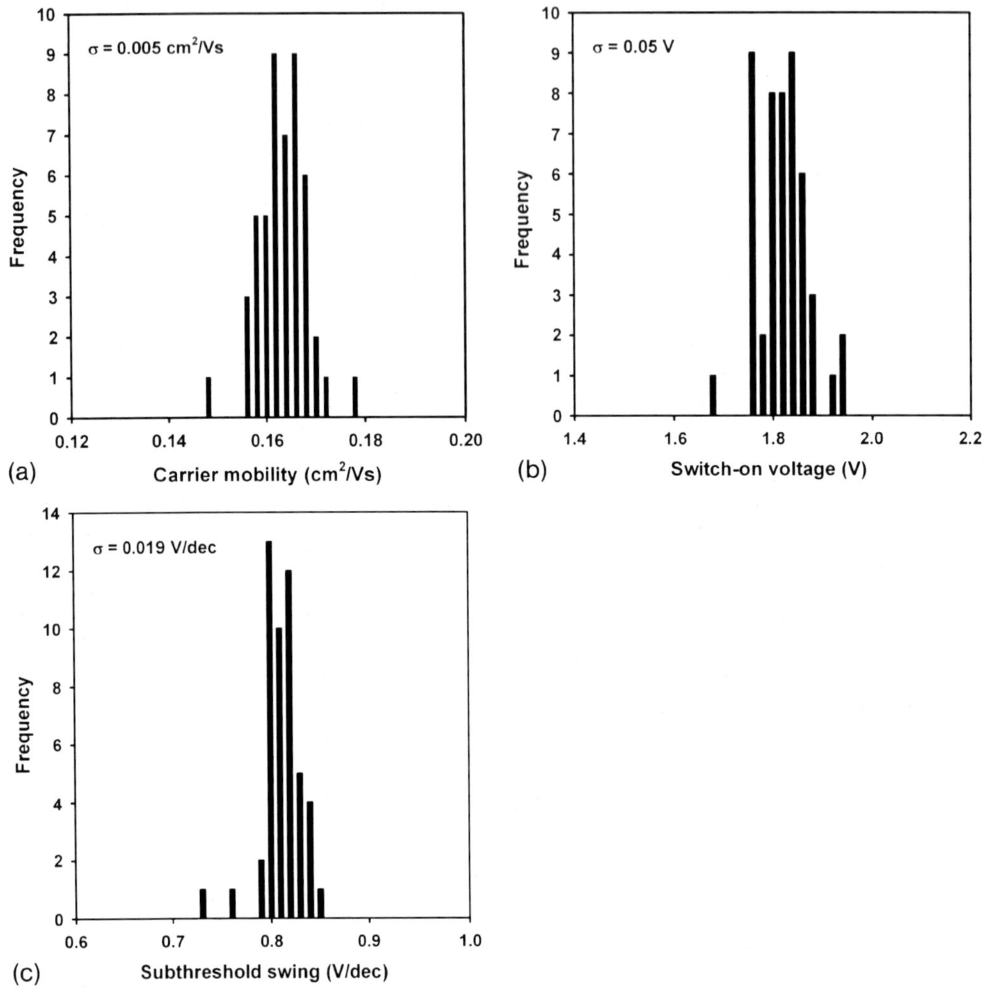

Fig. 6.9. Statistical distribution of (a) mobility, (b) switch-on voltage, and (c) subthreshold swing of 49 pentacene TFTs on the same substrate.

tion of sweep of the gate-source voltage. Whereas the hysteresis for inorganic gate dielectric TFTs is typically less than 1 V, the hysteresis for polymer gate dielectric TFTs can be as large as 15 V [25]. Hysteresis can usually be divided in two classes, permanent hysteresis and dynamic hysteresis. Although permanent hysteresis (no significant change in threshold voltage over time) can be used for data storage (Section 6.3.5) [57], dynamic or volatile hysteresis is a major problem in the design of integrated circuits. The successful integration of TFTs into functional digital and analog ICs depends critically on stable operating points (i.e. on very narrow operating windows) for each transistor. These operating windows define the bias condi-

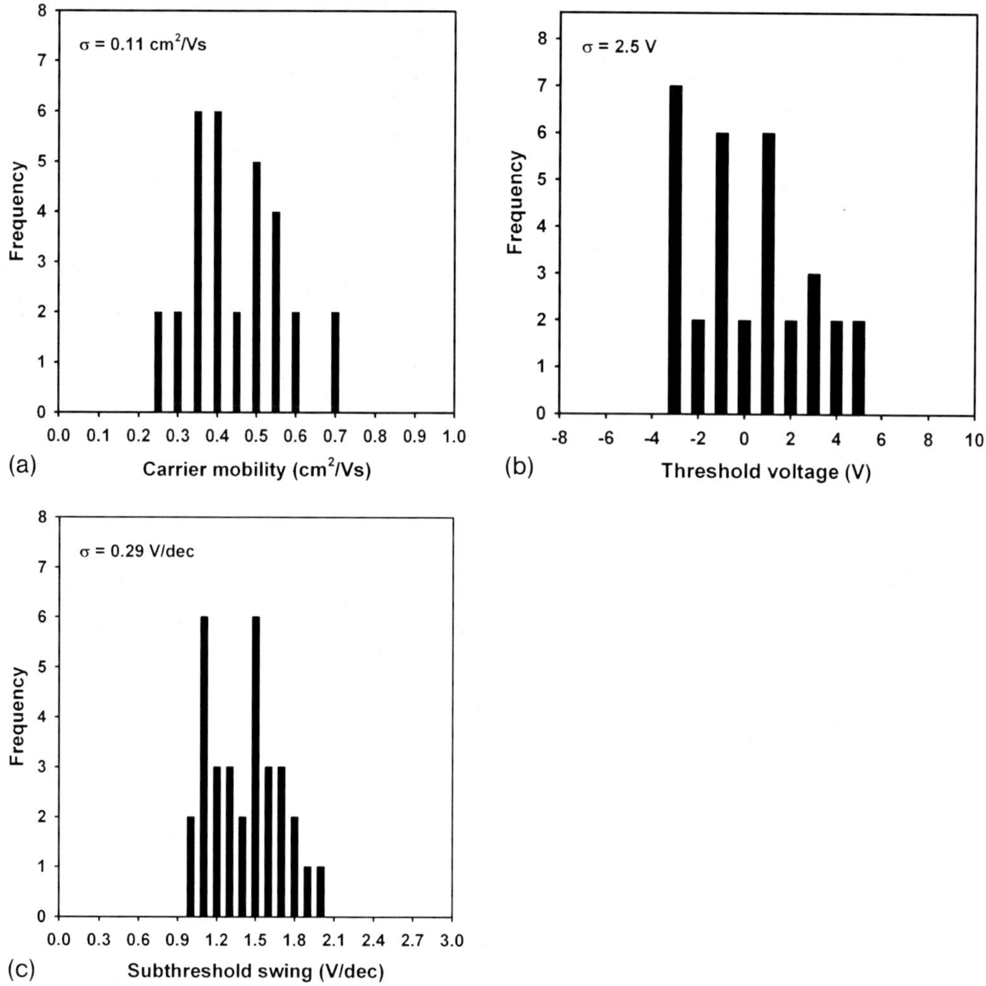

Fig. 6.10. Statistical distribution of (a) mobility, (b) threshold voltage, and (c) subthreshold swing of 32 pentacene TFTs on 32 individually and manually processed substrates.

tions under which the TFTs are in their "ON" and "OFF" states, which are related to the threshold or switch-on voltage. In integrated circuits or displays, in which the TFTs are driven between their "ON" and "OFF" states, a large shift in threshold voltage (i.e., a large hysteresis) could shift the point of response (starting interaction) outside the operating window and consequently the circuit will not operate. The impact of hysteresis on the function of circuits depends approximately on the number of interacting TFTs in a given design.

One dominating factor for hysteresis in organic polymers is related to the presence of mobile dipoles in the polymer bulk. Consequently the effect varies with film thickness for the same dielectric material. Thinner films with a smaller number of mobile dipoles per unit area have a smaller effect on the shift in threshold voltage than thicker films. For a given thickness of one material the hysteresis is critically dependent on the speed of the measurement (sweep rate) and on the applied field, indicating dynamic behavior of the material related to the decay in reorientation of the dipoles. Thus, the values measured for hysteresis in TFTs are usually smaller for high-frequency measurements with low applied fields than for slow measurement at large potentials.

To answer the question "What are the mobile dipoles in the polymer film and where do they come from?" it is necessary to follow the process flow of the polymer dielectric layer from basic polymer to finished film in the TFT device. Some of the answers will be speculative but could help to understand the complexity of hysteresis and some have been identified as key points in the suppression of this effect to produce excellent insulating layers with negligible hysteresis. At this point the nature of "mobile dipoles" should be classified. The first class covers all molecules or fragments that are related to external effects, including remaining solvents, impurities (e.g. traces of catalysts, oligomers) or additives in the polymer formulation. These molecules or fragments could remain in the polymer network (after deposition and drying) as true mobile species with low molecular weight compared with the basic polymer. The second class of mobile dipoles arises from the macroscopic behavior of the basic polymer. Depending on the chemical structure of the monomers, the molecular weight, and the solvent, the macromolecules are dissolved in different macroscopic shape in the polymer formulation (filament, mesh, etc.). Consequently, the network of the basic polymer in a dielectric film is affected by this macroscopic shape also, resulting in a more or less densely packed film with different degrees of freedom for movement of chain ends and backbone parts of the macromolecules. These movable parts in the basic polymer network would be affected by an external electric field. The third class of mobile dipoles is related to intramolecular dipoles, and depends on dipole moments of the monomers or building blocks in copolymers.

It should first be noted that hysteresis in TFTs is not an intrinsic property of organic polymer dielectrics. Results obtained on pentacene TFTs with Parylene gate-dielectric layers reveal electrical transfer characteristics without significant hysteresis. This could be explained by the "dry" CVD process used to produce the Parylene layer yielding a film without any remaining solvent impurities or any thermal damage from drying the film. The absence of solvents during processing may be an important aspect of these properties but would not, however, explain the significant hysteresis obtained for carefully dried films of other polymers deposited from solution, even with analytical evidence that no solvent remains. For Parylene the polymer network or the package density in the bulk may be affected by the bottom-up approach of surface polymerization compared with the top-down approach for solution-grafted films, in which the packing in the film is determined by the behavior in solution.

Most organic dielectrics are usually grafted from solution, so again we will consider PVP, which is suitable as blank polymer film and as cross-linked material in TFT devices. In pentacene TFTs with pure PVP dielectric films (without cross-link) significant hysteresis of around 15 V was seen that was virtually independent of the type of solvent and the conditions used to dry the polymer film. For pentacene TFTs with cross-linked PVP (generated from a spin-coated formulation of the same PVP basic polymer (M_w 20,000) with a low-molecular-weight cross-linking agent (poly(melamine-co-formaldehyde) methylated, M_n 511, as additive) dissolved in PGMEA, and thermally cured at 200 °C), however, the behavior was more complex.

Hysteresis of 12 V was measured in devices cross-linked at 200 °C for 10 min in a nitrogen atmosphere at ambient pressure (Fig. 6.11a). TFTs with hysteresis of 7 V were obtained by performing the cross-link reaction in nitrogen at reduced pressure (Fig. 6.11b). And, finally, negligible hysteresis of approximately 1 V was obtained when the cross-link reaction was performed for 45 min at reduced pressure (Fig. 6.11c). These observations, with results obtained by FTIR and thermogravimetric measurement of the polymers cross-linked under different conditions, suggest the hysteresis depends critically on the "degree" of cross-linking and on the amount of residue of the cross-linking agent in the polymer film. Thus, in the PVP system the hysteresis is increased by impurities mainly attributed to the cross-linking agent and movable parts in the basic polymer which can be "fixed" (made immobile) with the cross-link reaction. By tuning the reaction conditions to longer reaction times (presumably resulting in more cross-linking) and to reduced pressure (to out-gas by-products more efficiently) the two opposite effects of the cross-linking agent can be used to obtain dielectric films with low hysteresis. It should be noted that the carrier mobility in these pentacene TFTs is related to the hysteresis. With decreasing hysteresis from 12 V and 7 V to 1 V the mobility decreases from 0.8 and 0.6 $cm^2\ V^{-1}\ s^{-1}$ to 0.1 $cm^2\ V^{-1}\ s^{-1}$.

As mentioned previously, however, hysteresis is a complex property of polymer dielectrics. Besides the impact of the polymer material itself, other relationships have also been identified. For example, for the cross-linked PVP system increased hysteresis of around 6 V was obtained on soda lime glass substrates, even using an optimized process, whereas on other substrates, for example PEN, PET, and thermally oxidized silicon wafers, hysteresis was negligible (less than 1 V). This effect may depend on ions moving from the substrate into the dielectric layer during the cross-link temperature step. Although the exact cause of this effect has not yet been fully investigated, it has a large effect on the choice of substrate in the production of IC based on this technology. Another example is the observed increase in the hysteresis of pentacene TFTs with cross-linked PVP gate dielectric with exposure to ultraviolet light, as shown in Fig. 6.12.

This observation can be explained in terms of photo-induced damage in the PVP network, also indicated by FTIR measurements. It should be noted that exposure doses in this experiment are fairly large compared with typical environmental conditions, but the result should nonetheless be taken into account with the focus on long-term stability of devices.

Using TFTs with thermally cross-linked PVP as gate-dielectric layer a wide range of integrated circuits have been demonstrated. By building on coplanar bottom-

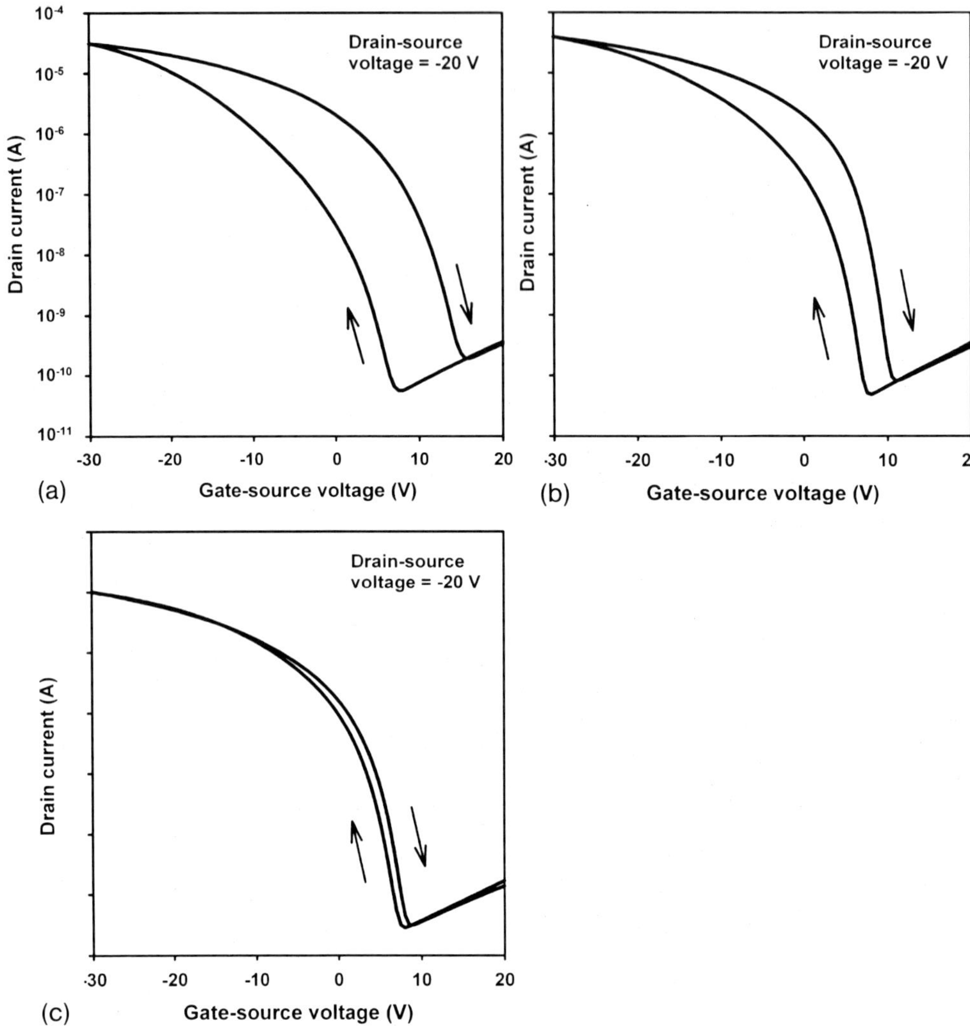

Fig. 6.11. Electrical transfer characteristics for pentacene TFTs ($W = 500$ μm, $L = 50$ μm) with 180 nm PVP dielectric layer cured at 200 °C for (a) 10 min in N_2 at 1000 mbar; (b) 10 min in N_2 at 400 mbar; (c) 45 min in N_2 at 400 mbar.

contact TFTs with vias and patterned pentacene layer simple circuits such as inverters and ring oscillators have been reported [61]. Owing to the high quality of the PVP dielectric material even more complex circuits have been realized, including various logic gates (NAND, AND, OR, XOR) and latches, flip-flops, and frequency dividers (e.g. a five-stage frequency divider "divide-by-32" that is composed of 225 pentacene TFTs).

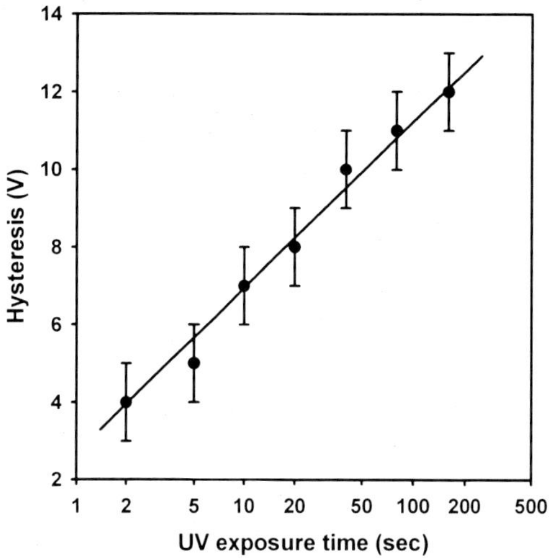

Fig. 6.12. Dependence of hysteresis in pentacene TFTs on exposure to UV light (Ushio model UXM-501XD lamp, 365 nm, 7 mW cm^{-2}).

To date, PVP (especially in the cross-linked form) is the most promising gate-dielectric material in organic electronics in the sense of realizing complex circuits on rigid and flexible substrates. Preparation and patterning of suitable dielectric films has, however, thus far usually followed the standard silicon technology using spin coating, photolithography, and plasma etching [61]. The dream of "fully printed devices" seems hardly realizable with this polymer formulation. Deposition itself might conceivably be performed by use of inexpensive methods, for example spray or dip coating, but a directly printed pattern (including vias as required to realize IC) of a polymer layer from a formulation of low viscosity needs further improvements.

6.4.2
The Self-assembled Monolayer Approach

The possibility of using organic thin-film transistors for a wide range of potential applications has been demonstrated. Now, commercialization is in focus. It is indisputable that organic electronics will not compete with standard silicon devices, because of the limited performance (e.g. charge-carrier mobility). As in state-of-the-art silicon technology, however, there is a trend in market applications toward battery-driven mobile solutions or completely internal power-independent applications (e.g. RFID), both of which require low power consumption by the active devices. Also, hybrid solutions of large-area organic-based arrays and silicon-based controller units are conceivable; these would combine the advantages of both technologies. So, besides the general trend of reduced power consumption in electronics, the need for organic devices to work with low supply voltage – in the range of standard silicon devices – is achieved by interaction with an existing silicon environment. The strategies used to achieve functional organic transistors with suitable performance at this supply voltage of 1–3 V are mainly predetermined by tuning the capacitance C_i of the dielectric layer (Eqs. 1 and 2) by alteration of their dielectric constant ε_I and/or their film thickness d. Consequently, two approaches are under investigation, use of dielectric materials with either high ε [6, 27, 31, 36–40] or reduced film thickness. The self-assembled monolayer approach targets the dielectric film thickness d. The key feature of this approach is the combination of nanometer scale film thickness with very small leakage current through the layer. The key challenge in using self-assembled monolayer dielectrics is the preparation of a dense pinhole-free robust layer to prevent electrical shorts and enabling further processing on top of the layers.

At first a suitable molecule is required, preferably with a linear structure in the insulating tail. The *n*-alkyl moiety is favored over conjugated systems, because it provides better insulation along the axis of the molecule [63]. To facilitate self-assembly a proper anchor group at the end of the molecular chain is essential. A wide range of anchor groups have been investigated for different substrates, generally with the focus on SAMs as etch mask or to match the work functions of different electrode materials [64–67]. The anchor groups also affect the packing density and the stability of a SAM, depending on the different interaction mechanisms between molecules and surfaces. The strength of interactions ranges from weak for hydrogen bonding (moving molecules at room temperature) to very strong for covalently bonded molecules (stable up to 350 °C) [51]. Because the thickness of a monolayer is approximately equal to the length of the molecule (slightly smaller, because of the effect of tilting depending on the type of substrate and the anchor group), the electrical insulation properties (especially the leakage currents) of a SAM also depend on the molecular length. An increased alkyl chain length leads to a reduced conductance along the molecule, as demonstrated in single-molecular measurements on different *n*-alkyldithiols [63]. The length of the molecules cannot, however, be extended endlessly without reducing the ability for self-assembly. Finally, the end group (head group) of the molecule affects the packing density and the morphology of the SAM and provides the surface for the molecular ordering of the organic semiconductor in the thin-film transistor.

The use of SAM gate dielectrics for organic TFTs was pioneered by the Vuillaume group [52]. They investigated the mechanism of carrier tunneling through SAMs and showed that current leakage through densely packed organic molecular monolayers with highly ordered aliphatic chains can be remarkably low, despite their thickness of only a few nanometers [68]. To prepare organic TFTs (and even silicon MISFETs) with SAM dielectrics, the Vuillaume group relied on carboxyl-terminated *n*-alkyltrichlorosilanes [51]. Using α-sexithiophene as the organic semiconductor they produced organic TFTs operating at 2 V with a field-effect mobility of 3.6×10^{-4} cm^2 V^{-1} s^{-1}, an on/off current ratio of 10^4, and a gate current density of 10^{-6} A cm^{-2} [10]. Recently, the Marks group has reported successful use of a thin organic multilayer as a dielectric in organic transistors [69]. In both examples the final insulating layer was created by self-assembly from a precursor followed by one or more chemical modifications. These bottom-up approaches furnish layers with excellent electrical properties but which require multiple chemical reactions.

Another approach in generating molecular insulating layers without the need of chemical conversion after deposition is the use of preliminarily modified molecules which can form dense self-assembled monolayers. To create dense self-assembled monolayers with sufficient robustness and insulating properties, a modified alkyltrichlorosilane with an aromatic end-group (18-phenoxyoctadecyl)trichlorosilane (PhO-OTS; chemical structure Fig. 6.15a) was synthesized and tested [50]. The SAMs were created in a one-step process from vapor phase or solution. On self-assembly on a natively oxidized silicon surface the π–π interaction between the phenoxy end-groups of adjacent molecules creates an intermolecular top-link, leading to a more closely packed surface compared to monolayer than when linear end groups are used.

It was shown by surface analysis (AFM and STM) that these films cover large areas (micrometer scale) without the occurrence of pinholes, furnishing a molecular flat amorphous surface with good short-range order but no long-range order, and thus free of domain boundaries (Fig. 6.13).

The insulating properties of the PhO-OTS layers were tested in capacitor devices. Figure 6.14 shows the breakdown characteristics of the PhO-OTS SAM gate dielectric, determined on twenty Si-SAM-Au capacitors measured on structures with a

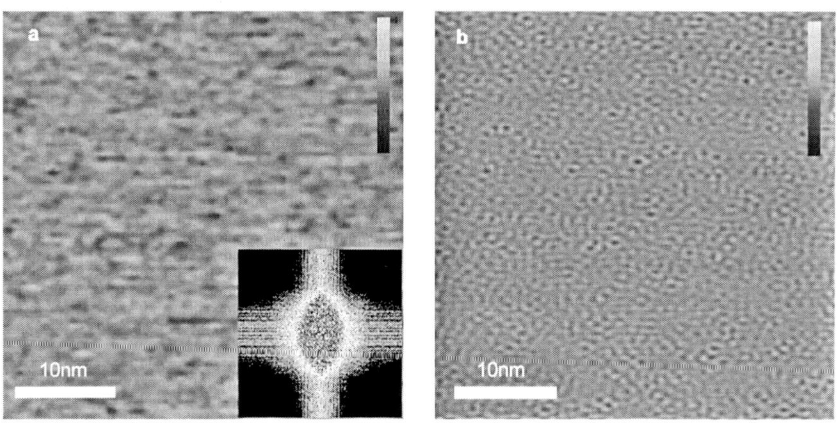

Fig. 6.13. STM images of a PhO-OTS monolayer: (a) shows the flatness of the monolayer, its high density and the absence of grain boundaries and defects. The inset in the bottom right corner is the Fourier transform of the image with a round broad peak for a spacing of ~6–8 Å. 5 Å is a typical spacing for benzene rings forming a herringbone structure in SAMs. The larger spacing is indicative of random orientation of the benzene rings in the monolayer. The absence of preferential directions in this peak proves that the monolayer is amorphous. (b) is the same image filtered by a band-pass filter collecting only the frequency components present in the ring peak shown in the inset. This image shows more clearly the organic "glassy" nature of the monolayer.

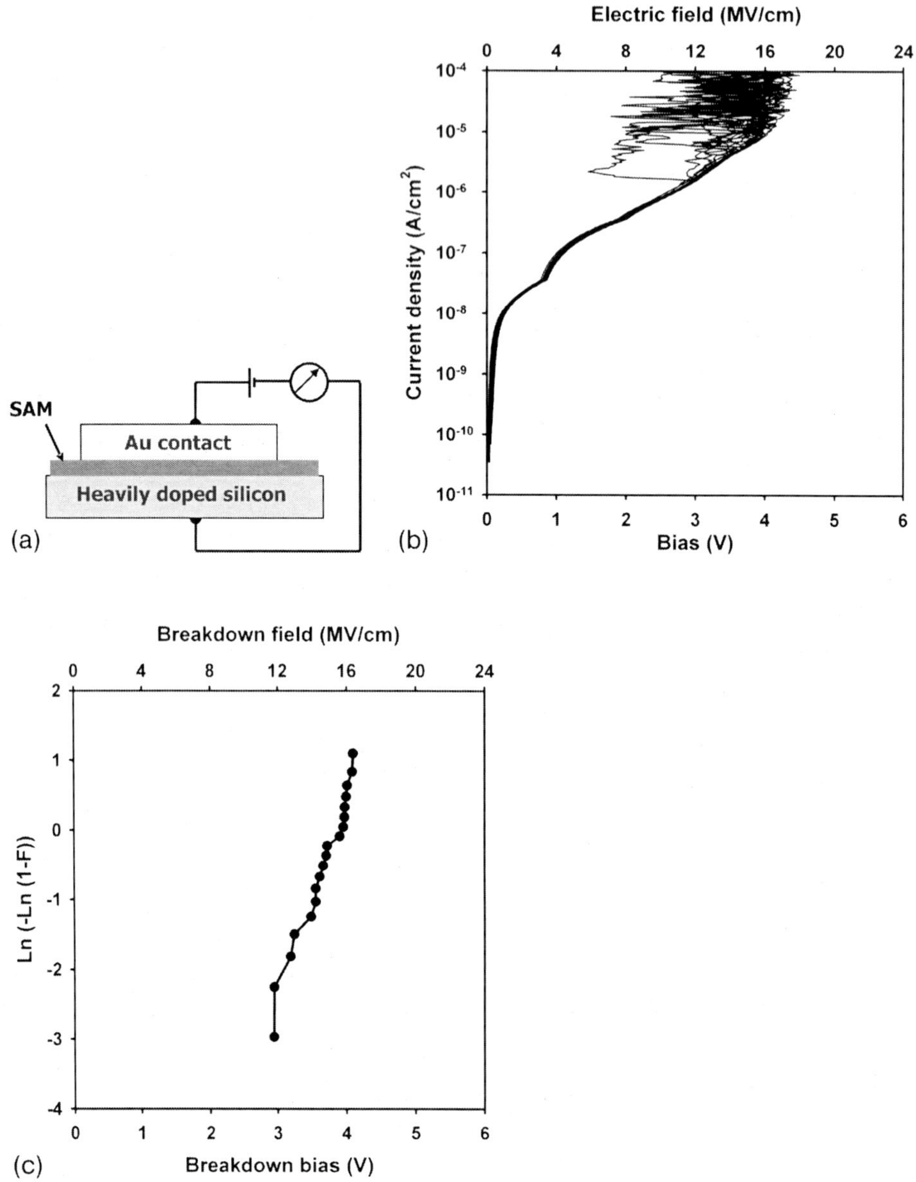

Fig. 6.14. Breakdown characteristics of PhO-OTS SAM capacitors. (a) Schematic cross section of a capacitor with PhO-OTS SAM dielectric. (b) Current density through the 2.5 nm thick molecular SAM dielectric in 20 devices. (c) Distribution of breakdown field for the 2.5 nm thick molecular SAM dielectric (20 devices).

contact area of 3×10^{-4} cm^2. At a potential of 1 V applied across the SAM (corresponding to an electric field of 4 MV cm^{-1}) a current density of $(8 \pm 1) \times 10^{-8}$ A cm^{-2} was measured. At 2.5 V the current density is $(8 \pm 0.5) \times 10^{-7}$ A cm^{-2}. Dielectric breakdown occurs at 14 ± 2 MV cm^{-1}. These results suggest that for very thin dielectrics, molecular monolayers may provide better performance than traditional inorganic oxides [70, 71].

Figure 6.15 shows the electrical characteristics of a pentacene TFT with top-contact configuration. The transistor has a carrier field-effect mobility of

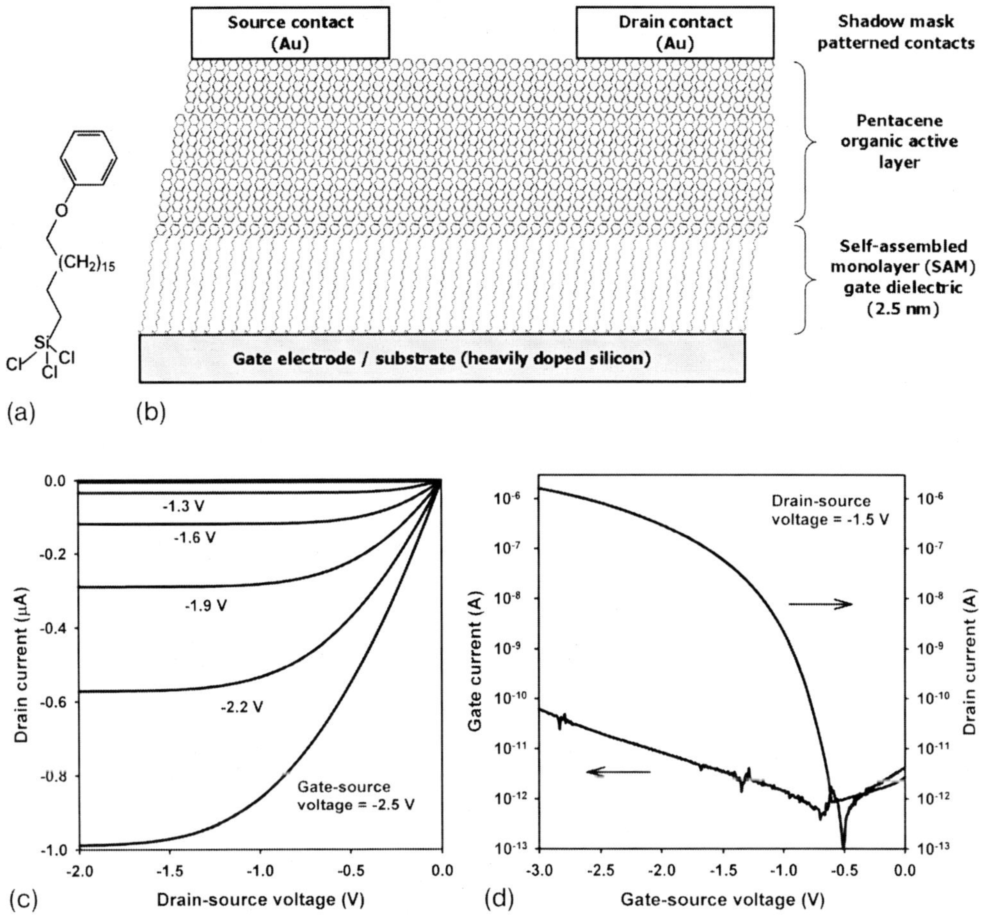

Fig. 6.15. (a) Chemical structure of (18-phenoxyoctadecyl)-trichlorosilane. (b) Cross-section of a pentacene TFT with SAM dielectric and source/drain contacts deposited through a shadow mask. (c) Output characteristics. (d) Transfer characteristics [50].

1 $cm^2\ V^{-1}\ s^{-1}$, a threshold voltage of −1.3 V, an on/off current ratio of 10^6, and a subthreshold swing of 100 mV $decade^{-1}$. For gate voltages larger than approximately 1 V the drain current exceeds the gate current by a factor of 4×10^4, confirming the excellent gate insulation provided by the SAM dielectric layer. The TFTs consistently have a negative threshold voltage (i.e. the transistors are normally-off devices), which is an important prerequisite for the straightforward implementation of digital circuits with these TFTs [72]. The excellent device characteristics confirm the high quality of the SAM gate dielectric; this is also apparent from the absence of hysteresis in the TFT characteristics. (Hysteresis is frequently observed in organic TFTs and is often attributed to mobile charges in the gate dielectric.)

Integrating TFTs into circuits requires that all functional device layers can be patterned. To create gate electrodes that can be lithographically patterned a thin layer of aluminum, deposited by evaporation in vacuum, is suitable. The natively oxidized Al surface has a large density of functional groups for molecular self-assembly, which is performed from solution or from the vapor phase with adapted molecules. Contact angle analysis reveals a large decrease in surface energy on self-assembly, confirming the formation of a dense monolayer. To measure the breakdown voltage of the 2.5 nm thick monolayers, capacitors with Al bottom electrodes, SAM dielectric, and Au top electrodes were prepared. The top electrodes were patterned either by thermal evaporation through a shadow mask (20 capacitors; electrode area $3 \times 10^{-4}\ cm^2$) or by use of photolithography and wet etching (20 capacitors; $8 \times 10^{-5}\ cm^2$). The results indicate that the impact of the wet chemical patterning process on the integrity of the molecular dielectric is insignificant. At an electric field of 10 MV cm^{-1} the current density is only 1 μA cm^{-2} and breakdown occurs at 17 ± 1 MV cm^{-1} [73]. For many applications it is also necessary to open vias in the dielectric and to pattern the organic semiconductor layer. Creating vias in the SAMs can accomplished by using a combination of standard photolithography and O_2 plasma etching or by defining conducting pattern (the later interconnects) from metals which will not be covered by the SAM, by using the substrate selectivity of the anchor group. Because of the larger contact resistance of the coplanar structure [14], the TFT has a slightly lower mobility of 0.15 $cm^2\ V^{-1}\ s^{-1}$, a threshold voltage of −1.4 V, an on/off ratio of 10^5, and a subthreshold swing of 110 mV $decade^{-1}$. Owing to the low volume and interface defect density of the molecular gate dielectric, the hysteresis in the TFT characteristics is less than 100 mV, much lower than for TFTs with a polymer gate dielectric (Fig. 6.16).

Building on coplanar TFTs with vias and patterned pentacene layer, inverters with saturated load, and five-stage ring oscillators with output buffer, is shown in Fig. 6.17. These circuits were made on glass substrates and on flexible, transparent poly(ethylene naphthalate) (PEN) film. Ring oscillators on glass operate with supply potentials as low as 1.5 V (Fig. 6.17b), and ring oscillators on PEN with supply voltage as low as 2 V (Fig. 6.17c). Because the supply potential never drops entirely across the gate dielectric, the circuits sustain supply potentials that are greater than the breakdown voltage of the SAM. At a supply potential of 6 V a signal delay of 230 μs per stage for ring oscillators on glass was measured. TFTs on

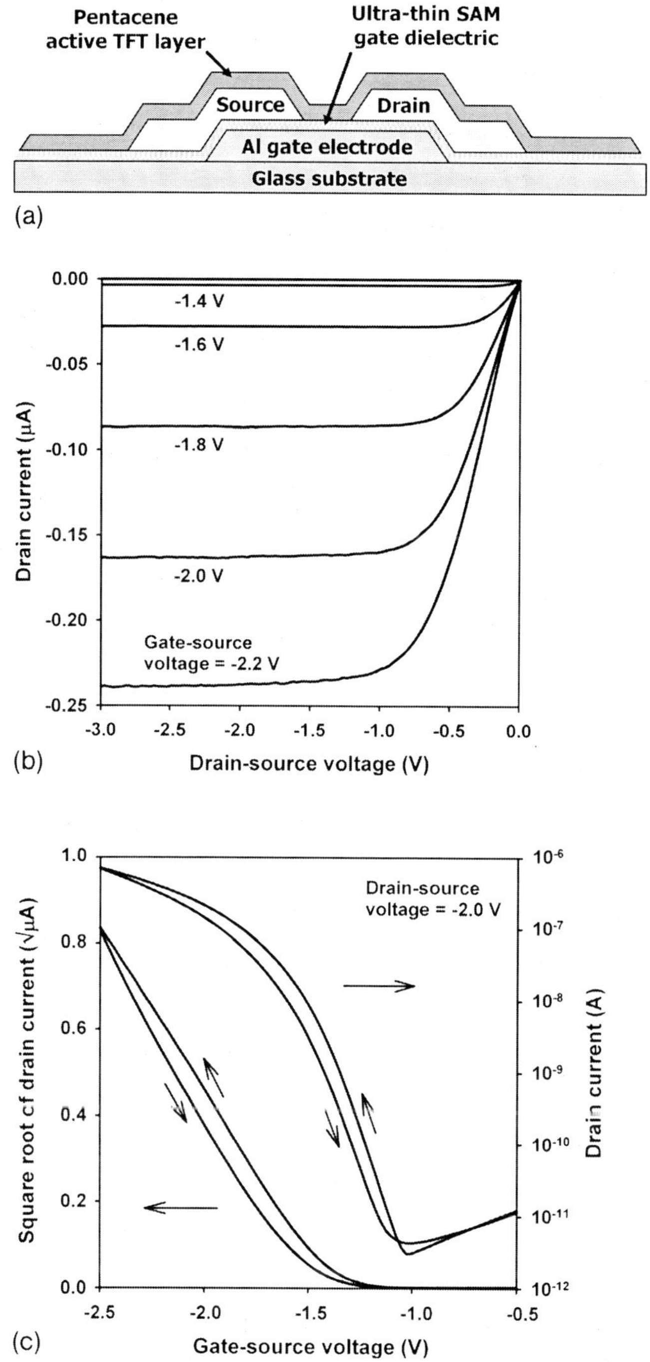

Fig. 6.16. (a) Cross-section of a coplanar pentacene TFT with patterned Al gate and SAM dielectric on a glass substrate. The mobility is 0.15 $cm^2\ V^{-1}\ s^{-1}$ and the hysteresis is less than 100 mV. (b) Output and (c) transfer characteristics.

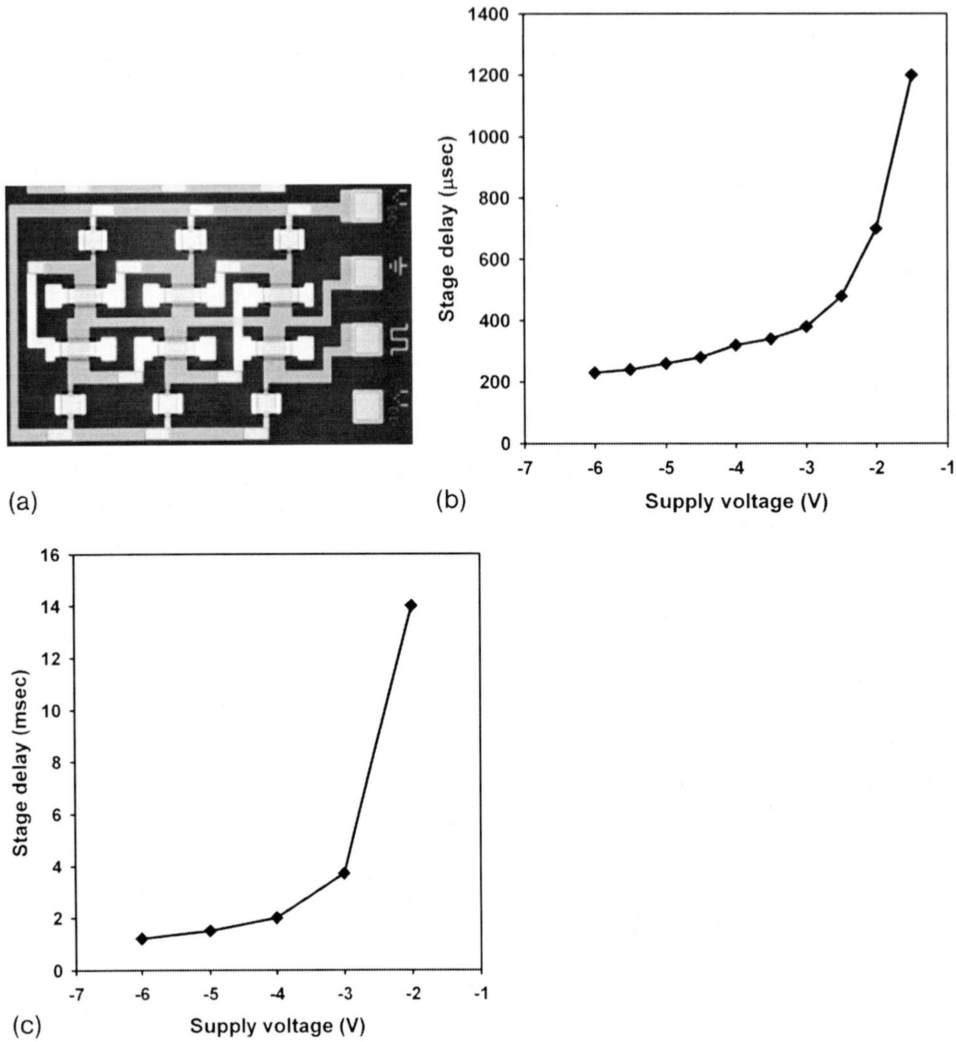

Fig. 6.17. (a) Photograph of a five-stage ring oscillators with output buffer with SAM gate dielectric on glass substrate. (b) Signal delay as a function of supply potential for glass substrate. (c) Signal delay as a function of supply potential for PEN substrate.

flexible PEN have somewhat lower mobility than TFTs on glass, and the lowest stage delay measured on PEN was 1.2 ms.

The device yield of TFTs with molecular gate dielectrics would be expected to be low, because of defects in the thin self-assembled monolayer arising from the imperfect surface of the evaporated aluminum bottom electrode. Results obtained on capacitor devices (20 devices with 3×10^{-4} cm^2 area each) indicate, however, that the self-assembly process can be used to cover even large areas with pinhole-free layers. A further indication is the function of TFT test arrays with 3000 TFTs with patterned Al gates, SAM dielectric, and patterned pentacene active layer all wired in parallel [73].

References

1 J.A. Rogers et al. *Proc. Natl. Acad. Sci. USA* **2001**, *98*, 4835–4840.
2 C.D. Sheraw et al. *Appl. Phys. Lett.* **2002**, *80*, 1088–1090.
3 H.E.A. Huitema et al. *Nature* **2001**, *414*, 599.
4 B. Crone et al. *Nature* **2000**, *403*, 521–523.
5 D.M. de Leeuw, D.M. et al. *IEDM – Technical Digest* **2002**, 293–296.
6 P. Baude et al. *Appl. Phys. Lett.* **2003**, *82*, 3964–3966.
7 L. Wang et al. *Appl. Phys. Lett.* **2004**, *85*, 1772–1774.
8 L.-L. Chua et al. *Nature* **2005**, *434*, 194–199.
9 W. Fix et al. *Appl. Phys. Lett.* **2002**, *81*, 1735–1737.
10 H. Klauk et al. *IEDM – Technical Digest* **2002**, 557–560.
11 G.H. Gelinck et al. *Appl. Phys. Lett.* **2000**, *77*, 1487–1498.
12 C. Reese et al. *Materials Today* (Elsevier) **2004**, 20–27.
13 M. Halik et al. *Adv. Mater.* **2003**, *15*, 917–922.
14 P.V. Necliudov et al. *J. Appl. Phys.* **2000**, *88*, 6594.
15 H. Klauk et al. *Solid-State Electronics* **2003**, *47*, 297–301.
16 A. Tsumura et al. *Appl. Phys. Lett.* **1986**, *49*, 1210.
17 H.E. Katz et al. *Acc. Chem. Res.* **2001**, *34*, 359–369.
18 C.D. Dimitrakopoulos et al. *IBM J. Res. Dev.* **2001**, *45*, 11–27.
19 C.J. Drury et al. *Appl. Phys. Lett.* **1998**, *73*, 108.
20 H. Sirringhaus et al. *Science* **2000**, *290*, 2123.
21 T. Kawase et al. *Adv. Mater.* **2001**, *13*, 1601.
22 F. Garnier et al. *Science* **1994**, *265*, 1684.
23 M. Halik et al. *Adv. Mater.* **2002**, *14*, 1717–1722.
24 F. Eder et al. *Appl. Phys. Lett.* **2004**, *84*, 2673–2675.
25 H. Klauk et al. *J. Appl. Phys.* **2002**, *92*, 5259–5263.
26 D.J. Gundlach et al. *IEEE Electron Device Lett.* **1997**, *18*, 87.
27 A.F. Strassen et al. *Appl. Phys. Lett.* **2004**, *85*, 3899–3901.
28 Y.Y. Lin et al. *IEEE Electron Device Lett.* **1997**, 18, 606.
29 C.D. Sheraw et al. *IEDM – Technical Digest* **2000**, 619.
30 H. Klauk et al. *IEEE Transactions on Electron Devices* **1999**, *46*, 1258–1263.
31 T.W. Kelly et al. *MRS Symp. Proc.* **2003**, *771*, 169–179.
32 B.K. Crone et al. *J. Appl. Phys.* **2001**, *89*, 5125.
33 H.E.A. Huitema et al. *Adv. Mater.* **2002**, *14*, 1201.
34 D. Knipp et al. *MRS Symp. Proc.* **2001**, 553.
35 F.-C. Chen et al. *Appl. Phys. Lett.* **2004**, *85*, 3295–3297.
36 L.A. Majewski et al. *Adv. Mater.* **2005**, *17*, 192–196.
37 C.D. Dimitrakopoulos et al. *Science* **1999**, *283*, 822–824.
38 R. Schroeder et al. *Appl. Phys. Lett.* **2003**, *83*, 3210–3212.
39 L.A. Majewski et al. *J. Appl. Phys.* **2004**, *96*, 5781–5787.
40 L.A. Majewski et al. *J. Phys. D: Appl. Phys.* **2004**, *37*, 21–24.
41 J.A. Rogers et al. *IEEE Electron Device Lett.* **2000**, *21*, 100.
42 Y. Kato et al. *Appl. Phys. Lett.* **2004**, *84*, 3789–3791.

43 Severt et al. *Chem. Mater.* **1994**, *6*, 1809–1815.
44 R. Parashkov et al. *J. Appl. Phys.* **2004**, *95*, 1594–1596.
45 Z. Bao et al. *Adv. Funct. Mater.* **2002**, *12*, 526–531.
46 L.-L. Chua et al. *Appl. Phys. Lett.* **2004**, *84*, 3400–3402.
47 C.D. Dimitrakopoulos et al. *Synth. Metals* **1998**, *92*, 47.
48 T. Yasuda et al. *Jpn. J. Appl. Phys.* **2003**, *42*, 6614–6618.
49 I.M. Rutenberg et al. *J. Am. Chem. Soc.* **2004**, *126*, 4062–4063.
50 M. Halik et al. *Nature* **2004**, *431*, 963–966.
51 J. Collet et al. *Appl. Phys. Lett.* **1998**, *73*, 2681–2683.
52 J. Collet et al. *Appl. Phys. Lett.* **2000**, *76*, 1941–1943.
53 H. Klauk et al. *IEDM – Technical Digest* **2004**, 369–372.
54 S. Kobayashi et al. *Nature Materials* **2004**, *3*, 317–322.
55 K.P. Pernstich et al. *J. Appl. Phys.* **2004**, *96*, 6431–6438.
56 J. Park et al. *Appl. Phys. Lett.* **2004**, *85*, 2283–2285.
57 R. Schroeder et al. *IEEE Electron Device Lett.* **2005**, *26*, 69–71.
58 M. Halik et al. *Appl. Phys. Lett.* **2002**, *81*, 289–291.
59 M. Halik, (unpublished results).
60 H. Klauk et al. *IEEE Transactions on Electron Devices* **2005**, *52*, 618–622.
61 H. Klauk et al. *Appl. Phys. Lett.* **2003**, *82*, 4175–4177.
62 G.H. Gelink et al. *Nature Materials* **2004**, *3*, 106–110.
63 B. Xu et al. *Science* **2003**, *301*, 1221–1223.
64 A. Kumar et al. *Langmuir* **1994**, *10*, 1498–1511.
65 P.E. Laibinis et al. *J. Am. Chem. Soc.* **1991**, *113*, 7152–7167.
66 Y.-T. Tao et al. *J. Am. Chem. Soc.* **1993**, *115*, 4350–4358.
67 A. Ulman et al. *Chem. Rev.* **1996**, *96*, 1533–1554.
68 C. Boulas et al. *Phys. Rev. Lett.* **1996**, *76*, 4797–4800.
69 M.-H. Yoon et al. *Proceedings of the National Academy of Sciences* **2005**, *102*, 4678–4682.
70 K. Sekine et al. *IEEE Trans. Electr. Dev.* **2000**, *48*, 1550–1555.
71 S.H. Lo et al. *IEEE Electr. Dev. Lett.* **1997**, *18*, 209–211.
72 H. Klauk et al. *2003 IEDM – Techn. Dig.* **2003**, 195.
73 H. Klauk et al. *2004 IEDM – Techn. Digest* **2004**, 369.

7
Advanced Flexible Polymeric Substrates

William A. MacDonald

7.1
Introduction

There is substantial interest in flexible electronics, and many electronics-based companies and a range of new start-up companies are actively conducting research into flexible electronics based on organic materials. Flexible electronics promises substantial rewards as a route to the development of displays that are thin, light, robust, conformable, and can, if required, be rolled away when not in use. In addition, the use of plastic-based substrates, coupled with recent developments in solution deposition and inkjet printing for laying down organic light emitting polymers and active matrix thin film transistor arrays, opens up the possibility of cost-effective processing in high volumes by use of roll-to-roll processing. When this area of technology first began to emerge, the substrate on which the electronic structures were being applied was not regarded as very important. This may have been acceptable for demonstrating the potential of the technology, but as this technology has moved closer to prototype devices and commercial production processes it has become clear that to make this technology viable, choice of substrates with the required set of properties is essential. This is also a very wide field covering "simple" organic circuitry through to active matrix backplanes for OLED displays. The requirements for the different applications are very different and will require substrates with different sets of properties. This review will concentrate on biaxially oriented polyester films and will cover the properties of these materials in the context of the set of properties required for printed organic circuitry.

7.2
Polyester Substrates

Poly(ethylene terephthalate) (PET), e.g. DuPont Teijin Films Melinex® polyester film, and poly(ethylene naphthalate) (PEN), e.g. DuPont Teijin Films Teonex® polyester film, are biaxially oriented semicrystalline films [1]. The chemical structures of PET and PEN are shown in Fig. 7.1.

Organic Electronics, Materials, Manufacturing and Applications. Edited by Hagen Klauk

ISBN: 3-527-31264-1

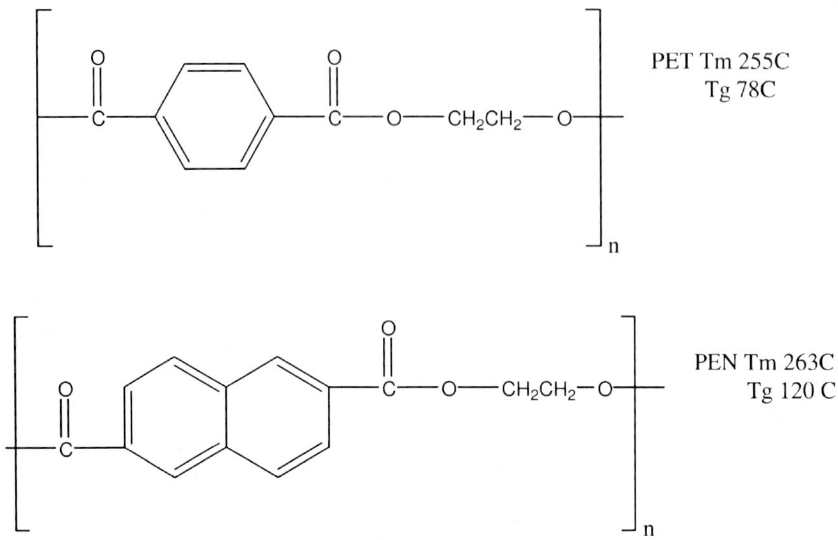

Fig. 7.1. Chemical structure of PET and PEN films.

Substitution of the phenyl ring of PET by the naphthalene double ring of PEN has very little effect on the melting point (*Tm*), which increases by only a few degrees but a substantial effect on the glass transition temperature (*Tg*, the temperature at which the polymer changes from a rigid glass to a rubber or polymer melt and the polymer molecules start to have significant mobility), which increases from 78 °C for PET to 120 °C for PEN [2]. PET and PEN films are prepared by a process in which the amorphous cast is drawn in both the machine direction and the transverse direction. The biaxially oriented film is then heat set to crystallize the film [3, 4].

The success of polyester film in general applications is because of properties derived from the basic polymer coupled with the manufacturing process of biaxial orientation and heat setting, described above. These properties include high mechanical strength, good resistance to a wide range of chemicals and solvents, low water absorption, excellent dielectric properties, good dimensional stability, and good thermal resistance in terms of shrinkage and degradation of the polymer chains. Fillers can be incorporated into the polymer to change the surface topography and opacity of the film. The film surface can also be altered by pretreatment to give a further range of properties, including enhanced adhesion to a wide range of inks, lacquers, and adhesives. These basic properties have resulted in PET films being used in a wide range of applications, ranging from magnetic media and photographic applications, in which optical properties and excellent cleanliness are of paramount importance, to electronics applications such as flexible circuitry and touch switches, for which thermal stability is key. More demanding polyester film markets, which exploit the higher performance and benefits of PEN include magnetic media for high-density data storage and electronic circuitry for hydrolysis-

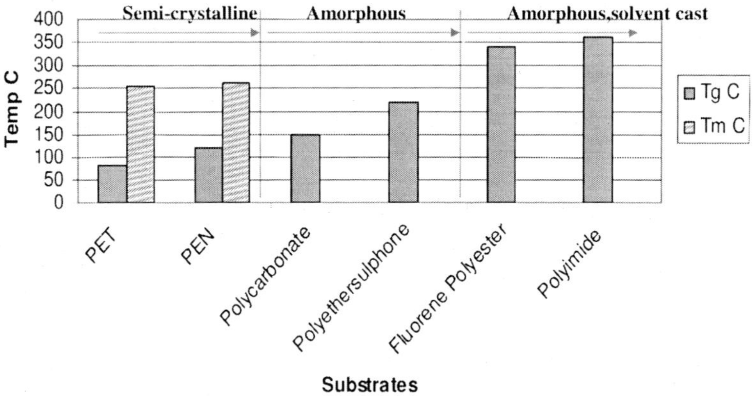

Fig. 7.2. Glass transition of film substrates of interest for applications in flexible electronics.

resistant automotive wiring. This set of properties provides the basis on which we can now build to meet the demands of the organic electronics market.

It is interesting to contrast these films with the other films being considered for flexible electronics especially for the higher performance flexible display market. The main candidates are shown in Fig. 7.2 which lists the substrates on the basis of increasing glass transition temperature (*Tg*) [5, 6].

The polymers can be further categorized into films that are semi-crystalline (PET and PEN as mentioned above), amorphous and thermoplastic, and amorphous but solvent-cast. Polymers with *Tg* values higher than 140 °C that are semi-crystalline tend to have melting points too high to enable the polymers to be melt processed without significant degradation – PEN is the highest-performance material commercially available as a biaxially oriented semicrystalline film. The next category includes polymers that are thermoplastic but noncrystalline; these range from polycarbonate (PC), e.g. Teijin's Pure-Ace® [7] and GE's Lexan® [8], with *Tg* of ~150 °C, to polyethersulfone (PES), e.g. Sumitomo Bakelite's Sumilite® [9], with a *Tg* of ~220 °C. Although thermoplastic, these polymers may also be solvent-cast, resulting in high optical clarity. The third category are high-*Tg* materials that cannot be melt processed; these include aromatic fluorene-containing polyarylates (PAR), e.g. Ferranias Arylite® [10], and polyimide (PI), e.g. DuPont's Kapton® [11].

7.3 Properties of Base Substrates

7.3.1 Optical Properties

The clarity of the film is important for bottom-emissive displays, when one is viewing through the film. A total light transmission (TLT) of >85% over 400–800 nm coupled with haze of less than 0.7% are typical requirements for this application – both PET and PEN films meet this requirement. For top emissive displays and for some nondisplay printed-circuitry applications, clarity in the base substrate is not essential.

7.3.2 Birefringence

Biaxially oriented films such as PET and PEN are birefringent. For LC displays which depend on light of known polarization this means that birefringent films, which would change the polarization state, are unlikely to be used as substrates. Films based on amorphous polymer are not birefringent and are more suitable for LC displays. Birefringence is not an issue with OLED, electrophoretic displays, or, indeed, some LC displays.

7.3.3 Thermal Properties

Dimensional and thermal stability and reproducibility are critical, for withstanding the high temperatures of deposition of ITO coatings or any barrier coatings and the curing of dielectric coatings, to ensure attachment of the different layers in the final device, and for the multilayer device to be able to withstand thermal cycling. This is manifested in two ways. First, the shrinkage a film undergoes when heated and then cooled to the starting temperature. Low levels of shrinkage are desired for accurate alignment on the substrates after each thermal processing step. The dimensional stability of both PET and PEN films can be enhanced by a heat-stabilization process in which the internal strain in the film is relaxed by exposure to high temperature while under minimum line tension [2, 5, 6, 12]. Both films are dimensionally reproducible up to the temperature at which they are heat stabilized and this capacity to undergo a heat stabilization process above the *Tg* is unique to the biaxially oriented crystalline films. The *Tg* process does not define their upper processing temperature, as it does for amorphous polymers. For PET this is typically 150 °C and for PEN it is 180–220 °C. Below these temperatures shrinkage of both films is of the order of 200–500 ppm.

The second factor that affects dimensional reproducibility is the natural expansion of the film as the temperature is cycled, as measured by the coefficient of linear thermal expansion (CLTE). (The CLTE defines an increase in length along one given axis – it is often abbreviated to CTE.) A low CLTE typically < 20 ppm per degree is desirable to match the thermal expansion of the base film to the layers which are subsequently deposited. A mismatch in thermal expansion means that the deposited layers become strained and cracked as a consequence of thermal cycling. In the temperature range from room temperature to the *Tg* the typical CLTE of PEN and PET are 18–20 ppm and 20–25 ppm, respectively (but note that the *Tg* of PEN is 40 °C higher than for PET). Above the *Tg*, the natural expansion of PET and PEN films which have not been stabilized is dominated by the shrinkage

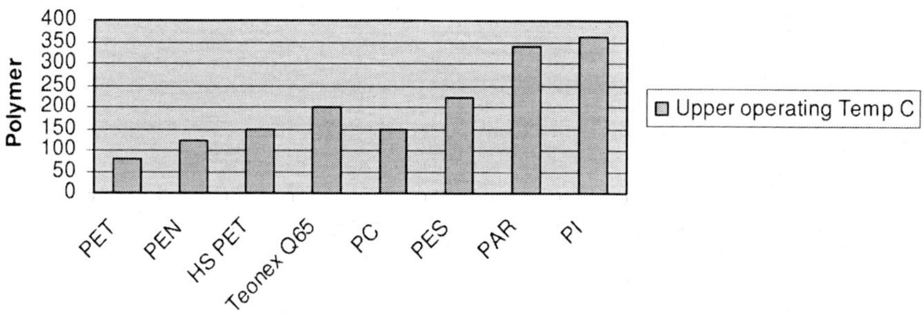

Fig. 7.3. Upper processing temperature of film substrates of interest for flexible electronics applications.

the films undergo as the internal strains in the film relax, as discussed above. For the heat-stabilized films discussed above, however, only a very small increase in CTLE is observed in the temperature range from the *Tg* to the temperature at which they were heat-stabilized. This contrasts favorably with the quoted coefficient of expansion of amorphous polymers which is typically 50 ppm per degree [6] below the *Tg* but can increase by a factor of three above the *Tg*.

In addition to dimensional stability another important factor to be considered is the upper processing temperature (T_{max}) at which a film can be used. Although, as outlined above, the *Tg* does not define T_{max} for semi-crystalline polymers, it largely does for amorphous polymers. Figure 7.3 shows the upper operating temperature if the effect of heat stabilization is taken into account.

7.3.4 Solvent Resistance

A wide range of solvents and chemicals can potentially be used when laying down the different layers in a display, depending on the processing steps involved. Amorphous polymers usually have poor solvent resistance compared with semi-crystalline polymers (Table 7.1).

This deficiency is overcome by application of a hard coat to the amorphous resins; this substantially improves resistance to the solvents and chemicals such as NMP, IPA, acetone, methanol, THF, ethyl acetate, 98% sulfuric acid, glacial acetic acid, 30% hydrogen peroxide, and saturated bases such as sodium hydroxide [13]. With poly(ethylene terephthalate) and of poly(ethylene naphthalate) films a hard coat is not required for solvent resistance.

The residual shrinkage of TeonexQ65 after 30 min at 150 °C is of the order of 500 ppm but can be reduced to below 200 ppm or better by careful process control. Dimensional reproducibility down to 25 ppm is being requested for the more demanding applications such as inorganic AM backplanes on flexible substrates and it has been shown that it is possible to achieve this level of shrinkage with Teo-

Tab. 7.1. Table of solvent resistance.

	Unit	Q65	PET	PC	PES
Ketone	Acetone	Good	Good	(Fair)	(NG)
	MEK	Good	Good	(Fair)	(–)
Alcohol	Methanol	Good	Good	(Good)	(–)
	Ethanol	Good	Good	(Good)	(Good)
	Isopropanol	Good	Good	(Good)	(–)
	Butanol	Good	Good	(Good)	(–)
Ester	Ethyl acetate	Good	Good	(Fair)	(Good)
Hydrocarbon	Formalin	Good	Good	(NG)	(–)
	Tetrachloroethane	Good	Good	(–)	(Good)
Acid	10% HCl	Good	Good	(–)	(Good)
	10% HNO_3	Good	Good	(–)	(Good)
	10% H_2SO_4	Good	Good	(–)	(Good)
	Acetic acid	Good	Good	(–)	(–)
Alkali	10% NaOH	Good	Fair	(–)	(Good)

nexQ65 [14]. In addition to this residual shrinkage, the processing environment, in effect the prevailing humidity, must be taken into account, because even crystalline TeonexQ65 will absorb up to 1500 ppm moisture under ambient conditions, yet readily lose this at higher processing temperatures. With a knowledge of the solubility of moisture in PEN film and its rate of diffusion as a function of temperature it is possible to model the impact of such environmental conditions on volumetric changes in the film. Figure 7.4 shows the dependence of moisture loss on time as 125-μm film is heated and held at 90, 100, 120, and 150 °C at an RH of 40% (at 20 °C).

At 150 °C the film takes 6 min to reach an equilibrium moisture level of 5 ppm but at 90 °C the film takes 30 min to reach an equilibrium level of 40 ppm.

On removal from the heated environment and return to ambient conditions, the film initially picks up to 50% of its equilibrium moisture level within a few minutes while cooling but the remaining moisture to reach equilibrium is picked up much more slowly. Films can typically take over 12 h to reach equilibrium. Figure 7.5 shows the effect of RH on this process and indicates that final (equilibrium) moisture levels will change significantly with different RH.

At an RH of 20% the film reaches an equilibrium level of ca 500 ppm but at 60% RH the equilibrium level is ca 1500 ppm.

Film thickness will also affect the rate of moisture pickup. Films of thickness up to 150 ppm take 12 h to reach equilibrium whereas films of thickness 175, 200, and 250 μm have not reached equilibrium in 15 h. Films of lower thickness

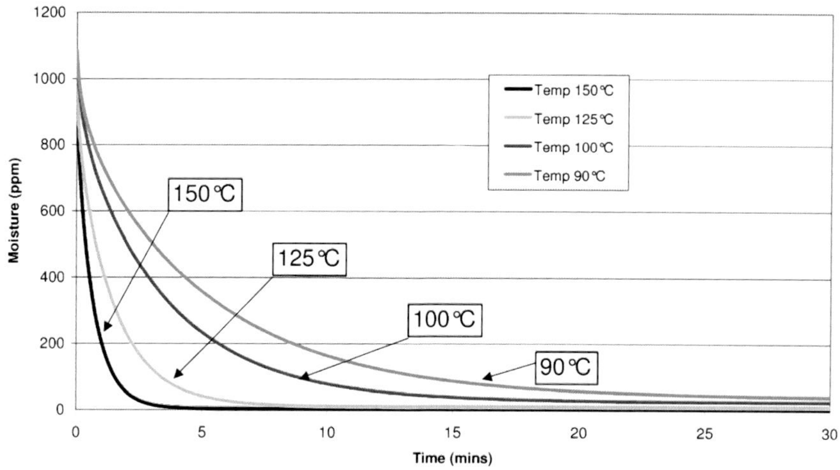

Fig. 7.4. Dependence of moisture loss on time for heating to different temperatures at RH 40% (at 20 °C).

absorb moisture more rapidly and reach equilibrium more quickly at elevated temperatures – this is apparent from Fig. 7.6 which shows the drying and subsequent moisture pickup of films from 100 μm to 200 μm at 25-μm intervals.

Measurements in our laboratory indicate that the film increases in length in each axis by approximately 44 ppm per 100 ppm of moisture absorbed. This is what would be predicted for such small changes in dimensions from a simple vol-

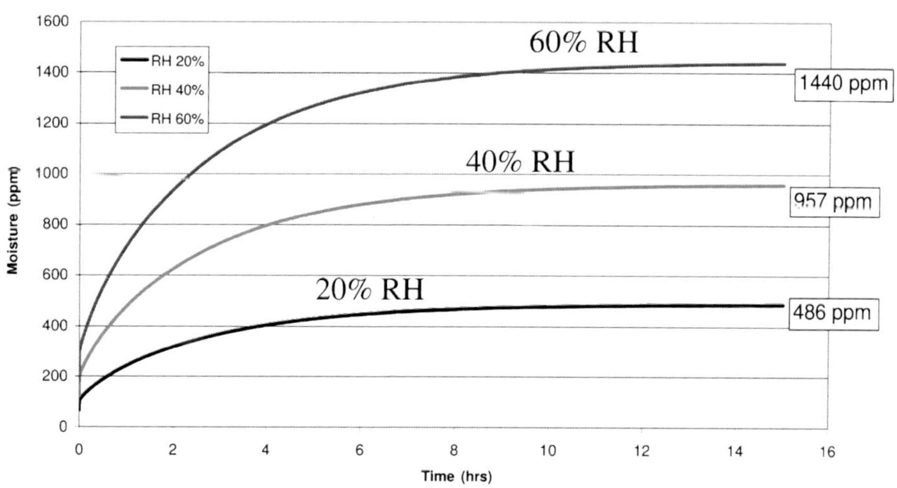

Fig. 7.5. Effect of RH on moisture pickup at 23 °C.

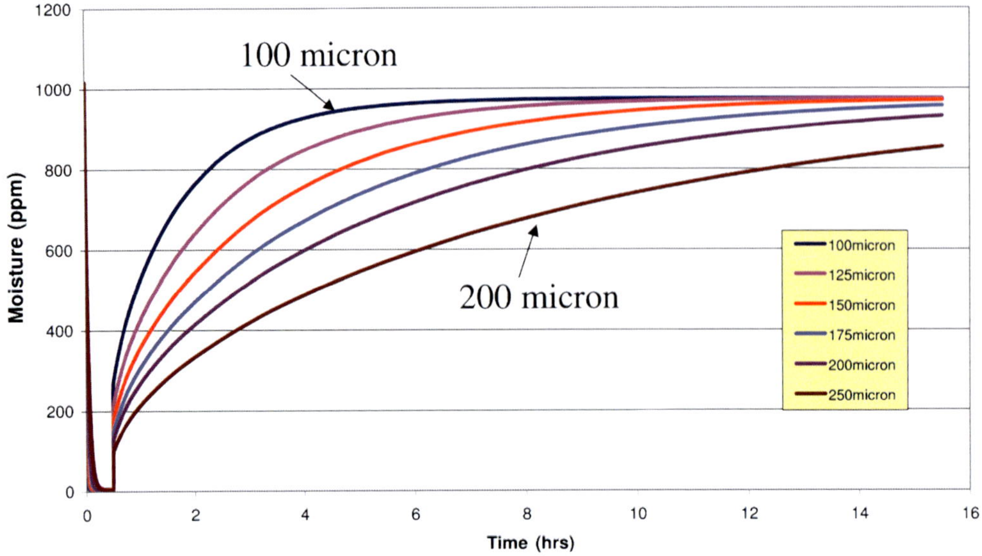

Fig. 7.6. Effect of film thickness on moisture pickup at 23 °C and 40% RH.

ume calculation. Taking into account the variation in moisture content that can result, depending on temperature, RH, and film thickness, which can be of the order of several hundreds of ppm, it can be seen that uncontrolled moisture pickup will have a significant effect on dimensional reproducibility. This will be particularly significant at lower processing temperatures, at which film shrinkage is minimum, but at which the film takes longer to reach equilibrium moisture level than the time at the temperature used to perform the processing step. By carrying out modeling studies of the type described above the effect of this moisture change, once understood, can be minimized.

PET and PEN films have an inherent advantage over the amorphous polymers films being both semi-crystalline and biaxially oriented and typically absorbing approximately 1400 ppm moisture at equilibrium (the exact figure depending on temperature and relative humidity). Polyethersulfone and polyimide films are particularly problematic and will absorb over 1% moisture at equilibrium.

7.3.5
Surface Quality

Surface smoothness and cleanliness of the flexible substrates are both essential to ensure the integrity of subsequent layers such as conductive coatings, TFT arrays, or barrier coatings; these properties can be considered separately.

PET and PEN have an inherent surface smoothness of less than 1 nm, in terms of both Roughness Average (*Ra*) and Root Mean Square Roughness (*Rq*), using

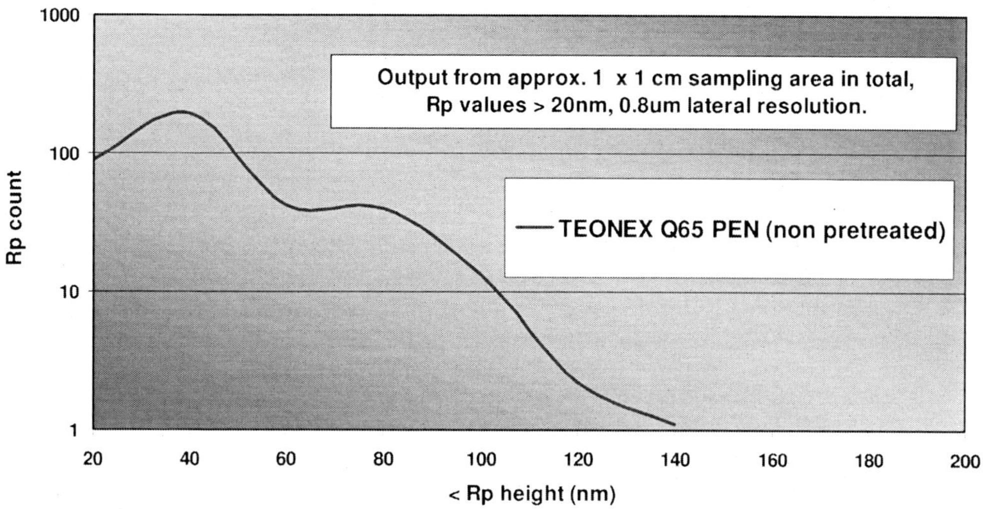

Fig. 7.7. Extreme surface peak 'Rp' (highest point) – frequency distribution for Teonex Q65 125 μm PEN.

white light interferometric methods of measurement. The surface smoothness depends, however, on the extent of "internal" contaminants such as catalyst residues, polymer degradants etc. Typically over a 1 cm^2 sample area these internal contaminants lead to a range of surface peaks less than 0.2 μm high. The frequency of such surface peaks (*Rp* values) can be shown in the graph below (Fig. 7.7).

The graph comprises numerous measurements, which in total make up a 1 cm × 1 cm sampling area. For each measurement the highest point (Rp value), is recorded and the following distribution observed. The particles are predominantly below 140 nm. The predominance of these particles can be influenced by the polymer recipe and by process control to minimize degradation.

Surface cleanliness is dominated by external contaminants such as airborne debris, scratches, etc. The size range of these contaminants is typically from submicron to tens of microns, both laterally and vertically. Polyester film is not manufactured under clean-room conditions and typically rolls of film are slit to size – dust is present or generated in these processes. These external contaminants are the major cause of surface related issues in down stream processing, but the bulk of these contaminants can be removed by some form of surface cleaning.

In demanding applications, for example barriers for OLED displays, control of polymer smoothness and cleanliness by the routes mentioned above may not be sufficient. In such circumstances it is necessary to lay down a further coating under clean-room conditions which acts both as a planarizing coating and as a hard coat to prevent scratching on subsequent processing. This has been discussed elsewhere [6, 12, 13].

7.3.6
Mechanical Properties

Production of glass-based displays currently involves moving batches of glass between the different processing stages. The mechanical difference, in particular stiffness, between rigid glass and flexible substrates will probably require very different methods of processing. PET and PEN films are inherently stiffer than amorphous films; their Young's Modulus is typically three times higher – an artifact of being semi-crystalline and biaxially oriented [15]. Young's modulus is independent of thickness and does not indicate how stiffness will change with thickness. Another way of expressing this is by defining the rigidity by use of the equation:

$$D = \frac{Et^3}{12(1-n)}$$

where E is the tensile or Young's modulus, t is the thickness, and n is Poisson's ratio (0.3–0.4). Assuming a Young's modulus of 2 Gpa for amorphous films and 6 Gpa for PEN film it can be seen from Table 7.2 that 200 μm poly(ethylene naphthalate) film is four times more rigid than 125 μm poly(ethylene naphthalate) film and twelve times more rigid than a 125 μm amorphous film. This stiffness may prove to be an advantage in a batch-based display manufacturing process.

One of the main drivers in moving to plastic substrates is that it opens up the possibility of roll-to-roll processing and the process and economic advantages that this brings. Under these conditions a winding tension will clearly be present and polymer film substrates with low moduli will be susceptible to internal deformation, particularly at elevated process temperatures. Figure 7.8 shows a comparison between poly(ethylene terephthalate) and poly(ethylene naphthalate) films.

The storage modulus, E', is obtained by dynamic mechanical thermal analysis and as the temperature is increased, the stiffness of both materials is seen to fall. In the region 120–160 °C, however, PEN is significantly stiffer and stronger, with a modulus almost twice that of PET.

Tab. 7.2. Comparative rigidity of amorphous and semicrystalline films.

Material	Thickness (μm)	Rigidity (Nm × 10^{-4})	Rigidity relative to 125 μm amorphous film
Amorphous	125	5	1
Amorphous	200	20	4
TeonexQ65	125	15	3
TeonexQ65	200	61	12

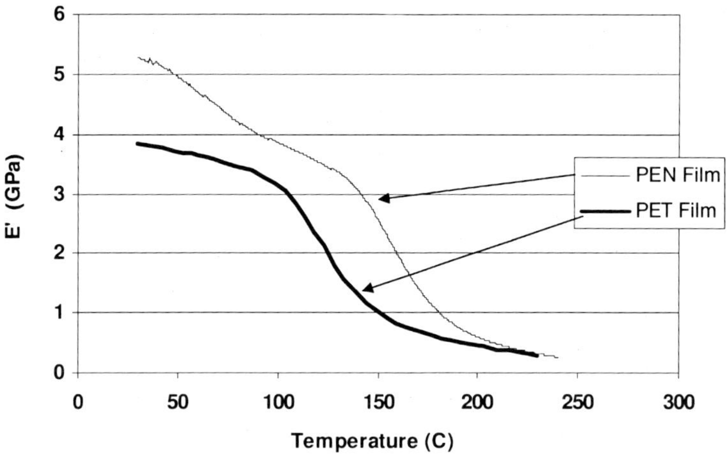

Fig. 7.8. Stiffness (E') of polyester film with temperature.

7.3.7 Summary of Key Properties of Base Substrates

The main properties of heat-stabilized PET and PEN relevant to flexible electronics are summarized in Fig. 7.9. Unstabilized PET and PEN films have the same set of

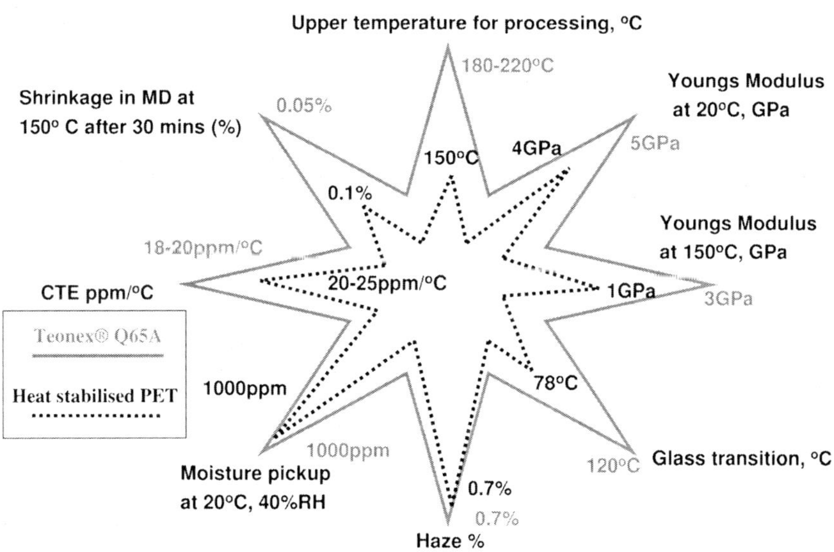

Fig. 7.9. Summary of properties of PET and PEN relevant to flexible electronics.

Tab. 7.3. Comparison of the key properties relevant to flexible electronics.

Property	Substrate					
	Heat-stabilized PEN	Heat-stabilized PET	PC	PES	Polyacrylate	Polyimide
CTE (−55 to 85 °C) ppm per degree	√√	√√	√	√	√	√√
% Transmission (400–700 nm)	√√	√√	√√√	√√	√√	×
Water absorption (%)	√√	√√	√	×	√	×
Young's modulus (Gpa)	√√	√√	√	√	√	√
Tensile strength (Mpa)	√√	√√	√	√	√	√
Solvent resistance	√√	√√	×	×	×	√√
Upper operating temperature	√√	√	√	√√	√√√	√√√

properties apart from the shrinkage and upper processing temperature, which for these films is largely dictated by the *Tg*. Table 7.3 contrasts the set of properties of the different substrates under consideration for flexible electronics.

These properties have been discussed in the text and elsewhere [5, 6]. This table shows that both heat-stabilized PET (e.g. Melinex ST504) and heat stabilized PEN (TeonexQ65A) have an excellent balance of the key properties required for flexible electronics. TeonexQ65A has a higher-temperature performance than Melinex (Fig. 7.9) and as a result of this set of properties TeonexQ65A is emerging as a leading material for the base substrate of OLED displays and active matrix backplanes.

7.4 Multilayer Structures

To replace glass, a plastic substrate must offer the properties of glass, i.e. clarity, dimensional stability, thermal stability, barrier, solvent resistance, low coefficient of thermal expansion, and a smooth surface. No plastic film has all these properties so any plastic based substrate will almost certainly be a multilayer composite structure. In addition to choosing the right materials for the different layers, one now has a new set of issues associated with the properties of multilayer structures. These issues include the adhesion of the different layers, the effect of thermal and environmental cycling, and the effect of flexing the structure, not only on specific properties such as barrier and conductivity, but also the robustness of the total structure. Understanding the impact of these effects and optimizing both the product structure and the processing steps involved in device manufacture will be critical to achieving the demanding property requirements being asked of substrates in flexible electronics-based applications..

The component layers in engineered substrates embrace a wide range of mechanical behavior. Polymeric substrate films are flexible and tough, whereas the active layers are soft organic polymers and the conducting and barrier layers are stiff and brittle inorganic materials and metals. The structures may be subjected to residual stresses from manufacture, differential thermal expansion, and bending during handling. There is a danger of damage in the layers resulting in a loss of function. The likelihood of failure in a layer depends on the arrangement, thickness, and properties of all layers.

Models of behavior based on mechanical properties, stack geometry, and mode of bending have been compiled from beam theory for linear isotropic elastic materials, and implemented for simple geometries by using spreadsheets. Bending tests subject the laminate to three-point or two-point bending. In each, tensile and compressive stresses are greatest at one particular location, and shear stresses develop throughout much of the laminate. Using the models as design tools will enable the structure to be modified to reduce the risk of failure. TeonexQ65 was used for this study. TeonexQ65 125 μm thick was coated after manufacture with a commercial-grade acrylate which was subsequently cross-linked in a thermal step. Two samples of laminate were manufactured, using the same TeonexQ65 substrate and coatings of thicknesses 1.6 and 2.4 μm.

An Instron extensometer was used to test the integrity of the coating under two different modes of deflection. In the first, specimens of laminate were stretched in uniaxial tension, subjecting both layers to the same tensile strain. In the second, the two-point bending test described above was performed. Samples were reexamined under optical magnification for evidence of fracture within the coating layer and for the amount of damage estimated by counting the density of cracks which had developed across the entire width of the specimen. Figures 7.10 and 7.11 show the results of simple extensional strain on the integrity of the planarizing coating on the polyester substrate.

Although the fracture strain of the polyester is known to be approximately 50%, the cross-linked coating was found to develop observable cracks at strain as low as 3%. Comparing Figs. 7.10 and 7.11 suggests the thicker coating is slightly more brittle but clearly neither material is ductile. Thus beyond 6% strain, which corresponds to crack separation distances less than 70 μm, crack density continues to increase with increasing strain.

The fracture stress, σ_f corresponding to failure at 3% strain is 150 Mpa. By applying laminate theory and the working spreadsheet model described earlier, the deformation required to reproduce the fracture stress under a different geometry can be easily calculated. Thus for the same laminate sample, the solution for a two-point bend deformation is that a plate separation of 6.5 mm will apply a strain of 3% and develop fracture stress in the coating layer.

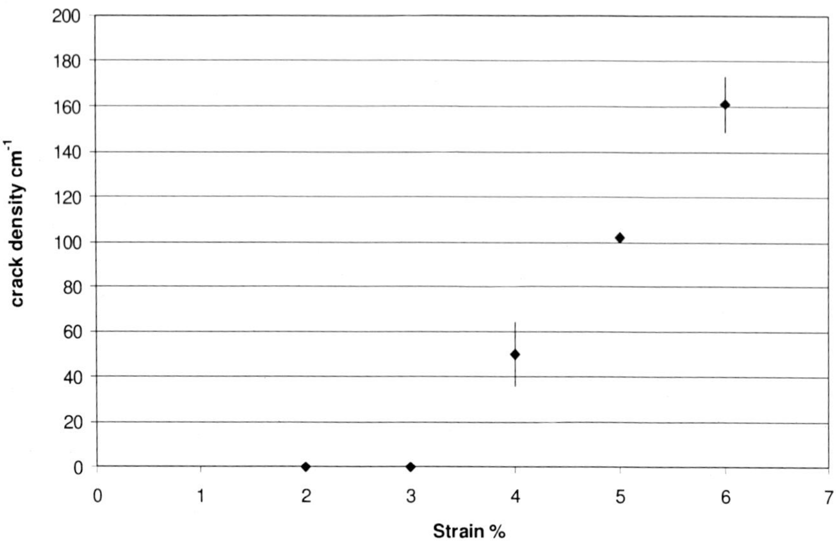

Fig. 7.10. Cracks versus % strain for a coat thickness of 1.6 μm.

Figure 7.12 shows the result of a fracture measurement using two-point bend geometry. The agreement between observed and predicted behavior is excellent – cracks began to develop in the coating layer at a plate separation between 6 and 7 mm and propagated across the width of the specimen. The point of maximum

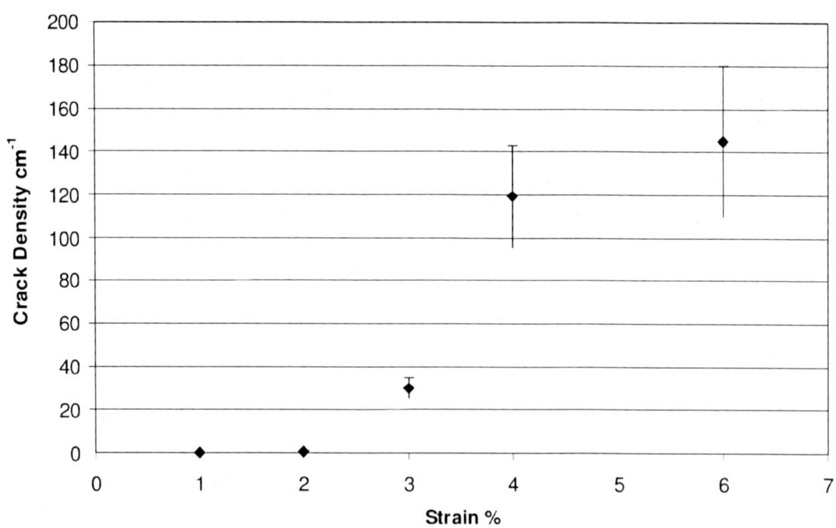

Fig. 7.11. Cracks versus % strain for a coat thickness of 2.4 μm.

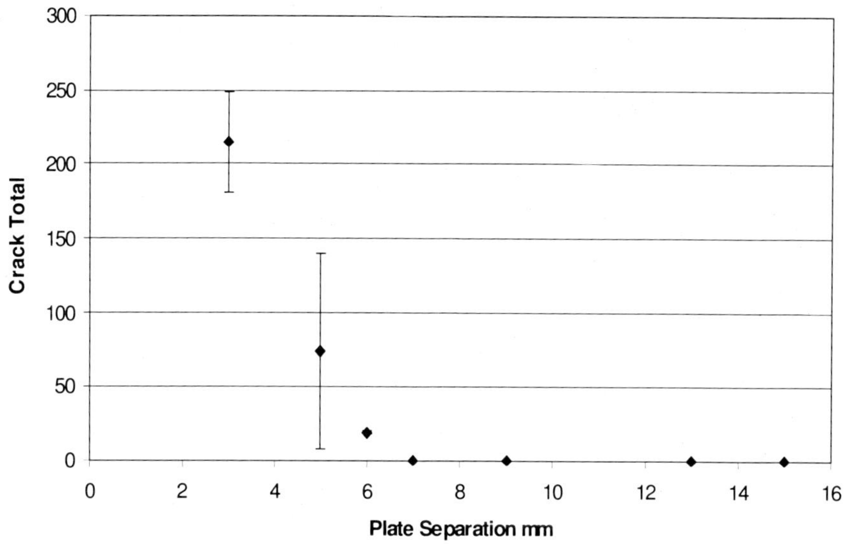

Fig. 7.12. Experimental measure of fracture under two-point bend.

stress and, therefore, maximum crack density is at the midpoint along the specimen length; because the measurable area of the specimen was small, however, it was not possible to resolve a distribution of crack density as a function of position along the specimen.

7.5 Film in Application

As discussed in the introduction requirements for the different applications envisaged for printable electronics are very different and will require substrates with different sets of properties. This is summarized in Fig. 7.13 and for this classification covers "simple" organic circuitry, e.g. RFID, organic based active matrix backplanes, OLED displays, but also includes the requirements of inorganic TFTs on flexible substrates.

The key properties used to illustrate this are smoothness, dimensional stability, barrier, and upper processing temperature. With "simple" organic-based printed circuitry for which processing temperatures are below 80 °C, surface cleanliness is an important factor but high-temperature performance is not required. Cost is likely to be a key consideration. For this application the thermal properties of PET film are adequate, although the surface smoothness and cleanliness requirements are such that special grades may be required. At the other extreme the film requirements for OLED displays, for which a flexible substrate with glass-like properties is required, are extremely demanding and the film structure will be more complex.

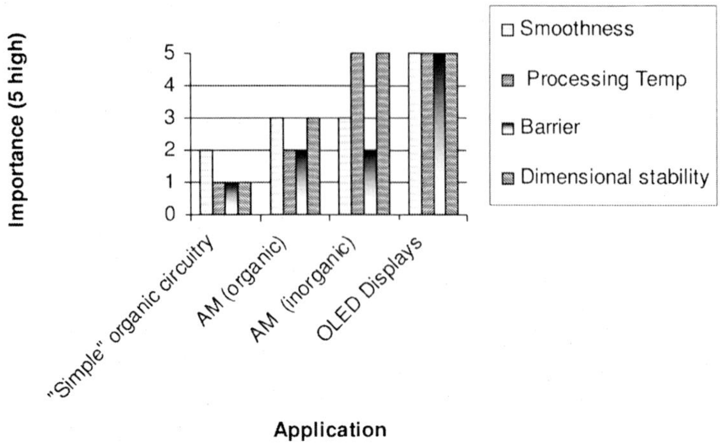

Fig. 7.13. Property requirements for different flexible electronics applications.

In this case the set of properties of heat-stabilized PEN (TeonexQ65A) is more appropriate. Organic-based printed AM backplanes sit between the performance requirements of simple organic circuitry and OLED displays, particularly with regard to temperature, and the choice of substrate will depend upon the choice of organic materials and also the processing route and conditions used. There is, therefore, no one film that will meet the requirements of all these applications and for the more demanding applications areas films may well be engineered for a specific application. The basic set of properties of PET and PEN films, their availability on a commercial scale, and the new developments discussed in this chapter means polyester films are the leading substrate candidates for applications based on organic electronics.

References

1 Melinex, Teonex, DuPont Teijin Films Luxembourg S.A., P.O. Box 1681, L-1016 Luxembourg.

2 W.A. MacDonald, J.M. Mace and N.P. Polack, New developments in polyester films for display applications, 45th Annual Technical Conference Proceedings of the Society of Vacuum Coaters, 2002, 482.

3 "PET Film and Sheet", MacDonald, W.A., Mackerron, D.H., Brooks, D.W., Chapter in "PET Packaging Technology" Ed Brookes, D.W., Sheffield Academic Press, 2002.

4 MacDonald, W.A., Polyester Film, Encyclopedia Polymer Science & Technology, 3rd Edition, John Wiley & Sons, Inc.

5 B.A. MacDonald, K. Rollins, R. Eveson, K. Rakos, B.A. Rustin and M. Handa, New developments in polyester films for flexible electronics, Mat. Res. Soc. Symp. Proc., 2003, **769**, paper H9.3.

6 MacDonald, W.A., Engineered Films for Display Application, J Mat Chem 14, 2004, 4.

7 PURE-ACE, Teijin Chemicals Ltd,

1-2-2 Uchisaiwai-cho, Chiyoda-ku, Tokyo, Japan.
8 Lexan, GE Plastics, One Plastics Ave, Pittsfield, MA 01201, USA.
9 Sumilite, Sumitomo Bakelite Co. Ltd, Ten-Nouzu Parkside Blgd., 5-8, 2-Chome, Higashi-Shinagawa, Shinagawa-Ku; Tokyo; 140-0002, Japan.
10 Arylite, Ferrania S.p.A., Centro Direzionale, Via Rivoltana, 2/d, Edificio A, I 20090 Segrate (MI), Italy.
11 Kapton, DuPont High Performance Films, P.O. Box 89, Route 23 South and DuPont Road, Circleville, OH 43113, USA.
12 W.A. MacDonald, K. Rollins, R. Eveson, R.A. Rustin and M. Handa, Plastic Displays – new developments in polyester films for plastic electronics, Society For Information Display, Digest of Technical Papers, 2003, 264–267.
13 S. Angiolini, M. Avidano, R. Bracco, C. Barlocco, N.G. Young, M. Trainor and X.-M. Zhao, High performance plastic substrates for active matrix flexible FPD, Society For Information Display, Digest of Technical papers, 2003, 1325–1327.
14 K.R. Sarma, C. Chanley, S. Dodd, J. Roush, J. Schmidt, G. Srdanov, M. Stevenson, R. Wessel, J. Innocenzo, G. Yu, M. O'Regan, W.A. Macdonald, R. Eveson, K. Long, H. Gleskova, S. Wagner, J.C. Sturm, Active matrix OLED using 150 °C a-Si TFT backplane built on flexible plastic substrate, Proceedings from SPIE Aerosense, Technologies and Systems for Defense and Security, April 22–25, 2003.
15 W.A. MacDonald, K. Rollins, D. MacKerron, R. Eveson, R.A. Rustin, K. Rakos, M. Handa, Society For Information Display, Digest of Technical Papers, 2004, 420–423.

III
Manufacturing for Organic Electronics

8
Reel-to-reel Vacuum Metallization

Roland Treutlein, Martin Bergsmann, and Carl J. Stonley

8.1
Reel-to-reel Vacuum Metallization

Vacuum metallization is the process of coating a substrate with a thin metal layer under high-vacuum conditions. This process is often simply referred to as "metallizing". The coating of web materials is referred to as vacuum web coating.

For more than 25 years, vacuum metallization of polymer webs with aluminum has been carried out on an industrial scale. Initially used to coat PVC for decorative purposes, the metallization of synthetic materials is today first and foremost used to produce functional coatings [1]. In the course of the past few years, metals other than aluminum have gained increasing importance in the metallization of polymer webs. The development of the metallization process was, and still is, subject to two requirements, namely *higher quality* and *lower cost*. Higher quality means:

- Higher conductivity (higher thickness and/or greater density of the metal film)
- Smoother surfaces (essentially because of the surface of the film substrate)
- Better homogeneity of metal film thickness
- Fewer defects in the metal layer (holes because of spitting during metallization)
- Fewer defects in the reel-winding process (creases)
- Improved barrier properties

Costs can be reduced by means of:

- Greater reel length/diameter
- Greater web width
- Higher web speed
- Shorter downtime

This chapter addresses, in particular, the process of vacuum metallization of flexible substrates with aluminum and copper and the respective machine requirements. Importance will also be attached to the properties of the layers produced and their applications, especially for organic electronics and packaging.

Organic Electronics, Materials, Manufacturing and Applications. Edited by Hagen Klauk

ISBN: 3-527-31264-1

8.1.1
The Metallization Process

Evaporation of pure metal layers from the vapor phase requires high-vacuum conditions (pressure $< 5 \times 10^{-4}$ mbar). To obtain good vacuum conditions the entire reel to be coated must be placed inside the vacuum chamber. The volume of vacuum chambers for industrial vacuum web coating must therefore be 15–20 m^3. To achieve the required vacuum conditions in such large vacuum chambers after short pumping time and to keep them for the whole process, high-performance pumping units with diffusion pumps and mechanical roughing pumps are required. It is essential to quickly achieve ideal and constant vacuum conditions to minimize downtime. In the process, the amount of residual gas is reduced by a factor of 10^7 to prevent its reaction with the metal vapor. A reaction between metal and oxygen or water vapor from residual gas would cause contamination of the evaporated layer, resulting in poor brilliance, lack of metallic character, poor barrier, poor adhesion, and high electrical resistance of the metal layer. Other residual gas components may pollute the layer and cause defects because of adsorption on the substrate.

Metallizers with external unwind and rewind facilities have also become known. Here the web is transported into and out of the vacuum chamber through locks. Because of demanding requirements of instruments and apparatus, these coaters have not gained acceptance for coating of polymer webs. They are successfully used in semi-continuous processes, for example glass coating.

The following section briefly describes the three most commonly used evaporation methods used in reel-to-reel metallization.

8.1.1.1 Evaporation Sources

Thermal Evaporation The easiest way of evaporating metal is by means of resistance evaporators known commonly as "boats". Boats, made of sintered ceramics, are positioned side by side at a distance of approximately 10 cm across the web width (Fig. 8.1). Titanium boride TiB_2 is used as an electrically conductive material with boron nitride BN (two-component evaporator) or BN and aluminum nitride AlN (three-component evaporator) as an insulating material [2]. By combination of conductive and insulating materials, the electrical properties of evaporators are adjusted.

The evaporators are heated resistively to approximately 1500 °C. A wire made from the material to be vaporized is brought into contact with each boat at which point it instantly melts and becomes gaseous. The metal vapor condenses on the cooled film passing by, and forms a thin metal coating on the film surface. This is the most commonly used method of evaporation.

E-beam Evaporation Modern machines increasingly use electron-beam evaporators. In this process an e-beam is focused on the evaporation material which, as a result of the energy input from the beam, is heated and evaporates (Fig. 8.2).

Fig. 8.1. Vacuum evaporator (Applied Films).

E-beam evaporators have the benefit of higher evaporation temperatures than resistance evaporators and it is also possible to evaporate materials, for example Ni, B, and Co, which alloy or react chemically when using resistance evaporators [3]. A disadvantage of e-beam evaporators is that substantially more complex instrumentation is required for process control.

Sputtering A third frequently used method is sputtering (Fig. 8.3), a technique in which metal atoms are removed from a solid target by means of ion bombardment

Fig. 8.2. E-beam evaporator (AP&T).

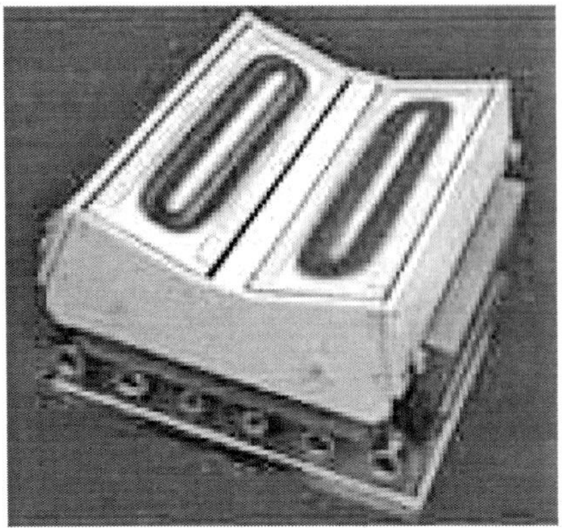

Fig. 8.3. Magnetron sputtering source (Von Ardenne).

and are then absorbed on the substrate surface facing the target. At normal web speeds, this process enables the formation of very thin metal layers only, these are prone to contamination by the residual gas. To avoid this contamination the process must be controlled precisely. Materials which are otherwise difficult to deposit, for example indium tin oxide (ITO), can be evaporated by sputtering [3].

8.1.1.2 Pretreatment and Cleaning of the Web Substrate

Modern metallizers are usually equipped with plasma pretreatment units (low-pressure plasma) for cleaning and pretreating the film. Flat cathodes are used for the purpose. By setting up an electric field and introducing a process gas (Ar), a plasma is generated. The high-energy ions are accelerated in the direction of the substrate, collide with the substrate, and remove adsorbed material from the surface, which results in a cleaning effect. The ions also break covalent bonds of organic contaminants in the surface, which facilitates evaporation [4].

By means of ion bombardment and the concurrent UV radiation, functional groups on the surface of the substrate are activated. This increases the polar portion of the surface energy, thereby improving the adhesion of the evaporated layer to the substrate (Fig. 8.4) [5].

Excessive pretreatment destroys the polymer surface and therefore reduces adhesion. In addition to the adhesion, the surface of the substrate affects the growth, and thus the structure, resistance, and barrier properties, of the metal layer.

8.1.1.3 PVD Process Flow

A schematic view of a web coater is shown in Fig. 8.5. The reel of film to be coated is placed in the upper part of the vacuum chamber, known as the winding chamber and, by means of suitable pumps, this chamber is evacuated to approximately 1×10^{-3} mbar. The winding chamber is separated from the lower part of the vacuum chamber, known as the evaporation chamber, by a differential pumping stage.

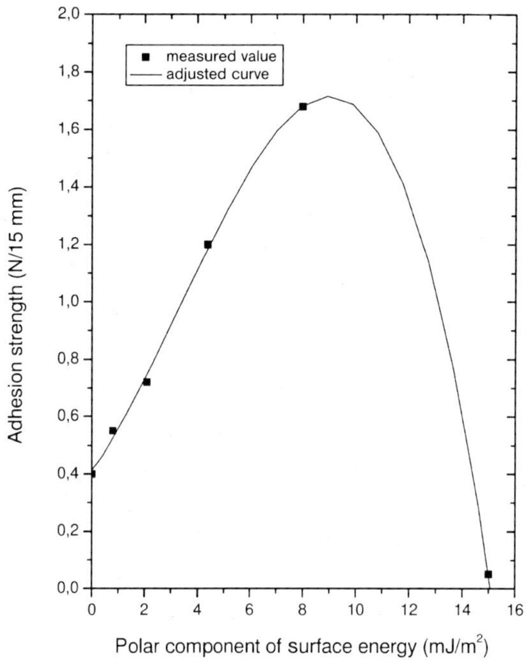

Fig. 8.4. Adhesion in the polar portion of the surface energy as a result of plasma treatment by BOPP [7].

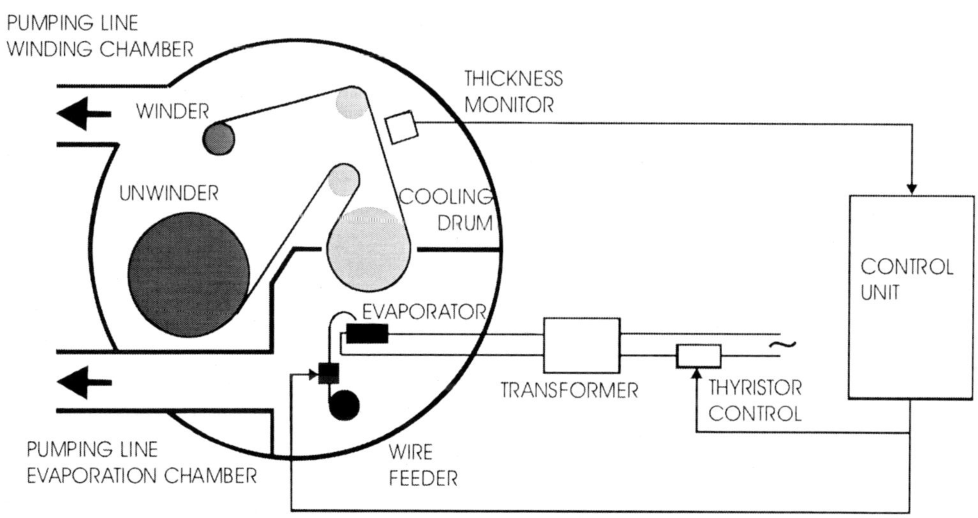

Fig. 8.5. Web coater, schematic view.

The evaporation chamber itself is evacuated to approximately 1×10^{-4} mbar by a separate pumping unit, usually by use of high-performance oil diffusion pumps. Separation of the winding and evaporation chambers is necessary because the web substrates (polymer, paper, nonwovens) outgas in vacuum, which means that they release adsorbed and absorbed components into the chamber. If the film were to be metallized under these bad vacuum conditions, a pure metallized layer would not be obtained. For this reason, only the part of the reel to be metallized is brought into the evaporation chamber, through a small slot. The web substrate then passes around a cooling drum, is metallized, and then is transported back into the winding chamber, and wound up.

In some machines the film is not passed over a cooling drum, but is directly transported over the slot and metallized ("free-span" metallization). This increases the efficiency of the evaporation sources and prevents the formation of creases initiated by the cooling drum [6]. The thermal load of the film also changes. On the one hand it increases, because there is no cooling by the drum; on the other hand it decreases, because there is also no heat reflection from the drum and the distance between evaporator and substrate can be increased. Whether the heat load increases or decreases depends very much on the configuration of the machine. For equal evaporator–substrate distances the thermal load, and thus the density of the metal structure, increases with free-span metallization (Fig. 8.6).

Each of the evaporators is electrically heated by use of a high current, some 100 A, which is fed at a low alternating potential of approximately 10 V. This is achieved by using high-current transformers with a thyristor-controlled primary current. Layer thickness monitors, positioned over the width of the web at the same dis-

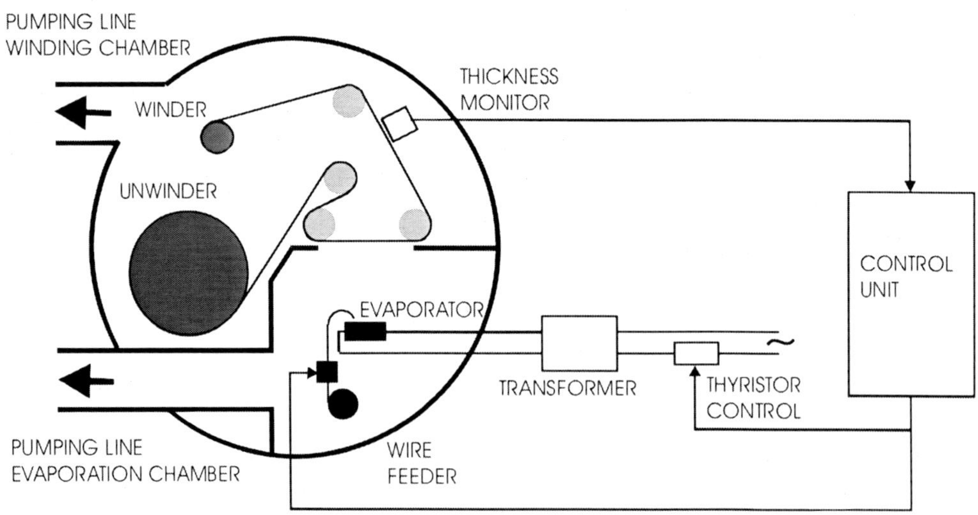

Fig. 8.6. Free span web coater, schematic view.

Fig. 8.7. TOPMET metallizer (Applied Films).

tance as the evaporators, measure the thickness and homogeneity of the evaporated layer. These measurable variables serve as input for control of the evaporator circuit and wire feeding [7]. Figure 8.7 shows a state-of-the-art metallizer (manufactured by Applied Films).

The Metallization Cycle A typical metallization cycle comprises the following steps:

1. Loading of the nonmetallized reel into the winding chamber
2. Closing and evacuation of the chamber
3. Heating of the boats
4. Unwinding of the reel
5. Opening the aperture above the boats
6. Metallization of the film reel
7. Cooling of the boats
8. Venting of the chamber
9. Opening of the chamber and removing the metallized film reel
10. Cleaning of the chamber

8.1.1.4 Typical Process Times, Rates, and Quantities

Depending on the length of the film reel, modern machines need between 30 min and 2 h per reel to complete these steps. Process rates depend on the method of metallization (e.g. sputtering or thermal evaporation) and the substrate, and also on the desired metal film thickness and, last but not least, the required quality. Typical process rates for metallization of poly(ethylene terephthalate) (PET) film are given in Table 8.1.

The quantities produced per metallization cycle depend on the type of substrate, e.g. polymer film type, paper, or nonwoven, its thickness, and the width and length of the reel. Table 8.2 provides a few typical examples:

Tab. 8.1. Process rates for metallization of poly(ethylene terephthalate) (PET) film.

Method/metal	Substrate	Layer thickness of metallization (nm)	Application/ requirements	Process speed
Thermal/Al	PET, 6 μm	20	Capacitors/low	Up to 1000 m min^{-1}
Thermal/Al	PET, 12 μm	30	Packaging/medium	Up to 800 m min^{-1}
Thermal/Cu	PET, 50 μm	800	Electronics/high	Up to 100 m min^{-1}
E-beam/Au	PET, 100 μm	50	Electronics/high	Up to 30 m min^{-1}
Sputtering/Al	PET, 50 μm	10	Window films/high	Up to 50 m min^{-1}
Sputtering/Cr	PET, 36 μm	20	Decoration/high	Up to 15 m min^{-1}

Tab. 8.2. Quantities produced per metallization cycle.

Method/metal	Substrate type/ thickness/ reel width/reel length	Layer thickness of metallization (nm)	Application/ requirements	Quantity per cycle (time)
Thermal/Al	PET/6 μm/8 m/ 48 000 m	20	Capacitors/low	384 000 m^2 (max. 2 h)
Thermal/Al	PET/12 μm/4 m/ 48 000 m	30	Packaging/medium	192 000 m^2 (max. 3 h)
Thermal/Cu	PET/50 μm/1 m/ 5000 m	800	Electronics/high	5000 m^2 (max. 2.5 h)
E-beam/Au	PET/100 μm/0.3 m/ 1000 m	50	Electronics/high	300 m^2 (max. 1.5 h)
Sputtering/Al	PET/12 μm/1 m/ 10 000 m	100	Window films/high	10 000 m^2 (max. 3 h)
Sputtering/Cr	PET/36 μm/1 m/ 3000 m	20	Decoration/high	3000 m^2 (max. 5 h)

8.1.1.5 Transfer Metallization

Apart from direct metallization of substrates as described above, transfer metallization is also possible. In this technology the metal layer is deposited on to a carrier film on which the metal adheres only very weakly, or which is coated in such a way that it is subsequently separable. The metallized side of the film is then laminated against another substrate and, after curing of the laminating adhesive, separated from the metal carrier film. During separation, the metal layer is transferred from the carrier film and remains on the second substrate. The carrier film can then be metallized again. This method is used for materials which cannot be directly metallized at all or which cannot be metallized without significant loss of visual and functional quality of the metal layer, e.g. metallization of paper or board.

8.1.1.6 Pattern-evaporated Layers

Pattern-evaporated layers can be produced by direct or indirect methods. In the direct route either a fixed mask is placed between the evaporation source and the web substrate or a co-traveling masking web is applied to the substrate web as it travels in the vacuum chamber. An alternative indirect method is a multi-step process including pretreatment of the film, metallizing and subsequent removal of unwanted sections. Applications of pattern evaporated layers include:

- thin-film solar cells
- security features
- flexible printed circuits (for signal-carrying circuits)
- RFID antennas (for HF short-range systems, unlimited in UHF range)

8.1.2 Properties of the Evaporated Layer

Despite the differences between each evaporation method, described above, evaporated metal layers produced by any method are regarded as relatively very thin with thickness ranging between several nanometers and one micrometer. The properties of these layers are therefore not comparable with those of rolled metals, for example Al foil, or with those of oxides with lacquer or adhesive coatings of several micrometers.

Evaporation produces an ultra-thin layer which, although covering the surface, does not change the basic properties of the underlying layer, for example roughness etc. Surface defects such as scratches and raised surfaces, etc., are apparent in the evaporated layer. The metal layer can be compared with thin snow cover – it changes the surface optically, but its topography remains essentially the same. The comparison works also with regard to the structure of the layers; neither is massive, both consist of connected crystallites.

8.1.2.1 Structure

The structure of a metallized layer is determined by the surface topography of the web substrate and by interaction of the involved condensation processes (shadowing, surface diffusion, volume diffusion, and desorption), which cause growth in the form of more or less densely packed pillars or conglomerates (Fig. 8.8) [8].

Depending on which of the four condensation mechanisms is predominant, the structure will be more or less dense (Zones I–III). The activation energy for the surface and volume diffusion of metals is proportional to their respective melting

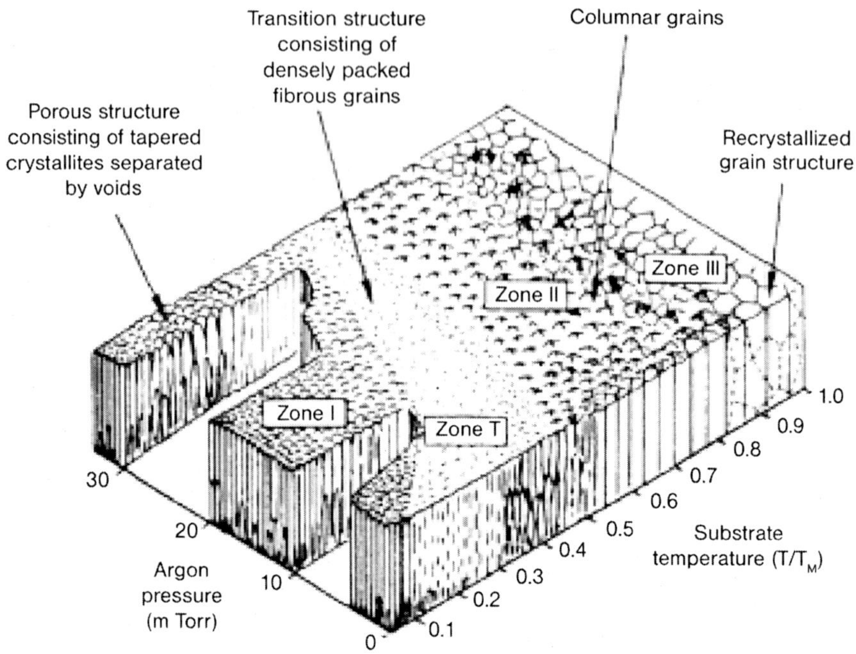

Fig. 8.8. Types of growth of metallized layers [8].

points. The structure of the film is therefore essentially determined by the relationship between substrate temperature and melting point of the metal. It is also affected by the relevant vacuum conditions.

Taking, for instance, Al, with a melting point of 660 °C and a web substrate temperature of 50 °C, zone I formations will be created (porous structure, pointed crystallites, large voids) and up to 250 °C, formations in the transitional area (densely packed fibers) will appear. Up to 450 °C zone II (pillar-shaped crystallites), and above this temperature zone III (conglomerate-type crystallites) formations will be seen. Because of the relatively low maximum thermal stress that may be applied to polymer webs, the growth in metallized layers on polymer webs mainly occurs in Zone I or in the transitional zone. The different growth is also evident from comparison of cooling drum and free-span coater methods.

8.1.2.2 **Layer Thickness (Conductivity)**

The typical thickness of metallized Al layers ranging from few nm to several hundred nm corresponds to an area weight of approximately 100 mg m^{-2}. Such film thicknesses cannot be reliably measured by mechanical meters or by weighing. Other laboratory measurement techniques are used, for example X-ray fluorescence (XRF) spectroscopy, ellipsometry, or ion scattering (RBS). All these techniques are unsuitable for in-line control during metallization, because they require highly sophisticated technical equipment. For industrial applications, other indirect measurement techniques are used, for example *optical density* and *resistance.*

Optical Density For aluminum, the most popular method of determination of the quantity metallized is measurement of "optical density" (O.D.). The vacuum-evaporated layer is exposed to light of intensity I_0 and the part of the light transmitted I is measured. Optical density (O.D.) is a function of the light transmission, T, and defined as [9, 10]:

$$\text{O.D.} = \log\left(\frac{I_0}{I}\right) - \log\left(\frac{100}{T(\%)}\right)$$

A typical Al layer with an O.D. of 2.0 has a visible light transmission of 1%. Lambert's Law:

$$I = I_0 \times 10^{-\alpha d}$$

indicates that in the simple relationship with the film thickness, d:

$$\text{O.D.} = \alpha \times d$$

where α is the absorption coefficient of the material. The absorption coefficient for metallized layers must be determined empirically (growth, density). For aluminum, α is ~0.069 nm^{-1}.

Surface Resistance For opaque substrates or very thick metal layers, measurement of the thickness optically is not feasible. For such materials eddy currents (induced current) are used to measure the electrical surface resistance of the metal layer and, consequently, the thickness of the metallized layer. The thinner the metal layer, the higher its resistance. Surface resistance is inversely proportional to the thickness of the layer (proportionality factor depending on the material).

In this technique a high-frequency magnetic field is created by a solenoid in a probe at a fixed distance above the layer. The high-frequency inductive current generates eddy currents in the layer to be measured; these change the impedance of the solenoid generating the high-frequency alternating field. According to Lenz's Law, the magnetic field set up by the induced current opposes the energizing field and weakens it. The change in impedance depends on the evaporated material, the film thickness, and the sample–probe distance. Given the material is known and a fixed distance is maintained between probe and film, the change in impedance serves as the measurable variable for film thickness.

The surface resistance R_{fl} is defined for square test surfaces. If the resistance R is defined via the specific resistance ρ_{el} (matter constant):

$$R = \rho_{el} \times \frac{1}{A} = \rho_{el} \times \frac{1}{bd}$$

Tab. 8.3. Specific resistances for evaporated aluminum and copper.

Material	Method applied	Specific resistance (10^{-9} Ωm)
Al	Thermal evaporation	39
Cu	Thermal evaporation	32
Cu	Sputtering	21

where l is the length of the conductor, $A = b \times d$, is the cross-sectional area of conductor, b is the width, d the thickness of the layer, and a square test surface is inserted ($l = b$):

$$R_{\text{fl}} = \frac{\rho_{\text{el}}}{d}$$

Because of their specific structure (Section 8.1) evaporated layers have only approximately 70% of the conductivity of the solid metal [11]. Depending on the method of application this factor may vary, because of the different densities of the layer that can be achieved. An overview of specific resistances for evaporated aluminum and copper is given in Table 8.3.

Figures 8.9 and 8.10 illustrate the correlations between the thicknesses of layers and measurable variables for aluminum and copper.

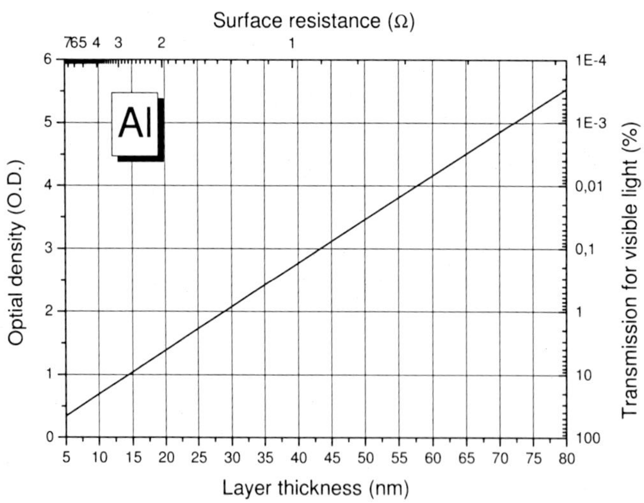

Fig. 8.9. Correlation of the different measurable variables for the film thickness of aluminum.

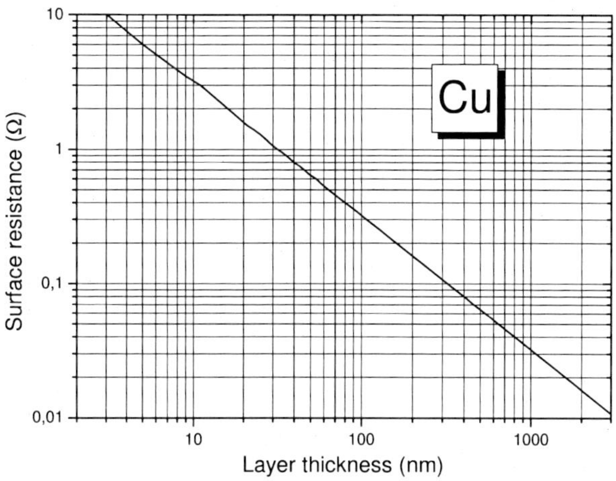

Fig. 8.10. Correlation of film thickness and surface resistance for copper.

8.1.2.3 Barrier

Al is frequently used to obtain good barrier properties with polymer webs and, despite the very thin metal layers produced, the barrier of the metallized film to water vapor (WVTR) and oxygen (OTR) is markedly improved compared with nonmetallized film. This enhancement of the barrier is the reason aluminum metallized films are used in packaging.

The barrier quality is determined by the surface properties of the polymer substrate and the properties of the metal layer [12]. The most crucial factor is the *surface quality of the substrate.* If there is an adsorbate on the surface, it may be desorbed during the evaporation process, which leads to pinholes and poor adhesion of the metal layer. If the adsorbate is water, Al may react with the water and lead to the formation of Al oxide which has barrier properties substantially inferior to those of metallic Al. Also the lack of functional groups on the polymer surface causes poor adhesion and thus a change in the structure of the layer.

Poor quality of the web substrate surface leads to:

- a greater number of pinholes;
- poor adhesion and consequently lower optical density; and
- poor nucleation and less dense structure.

These factors lead to an inferior barrier and a more rapid reduction of the barrier effect. For a good barrier effect, co-extrusion or coating of the substrate, corona discharge, or plasma pretreatment, are, therefore, usually necessary. Vacuum metallization conditions, thermal stress of the substrate, and mechanical stress of the metal layer also play a fundamental role:

- **Vacuum** Contamination of the Al layer by residual gas components (mainly water vapor) affects the barrier substantially; good vacuum ($<5 \times 10^4$ mbar) in the evaporation chamber is therefore required for barrier applications. Contamination from the evaporation material and the evaporators must also be avoided as much as possible.
- **Thermal stress** For PP, increasing the distance between the evaporator and the substrate, or increasing the web speed, reduces thermal stress on the polymer web and, consequently, the number of pinholes. For heat resistant polymers such as PET thermal stress is less significant.
- **Mechanical stress** High web tension causes micro-cracks within the metal layer. Reduction of the barrier in thicker layers is mentioned in the literature, but is probably a result of the stress to which the metal layer is exposed in subsequent process steps (e.g. lamination).

Microscopic pinholes primarily affect water-vapor barrier properties. The oxygen barrier is mainly affected by macroscopic defects.

In summary, good barrier properties can only be achieved by optimizing all the factors affecting the metallization process. Under these conditions, metallized layer barrier properties will be excellent. Superior barrier properties may only be achieved by using multi-layer structures. Such ultra high barrier films, formed from metallized Al sandwiched between two acrylate layers, were first reported in 2000 [13].

Table 8.4 shows clearly how metallization significantly improves the barrier qualities of polyethylene terephthalate (PET) and polypropylene (PP) films.

The excellent barrier properties of metallized films are particularly evident from direct comparison with other barrier materials. In general, the barrier to oxygen and water vapor is taken as a measure of suitability for food packaging. As can be seen from Table 8.5, metallized films achieve a very high rating in this comparison. It is worthy of note that metallized PET films constitute an excellent barrier both to water vapor and to oxygen, whereas an OPP film with the same metallization pro-

Tab. 8.4. Improvement of the barrier qualities of poly(ethylene terephthalate) (PET) and polypropylene (PP) films by metallization.

Film	Thickness (μm)	OTR (cm^3 m^{-2} d^{-1}) [14]	WVTR (g m^{-2} d^{-1}) [15]
PET film	25	31–93	16–20
PET film, metallized	25	0.16–1.7	<1.0
OPP film	25	1550–2500	3.9–6.2
OPP film, metallized	25	19–160	<0.6

Tab. 8.5. Comparison of different barrier materials (typical values).

Material	Light transmission (%)	OTR ($cm^3\ m^{-2}\ d^{-1}$)	WVTR ($g\ m^{-2}\ d^{-1}$)
Metallized PET film	<1	<2	<1
Metallized OPP film	<1	<160	<0.6
Ultra-high-barrier film	<1	0.02–0.08	0.02–0.09
Aluminum film, >20 μm	0	0	0
AlO_x coating	>80	<5	<2
SiO_x coating	>80	<2	<1
ITO coating	>80	Approx. 0.7	Approx. 0.6
Top-coated AlO_x layer	>80	Approx. 0.35	Approx. 0.12
PVDC coating[a]	>90	Approx. 8	Approx. 3
PAN film (Barex)[b]	>90	Approx. 13	Approx. 50
EVOH co-extrusion[c]	>90	Approx. 1	<5

[a] PVDC is poly(vinyl dichloride)
[b] PAN is polyacrylonitrile
[c] EVOH is ethylene vinyl alcohol

vides a good barrier to water vapor only. The oxygen barrier is significantly better than that of the nonmetallized film but does not meet barrier requirements for specific food packaging. Even after metallization, densities of other synthetic films, for example poly(vinyl chloride) (PVC), are not sufficient for use as food packaging material. Metallized paper has hardly any importance as a barrier layer to oxygen and water vapor.

If we compare the data in Table 8.5 with the barrier requirements set in polymer electronics (Fig. 8.11), it is evident they cannot be met with metallized films, not even with ultra-high-barrier films, multi-layer structures from metal evaporation, and polymeric layers. For transparent barriers, as required for OLEDs, displays, organic solar cells, etc., evaporated oxide layers are even further from meeting the values required.

For barriers in polymer electronics, evaporated layers cannot be used alone. Specific combinations of layers must be developed for that purpose.

8.1.2.4 Light Barrier

In addition to their outstanding properties as barriers to oxygen and water vapor, metallized films have another advantage – as a barrier to light. Because the appearance (e.g. yellowing) and/or function of some substances are affected by light, they must be protected appropriately. Food deteriorates much more rapidly when exposed to light; fatty food becomes rancid; with dairy products light causes undesired odor and vitamins decompose more quickly.

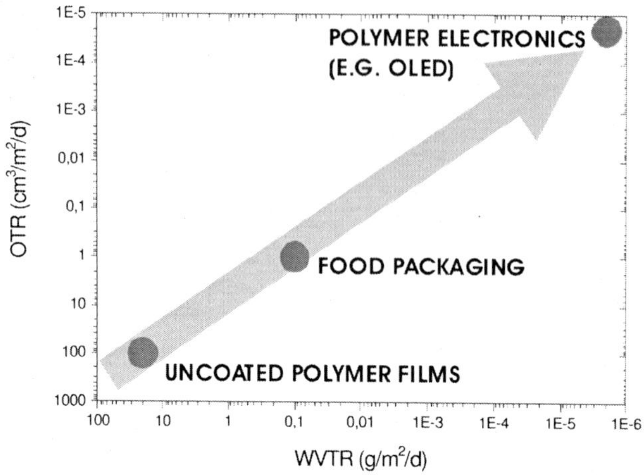

Fig. 8.11. Barrier requirements for polymer electronics compared with those for food packaging.

Aluminum metallized films provide a barrier to light which cannot be achieved with conventional lacquering or coating. For instance, a 2 O.D. metallization has a light transmittance of only 1%. In this respect, only aluminum foil is superior to metallized films. Barrier films or coatings such as PVDC, EVOH, and SiO_x are relatively clear and therefore have a clear disadvantage in application when both barrier and light protection are necessary.

8.1.3
Environmental Benefits of Vacuum Evaporated Layers

Vacuum metallized layers have environmental and energy advantages over metal foils. Material and energy demand are considerably lower than when using a foil of the same material. For example, with one kilogram of aluminum approximately 5000 m^2 can be vacuum metallized; if the same quantity of aluminum is used to produce a 9 μm thick foil the yield is just 41 m^2.

This significantly enhanced efficiency of vacuum metallization is also obvious from comparison of energy consumption. Production of aluminum is very power-consuming. Evaporation of aluminum for one square meter of film requires only 92 kJ whereas for production of 9 μm thick aluminum foil approximately 3000 kJ are needed. This is approximately a thirtyfold difference in energy consumption per unit area.

Patterned metallic layers, like those used for conducting tracks, not only have this obvious economic advantage but also have marked ecological benefits over conventional structuring techniques. In contrast with etching technologies, no etching media, for example $FeCl_3$ or NaOH are required; this substantially reduces the environmental impact of structuring.

8.1.4 Applications of Metallized Films

8.1.4.1 Barrier Packaging

In food packaging applications the packed goods are particularly sensitive to environmental conditions such as oxygen, moisture, flavor, or light. A typical example is the packaging of potato chips. In general the packaging consists of a laminate of metallized OPP film (usually 20 µm) and a printed OPP film of the same thickness. The metal layer and the print are positioned between the two films and are protected against abrasion and damage. The important aspect of packaging potato chips is to enhance the barrier to moisture to prevent loss of "crispiness".

When food products must be protected against both moisture and oxygen a laminate of metallized PET film, typically 12 µm thick, and PE-LD is usually used. The metallized PET film provides the excellent barrier properties and the PE film serves as sealing medium. A classical example is packaging of cakes with a fat-containing chocolate covering. The chocolate would become rancid if in contact with atmospheric oxygen, which would considerably reduce shelf life. By using a metallized packaging film, loss flavor from packed products (e.g. confectionery products or chocolates) is reduced or prevented, thus ensuring the content will taste consistent for longer.

8.1.4.2 Decorative Applications

In addition to these applications in the food industry which are first and foremost a result of the improved barrier properties, there are other applications in the "consumer sector" in which the aesthetic value of products is enhanced by use of metallized films. A typical example is the use of metallized paper for drinks bottles. Numerous brands of bottled beer are promoted by use of metallized paper labels. This design of the bottle primarily serves the purpose of producing an "eye-catching" effect to capture consumers' attention; the bottles on the shelf are designed to stand out from other beers and give the impression of a "superior" and "more precious" product. This is also the reason for using metallic elements on brochures and promotional materials.

Another application of metallized films for design purposes is the generation of metallic effects (car paneling, mobile phones, furniture, architecture, gift wrapping) which can be produced economically and without significant additional weight. The benefits of the synthetic material, for example durability and flexibility, are also maintained.

8.1.4.3 Functional Layers

Because of their conductivity, reflecting effect, and light barrier properties, metallized films are increasingly used as functional coatings in technical applications. Examples include battery cell cases, prepaid telephone cards, self-adhesive labels, insulation foil in the building industry, decorative foil for design elements, security features for bank notes, and tear tapes, etc.

In the electronics sector, metallized films and nonwovens are used as EMI

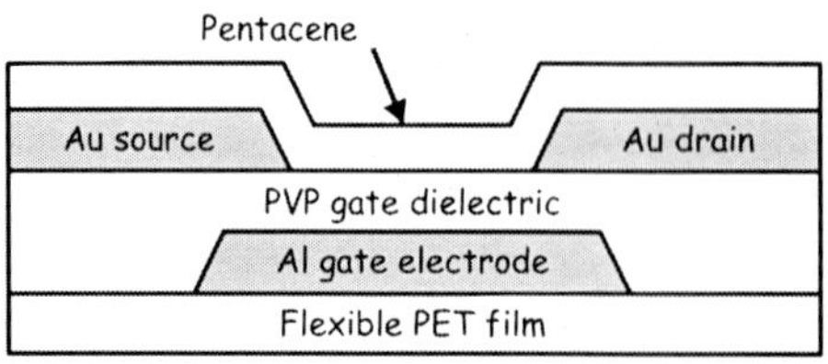

Fig. 8.12. Schematic cross-section of an organic pentacene TFT on metallized PET film.

shielding and capacitor foils, as a basic material for flexible printed circuit boards, and for RFID antennas. With the development of polymer electronics, metallized films are gaining importance as electrodes and conductive tracks (Cu, Au, Al) and for thin-film insulation (oxides). More information on this subject is given in Section 8.1.4.4.

8.1.4.4 **Polymer Electronic Substrates**

Metallized polymeric films are also of great interest for use as lightweight, flexible, cost-effective substrates for organic electronics. To demonstrate that the quality of reel-to-reel metallized PET is already sufficient for organic electronics, we have produced organic thin-film transistors (TFTs) and integrated ring oscillators on PET [16]. The cross-section of these organic TFTs is shown schematically in Fig. 8.12. Commercial PET film with a thickness of 36 μm was aluminized using an industrial web coater in a reel-to-reel process with a speed of 4 m s^{-1}, giving an optical density of 2.2. The metallized PET was cut into sheets and the Al film was patterned by photolithography and wet etching to define Al gate electrodes for the transistors and interconnect lines for the integrated circuits. A 270 nm thick polyvinylphenol (PVP) gate dielectric layer was deposited by spin-coating, gold source/drain contacts were evaporated and photolithographically patterned, and pentacene was evaporated as the organic semiconductor. Figure 8.13 shows a photograph of a

Fig. 8.13. PET test chip with pentacene transistors and circuits.

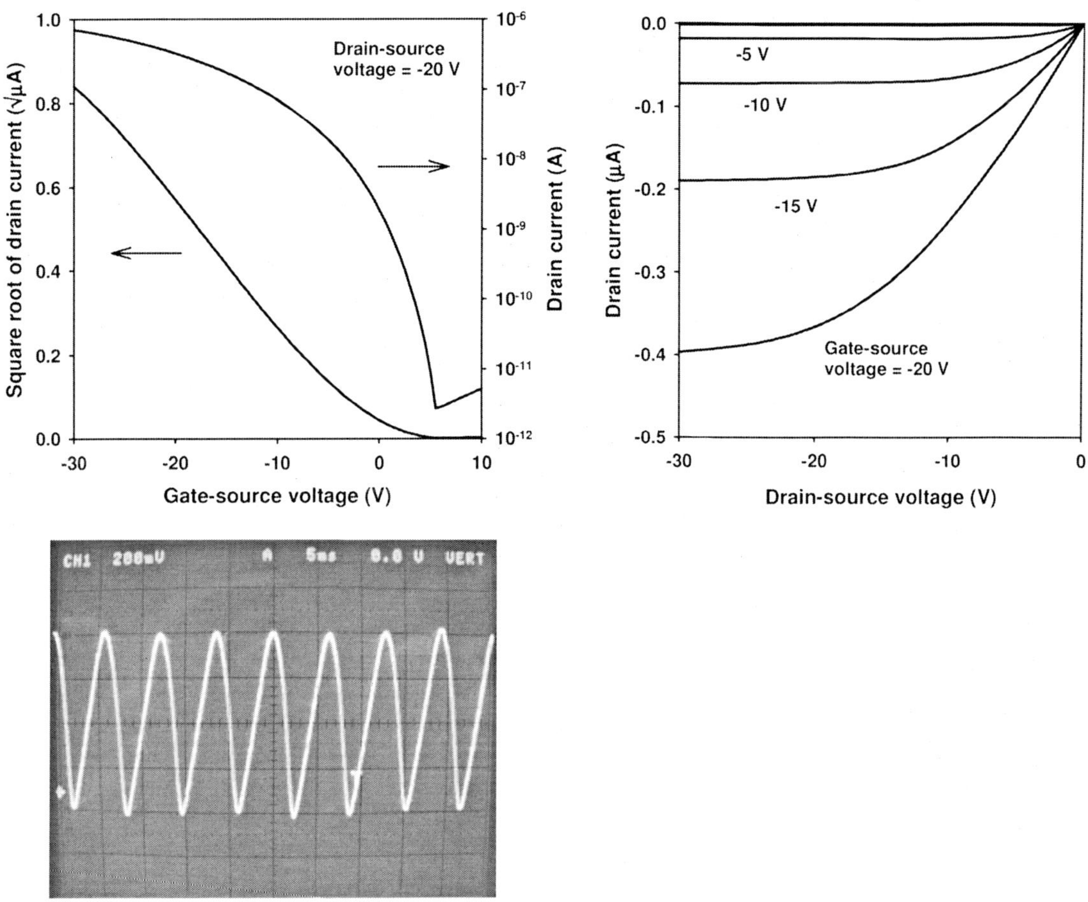

Fig. 8.14. Electrical characteristics of a pentacene TFT and output signal from a five-stage pentacene ring oscillator, both fabricated on reel-to-reel aluminized PET.

completed PET test chip with a variety of pentacene transistors and circuits. Figure 8.14 shows the electrical characteristics of the TFTs and circuits on reel-to-reel aluminized PETs. The performance of these transistors and ring oscillators is comparable with that of organic TFTs and circuits on other, more expensive types of substrate.

8.1.5 Market Analysis

Because of the many possible applications, the total market size for metallized films is very large. Because the individual applications are distributed to most different market segments, exact determination of the total market size is difficult to confirm. Most metallization is with Al, which accounts for in excess of 98% of all metallization conducted. The remainder is shared between metallization with Cu, Al_2O_3, and SiO_x.

In 2000, 162 000 000 kg metallized web material (10–15 billion m^2) were used for packaging applications in Europe; of these 42% were OPP, 12% PET, and 40% metallized paper. The remaining 6% is accounted for by other metallized films, for example PE, PVC, PA, or cellophane, and metallized boards or tissues. An annual growth rate of 3–4% is forecasted [17, 18]. An approximately equally quantity of metallized film is used for decorative and technical applications. For these, however, PET is the dominant substrate. In technical applications, capacitor foil currently has a large share. RFID and EAS technology and polymer electronics are wide fields of potential applications and future growth for metallized substrates.

References

1 R. Mitchell et al., Tech. Conf. Proc. Soc. Vacuum Coaters (1998) 472.
2 M. Seifert, Proc. Tech. Seminar Evaporation Boats (1997).
3 A.H.F. Muggleton, Vacuum 37 (1987) 785.
4 E. Finson et al., Tech. Conf. Proc. Soc. Vacuum Coaters (1995) 52.
5 G. Löbig et al., Tech. Conf. Proc. Soc. Vacuum Coaters (1998) 502.
6 F. Casey et al., Tech. Conf. Proc. Soc. Vacuum Coaters (1999) 480.
7 F. Casey, Tech. Conf. Proc. Soc. Vacuum Coaters (1991) 124.
8 S. Craig et al., J. Vac. Sci. Technol. 19 (1981) 205.
9 ExxonMobil Chemical #456 (www.exxonmobil.no).
10 ExxonMobil Chemical #474 (www.exxonmobil.no).
11 U. Leute, Kunststoffe und EMV, Carl Hanser, München (1997) 143.
12 A. Yializis et al., Tech. Conf. Proc. Soc. Vacuum Coaters (1997) 371.
13 A. Yializis et al., Tech. Conf. Proc. Soc. Vacuum Coaters (2000) 404.
14 ExxonMobil Chemical #475, 476, 478 (www.exxonmobil.no).
15 ExxonMobil Chemical #518 (www.exxonmobil.no).
16 H. Klauk, F. Eder, M. Halik, U. Zschieschang, G. Schmid, C. Dehm, M. Bergsmann, R. Einsiedler and R. Treutlein, "Organic Electronics on Flexible Substrates – Paper or Plastic?" *61st Device Research Conference Digest* (2003) 183.
17 W. Geitner, Tech. Conf. Proc. Soc. Vacuum Coaters (2000) 414.
18 P. Gaster, EMA Proc. Tech. Seminar 2001.

9
Organic Vapor Phase Deposition

Michael Heuken and Nico Meyer

9.1
Introduction

In recent years organic electronic devices such as thin-film transistors (TFTs), solar cells, photo detectors, and, especially, OLEDs (organic light-emitting diodes) for display applications and solid-state lighting applications have been intensively investigated [1–6]. As a result of these efforts OLED displays have already reached commercial significance, for example in cell phones and MP3 players, because of their intrinsic properties such as high brightness, high resolution, low energy consumption, and wide viewing angle characteristics realized in thin devices on glass or even flexible substrates. Further research and development focuses on device performance improvements for OLEDs, on improving our basic understanding of organic TFTs, solar cells, and photo detectors, and on integrating OLEDs with organic TFTs to create all-organic active-matrix OLED displays.

Today OLEDs and organic TFTs based on small molecular weight organic materials are mainly manufactured by using vacuum thermal evaporation (VTE). Conventional VTE is similar to molecular beam epitaxy (MBE), which was used in early production of III–V semiconductors. These high-vacuum evaporation techniques have limitations in process control, flexibility, scalability and cost effective manufacturing capabilities, however. Meanwhile, metal organic vapor phase deposition (MOCVD), a gas-phase transport technique, has replaced MBE for manufacturing III–V semiconductors. Inorganic ultra high brightness (UHB) LEDs have become very complex and can only be realized by MOCVD for industrial production.

Organic vapor phase deposition (OVPD) was invented by S. Forrest at Princeton University to transfer all the benefits of the gas phase process to applications in organic electronics. OVPD technology has the potential to overcome the limitations of VTE, similar to the replacement of MBE by MOCVD.

9.1.1
The Principle of OVPD

In 1995 Forrest introduced the dry OVPD principle and has since investigated the benefits of gas-phase transport by using a simple flow reactor to overcome many

Organic Electronics, Materials, Manufacturing and Applications. Edited by Hagen Klauk

ISBN: 3-527-31264-1

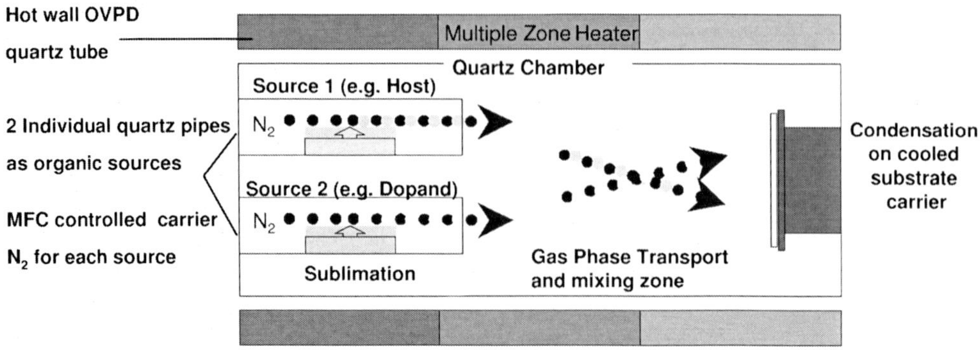

Fig. 9.1. OVPD principle and quartz flow deposition module invented by Forrest.

limitations of standard VTE [7–12]. Because of the gas-phase transport principle of OVPD, the arrangement of evaporation and condensation is decoupled, in contrast to VTE. Thus, the evaporation design can be optimized to any chemical needs without affecting the design for uniform and controlled large-area organic film deposition.

As shown in Fig. 9.1, evaporation of the organic material occurs in individual and decoupled quartz pipes. A precise amount of carrier gas, for example nitrogen, is added into each quartz pipe by standard mass-flow controllers (MFC) to pick up the organic molecules. The organic molecules are transported by the carrier gas into the hot wall deposition chamber and are uniformly mixed if two organic materials or more, for example host and dopant, are evaporated simultaneously. Finally the organic molecules in the gas phase diffuse through the boundary layer on to the cooled substrate where they condense. One consequence of this controlled gas-phase transport technique is that no unintended deposition of expensive organic molecules occurs in the hot-wall OVPD deposition chamber itself – deposition is mainly on the substrate. This enables high material utilization.

Using a carrier gas in OVPD enables the deposition of organic materials at a controlled pressure of 10^{-3}–10 torr. Thus the OVPD module does not have to be pumped down to high-vacuum conditions, as in VTE, which consequently increases the uptime of the OVPD deposition system. Also, the continuous purge of carrier gas in OVPD prevents any contamination of parasitic surfaces, which increases the reproducibility of the deposited organic film quality. Therefore Forrest envisaged with the flow deposition module that OVPD is an ideal solution for industrial mass production.

9.1.2 Close Coupled Showerhead Technology

The inherent advantages of OVPD can be exploited on an industrial scale when combined with close coupled showerhead (CCS) technology. With CCS the carrier

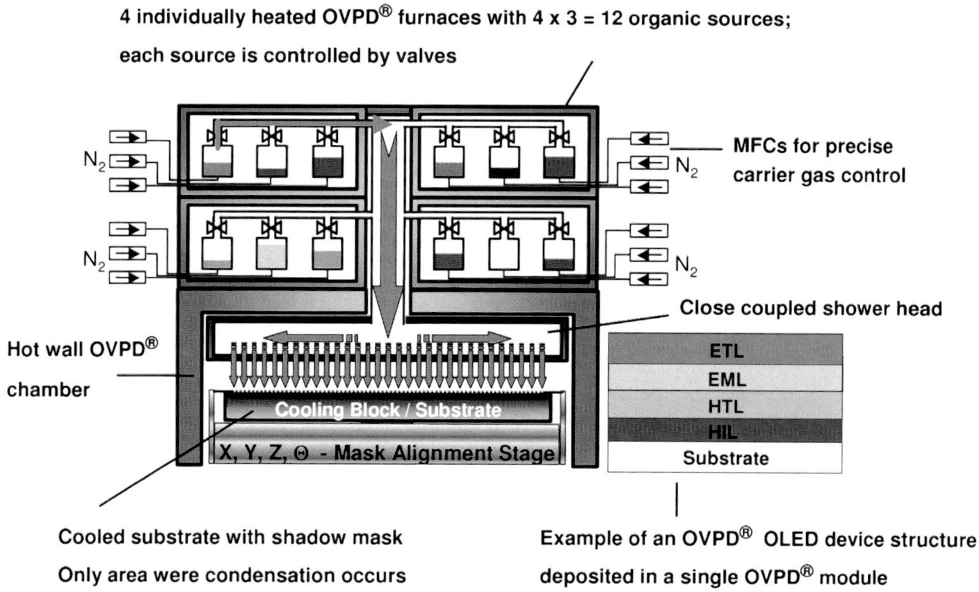

Fig. 9.2. Close-coupled showerhead technology for industrial-scale production of multiple layers and OVPD. OLED example deposited in a single OVPD module.

gas and organic materials are introduced uniformly through a showerhead injector across the entire deposition area. As the injection is in close proximity to the substrate surface the fluid dynamics enforce homogenous distribution of the species, enabling uniform deposition across the entire area without a need for substrate rotation (*close coupling effect*) [13–19]. In addition to intrinsic uniformity CCS enables deposition with high material utilization efficiency.

Figure 9.2 shows a schematic diagram of AIXTRON's Gen 2 OVPD equipment employing a temperature-controlled deposition chamber and the close coupled showerhead design with orifices acting as a multiple gas distributor. The organic source materials are heated in physically separated source containers, which are arranged in separate furnaces remote from the deposition chamber, and transported by an inert carrier gas into the hot-wall deposition chamber. As shown in this scheme the Gen 2 OVPD design was realized with twelve organic sources in four individual furnaces. The number of organic sources and furnaces is flexible and can be adjusted to the desired processes on request.

Individual valves switch the source flows to enable rapid on/off control of the respective deposition, which offers high precision control of layer interfaces and minimization of material waste. The organic molecules are homogeneously mixed in the gas phase before being introduced uniformly through the heated showerhead injector across the entire substrate surface where they condense to form the desired organic film.

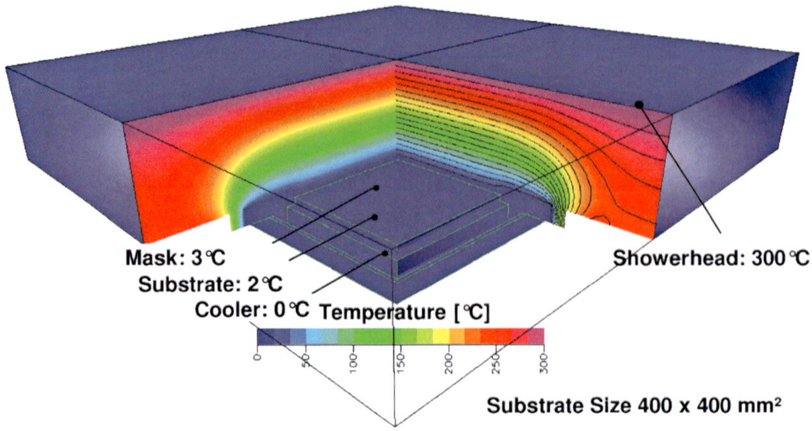

Fig. 9.3. Temperature distribution in the OVPD chamber and of the actively cooled substrate in Gen 2.

An important thermodynamic requirement for the homogeneous distribution of the organic gas phase in the close coupled showerhead is temperature uniformity, which also guarantees no unintended condensation of the organic material. Figure 9.3 depicts the temperature simulation of an Gen 2 OVPD showerhead suitable for substrate sizes of 400 mm × 400 mm. The showerhead is uniformly heated to 300 °C, whereas the chiller actively cools the substrate and the mask to 2 °C and to 3 °C, respectively, across the whole area. Both the mask and the substrate are in close contact and thereby actively cooled down across the whole area.

Experimental proof of control of the mask temperature with the chiller in a Gen 2 OVPD module under process conditions (showerhead heated to 325 °C) was achieved by *in situ* temperature measurement, as shown in Fig. 9.4. The experiments were performed at atmospheric pressure and at a deposition pressure of 0.9 mbar typical for OVPD, and for chiller temperatures between 5 and 30 °C. The mask temperature can be linearly controlled by the chiller temperature. The observed ΔT of 6.5 degrees is in good agreement with modeling prediction of 3 degrees in Fig. 9.3. In addition, measurements during a typical OVPD deposition time of 2 to 6 min confirmed there is no temperature drift under process conditions over time. The data prove that heat conductance and radiation is perfectly compensated by the chiller capacity.

In conclusion, stable and reproducible temperature conditions for the mask and the substrates, essential requirements for a stable production process, have been demonstrated for OVPD.

This CCS technology with its intrinsic laminar flow condition enables uniform deposition for a wide range of deposition pressure, carrier gas flow rate, and substrate temperature. The deposition rates and layer composition such as co-hosting or doping are adjusted by means of mass flow rate of the carrier gas at constant source temperature using individual electronic MFCs for each source. No recircu-

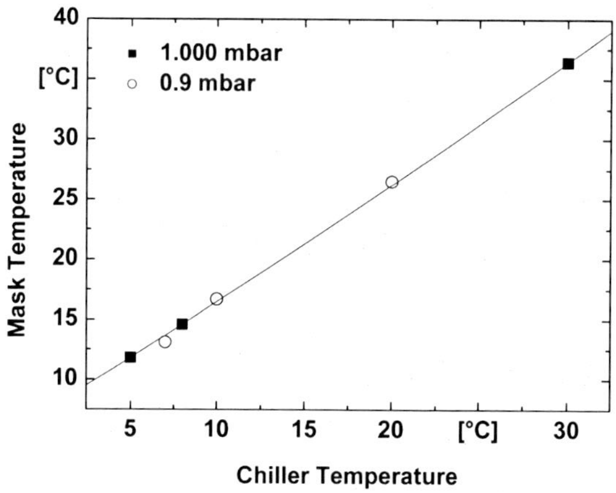

Fig. 9.4. Linear mask temperature dependence in OVPD as function of chiller temperature proves heat flux is controllable in OVPD.

lation of the carrier gas occurs, offering a high precision and reproducibility combined with flexibility in multilayer deposition sequences and accurate composition of multi-component layers (co-hosting/co-doping) and ramping of material flow for graded compositions.

Gas-phase transport enables placement of multiple organic sources remote from the deposition chamber, thus preventing cross-contamination between source materials. The hot-wall deposition chamber prevents parasitic condensation of the organic material. Consequently OVPD combined with the close coupled showerhead has the potential for low maintenance cycles, high material yield, high reproducibility, well-defined doping with multiple dopants, and high throughput, which are key factors for industrial mass production at low cost. Finally scaling of the technology of OVPD and CCS to any size is not limited and can be realized vertically or horizontally.

9.2 Deposition of Organic Thin Films

9.2.1 Process Control in OVPD

Key requirements for industrial deposition of single films are the rate of deposition achievable, controllability and reproducibility, and film quality, which is important for high uptime and production yield. In contrast to VTE, in which the deposition rate is controlled by the evaporation temperature in the crucible, OVPD is kept under steady-state temperature conditions and the deposition rate is adjusted only

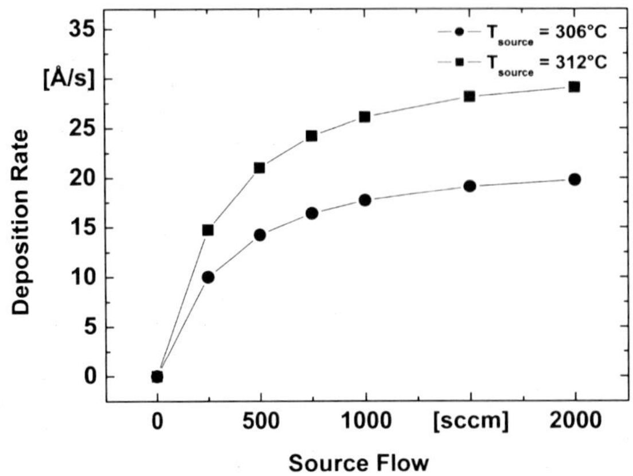

Fig. 9.5. Typical calibration curves for Alq_3 evaporated at 306 °C (circles) and 312 °C (squares).

by the source flow of carrier gas through the source container. The source flow is precisely controlled by standard mass-flow controllers and actively added or removed to the total flow by switching valves into the run/off mode. Depending on its specific vapor pressure transport of the material can be controlled by the amount of carrier gas flowing through the source. Thus, the rate of deposition of each material can be adjusted as function of the carrier flow for individual evaporation temperatures and container pressures, resulting in exact prediction of the deposition rate. Figure 9.5 shows two typical calibration curves for Alq_3 evaporated at 306 °C and 312 °C. Increasing the source flow to 2000 sccm resulted in a deposition rate of 30 Å s^{-1}.

These calibration curves also reveal the huge increase of deposition rate with temperature, as expected from the Clausius–Clapeyron equation. For small flows the deposition rate increases linearly with source flow whereas at flows > 500 sccm the increase is sub-linear. With higher flows the deposition rate is dominated by the flow restriction from the process chamber to the source container. If the evaporation temperature is increased from 306 to 312 °C the deposition rates increases approximately 47%, from 19.2 to 28.2 Å s^{-1}.

By combining multiple separate source containers of the same organic material in a parallel configuration the deposition rate can be linearly increased, leading, for example, to deposition rates of up to 60 Å s^{-1} for two Alq_3 sources.

The uniformity of such an OVPD film of Alq_3 is shown in Fig. 9.6. Analysis by variable angle spectroscopic ellipsometry (VASE) confirmed the surface was smooth across the entire substrate area with thickness deviation of $\pm 1.7\%$, a standard deviation, σ, of 1.0% only. Atomic force microscopic analysis of such a typical film revealed RMS values to be $\sim$6 Å, i.e. thickness differences in the range of a single monolayer only, irrespective of deposition rate [20–22].

Figure 9.7 illustrates the controllability of the OVPD process for achieving sharp

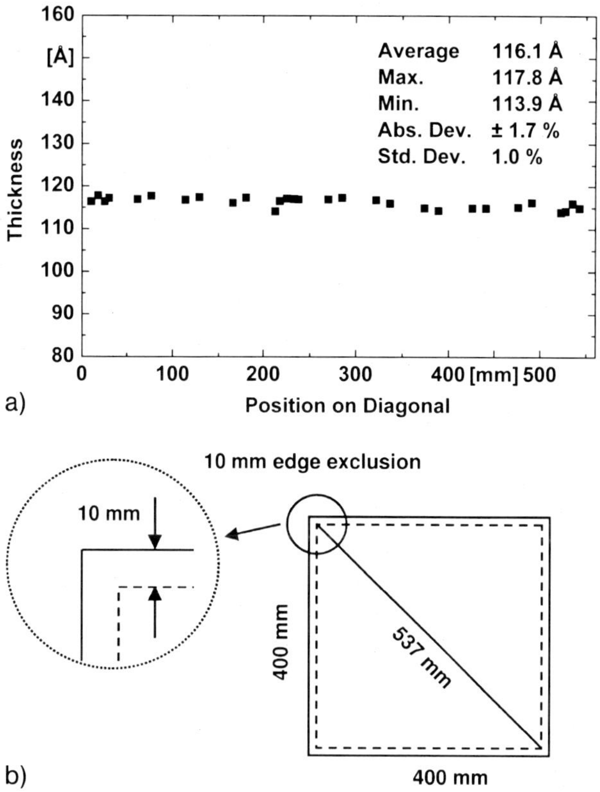

Fig. 9.6. (a) Film-thickness uniformity of Alq_3 deposited on a $400 \times 400\ mm^2$ glass substrate measured by variable angle spectroscopic ellipsometry (VASE) across the diagonal; (b) 10 mm edge exclusion of the $400 \times 400\ mm^2$ glass substrate.

interfaces, by measurement of the deposition rate using an in-situ quartz-crystal monitor. With a constant source flow of carrier gas through an Alq_3 container we demonstrated reproducible deposition rates of 12.28 and 12.27 Å s^{-1} for seven sequential depositions in two consecutive runs. For demonstration of a thick bulk layer we deposited for approximately 120 s with a constant deposition rate of 12.28 Å s^{-1} resulting in a total thickness of 1473 Å (1 in Fig. 9.7). Reducing the deposition time to 20 s resulted in a thickness of 245 Å (2 in Fig. 9.7). These single-layer results reveal that abrupt interfaces and reproducible thicknesses can be achieved by control of the pneumatic source valves.

Besides thickness reproducibility a basic requirement for an industrial production process is process stability. Figure 9.8 shows the long-term source stability of Alq_3 tested in an Gen 2 OVPD module. Using identical process conditions on days 1 and 21, more than 500 h continuous operation, we observed a deposition rate of 12.28 Å s^{-1} with a standard deviation of only 1.93%. Even an extensive equipment

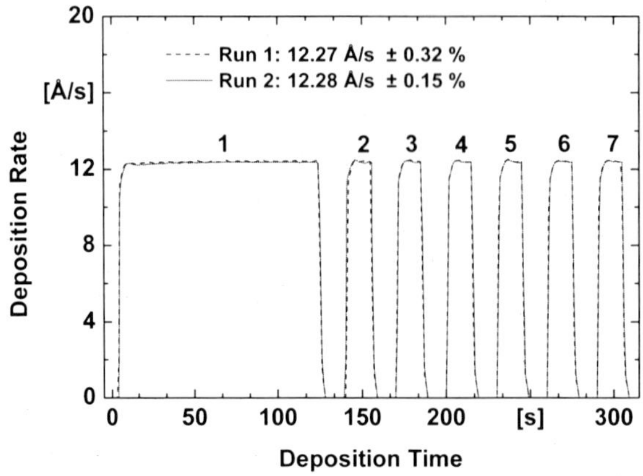

Fig. 9.7. Precise control of the deposition rates for Alq_3 in two separate runs with seven sequential depositions.

check and different OVPD depositions had no impact on organic film thickness and proves the industrial reliability for process stability and thickness reproducibility.

We subsequently demonstrated that the combination of close coupled showerhead and OVPD enables highly reproducible deposition rates with remarkable thickness uniformity. Alq_3 calibration curves revealed that deposition rates of 30 Å s^{-1} for a single container arrangement or up to 60 Å s^{-1} for a dual container configuration could be achieved. Because of the universal OVPD design principle, these high deposition rates are not limited to Alq_3 but could also be achieved for other typical organic materials. The standard deviation for Alq_3 in run-to-run and day-to-day thickness reproducibility was only 0.3 and 1.9%, respectively, demonstrating process stability and thickness reproducibility over a period of 500 h.

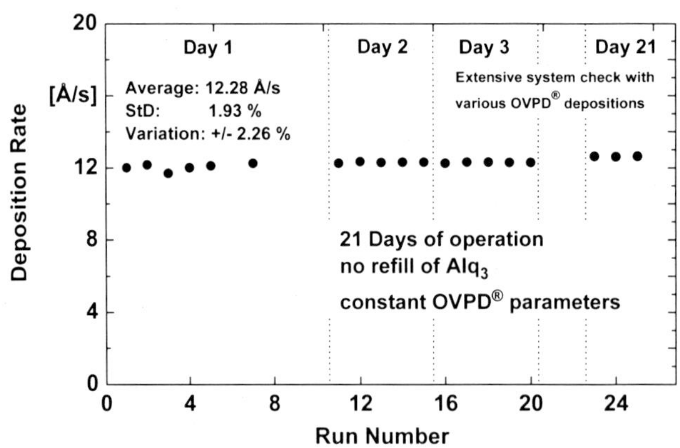

Fig. 9.8. Long-term source stability for Alq_3 in OVPD.

Spectroscopic ellipsometric characterization of OVPD-deposited films of Alq_3 and α-NPD by Himcinschi, Zahn and AIXTRON confirmed uniform layer thicknesses with standard deviations of 0.7% for Alq_3 and 0.9% for α-NPD across an 8-inch silicon wafer [20]. Atomic force microscopy images of such typical OVPD-deposited Alq_3 layers confirmed smoother surfaces characteristics, with RMS values of ~6 Å for the OVPD films; this corresponds to the thickness of a single monolayer for Alq_3 [21]. During determination of optical constants up to the ultraviolet range of 9.5 eV they reported significantly higher stability of OVPD-processed Alq_3, which is usually degraded by contact with moisture and air. This observation was explained by the smoother surface and indicates that molecular packing of Alq_3 in the single layer may play a role in environmental stability. Latest results using X-ray reflection (XRR) analysis of single films deposited by OVPD with deposition rates up to 12.7 Å s^{-1} for α-NPD and 23.6 Å s^{-1} for Alq_3 confirmed the VASE thickness determination. XRR determined density also was unchanged and slightly lower than the density of crystals for Alq_3 and α-NPD, as expected [22].

These data prove that thin layer multi-heterojunction devices are enabled by OVPD with accurate thickness reproducibility and layer performance, which are essential manufacturing requirements in industrial mass production. The OVPD technology also enables controlled material transport and a high material utilization efficiency of 50–70%, based on condensation on to the substrate only. For state-of-the art VTE manufacturing technology material utilization efficiencies are in the range of 1 to 6% [23].

For organic TFT, relatively low deposition rates may be desirable, in the interest of improved molecular ordering and higher carrier mobility in the plane parallel to the substrate surface. Unlike VTE, in which deposition rates below approximately 0.1 Å s^{-1} are often difficult to control, OVPD technology enables easy and reproducible control of low deposition rates down to 0.01 Å s^{-1}. The longer deposition time at such low deposition rates is less critical, because total film thickness is often much lower in organic TFTs compared with OLEDs. OVPD, however, further enables morphology tuning and control by variation of proven process conditions, as described below.

In conclusion we have illustrated material-specific calibration plots for individual sets of process temperatures and other conditions. Extending this precise process controllability of a single material to several materials offer a valuable technology for coevaporation of several materials, for example mixed hosts or accurate doping of dyes.

9.2.2 Co-deposition and Doping in OVPD

Research on organic semiconductor materials has resulted in remarkable progress and the introduction of material combinations using hosts and single or multiple guests or additional host materials [24–30]. This demand for simultaneous coevaporation or doping is often difficult to achieve in VTE, but easily achieved by OVPD. In the previous section we explained the OVPD process conditions essen-

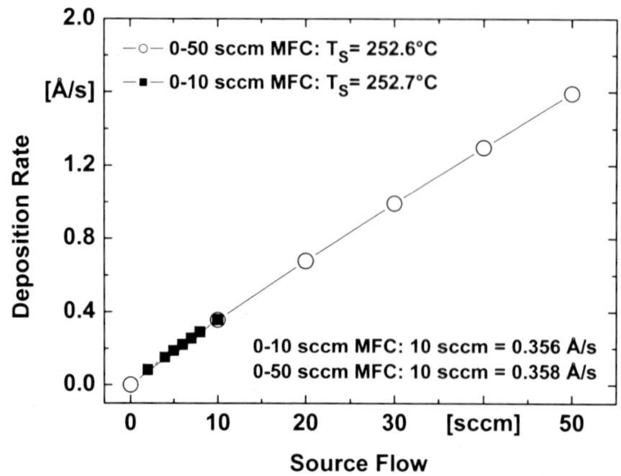

Fig. 9.9. Linear calibration curves for the dopant rubrene with a 10 sccm MFC (squares) and a 50 sccm MFC (circles). Both MFC achieved a 10 sccm deposition rate of 0.356 and 0.358 Å s^{-1}, a relative error of only 0.48%.

tial for deposition of single layers. Applying this concept of precise source flow controllability and long-term source stability to additional sources enables precise deposition, co-deposition, and doping of multi-component layers.

Almost linear OVPD calibration curves of the typical dopant rubrene for a variety of source flows up to 10 sccm and up to 50 sccm are presented in Fig. 9.9, which shows that the deposition rate can be precisely adjusted from 0.06 to 1.6 Å s^{-1}. Both curves are an ideal fit and reveal a linear relationship between deposition rate and source flow; they were collected with two mass-flow controllers of different capacity ranges (10 sccm and 50 sccm). Ellipsometric thickness analysis confirmed for both experiments a deposition rate of 0.3564 and 0.3582 Å s^{-1}, which is a relative error of only 0.48% and is identical with our prediction of dopant controllability (Table 9.1). Using a standard OVPD deposition rate of 10 Å s^{-1} for a hosts the doping range of rubrene can be very precisely adjusted in the range of 0–16%.

As we have seen in Fig. 9.7 the use of a precise mass-flow controller guarantees accurate and reproducible controllability for deposition of single layers. By use of physically separated sources in OVPD, which avoids any cross talk, this design of carrier transport can be extended from a single source to a high number of sources leading consequently to the deposition of any mixtures or doped layers, known as coevaporation or cohosting. By controlling the organic flux of two or even more organic materials the concentration of each organic species can be homogenously adjusted to any desired ratio in an abrupt or an ramping mode offering the potential to create precise stoichiometric mixtures in a single layer. Doping ranges from 0.1–50% are of interest in device development. All can be realized by OVPD. Besides this huge doping range the standard mass-flow controllability enables dynamic variation of each individual source flow. Ramping the carrier gas flows during the deposition of host materials and/or dopants enables deposition of even more complex layer compositions with gradually adjusted material mixtures.

Thin organic Alq_3/α-NPD heterostructures deposited by OVPD were investigated by use of spectroscopic ellipsometry by Himcinschi, Zahn and AIXTRON [31]. The analysis revealed that simultaneously mixed layers of Alq_3 and α-NPD consisting of 86.6% Alq_3 and 13.4% α-NPD were obtained; this is in good agreement with the target value of 15% for α-NPD and ellipsometry can be used for determination of multilayer compositions. These first results demonstrate that the easily controllable OVPD enables deposition of mixed layers or individual doping.

Consequently, this straightforward realization of vertical doping profiles and material mixtures enables precise layer fine-tuning to optimize device performance, for example power efficiency or lifetime [25]. For current OLED research and development and production [32, 33] doping and coevaporation are fundamental for device optimization and longer lifetimes.

9.2.3
Controlled Morphology and Layer Interfaces in OVPD

As we have described, OVPD technology enables the deposition of layers of pure materials or precise compositions of different materials, for example hosts, cohosts, and dopants. Besides layer composition, layer morphology [34], deposition rate [35, 36] and the substrate temperature [37–40] also affect device performance. Layer morphology is strongly affected by the substrate itself, but also by the deposition process and the deposition conditions.

Morphological changes of single layers or devices processed by VTE are known and have been widely discussed. Annealing procedures of fully processed and encapsulated devices are a known method for enhancement and stabilization (standardization) of device performance [37, 41]. In 2004 Djurisic and Chan reported the annealing and environmental stability of Alq_3 films deposited by VTE under different conditions [42]. They observed that annealing improves environmental stability, with substantially increased and stabilized PL intensity. Their observation indicates that morphological effects may play a role and is in accordance with the observation by Himcinschi and Zahn of higher stability of OVPD-deposited Alq_3 [21]. As discussed by Zahn, substrate temperature effects result in domains with different ratios of α- and β-PTCDA with increased surface roughness [43]. Cheng investigated single films of Alq_3 deposited by VTE with deposition rates of 0.01–1.03 Å s^{-1} [35]. Analysis by atomic-force microscopy revealed increasing film roughness of 12.0, 32.0, and 36.6 Å RMS with decreasing deposition rate. All these observed morphology effects had to be more or less tolerated in VTE, because process parameters other than deposition rate were not controllable. The crystalline growth mode in VTE was recently reported by Yang and Park and shown to lead to high performance in organic TFTs [78].

Forrest stressed as early as 1998 that two different types of growth regimes should be realized in this unique deposition process by variation of unique OVPD deposition parameters, for example substrate temperature, chamber pressure, and the concentration of organic molecules in the carrier vapor phase [8]. Yang and coworkers recently observed a strong effect of morphology control and material mixing on the power conversion efficiency of organic photovoltaic cells [40, 44].

A typical OVPD process with the substrate actively cooled to room temperature in combination with an increased chamber pressure and moderate deposition rate creates a deposition process which is diffusion-limited. The organic molecules are transported by diffusion into the boundary layer and are randomly distributed before temperature-induced condensation. Reaching the substrate the organic molecules can freeze as deposited, leading finally to amorphous films. With elevated substrate temperatures and reduced deposition rates the condensing molecules have sufficient energy and time to diffuse on the substrate surface before crystallization, creating three-dimensional structures. Here the intermolecular adhesion of the molecules exceeds that of molecular–substrate bond interaction, resulting in crystalline island growth. The gap-filling characteristics of this deposition regime can be also used for surface planarization, which is important if smooth layer uniformities are desired. This characteristic of OVPD enables the precise formation of large interface areas and interpenetration networks, which are essential for high photovoltaic cell power efficiencies. This gap-filling characteristic of OVPD also tolerates all particles, which are covered and planarized without any voids. This beneficial side-effect is important in mass production on large substrate sizes, in which particle contamination is critical.

Changing the OVPD conditions toward a lower substrate temperature in combination with a lower deposition chamber pressure and a higher deposition rate results in kinetic deposition. The number of intermolecular collisions of the organic molecules is reduced and the condensed molecules have less energy to crystallize. Because of their higher deposition rates the organic molecules directly freeze into an amorphous mesh or have only sufficient time to form much smaller crystals. Long-range diffusion of the organic molecules on the substrate is thus prevented and, finally, the deposited layer conforms to the substrate topology, without any voids in the layer deposited.

By enabling these two different intrinsic deposition regimes OVPD enables rapid switching between morphology-specifying deposition conditions. Depending on the desired device properties the morphology of organic layers may vary from amorphous to polycrystalline (in which the crystal dimensions are well defined) or, finally, to a perfectly crystalline layer. Morphology effects in ballistic VTE are very difficult to examine whereas the OVPD technology enables precise controllability in performance-driven device optimization.

This precise controllability of film morphology is requirement for adjustment of the interfaces between the substrate and the first organic layer and subsequent organic–organic interfaces. This feature is unique to gas-phase transport in OVPD and is not realizable in the ballistic transport principle of VTE. As described later for organic solar cells, controlled morphology of the organic thin films and their interfaces has huge potential for device and performance optimization. In conclusion, it is obvious that a better-controlled interface will be a means of improving device performance and operating lifetime. OVPD, with its two tunable growth modes, is a valuable and unique tool for exploiting this benefit.

9.3 Electronic Devices by OVPD

The most advanced organic electronic systems already in commercial production are high-efficiency, very bright and colorful thin displays based on organic light-emitting devices [45]. Significant progress is also being made in the realization of organic TFTs and low-cost organic photovoltaics (PV) for solar energy generation. The successful application of organic electronics will depend on capturing its low-cost potential by use of innovative technology for fabrication of devices on flexible and inexpensive large-area substrates [46, 47]. Starting in niche markets, organic electronics also have the potential to replace existing products in flat panel displays, solid-state lighting, photodetection, low-cost integrated circuits, and low-cost power generation. These different applications of organic electronics will be discussed briefly.

9.3.1 OLEDs Made by OVPD

The starting strategy to prove the unique advantages of OVPD is to demonstrate OLED devices by translating the VTE device structure into OVPD process parameters and obtain VTE-identical or better OLED device performance. With this VTE-based device structure OVPD process parameters are carefully screened and optimized to improve the performance and lifetime at device level. These OVPD parameters include deposition rate and the deposition time, for adjusting the thickness and morphology of each layer, and the doping range and profile of dopants. Use of the many combinations of OVPD process parameters (deposition rate, deposition pressure, and substrate temperature) to adjust the morphology in addition to variation of layer thicknesses and doping profiles results in a huge possibility of improved OVPD device performance. Finally the next step in the OVPD process strategy is to use these unique OVPD advantages as an option to extend device performance beyond VTE-based reference devices using new, maybe more complex, device structures which cannot be realized by VTE.

A first demonstration of phosphorescence-doped OVPD–OLEDs with identical VTE performance was achieved at Universal Display Corporation (UDC) and at TU Braunschweig [48] by use of PtOEP. The electroluminescence spectrum, with emission at 651.1 nm, and the structure of the device are shown in Fig. 9.10. The external quantum efficiency of this OVPD device reached 3.88% at 3.7 V forward potential (Fig. 9.11); this is identical to the external quantum efficiency of the nearly identical VTE device reported by Baldo and Forrest [49].

In 2005 Universal Display Corporation published OVPD results of an advanced

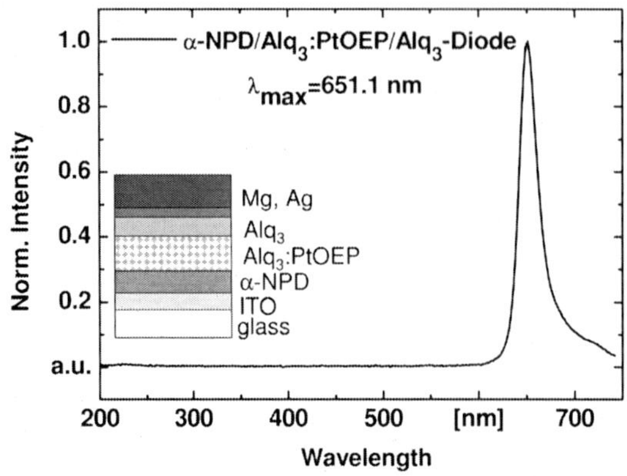

Fig. 9.10. Electroluminescence spectrum of the OVPD–OLED. The structure of the device is shown in the inset.

phosphorescent OLED (PhOLED) using the highly efficient green dopant $Ir(ppy)_3$ [50]. The structure of the device and the OVPD deposition rates are shown in Fig. 9.12.

Detailed performance analysis of the OVPD device again revealed comparable or even slightly better device performance of the OVPD processed device compared with the VTE reference. Figure 9.13 shows the luminous efficiency of the PhOLEDs processed by OVPD and VTE. The OVPD-PhOLED reached 25 cd A^{-1} and the VTE device approximately 24 cd A^{-1}, equivalent to a maximum external quantum efficiency of $7.0 \pm 0.1\%$, which is desirable for active matrix applications. Besides these comparable luminous efficiencies, the spectra differ slightly in their shape

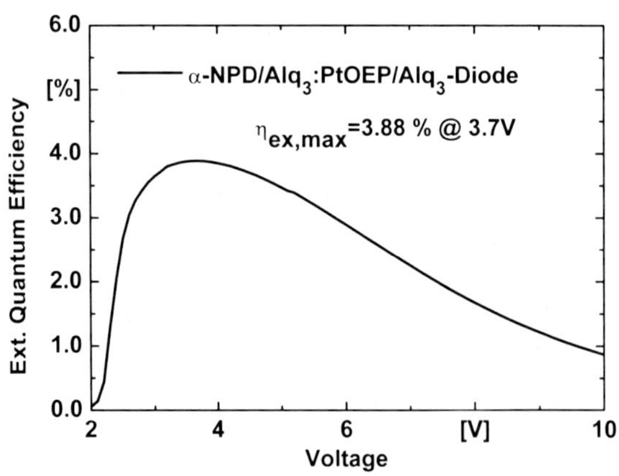

Fig. 9.11. External quantum efficiency of 3.88% for the PtOEP-doped OVPD–OLED, demonstrating that OVPD and VTE devices enable identical performance.

Layer	OVPD® Dep.-Rate [Å/s]	Dep. Time [s]
100 nm Al 1 nm LiF	deposited by VTE	deposited by VTE
40 nm Alq_3	3	133
10 nm BAlq	5	20
30 nm CBP with 4.3 vol.% $Ir(ppy)_3$	4.8 + 0.2	60
30 nm α-NPD	1.4	213
10 nm CuPc	1 by VTE	1 by VTE
Anode		

Fig. 9.12. Device structure and OVPD deposition rates for the phosphorescent OLED demonstrated by UDC. (Figure courtesy of UDC.)

and width, indicating that morphologically different interfaces may play a role. The inset in Fig. 9.13 shows the current density characteristics for the OVPD and the VTE processed devices; the devices have almost identical performance.

Finally, the operational lifetimes of both devices were analyzed. Figure 9.14 shows the normalized luminance of the devices tested under accelerated test conditions of high-dc current injection (40 mA cm^{-2}), corresponding to an initial

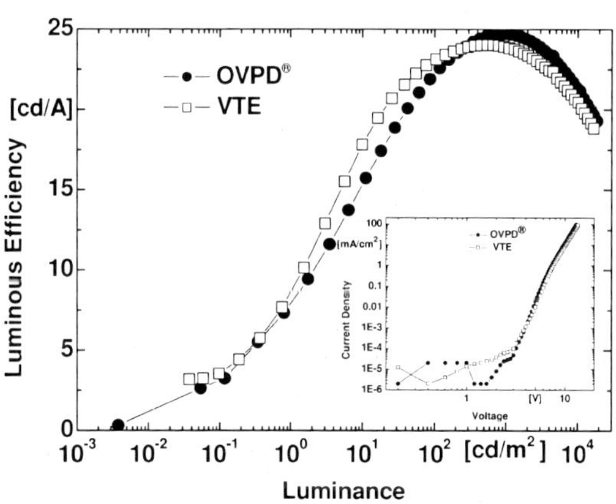

Fig. 9.13. Luminous efficiencies for the UDC PhOLED deposited by OVPD (circles) and VTE (squares). The VTE device reaches a maximum efficiency of 24 cd A^{-1} whereas that for the OVPD device is slightly better – 25 cd A^{-1}. The inset shows almost identical current density–voltage characteristics of the PhOLED deposited by OVPD and VTE. (Figure courtesy of UDC.)

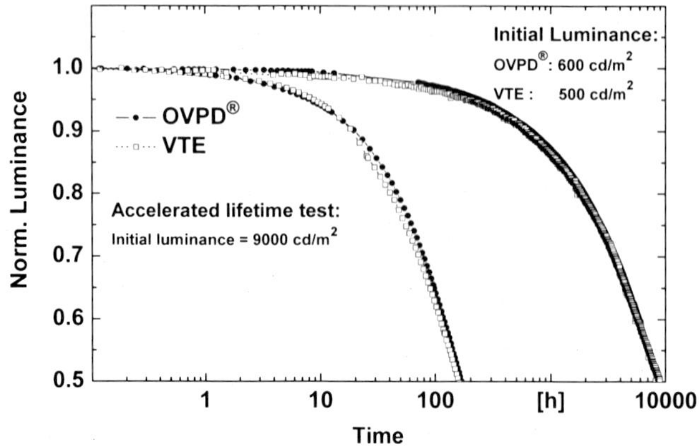

Fig. 9.14. Normalized luminance of PhOLEDs deposited by OVPD (circles) and VTE (squares) as a function of time, showing the operational lifetime under highly accelerated aging condition (initial luminance $L_0 \approx 9000$ cd m^{-2}) and under typical display luminance conditions (initial luminance $L_0 \approx 500$ cd m^{-2} and $L_0 \approx 600$ cd m^{-2}). (Figure courtesy of UDC.)

luminance of $L_0 = 9000$ cd m^{-2}. Both devices have identical lifetimes and reached 50% of initial luminance after approximately 160 h of continuous operation at room temperature. In addition to these highly accelerated test conditions the OVPD device was also tested with typical display luminance conditions of 600 cd m^{-2} and had 50% of the initial luminance after continuous operation for almost 10 000 h. The VTE device reached this lifetime of 10 000 h only under luminance test conditions of 500 cd m^{-2}.

With these data Universal Display Corporation proved that OVPD-PhOLEDs achieved comparably good results as VTE-processed devices. After the improvement strategy Universal Display Corporation began to modify the OVPD process conditions and doubled deposition rates for the emissive layer of CBP doped with $Ir(ppy)_3$ from approximately 4.8 to 10.3 Å s^{-1} by increasing the source flow and reducing the deposition time for the emissive layer from approximately 60 s to 29.1 s (see table in Fig. 9.12 for OVPD deposition rates and deposition times). This optimized device was analyzed at device level and also tested under lifetime conditions. Fig. 9.15 shows the luminous efficiencies for both OVPD devices. The shapes of the curves are almost identical and luminance efficiencies were even slightly higher for OVPD devices deposited at high deposition rates. These results show that high deposition rates of more than 10 Å s^{-1} for the emissive layer have no negative effect on device performance. Lifetime testing of both devices under accelerated test conditions showed operational performance to be almost identical, as shown in Fig. 9.16.

These results for OLED display applications show that OVPD-deposited devices achieve at least the same performance as the VTE-based devices. Deposition rates are, furthermore, higher for OVPD and are suitable on the device level (Fig. 9.16).

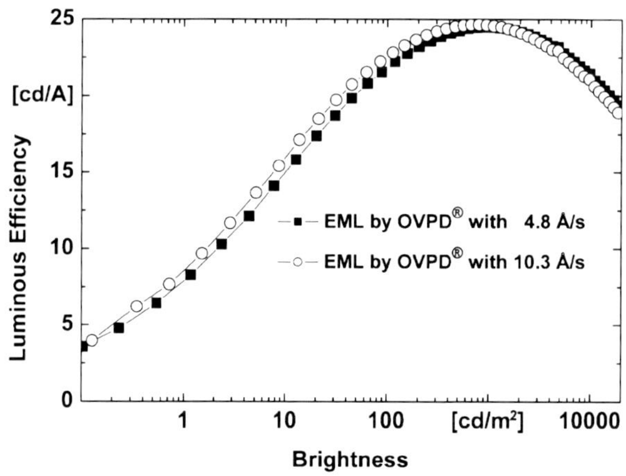

Fig. 9.15. Luminous efficiencies of PhOLEDs with CBP:Ir(ppy)$_3$ emissive layers deposited at 10.3 Å s^{-1} (circles) and 4.8 Å s^{-1} (squares). (Figure courtesy of UDC.)

On the basis of these initial results passive matrix OLED displays were realized in a German government funded project, headed by AIXTRON, to investigate the potential of OVPD for display applications. Figure 9.17 shows low information-content displays produced on passive matrix substrates and large-area OLEDs.

AIXTRON actively supports the research and development activity on OLEDs to improve their lifetime [51] and to explore and optimize the OVPD process in several

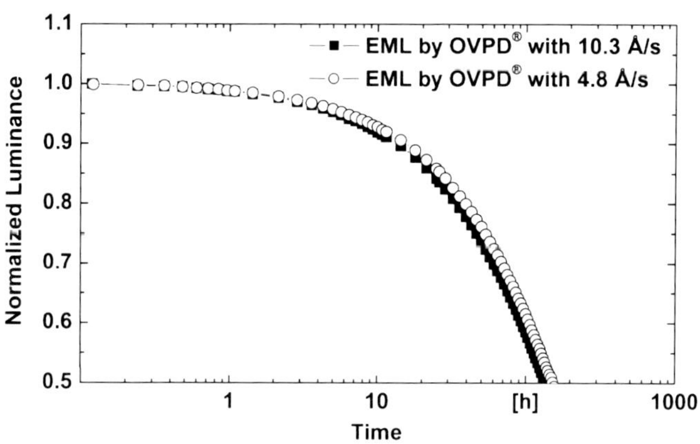

Fig. 9.16. Normalized luminance as a function of time of PhOLEDs deposited at 10.3 Å s^{-1} (squares) and 4.8 Å s^{-1} (circles) by OVPD, showing operational lifetime under highly accelerated ageing conditions (initial luminance ~9000 cd m^{-2}). (Figure courtesy of UDC.)

Fig. 9.17. Passive matrix display and large-area OLED logos showing RGB and AIXTRON obtained by OVPD at TU Braunschweig.

funded projects. In addition to OLEDs for display applications and white light OLEDs for indoor illumination, targets for device application also include organic photovoltaics. Each project combines the fundamental elements for process development – expertise in the field of organic materials and chemistry, in OVPD hardware and process conditions, and on the device level for individual application. We are thereby creating a strong network of OVPD process experience at universities and our industrial project partners.

9.3.2
Organic Photovoltaics by OVPD

Application of organic films for organic photovoltaic cells is being actively investigated in current research and development [5, 52–56]. Forrest demonstrated improved photovoltaic power conversion as a result of controlled growth of molecular bulk heterojunction (HJ) photovoltaic cells using the unique OVPD technology by applying two different growth modes [40, 44]. The power conversion of the same organic PV was analyzed in respect of its deposition process and deposition regimes. The planar heterojunction of the organic PV, deposited by VTE and by OVPD, achieved a power conversion of 1.1%. Post-growth annealing improved power conversion to 1.4% and was explained in terms of thermodynamically driven phase segregation, also forming inclusions and cul-de-sacs, but which also resulted in space charge effects reducing the fill factors of the organic photovoltaic device. Using the two different growth conditions in OVPD Forrest demonstrated the controlled growth of a thin but continuous wetting layer attached with short polycrystalline CuPc needles, all of similar dimensions. This controlled OVPD resulted in an increased interface surface area without formation of inclusions or cul-de-sacs reaching a power conversion of 2.7%, an increase by a factor of 2.5 with the highest fill factor of 58% in this examination. These results of controlled morphology and layer interfaces by application of OVPD in a variety of organic PV cells and direct comparison of their efficiency performance data with that for identical devices processed in VTE are explained in terms of reduction, or even elimination, of the bottle-neck of exciton diffusion length. A controlled CuPc/PTCBI HJ cell prepared by OVPD with a exciton diffusion length, L_D, of 3 nm for PTCBI reached an efficiency, η_P, of 2.7%, comparable with that of typical CuPc/C_{60} cell efficiencies of $\eta_P \approx 3\%$, with an exciton diffusion length of 40 nm for C_{60}. Consequently power conversion efficiencies of organic PV are probably more influenced by device morphology and its controlled interface than by their organic material composition [40].

AIXTRON already joined a German government-funded project, headed by the Hahn–Meitner Institute to explore industrial production of organic solar cells by OVPD.

9.3.3 Organic Thin-film Transistors by OVPD

In addition to OLEDs and organic photovoltaics, thin-film transistors (TFTs) are another research topic in organic electronics already attracting the interest of industry, because of their flexibility and low-cost potential.

Basic requirements for organic TFTs are high carrier mobility, high on/off current ratio, high current output and a low operating voltage. Pentacene is one of the most promising organic materials employed as the active material in organic TFTs. Thin-film pentacene transistors have been fabricated by solution precipitation, organic molecular beam deposition, VTE, and OVPD, all resulting in comparable device performance [10].

In 2000 Klauk et al. reported a TFT structure with reduced complexity [57]. They also described substrate-surface treatment with organosilane before pentacene deposition, which significantly increased hole mobility [58]. This observation of increased mobility was explained by improved molecular ordering of the molecules in the pentacene layer. Using the two different deposition regimes in OVPD Shtein et al. demonstrated that the crystal size of pentacene could be controlled and, obviously, that the morphology of pentacene layers is tunable. As a result of the additional organosilane treatment of the substrate surface and by using the diffusion-limited deposition regime in OVPD they achieved hole mobilities of up to 1.6 $cm^2\ V^{-1}\ s^{-1}$, comparable with electron mobilities in amorphous silicon TFTs [10].

In 2004 Ma and Yang introduced unique device architectures for organic TFTs by VTE with a vertically stacked layer arrangement for the gate, source, and drain electrodes [59]. They reported a low operating voltage (less than 5 V), high current output (up to 10 mA or 4 A cm^{-2}), and an ON/OFF ratio of 4×10^6, and pointed out that a thin and rough source electrode is essential for the performance. Obviously, the high controllability of the surface morphology in OVPD may result in additional improvements for such vertically stacked device arrangements. Ma and Yang also combined the organic TFT and an OLED in a single device and demonstrated a brightness of 1000 cd m^{-2} for the organic TFT-driven OLED. This indicates the feasibility of active matrix-driven OLEDs by organic TFTs enabling potentially low-cost production of the organic display backplane.

Consequently, organic electronics in combinations of several applications such as organic TFTs and OLEDs have huge potential for low-cost production and large-volume products using a variety of substrates, for example glass, metal foil, or plastics.

9.4 Full-color OLED Displays

OLED displays are the most promising technology for low-power high-quality flat-panel displays, although there is much competition, especially from liquid crystal displays (LCDs), which currently dominate market share. Because the viewing characteristics of LCDs have been substantially improved within the last decade, the demands to be fulfilled by OLED displays have also substantially increased. In the long-term only full-color high-resolution AM-OLED displays will be successfully introduced to the market on a large scale. Furthermore, competitive costs in the long term can be achieved only by mass production with substrate sizes comparable with those of LCD production lines.

Three main approaches are used to realize full-color OLED displays. One of these is color from white by applying a color-filter-matrix to a white backlight-emitting OLED display. Another approach is to convert blue light into green and red light by use of a color-converting phosphor matrix. These first two alternatives enable RGB-displays but also result in loss of power efficiency because of the color filters/converters. As a consequence the micro patterning approach of red, green, and blue OLEDs is favored and offers high efficiency combined with low-cost without any color filters.

An individually addressable double-sided 1.5 inch AM-OLED display was recently presented by AU Optronics; these are the thinnest OLED devices for mobile applications [60]. Samsung SDI has demonstrated a 17 inch UXGA-AMOLED display suitable for HDTV application in home entertainment [61].

9.4.1 Micropatterning by use of Shadow Masks

High-precision shadow masking is the most common patterning technique for vacuum-deposited OLEDs, with typical pixel apertures in the range 10–100 μm. Shtein et al. investigated the micropatterning of organic semiconductors using OVPD [11]. In VTE the substrate is placed within a distance less than the molecular mean free path (mfp) from the source and the film patterning through a shadow mask depends on the crucible and the mask geometry defining the trapezoidal profile of the deposit by the ballistic trajectories. VTE also uses substrate rotation to reduce non-uniformities of the deposited film originating from the source crucible dimensions and uneven heat conductivity, which result in deposition rate

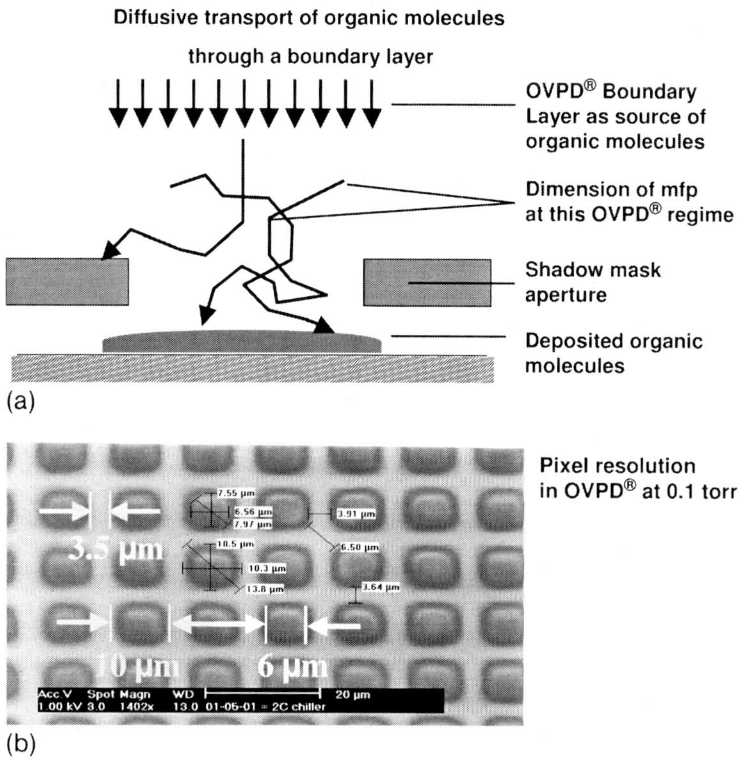

Fig. 9.18. (a) Conceptual schematic diagram of the OVPD process through a shadow mask. (b) SEM showing Alq_3 pattern dimensions deposited by OVPD at 0.1 torr on a silicon wafer with a shadow mask (mask thickness 3.5 µm, aperture 7.5 µm, mask-to-substrate separation < 1 µm) (Figure courtesy of S. R. Forrest).

fluctuations. Substrate rotation also promotes uniform film deposition to achieve the same shape even for pixels not directly located above the VTE crucible. In OVPD the locations of evaporation and condensation are not connected and source geometry has no impact on the aperture resolution. Monte Carlo simulations and experimental results by Shtein et al. indicate that the deposited shape in OVPD is controlled by the mfp, aperture geometry, and mask-to-substrate separation. Figure 9.18 shows a schematic diagram of the diffusion length of organic molecules in the OVPD process through a shadow mask with a side-wall-angle of 90°. The mask aperture dimensions were 7.5 µm, the mask thickness 7 µm, and the mask-to-substrate separation 3 µm. Using this mask geometry in OVPD Shtein et al. produced bell-shaped pixels of Alq_3 approximately 6 µm wide for the thick Alq_3 deposit and approximately 10 µm wide for the diffuse tailing effect.

In general the mfp is mainly determined by the pressure. In OVPD, typically operating at higher pressures than VTE, the gas phase also builds up a hydrodynamic boundary layer whose thickness is precisely adjusted by the total flow and the pressure in the deposition chamber. Organic molecules must diffuse through the boundary layer and are randomly distributed across the shadow mask. With a typical deposition pressure in the range 0.1–13 mbar the mfp of nitrogen in OVPD reaches 100–1 μm, which defines the resolution in OVPD. It is the boundary layer with its randomly distributed molecules above the shadow mask acting as the source of organic molecules. Because of this random diffusion transport, the pixel shape can be controlled to a bell-shaped profile. Here the mask thickness and the aperture wall design affect the deposition efficiency and the shape of the deposited film. The sharpest patterns are achieved for the thinnest mask and for apertures with side-wall-angles of 45°. Finally, the mask-to-substrate separation must be smaller than the mfp to reduce diffuse pattern edges.

For many future organic TFTs applications, for example large-area sensor or active-matrix display panels, in-line high-resolution patterning of the organic semiconductor layer will be crucial. Organic layer patterning for organic TFT with VTE-deposited organic layers has been demonstrated using photolithographic methods [79, 80] and shadow-masking [81]. Compared with these conventional techniques, OVPD with high-resolution in-line micropatterning is simpler and process control is better; this may greatly improve system performance.

AIXTRON is investigating the micropatterning of organic films through a variety of shadow masks with different apertures by modeling and by experimental screening of typical OVPD process parameters for large-area application. In conclusion, the mask design can be adjusted to the OVPD process conditions resulting in pixel shape comparable with VTE. The top-down deposition geometry in OVPD, in contrast with bottom-up deposition in VTE, can enable a further decrease in mask thickness and mask-to-substrate separation for high-resolution patterns.

9.4.2
Mask-less Processes

The deposition of organic thin films and the micropatterning step into pixel dimensions are two different processes and do not have to be performed principally at once in a single step, for example by using high-precision shadow masks. Also, the major disadvantage of high-precision shadow masking is the limited scalability of the mask itself.

A variety of technological solutions have been developed to decouple the deposition and the structuring. Samsung SDI developed a 2.6-inch full color VGA AM-OLED with a resolution of 302 ppi by use of laser-induced thermal imaging (LITI) [62]. The principle is based on a vacuum-deposited organic film approximately 30 nm thick on a donor film. The donor film is placed in intimate contact with the poly-Si-TFT substrate and then exposed to the LITI patterning process. Using a high-precision position accuracy of ±3.5 μm and an LITI process width variation of ±2.0 μm they demonstrated an aperture ratio of 40–50% with a fine pattern of 28 μm pixel pitch. To circumvent the potential particle generation in the solid-to-solid contact of donor film and substrate of the LITI Process, Eastman Kodak developed the radiation-induced-sublimation transfer process (RIST) [63]. Here a spacer is placed between donor film and substrate, maintaining a transfer gap of 1–10 μm, which secures pixel resolution and avoids any particle contamination. Kröger and Kowalsky also developed a very similar non-contact laser induced local transfer, LILT [64]. Using the LILT process and typical scanning speeds of up to 1.0–1.7 m s^{-1} they demonstrated RGB OLEDs with lateral resolution of 40 μm.

These latest results demonstrate that decoupling deposition and micropatterning can be performed on industrial scale leading to sophisticated full-color displays.

9.5 Material Properties of Organic Molecules for Use in OVPD

As we have described in the previous sections the OVPD process differs in its intrinsic properties from the VTE deposition principle. Differences of deposition technologies for the processed organic materials must also be considered for successful processing. A general advantage of organic electronics is that chemical synthesis provides organic materials with well-defined molecular properties for the desired application. Current trends in materials design focuses on improved electro-optical material properties [65–67], improved charge-blocking materials [68], optimized guest–host materials [69], improved phosphorescent dopants [70–72], improved lifetimes by doping [73], improved understanding of thermal, structural and photo-physical properties of organic material [74], and molecular glasses with higher glass-transition temperatures [75] for improved morphological stability of devices.

A consequence of this chemical design freedom is that multiple materials are screened for their performance and many materials will be used in industrial manufacturing for individual applications. As a result all these materials are not a chemical commodity and are therefore expensive. In addition each material must be suitable for the desired deposition technology. With its high material utilization efficiency of 50–70% the OVPD process offers high cost-reduction potential. This high material utilization efficiency in OVPD may, moreover, make research and development for new organic materials in the chemical industry more cost-efficient. This trend is already apparent for the highly efficient triplet metal–organic complexes based on iridium [70–72, 76] or other transition metals complexes, which are driven by OLED device performance.

Besides chemical synthesis, purity of the employed chemicals is essential for the organic device performance desired. Small-molecule materials, which are only poorly soluble, are usually purified by repeated train sublimation, which results in substantial losses of material and additional purification costs. These material requirements are almost identical for OVPD and VTE. Because of the different deposition processes, the thermal stress on the organic materials is different in OVPD and in VTE. OVPD operates under steady-state temperature conditions without ramping of the evaporation temperature during processing, as is typically applied in VTE. Depending on the vapor pressure of the organic materials used and the carrier gas transport the steady-state container temperature in OVPD can be significantly lower than in VTE. Even at lower process temperatures the thermal stress over time may, furthermore, differ for OVPD and VTE, if VTE is not operated in continuous operation mode.

9.6 Summary

There are similarities but also many differences between the deposition technologies VTE and OVPD. On the basis of a detailed technical evaluation of both technologies OVPD, as the new deposition technique, must produce results comparable with those of VTE and, additionally, advantages, to attract industrial attention. Besides overall hardware equipment differences, for example source containers and showerhead, these two technologies differ also in the principle of deposition on the molecular level. The individual process characteristics of OVPD and conventional VTE, which will be discussed below, are listed in Table 9.1.

Use of a carrier gas in OVPD enables deposition of the organic materials at a pressure of 10^{-3}–10 torr (Table 9.1, no. 1) whereas the VTE system must be pumped down to high-vacuum conditions, which is a time-consuming and critical

Tab. 9.1. Intrinsic process characteristics of close coupled showerhead OVPD technology compared with conventional VTE.

No.	Process properties	OVPD	VTE
1	Pressure	10^{-3}–10 Torr	$<10^{-6}$ Torr
2	Scalability	Two-dimensional	Three-dimensional
3	Layer uniformity	(~1%)	(~3%)
4	Doping control	<0.5%	>2.0%
5	Cross contamination	Low	High
6	Thickness control	~5–10 Å	~5–10 Å
7	Deposition efficiency (material utilization)	50–70%	<15%
8	System downtime	Low because of hot wall design	High because of particle generation
9	Deposition rate	~30 Å s^{-1}	~3–10 Å s^{-1}
10	Morphology control	Tunable	Difficult to control

procedure if, for example, moisture must be baked out from all the parasitic surfaces of the VTE module.

As already described, the close coupled showerhead in OVPD enables homogeneous distribution of the organic gas phase and two-dimensional scalability (Table 9.1, no. 2) suitable for deposition on larger substrates. Conventional VTE operates with a point source, which is placed about the same distance from the substrate as the substrate diagonal to achieve the desired layer uniformity. This three-dimensional approach in VTE dramatically increases the equipment size and pump-down time to high vacuum, obviously limiting the scalability of VTE equipment.

Scaling was introduced to VTE by using linear sources [23, 77]. Nevertheless, realization of high deposition rates in VTE needs high evaporation temperatures to enable the material flux. In OVPD evaporation and condensation of the organic materials are separated independent processes. Transport of material from the evaporation zone to the deposition zone is achieved by using a carrier gas to realize molecular mass transport. High deposition rates are thereby enabled under these steady-state temperature conditions by continuously improved source design and by controlling the flow of carrier gas, not by increasing the evaporation temperature.

The uniformity of the deposited layer (Table 9.1, no. 3) also differs in both deposition technologies. In OVPD the organic molecules are randomly distributed by intermolecular collisions with carrier gas molecules which results in a very uniform and quantitative coverage of the substrate. OVPD thus also has the potential to cover unintended substrate non-uniformities, for example defects or particles. Consequently OVPD can also be applied to complex three-dimensional structured substrates. A single layer of Alq_3, deposited by OVPD on a silicon wafer had a thickness uniformity of only 0.6% standard deviation, and surface roughness analysis by AFM confirmed, with an RMS value of 0.6 nm, that the thickness deviation of the Alq_3-layer is already in the molecular dimension [20].

In VTE arrival of the individual molecules can be described as ballistic transport, thus the mfp is comparable with the crucible–substrate distance. Consequently the symmetry of the crucible, for example point source, multiple point source or linear source, and the texture of the organic materials loaded, is reflected in the thickness uniformity and layer coverage of the substrate. This explains the shadowing effects observed for structured substrates [12, 40, 44]. As a result, coverage of the substrate by VTE is less uniform and may lead to pin-holes and is obviously not as perfect or quantitative as in OVPD. To reduce this disadvantage in VTE and to improve layer uniformity and coverage VTE uses, for example, substrate rotation to randomize the ballistic trajectories.

By actively switching the organic source, which is kept at accurate evaporation temperatures, evaporation, and thus the amount of organic material transported from the source, is precisely controlled in OVPD. The organic material is transported through run lines into a mixing unit and through the showerhead onto the substrate without any loss of material. This enables high thickness reproducibility of the deposited layers. Similarly precise handling of multiple sources, for example host and dopant, also enables precise coevaporation or doping control (Table 9.1, no. 4) of $\sigma < 0.5\%$. Single films and hetero structures of OVPD-processed Alq_3 and α-NPD have been investigated by VASE; this revealed that with different optical constants the dopant concentration can be monitored precisely [31]. In VTE precise coevaporation or doping is realized by ramping the crucible temperature, which results directly in exponential changes of the deposition rate. Precise doping in VTE is therefore difficult, with $\sigma > 2.0\%$.

Transport of all the organic material into the deposition chamber prevents any cross contamination (Table 9.1, no. 5) of the organic sources and enables the deposition of multiple layers in a singular OVPD module. The OVPD module can thus be extended to the desired number of sources for individual manufacturing needs.

Thickness controllability (Table 9.1, no. 6) and reproducibility in OVPD is achieved by accurate adjustment of the flow of carrier gas by means of mass-flow controllers whereas in VTE quartz crystal monitors are used to control the rate of deposition by adjustment of the evaporation temperature. In VTE small deviations of the evaporation temperature are known to affect the stability of the deposition rate and consequently the layer thickness, which may also affect the roughness and morphology of the VTE-deposited layer.

Because there is no loss of organic material during carrier gas transport in OVPD, material utilization efficiency is very high (Table 9.1, no. 7). Depending on hardware geometry and individual process conditions material utilization efficiencies of 50–70% are achieved. In VTE material utilization efficiency is only approximately 1 to 6% [23], because of huge parasitic surfaces. Here OVPD has the large advantage of cost reduction, especially important for extremely expensive organic compounds designed for product-specific applications of small amounts of material. The high material utilization efficiency in the hot wall OVPD module also reduces particle formation and its carrier gas purges the module continuously to maintain stable process conditions. Obviously expensive maintenance cycles are less frequent, enabling increased standby times for production with significant reduced system downtime (Table 9.1, no. 8).

In OVPD the deposition rate is controlled by the amount of carrier gas passed through individual source containers and the vapor pressure of the organic material, and reproducible high deposition rates of $30\,Å\,s^{-1}$ (Table 9.1, no. 9) have been demonstrated. In VTE typical deposition rates for OLEDs are adjusted by controlling evaporation temperatures, and are in the range $1–10\,Å\,s^{-1}$. The effect of deposition rate on surface roughness and on the electrical and luminance performance of OLED devices is known in VTE [36]. The high deposition rates possible with OVPD are currently an important topic in industrial research and development, because they lead to substantially reduced tact times, which are essential for low-cost production.

Another advantage of OVPD over VTE is the ability to control surface morphology (Table 9.1, no. 10). Use of two different deposition modes in OVPD enables active design of layer morphology and interfaces with very valuable properties for device improvements; this is of particular importance for high-performance organic TFTs.

In conclusion OVPD technology is a powerful mass-production method with unique advantages for low-cost production of organic electronics with the highest performance and improved lifetimes in devices in daily use.

Acknowledgment

The authors gratefully acknowledges the support and valuable contributions of M. Gersdorff, H. Blei, D. Keiper, M. Schwambera, M. Reinhold, K.-H. Büchel, F. Kittel, M. Dauelsberg, C. Martin, D. Brien, G. Strauch, and B. Marheineke at AIXTRON AG, Germany; J.J. Brown, T.X. Zhou, and T. Ngo at Universal Display Corporation, USA; S.R. Forrest, M. Shtein, and F. Yang at Princeton University, USA; S. Hartmann, A. Janssen, H.-H. Johannes, R. Caspary, and W. Kowalsky at TU Braunschweig, Germany; C. Himcinschi and D. Zahn at TU Chemnitz; K. Fostiropoulos, M. Rusu, S. Wiesner and M.C. Lux-Steiner at Hahn–Meitner Institut, Berlin, Germany; M. Beigmohamadi, P. Niyamakom, A. Farahzadi, S. Kremers, and M. Wuttig at RWTH Aachen; P. v. Gemmern, F.O. Jessen, and H. Kahlisch at RWTH Aachen; and S. Brand, W. Müller, H. Schwab, and D. Bertram at Philips, Aachen who performed most of the work reported in this article.

Part of this work was financially supported by the Federal Ministry of Education and Research in Germany (BMBF, No. 001BD153; No. 13N8650 and BMU 0329927B).

OVPD technology has been exclusively licensed to AIXTRON from Universal Display Corporation, Ewing, NJ, USA, for equipment manufacture. OVPD technology is based on an invention by Professor Stephen R. Forrest et al. at Princeton University, USA, which was exclusively licensed to UDC. AIXTRON and UDC have jointly developed and validated OVPD pre-production equipment. OVPD is a registered trademark.

References

1 J. Shinar in *Organic Light-Emitting Devoices: A Survey*, Springer **2004**.

2 L. S. Hung, C. H. Chen, *Materials Science and Engineering*, **2000**, *R 39*, 143.

3 S. Forrest, *MRS Bulletin*, **2001**, 108.

4 S. R. Forrest, *Chem. Rev.*, **1997**, *97*, 1793.

5 S. R. Forrest, *MRS Bulletin*, **2005**, *30*, 28.

6 B. D'Andrade, V. Adamovich, R. Hewitt, M. Hack, J. J. Brown, *SPIE*, **2005**, *5937*, 87.

7 P. E. Burrows, S. R. Forrest, L. S. Sapochak, J. Schwartz, P. Fenter, T. Buma, V. S. Ban, J. L. Forrest, *J. Crystal Growth*, **1995**, *156*, 91.

8 M. Baldo, M. Deutsch, P. Burrows, H. Gossenberger, M. Gerstenberg, V. Ban, S. R. Forrest, *Adv. Mater.*, **1998**, *10*, *18*, 1505.

9 M. Shtein, H. F. Gossenberger, J. B. Benzinger, S. R. Forrest, *J. Appl. Phys.*, **2001**, *89*, *2*, 1470.

10 M. Shtein, J. Mapel, J. B. Benzinger, S. R. Forrest, *Appl. Phys. Lett.*, **2002**, *81*, *2*, 268.

11 M. Shtein, P. Peumans, J. B. Benzinger, S. R. Forrest, *J. Appl. Phys.*, **2003**, *93*, 7, 4005.

12 S. R. Forrest, *Nature*, **2004**, *428*, 911.

13 M. Schwambera, N. Meyer, S. Leder, M. Reinhold, M. Dauelsberg, G. Strauch, M. Heuken, H.

JUERGENSEN, T. ZHOU, T. NGO, J. BROWN, M. SHTEIN, S. R. FORREST, *SID Digest*, **2002**, 894.

14 J. K. MAHON, T. ZHOU, S. R. FORREST, M. SCHWAMBERA, N. MEYER, *Solid State Technology*, **2002**, *July*, 131.

15 M. SCHWAMBERA, N. MEYER, M. GERSDORFF, M. REINHOLD, G. STRAUCH, R. BECCARD, M. HEUKEN, *SID Digest*, **2003**, 1419.

16 D. R. T. ZAHN, C. HIMCINSCHI, M. FRIEDRICH, A. PARAIAN, M. HEUKEN, *Vakuum in Forschung und Praxis*, **2003**, *15*, *6*, 312.

17 S. HARTMANN, A. JANSSEN, M. GERSDORFF, N. MEYER, M. REINHOLD, M. SCHWAMBERA, G. STRAUCH, B. MARHEINECKE, M. HEUKEN, H.-H. JOHANNES, W. KOWALSKY, C. HIMCINSCHI, D. ZAHN, *SID Digest*, **2004**, 1388.

18 M. SCHWAMBERA, M. GERSDORFF, M. REINHOLD, N. MEYER, G. STRAUCH, B. MARHEINECKE, M. HEUKEN, T. X. ZHOU, T. NGO, J. J. BROWN, M. SHTEIN, S. R. FORREST, *IMID*, **2004**, 824.

19 B. MARHEINEKE, *SPIE Proceedings*, **2005**, *5961*, 3.

20 C. HIMCINSCHI, N. MEYER, S. HARTMANN, M. GERSDORFF, M. FRIEDRICH, H.-H. JOHANNES, W. KOWALSKY, M. SCHWAMBERA, G. STRAUCH, M. HEUKEN, D. R. T. ZAHN, *Appl. Phys. A: Mater. Sci. Process.*, **2005**, *80*, 551.

21 C. HIMCINSCHI, O. GORDAN, G. SALVAN, F. MÜLLER, D. R. T. ZAHN, C. COBET, N. ESSER, W. BRAUN, *Appl. Phys. Lett.*, **2005**, *86*, 111907.

22 M. BEIGMOHAMADI, P. NIYAMAKOM, A. FARAHZADI, S. KREMERS, M. WUTTIG, P. v. GEMMERN, F. O. JESSEN, H. KAHLISCH, S. BRAND, W. MÜLLER, H. SCHWAB, H. BLEI, N. MEYER, M. HEUKEN, D. BERTRAM, to be published.

23 D. M. WATERS, *Production Systems and Technologies for OLED Displays*, presented at OLED 03, San Diego, USA, *19–21 November*, **2003**.

24 S. NAKA, K. SHINNO, H. OKADA, H. ONNAGAWA, K. MIYASHITA, *Jpn. Appl. Phys.*, **1994**, *33*, L1772.

25 Z. D. POPOVIC, S. XIE, N. HU, A. HOR, D. FORK, G. ANDERSON, C. TRIPP, *Thin Solid Films*, **2000**, *363*, 6.

26 V.-E. CHOONG, J. SHEN, J. CURLESS, S. SHI, J. YANG, F. SO, *J. Phys. D. Appl. Phys.*, **2000**, *33*, 760.

27 A. B. CHWANG, R. C. KWONG, J. J. BROWN, *Appl. Phys. Lett.*, **2002**, *80*, *5*, 725.

28 B. W. D'ANDRADE, S. R. FORREST, A. B. CHWANG, *Appl. Phys. Lett.*, **2003**, *83*, *19*, 3858.

29 G. HE, O. SCHNEIDER, D. QIN, X. ZHOU, M. PFEIFFER, K. LEO, *J. Appl. Phys.*, **2004**, *95*, *10*, 5773.

30 L. S. LIAO, K. P. KLUBEK, C. W. TANG, *Appl. Phys. Lett.*, **2004**, *84*, *2*, 167.

31 C. HIMCINSCHI, S. HARTMANN, A. JANSSEN, N. MEYER, M. FRIEDRICH, W. KOWALSKY, D. R. T. ZAHN, M. HEUKEN, *J. Cryst. Growth*, **2005**, *275*, e1035.

32 J.-H. LEE, S. W. LIU, C.-A. HUANG, K. H. YANG, Y. CHANG, *Proc. of SPIE*, **2004**, *5464*, 434.

33 S. W. LIU, C. A. HUANG, J. H. LEE, K. H. YANG, C. C. CHEN, Y. CHANG, *Thin Solid Films*, **2004**, *453*, 312.

34 T. KATO, T. MORI, T. MIZUTANI, *Thin Solid Films*, **2001**, *393*, 109.

35 L. F. CHENG, L. S. LIAO, W. Y. LAI, X. H. SUN, N. B. WONG, C. S. LEE, S. T. LEE, *Chem. Phys. Lett.*, **2000**, *319*, 418.

36 C. B. LEE, A. UDDIN, X. HU, T. G. ANDERSSON, *Mater. Sci. Eng. B*, **2004**, *112*, 14.

37 C. Y. KWONG, A. B. DJURISIC, V. A. L. ROY, P. T. LAI, W. K. CHAN, *Thin Solid Films*, **2004**, *458*, 281.

38 A. Y. ANDREEV, C. TEICHERT, G. HLAWACEK, H. HOPPE, R. RESEL, D.-M. SMILGIES, H. SITTER, N. S. SARICIFTCI, *Org. Electronics*, **2004**, *5*, 23.

39 R. G. D. VALLE, A. BRILLANTE, E. VENUTI, L. FARINA, A. GIRLANDO, M. MASINO, *Org. Electronics*, **2004**, *5*, 1.

40 F. YANG, M. SHTEIN, S. R. FORREST, *J. Appl. Phys.*, **2005**, *98*, 014906.

41 J. LIU, T.-F. GUO, Y. YANG, *J. Appl. Phys.*, **2002**, *91*, *3*, 1595.

42 A. B. DJURISIC, T. W. LAU, L. S. M. LAM, W. K. CHAN, *Appl. Phys. A. Mater. Sci. Process.*, **2004**, *78*, 375.

43 G. Salvan, D. A. Tenne, A. Das, T. U. Kampen, D. R. T. Zahn, *Org. Electronics*, **2000**, *1*, 49.
44 F. Yang, M. Shtein, S. R. Forrest, *Nature Materials*, **2005**, *4*, 37.
45 J. N. Bardsley, *IEEE J. sel. Top. Quantum. Electronics*, **2004**, *10*, 1.
46 S. E. Shaheen, D. S. Ginley, G. E. Jabbour, *MRS Bulletin*, **2005**, *30*, 10.
47 C. J. Brabec, J. A. Hauch, P. Schilinsky, C. Waldauf, *MRS Bulletin*, **2005**, *30*, 50.
48 S. Hartmann, A. Janssen, H. Krautwald, H.-H. Johannes, W. Kowalsky, N. Meyer, M. Schwambera, M. Reinhold, M. Heuken, G. Strauch, *ITG Conference*, **2004**.
49 M. A. Baldo, D. F. O'Brian, Y. You, A. Shoustikov, S. Sibley, M. E. Thomson, S. R. Forrest, *Nature*, **1998**, *395*, 151.
50 T. Zhou, T. Ngo, J. Brown, M. Shtein, S. R. Forrest, *Appl. Phys. Lett.*, **2005**, *86*, 021107.
51 J.-H. Lee, J. J. Huang, C.-C. Liao, P.-J. Hu, Y. Chang, *Chem. Phys. Lett.*, **2005**, *402*, 335.
52 P. Peumans, S. Uchida, S. R. Forrest, *Nature*, **2003**, *425*, 158.
53 J. Xue, S. Uchida, B. P. Rand, S. R. Forrest, *Appl. Phys. Lett.*, **2004**, *84*, *16*, 3013.
54 S. Uchida, J. Xue, B. P. Rand, S. R. Forrest, *Appl. Phys. Lett.*, **2004**, *84*, *21*, 4218.
55 S. Yoo, B. Domercq, B. Kippelen, *Appl. Phys. Lett.*, **2004**, *85*, *22*, 5427.
56 J. Xue, S. Uchida, B. P. Rand, S. R. Forrest, *Appl. Phys. Lett.*, **2004**, *85*, *23*, 5757.
56a J. Xue, S. Uchida, B. P. Rand, S. R. Forrest, *Appl. Phys. Lett.*, **2004**, *85*, *23*, 5757.
56b B. Johnev, M. Vogel, K. Fostiropoulos, B. Mertesacker, M. Rusu, M.-C. Lux-Steiner, A. Weidinger, *Thin Solid Films*, **2005**, *488*, 270.
56c M. Vogel, J. Strotmann, B. Johnev, M.-C. Lux-Steiner, K. Fostiropoulos, *Thin Solid Films*, accepted
57 H. Klauk, D. J. Gundlach, M. Bonse, C.-C. Kuo, T. N. Jackson, *Appl. Phys. Lett.*, **2000**, *76*, *13*, 1692.
58 D. J. Gundlach, H. Klauk, C. D. Sheraw, C.-C. Kuo, J. R. Huang, T. N. Jackson, *Dig. Int. Electron Devices Meet.*, **1999**, 111.
59 L. Ma, Y. Yang, *Appl. Phys. Lett.*, **2004**, *85*, *21*, 5084.
60 C.-W. Ko, S.-H. Hu, S.-H. Li, T.-H. Hsiao, K.-S. Lee, C.-J. Chen, J.-J. Lih, *Sid Digest*, **2005**, 961.
61 H. K. Chung, K. Y. Lee, *Sid Digest*, **2005**, 956.
62 K.-J. Yoo, S.-H. Lee, A.-S. Lee, C.-Y. Im, T.-M. Kang, W.-J. Lee, S.-T. Lee, H.-D. Kim, H.-K. Chung, *SID Digest*, **2005**, 1344.
63 M. Boroson, L. Tutt, K. Nguyen, D. Preuss, M. Culver, G. Phelan, *SID Digest*, **2005**, 972.
64 M. Kröger, M. Hüske, T. Dobbertin, J. Meyer, T. Riedl, H.-H. Johannes, W. Kowalsky, *Proc. of SPIE*, **2005**, 5840.
65 L. S. Sapochak, A. Padmaperuma, N. Washton, F. Endrino, G. T. Schmett, J. Marshall, D. Fogarty, P. E. Burrows, S. R. Forrest, *J. Am. Chem. Soc.*, **2001**, *123*, 6300.
66 T. Fuhrmann, J. Salbeck, *MRS Bulletin*, **2003**, 354.
67 T. Spehr, R. Pudzich, T. Fuhrmann, J. Salbeck, *Organic Electronics*, **2003**, *4*, 61.
68 V. I. Adamovich, S. R. Cordero, P. I. Djurovich, A. Tamayo, M. E. Thompson, B. W. D'Andrade, S. R. Forrest, *Org. Electronic*, **2003**, *4*, 77.
69 I. Tanaka, Y. Tabata, S. Tokito, *Chem. Phys. Lett.*, **2004**, *400*, 86.
70 S. Lamansky, P. Djurovich, D. Murphy, F. Abdel-Razzaq, H.-E. Lee, C. Adachi, P. E. Burrows, S. R. Forrest, M. E. Thompson, *J. Am. Chem. Soc.*, **2001**, *123*, 4304.
71 S.-J. Yeh, M.-F. Wu, C.-T. Chen, Y.-H. Song, Y. Chi, M.-H. Ho, S.-F. Hsu, C. H. Chen, *Adv. Mater.*, **2005**, *17*, *3*, 285.
72 Y. Kawamura, K. Goushi, J. Brooks, J. J. Brown, H. Sasabe, C. Adachi, *Appl. Phys. Lett.*, **2005**, *86*, 071104.

73 H. Aziz, Z. D. Popovic, N-X. Hu, A-M. Hor, G. Xu, *Science*, **1999**, *283*, 1900.

74 M. Cölle, W. Brütting, *Phys. Stat. Sol. A*, **2004**, *201*, 6, 1095.

75 T. Fuhrmann, J. Salbeck, *Advances in Photochemistry, Vol. 27*, Edt.: D. C. Neckers, D. H. Volman, G. von Bünau, W.S. Jenks, **2002**, 83.

76 E. Holder, B. M. W. Langeveld, U. S. Schubert, *Adv. Mater.*, **2005**, *17*, 1109.

77 T. Dobbertin, E. Becker, T. Benstem, G. Ginev, D. Heithecker, H.-H. Johannes, D. Metzdorf, H. Neuner, R. Parashkov, W. Kowalsky, *Thin Solid Films*, **2003**, *442*, 132.

78 S. Y. Yang, K. Shin, and C. E. Park, *Adv. Funct. Mater.*, **2005**, *15*, 1806.

79 C. D. Sheraw, L. Zhou, J. R. Huang, D. J. Gundlach, T. N. Jackson, M. G. Kane, I. G. Hill, M. S. Hammond, J. Campi, B. K. Greening, J. Francl, and J. West, *Appl. Phys. Lett.*, **2000**, *80*, 1088.

80 I. Kymissis, C. D. Dimitrakopoulos, and S. Purushothaman, *J. Vac. Sci. Technol. B*, **2002**, *20*, 956.

81 H. Kawaguchi, T. Someya, T. Sekitani, and T. Sakurai, *IEEE Journal of Solid-State Circuits*, **2005**, *40*, 177.

10 Thermal Imaging and Micro-contact Printing

Hee Hyun Lee, John Rogers, and Graciela Blanchet

10.1 Introduction

The most widely used techniques for micro and nanofabrication – photolithography, electron beam lithography, and related methods – are extremely well suited to the types of application in microelectronics and display systems for which they were principally designed. There are substantial challenges, however, in adapting these techniques to new applications in fields such as plastic and molecular electronics. In these applications the complexity, the high capital and operating costs, and the difficulty in patterning large areas are significant disadvantages. In addition, the chemicals required for these methods – the resists, etchants, developers, solvents, etc. – are incompatible with many of the organic electronic materials of interest. As a result, some of the oldest, and conceptually simplest, forms of lithography – embossing, molding, stamping, writing, etc. – are now being examined as tools for forming micro or nanostructures [1] that can be implemented in these and other applications. Substantial progress has been made in the last few years, mainly by combining these approaches, or variants, with new materials, chemistries, and processing techniques. This chapter emphasizes three high-resolution patterning methods that with demonstrated applications in the building of flexible, plastic circuits. This emerging type of electronics is of interest because it enables important classes of consumer devices – electronic paper, wearable computers or sensors, disposable wireless ID tags, etc – that would be difficult to realize with conventional electronics. Low cost, large-area patterning techniques are essential for commercialization of such systems.

10.2 Building Blocks

Thin-film transistors (TFTs) are among the key building blocks of these circuits [2]. Figure 10.1 shows a cross-sectional view of a TFT. A thin insulating film (i.e. the gate dielectric) isolates source and drain electrodes and a semiconductor layer

Organic Electronics, Materials, Manufacturing and Applications. Edited by Hagen Klauk

ISBN: 3-527-31264-1

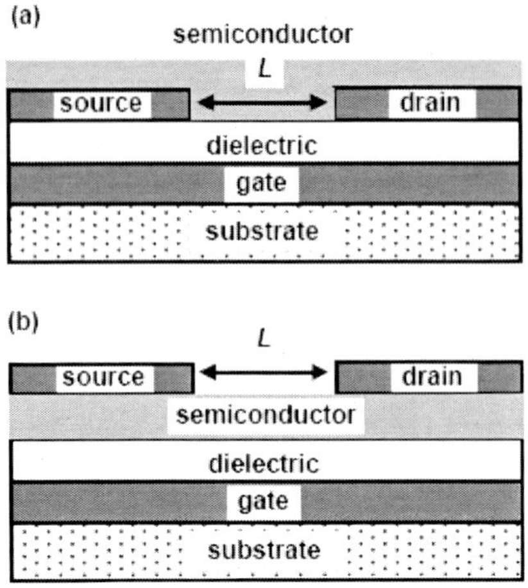

Fig. 10.1. Schematic illustrations of two device geometries for organic thin-film transistors. Part (a) shows the layout of a "top contact" device in which the source and drain electrodes are deposited on top of the semiconductor. Part (b) shows the "bottom contact" geometry in which the semiconductor is deposited on top of the electrodes. The distance between the electrodes defines the channel length, *L*.

from an underlying gate electrode. The region between the source/drain electrodes is known as the channel. The separation between these electrodes defines the channel length, *L*; their lengths define the channel width, *W*. In an accumulation-mode device, the channel has a high effective resistance when no potential is applied to the gate – the transistor is in its "off" state. When a voltage is applied to the gate, current flows between the source and drain electrodes if there is a potential difference between them – the transistor is in its "on" state. For many systems, the transistor must produce enough "on" current to activate or switch another part of a circuit or device (e.g. a pixel in a display) without producing "off" currents that could induce unwanted switching. These characteristics are governed by the mobility of the semiconductor and its doping level, the efficiency with which it is coupled to the source/drain electrodes (i.e. the resistances associated with the contacts), the capacitance of the gate dielectric, and the *W*/*L* ratio. Reducing the channel length increases the "on" currents. The ability to print (at low cost, over large areas, on plastic sheets) the source/drain electrodes at high resolution and with low contact resistance to the semiconductor is therefore crucial to achieving good performance. Several techniques have been demonstrated for this patterning task, including a type of photolithographic process that relies on photochemical conversion of polymers from nonconducting to conducting states [3, 4], specialized adaptations of inkjet [5] and screen printing [6, 7], and certain types of molding [8] and imprinting [9] techniques. The patterning approaches described are two forms of contact printing with a stamp [10–14]; and thermal transfer patterning [15]. We review these processes and their applications, capabilities, and limitations.

The first half of this chapter focuses on thermal imaging, describing the technique and the strenuous requirements of suitable laser-printable materials. We illustrate the latter with polyaniline composites specifically tailored for high-resolution printing over large areas with moderate throughput. Their use in active matrix backplane circuits for large, mechanically flexible sheets of electronic paper is then described.

The second half of the chapter focuses on micro-contact printing, perhaps the only process that can realistically lead to roll-to-roll manufacturing of organic electronics devices. We first describe the technique and its use in forming transistors and circuits with a range of organic semiconductors. Fabricating large area plates and evaluating possible routes to layer-to-layer registration address the feasibility of roll-to-roll electronics. The capabilities and limitations encountered and foreseen with such approach are discussed.

10.3 Printing and Patterning Techniques

10.3.1 Thermal Imaging

Although functioning plastic transistors have been produced using widely different approaches [3–16], nobody has yet fabricated a one-transistor layer using techniques that would be appropriate for commercial manufacture. Thus, it is important to identify techniques that fulfill the speed, resolution, and cost required for commercial fabrication of organic devices. Printing presses, with flexographic or offset plates, enable high speed and reel-to-reel fabrication, desirable features in the fabrication of very inexpensive, throwaway plastic electronics components. They are not, however, equipped to handle flammable solvents, nor do they meet the resolution or registration required for fabrication of plastic electronics devices. Later we will describe possible means of overcoming these difficulties but printing electronics in a press is, at the very least, a few years away.

It is perhaps more practical today to consider initial fabrication of organic electronic devices by using printing techniques that enable circumvention of the many serious materials and process issues currently associated with printing in a press. Thermal imaging, a novel digital technique used by the proofing industry, provides an attractive alternative path. Printing by thermal imaging proceeds via the ablative transfer of solid layers. Therefore, all issues of chemical compatibility among layers faced when printing sequentially from solution are entirely avoided. Also, thermal imaging speed, resolution, and registration are sufficient for commercialization of large-area plastic electronics devices.

Printing by laser ablation involves transfer of a digital image from a donor film on to a flexible receiver, with the electronics device being built by sequential transfer of separate solid layers. The architecture of these films is illustrated in Fig. 10.2.

The donor film comprises a multi-layer structure on a flexible polymeric sub-

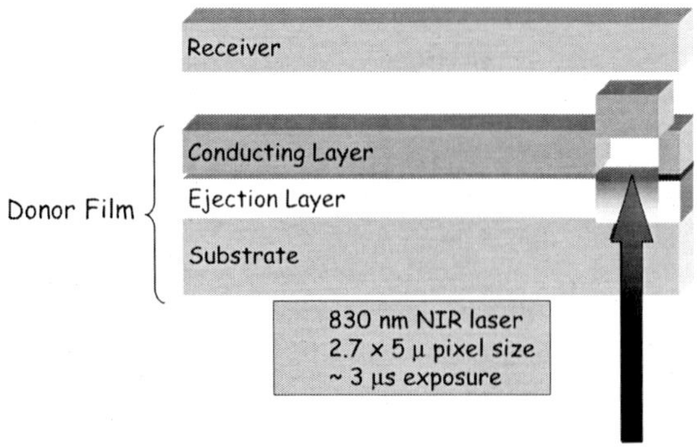

Fig. 10.2. Illustration of the thermal imaging process.

strate. A thin organic layer coated directly on the donor substrate provides the necessary propulsion forces. The light-to-heat conversion layer, placed directly on top of the propulsion layer, is then coated with the material to be printed, i.e. a conducting layer. The donor film vacuum held against the receiver is then exposed using a 40-W, 780-nm diode infrared laser. The laser beam is split into two-hundred-and-fifty 2-μm × 5-μm individually addressable spots.

The mode of operation of the material is as follows. The beam is focused through the donor base at the light-to-heat conversion interface. Efficient absorption of the beam generates heat efficiently, decomposing the surrounding organic materials into gaseous products. A high-temperature bubble trapped beneath the surface of the conducting layer thus propels the conducting layer on to the receiver film. Source and drains are printed by selectively exposing and transferring 5 μm × 2 μm pixels of the conducting layer on to the receiver. Because the beam is fully addressable, any pattern can be formed. In particular, multi-layer electronic circuitry could be built in this fashion.

10.3.2
Printed Devices: From TFTs to Large-area Backplanes

We illustrate the use of thermal imaging, a standard printing technique currently use in proofing shops, to fabricate organic electronic devices over large areas. The technique enables printing of areas up to ~1 m^2 and on to flexible substrates. The current resolution of approximately 5 μm corresponds to one pixel, the spot size of the focus laser beam. This is clearly adequate for early commercial production of the first generation of thin-film transistors (TFTs). Perhaps most important is that the data presented here reveal that thermal imaging enables manufacture of complex multi-layer circuits by sequential printing of solid films. Thus, it eliminates the solvent-compatibility issues faced by techniques that require the building organic multi-layers by the sequential application of liquid layers.

Figure 10.3 shows the I–V characteristics of transistors printed with a polyani-

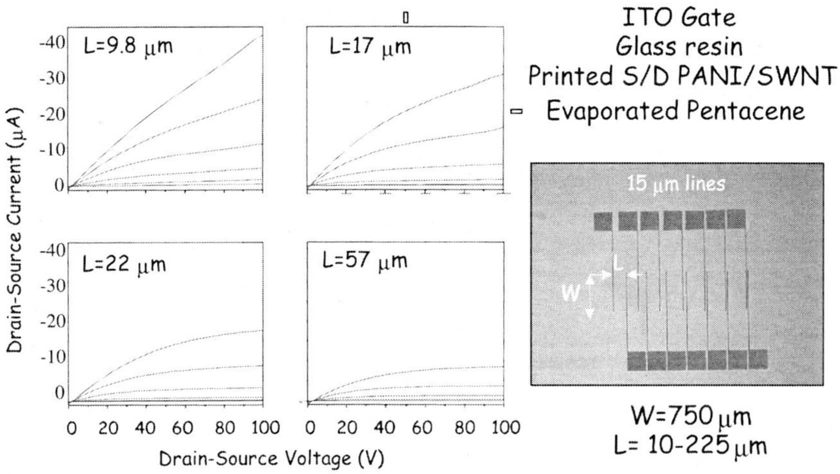

Fig. 10.3. I–V characteristics of TFTs with different channel lengths. The source and drains patterns were printed by thermal transfer with a polyaniline composite.

line composite source and drains [17, 18]. A 7-mm poly(ethylene terephthalate) (Mylar) sheet coated with 100 nm ITO was used as the gate. A glass resin was then spun on to the ITO to approximately 1.5 μm thick, insulating the gate from the other circuit elements. Source and drains with the pattern shown in the right hand side of Fig. 10.3 were printed by thermal imaging using DNNSA–PANI/SWNT [17], a polyaniline–carbon nanotube composite that will be described in the next section. Each set of source and drains have a fixed channel width (W) of 725 μm and channel lengths varying from 9.8 to 225 μm. After the source and drain level was printed, pentacene (Aldrich) was deposited by thermal evaporation without further purification. A set of I–V curves collected from a single array of transistors with 15 μm conducting lines is shown in Fig. 10.3. The gate voltage was varied from 0 to −100 V in −20 V steps. The channel-length values (L) are indicated on the right. Printed transistors yield stable bottom contact devices with on/off ratios (>1000) compatible with spin-cast dielectrics. The effective mobility was calculated at the saturation region and compared with control devices with Au source and drains fabricated by shadow masking. The calculated effective mobility of the device with 22 μm channel and printed DNNSA–PANI/SWNT electrodes was 0.3 $cm^2\ V^{-1}\ s^{-1}$. In contrast, the mobility of the control transistor with Au source and drain and similar W/L fabricated by shadow masking was only 0.15 $cm^2\ V^{-1}\ s^{-1}$. The threshold voltages and "off" currents are nearly identical in both. The differences between calculated mobility perhaps reflect differences between contact resistance for the DNNSA–PANI/pentacene and Au/pentacene interfaces or, more specifically, difference between the charge injection mechanism at the interface [18].

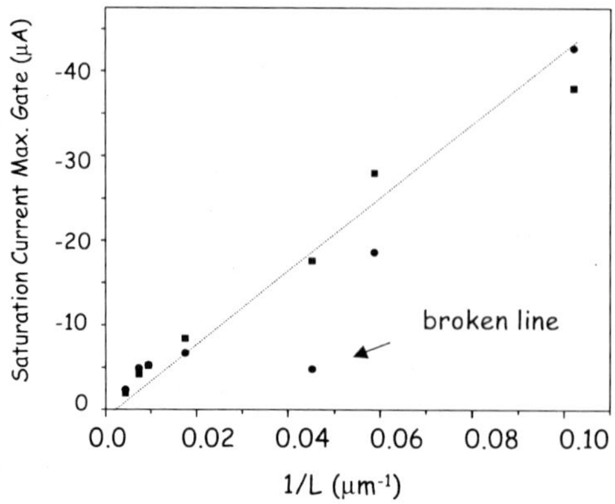

Fig. 10.4. Saturation current for two adjacent sets of TFT as a function of $1/L$.

Figure 10.4 shows the saturation current at maximum gate voltage for two adjacent sets of transistors, both with 15 µm lines, as a function of $1/L$. The results are internally consistent – e.g. maximum "on" currents scaling roughly linearly with $1/L$ for a given array of transistors. The one data point outside linear behavior was from a device that had a visible break in one of the source/drain electrodes. The linear dependence extrapolated to zero shows a relatively low contact resistance at the DNNSA–PANI/SWNT pentacene interface [15].

The possible impact of laser ablation printing in printable electronics is best illustrated by the manufacture of a 32-inch diagonal TFT display backplane [15]. This panel comprises over 4000 organic transistors printed on a mechanically flexible substrate with 20-µm channel lengths. The 32-inch diagonal electronic paper display backplane is extremely thin, flexible, and ultra-lightweight. The gate and source/drain layers isolated by a thin dielectric, were printed, in registry, using a dinonylnaphthalene-doped polyaniline (DNNSA–PANI)/single wall carbon nanotubes (SWNT) composite specifically tailored for laser ablative printing [17, 18]. The display backplane was completed by evaporation of pentacene through a shadow mask. An enabling element for fabrication of this panel, and an important outcome from the work, was the development of a printable conductor, a family of polyaniline/carbon nanotube composites that fulfills the resolution and conductivity requirements of TFT applications. The electronic properties of these composites will be described in detail in the next section.

The panel was produced as follows (Fig. 10.5). A donor film with the structure previously described and coated with a 1-µm DNNSA–PANI/SWNT composite [15] was used to print all conducting circuit elements. First, the large donor film was held by vacuum on the flexible receiver. The gate layer was then printed by selectively exposing the DNNSA–PANI/SWNT donor film as previously described. When exposure was complete the receiver was removed and a 1-µm dielectric layer was applied over the whole area. The receiver is then repositioned on to the drum in registry for laser printing of the source/drains and interconnects. The widths of

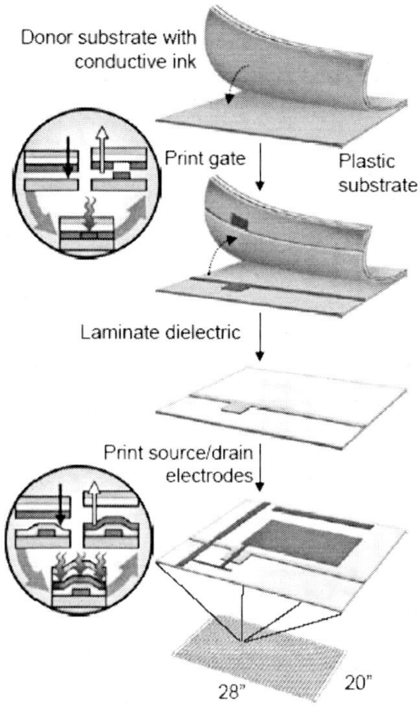

Fig. 10.5. Illustration of the printing process. Sequence for patterning structures using thermal transfer printing. A donor sheet coated with a layer of the material to be printed is placed against a substrate. The inset shows the process by which local heating with a laser causes the transfer of some part of the material from the donor to the substrate. Repeating these lamination, laser transfer printing, and removal steps provides an additive, dry, multilayer route to patterning for plastic electronics.

the source, drain, and channels were 20 μm. After the printing of the source and drains layer is complete the receiver is removed and pentacene is evaporated through a shadow mask.

The panel comprises about 5000 transistors. The higher magnification photographs show that printed lines have sharp edges and that transistors have clean, uniform channels. The high edge definition observed for printed features is not easily achievable when printing liquid layers. As seen in the magnified photograph, the source/drains are not perfectly centered on the gate. This is because of the manual repositioning of the receiver after the application of the dielectric layer.

We are currently in the process of developing printable dielectrics such that all layers can be sequentially applied without the need for re-registration. The source/drain and interconnect layer of the 20-inch × 28-inch source/drain panel shown in Fig. 10.6 was printed in 4 min, speed certainly appropriate for fabrication of electronic devices.

Figure 10.6 illustrates the printing of large-area electronics devices in registry

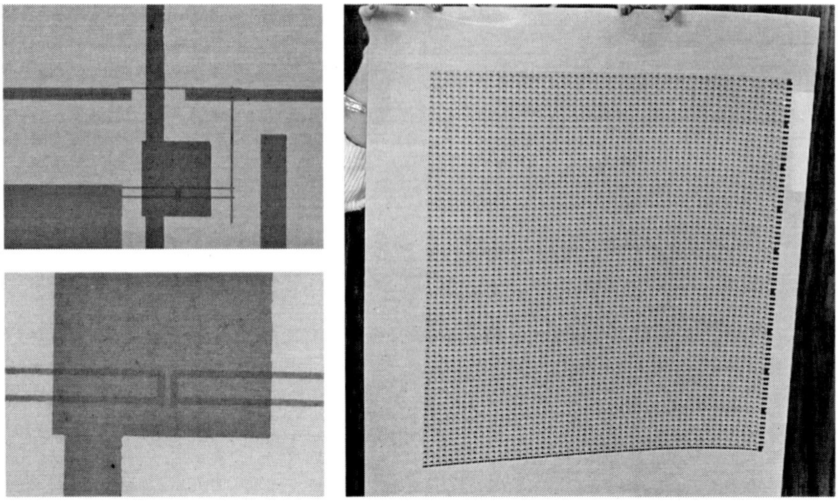

Fig. 10.6. Active matrix backplane circuit formed by thermal transfer. A photograph of a 50 cm × 75 cm printed panel is shown at the top. The micrograph in the middle at the left illustrates a unit cell of the circuit. The scanning electron micrograph on the bottom left provides a high resolution view of the source/drain electrodes and the channel region. The graph at the bottom right shows the electrical characteristic of one of the transistors in the large printed panel. The gate potential varies from 0 to −100 V in −20 V steps.

with high resolution and at speeds suitable for early commercialization of plastic electronics. The organic composites used to print source, drains and gates, were formulated as high resolution, high conductivity conductors printable via laser ablation for applications in organic electronics. We demonstrated here that conducting narrow sources and drains lines with 20 μm channels could be printed over large areas and at high speeds, therefore eliminating the need for solvents, masking, or lithography making the feasibility of organics electronics a step closer to reality.

10.4 Printable Materials

10.4.1 Polyaniline Nanotube Composites: A High-resolution Printable Conductor

Printing processes require candidate materials to be robust and to retain their electrical properties through the various steps. In addition, the material must be suitable for delivery to the substrate in a patterned high-throughput mode eliminating many slow serial delivery methods and most post-imaging processing. The material set required for fabrication of electronic devices is straightforward – dielectrics, conductors, and semiconductors along with appropriate packaging. While it is intuitively obvious to most that printing polymeric dielectrics by a laser ablative process is conceptually possible; using the same process for the delivery of either the conductor or semiconductor is much less obvious. In this section we show that conducting composites can be printed by laser ablation with sufficiently high resolution and conductivity appropriate for their application in plastic transistors. We further illustrate the potential resolution of the materials and the technique by printing polyaniline/nanotube structures, which could be used as source and drain with a channel 7 μm in length and a conductivity of 2 S cm^{-1}. Because doped conjugated polymers are the prototypical organic conductors, their choice as a starting point for conducting materials in organic electronics is apparent. Although heavily doped they have properties characteristic of the metallic state, including high electrical conductivity [19], their main drawbacks are insolubility and ease of oxidation. Polyaniline (PANI) is unique among conducting polymers, because it is environmentally stable and its conductivity can be reversibly controlled both by charge-transfer reaction with the conjugated backbone and by protonation of the nitrogen [19]. The conductivity of PANI is controlled by complete protonation with functionalized sulfonic acids [20]. The wide range of electrical properties [21] coupled with thermal and environmental stability makes PANI potentially attractive as a printable conductor.

The polyaniline was prepared by emulsion polymerization following the procedure outlined in US patent 5,863,465 with DNNSA used as a dopant [22]. Carbon Nanotubes (CNI) manufactured the nanotubes used in the work either by a high-pressure fabrication method (Hipco SWNT) or via laser ablation (Laser SWNT). We found that excellent dispersions of the nanotubes in PANI could be produced by two different procedures. The nanotubes could either be directly sonicated into the PANI solution or first sonicated into xylene and that dispersion afterwards sonicated into the DNNSA–PANI solution.

Conductivity of the composite was measured by coating films approximately 1 μm thick on to a glass slide upon which Ag conducting contacts had previously been deposited. The conductivity was measured using a standard four-probe method. Figure 10.7 shows a log–log plot of the conductivity of DNNSA–PANI SWNT composite thin films for nanotube concentrations ranging from 0.1% to 40% by weight. The data of Fig. 10.1 show that the conductivity of DNNSA–PANI increases over five orders of magnitude as the nanotube content is varied and extrapolates to report values of SWNT mats values at 100% loading [23–25]. Figure 10.7 also shows that the conductivity of the composite is highly dependent on the dispersion procedure but fully independent of the nanotube fabrication method. This result suggests that neither the graphitic component of the nanotube material (higher graphitic content in laser material) nor the metal catalysts associated with the differing fabrication methods affect composite conductivity. In contrast, the effect of the nanotube dispersion procedure on composite conductivity, which, in turn, reflects differences in contact resistance and charge transport, is large. Figure 10.7 also shows that the percolative behavior of nanotubes composites in a conducting (PANI) and a nonconducting (ethylcellulose) matrix is also dramatically

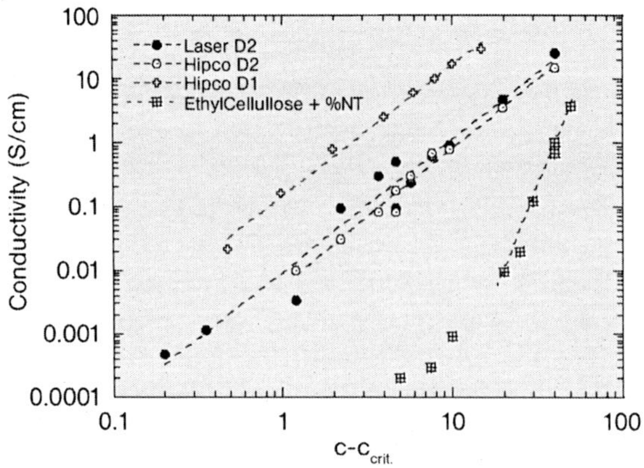

Fig. 10.7. Scaling behavior of DNSSA–PANI/SWNT and ethyl cellulose composites with single wall carbon nanotubes.

different. These data indicate the uniqueness of polyaniline as a conductive binder leading to superior properties in a composite.

Synthesis of a material satisfying the transport requirements for a polymeric device is not difficult and does not require a nanotube composite. Maintaining the conductivity of PANI throughout the laser driven printing process is a far more difficult challenge, however, because deprotonation, with loss of conductivity, occurs at fairly modest temperatures [26]. The more highly conducting PANI, for example dodecyl benzenesulfonic acid (DBSA) and camphor sulfonic acid (CSA) doped PANI, seem to be the most susceptible to deprotonation when laser imaged. Therefore, it was of little surprise that the archetypal conducting PANI are not able to withstand the temperature profile of the imaging process without significant degradation of properties. Thusly, we were forced to evaluate higher temperature stability alternatives. In particular, we found that PANI doped with dinonylnaphthalene sulfonic acid (DNNSA–PANI) is sufficiently robust to withstand the heat generated in the imaging process without degradation in conductivity. Although, its inherent conductivity of 10^{-4} S cm^{-1} makes it unsuitable as a printable conductor [22], we found that the conductivity can be increased by the addition of single-wall carbon nanotubes (SWNT) [23–25]. In Fig. 10.8 data are presented showing the conductivity of DBSA–PANI, DBSA–PANI with increased acid for high-resolution imaging, and DNNSA–PANI. Although the DBSA–PANI has higher conductivity, it does not print with fidelity. Although addition of acid or plasticizers led to higher-quality images, the conductivity degraded by two orders of magnitude.

In contrast, the 3% nanotube/DNNSA–PANI composite shown in the figure had both sufficient conductivity and resolution. As one would expect, a decrease in conductivity of all materials is eventually observed at sufficiently high fluence. Whereas DBSA materials have a roll off from even the initial incident energies, the PANI/SWNT composite exhibits instead an increase in conductivity until ap-

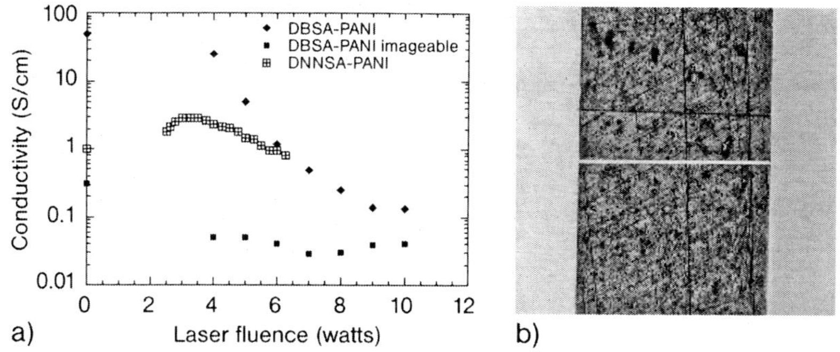

Fig. 10.8. Conductivity of transfer polyaniline as a function of laser fluence for three different polyaniline compositions: DBSA/PANI doped in a 1:1 and 1.25:1 DBSA:aniline ratio and DNNSA/PANI/SWNT composite. Source and drain lines with 7 μm channel length and 1 mm width printed via laser ablation from a DNNSA–PANI composite with 3% SWNT concentration are shown in the photograph.

proximately 4 W after which it also decreases. Although the exact temperatures of the laser process is not known, it is known that the PANI begins to deprotonate at temperatures in excess of 200–250 °C [26]. From this one might be led to conclude that the maximum in the conductivity roughly corresponds to a point where the peak temperature of the process reaches that point.

Nevertheless, this does allow for a reasonable operating window of exposure. An example of a printed structure is shown in the inset. This is simply a 500 μm wide line with a 7 μm gap which could potentially be used as a source drain of a printed field-effect transistor [27]. We have printed a variety of structures using this polymer as an imaging medium with acceptable results. Furthermore, the extremely high aspect ratio of the nanotubes result in the percolation effects extending to extremely low concentrations of nanotubes conferring several significant advantages. The desire for a polymeric or plastic device is largely driven by the desire for lightweight and flexibility. In addition, the low nanotube content required for printable composites assures that the mechanical properties of the neat polymer are retained throughout. We have not observed any degradation of the printed conductors as a result of simple flexing of the substrate (polyester), although detailed studies have not yet been attempted. The extremely low concentration for percolation of these highly anisotropic materials is a positive when tailoring the conductivity of DNNSA–PANI with SWNT composite materials for printing.

Clear quantitative descriptions of the transport in poorly defined material systems such as acid-doped PANI or mats and ropes of nanotubes are difficult to develop. Even more problematic is the description of a composite of the two. Nevertheless we do believe that there are some conclusions that can be drawn from consideration of the data. A large amount of work has gone into understanding the transport in heterogeneous systems such as conducting polymers and nanotube mats [25, 27–31]. However, that work has focused on understanding each separately rather than as a combined composite material. Interestingly, when combined the properties seem to have important differences from either by itself. The natural approach is to consider the (SWNT ropes $\sigma \approx 10^2$ S cm^{-1}) percolating in a nearly insulating matrix (PANI–DNNSA $\sigma \approx 10^{-4}$ S cm^{-1}). The expected scaling conductivity of the composite σ typically expressed as $\sigma = \sigma_0(c - c_c)^\gamma$ [28], where σ_0 is the conductivity of the-high conductivity phase, $c - c_c$ is the difference between the concentration of SWNT and the percolation concentration, and the critical exponent γ is a universal term reflecting the dimensionality of the network above percolation. The critical concentration ($c_c = 0.3\%$) was determined as the concentration of nanotubes at the intersection of the line through the data points and one with zero slope through the conductivity of neat PANI, σ_0 (10^{-4} S cm^{-1}). An exponent of 2.06 and 2.13 for Hipco D1 and D2, respectively and 2.11 for Laser SWNT (D2) were extracted from a power fit of the conductivity versus concentration and shown in Fig. 10.5. The scaling exponent of approximately two is consistent with the expected behavior of a 3D system as the system evolves from a neat PANI conducting polymer to a percolated network of nanotubes perhaps similar to the as produced mats of nanotubes. However, it is interesting to compare this behavior with that of the percolation of the nanotubes in polyethylcellulose also shown in Fig. 10.5. In this case $c_c = 3.0\%$ nor was the scaling behavior seen as cleanly with the conductivity rising much more slowly than the PANI samples at low concentrations and then rising at a higher rate at heavier loading of the nanotubes. This is interesting in two respects. Firstly if one is considering the percolation on a highly anisotropic shaped object in a matrix, the critical concentration should not depend upon the nature of the matrix assuming that the anisotropic material is well dispersed. Nor should the scaling behavior be affected. At the same time intuitively one is not surprised to see a difference. In principle one expects the inter-particle contact resistance to be far lower with the PANI matrix than in the case of an insulator like the cellulose. This assumption would seem to be supported by the observed data, despite the excellent dispersability and wetting of nanotubes in cellulose. This small layer of polymer between the tubes then acts as a large tunneling barrier, which is difficult to eliminate until rather high concentrations of the nanotubes.

Figure 10.9 shows the temperature-dependence of the normalized resistivity for DNNSA–PANI at 0, 3 and 5% SWNT loading spanning the percolation range. In addition, we have included Fisher et al. [27] normalized resistance of single wall nanotube pressed mats. This figure shows the effectiveness of the PANI in generating the inter rod contacts as described above. As shown in the figure, the resistivity of the 0% SWNT composite diverges for decreasing temperature as expected from doped polyaniline. In contrast, the resistivity of the 3% and 5% nanotube composites both above the critical concentration shows, to at least as pressed mats [27].

One can then propose that, as for polyaniline, a heterogeneous model for conduction [27, 28] can also describe transport in PANI/SWNT below percolation. In contrast, Kaiser model describes transport above percolation; with the system's metallic character reflected in a linear temperature dependence while lacking the exponential term associated with tunneling through conduction barriers [25]. The

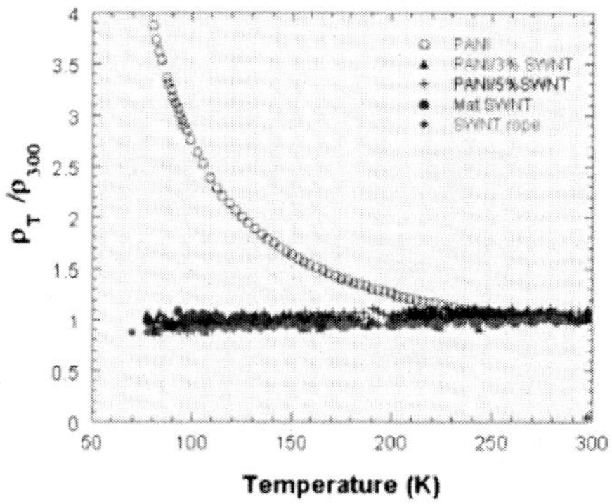

Fig. 10.9. Temperature dependence of the normalized resistivity for DNNSA–PANI, DNNSA–PANI/SWNT composites and SWNT ropes [25].

data then suggests that, charge transport must occurs mainly along a network of metallic tubes with PANI providing a path of low contact resistance for inter-tube charge transfer.

We showed in this section that the addition of SWNT in DNNSA–PANI creates a highly conducting three-dimensional percolating network, the linear temperature dependence above percolation reflecting their metallic character. These composites were formulated as high-resolution printable conductors for applications in organic electronics. Unlike solution printing techniques printing via laser ablation entails transferring a solid layer on to an existing circuit element. Thus, lifting material constrains because of solvent compatibility between sequential layers. We demonstrated the capability to produce narrow channels and large areas at reasonable speeds, which is crucial for plastic transistors and has been quite a challenge in the past. Thus, by printing micron-size channels at high speeds and without the need for wet chemistry, masking or photolithography printing via laser ablation brings organic electronics a step closer to reality.

10.5 Micro-contact Printing

10.5.1 Contact Printing with High-resolution Stamps

Contact printing involves the use of an element with surface relief (i.e. the "stamp") for patterned transfer of material (i.e. the "ink") to a substrate. Historically this approach has been used primarily to produce printed text or images with features that are 100 μm or larger in their minimum dimension. Its use for patterning functional materials in active systems which incorporate micron or nanometer-sized features is much less well explored. The nature of the inks and their role in process steps after the printing define the classes of materials that can be patterned. The resolution is determined by the nature of the ink and its interaction with the stamp and/or substrate, the resolution of the stamp, and the conditions that convert the pattern of ink into a pattern of functional material. This section describes printing methods that have sub-micron resolution and which can define electrodes for high performance plastic electronic devices.

10.5.1.1 **High-resolution Stamps**

The printing process can be separated into two parts: fabrication of the stamp and use of this stamp to pattern features defined by the relief on its surface. These two processes are typically quite different, although it is possible in some cases to use patterns generated by a stamp to produce a replica of that stamp. The structure from which the stamp is derived, which is known as the "master", can be fabricated with any technique that is capable of producing well-defined patterns of surface relief. This master can then be used directly as the stamp, or it can be used to produce stamps via molding or printing procedures.

It is important to note that the technique for producing the master does not need to be fast or low in cost. It is also not essential for it to have other characteristics that might be desirable for a given patterning task: it is used just once to produce a master, which is directly or indirectly used to fabricate stamps.

Each of these stamps can then be used many times for printing. In a common approach for the high-resolution techniques that are the focus of this chapter, an established lithographic method, such as one of those developed for the microelectronics industry, defines the master. Figure 10.10 schematically illustrates two possible routes to stamps. Both use photolithography to define a pattern of resist on a silicon wafer.

Stamps are generated from this structure in one of two ways: by casting against this master or by etching the substrate with the patterned resist as a mask. In the first approach, the master itself can be used multiple times to produce many stamps, typically using a light or heat-curable pre-polymer. In the second, the etched substrate serves as the stamp; additional stamps can be generated either by repeating the lithography and etching, or by using the original stamp to print replica stamps. For feature sizes larger than ~1–2 μm, contact or proximity mode photolithography with a photomask produced by direct write photolithography represents convenient methods to fabricate the master. Other techniques must be used for features smaller than ~2 μm. Established techniques include projection mode photolithography [32], direct write electron beam (or focused ion beam) lithography [33, 34], scanning probe lithography [35–38], and laser interference lithography [39]. The first two have features (i.e. flexibility in pattern design; well developed technology base, etc.) that make them useful for generating masters for plastic electronic systems. The circuit applications described below used masters formed by contact or projection mode photolithography. Some stamps that were designed to demonstrate patterning capabilities and resolution use masters formed by electron beam and laser interference lithography.

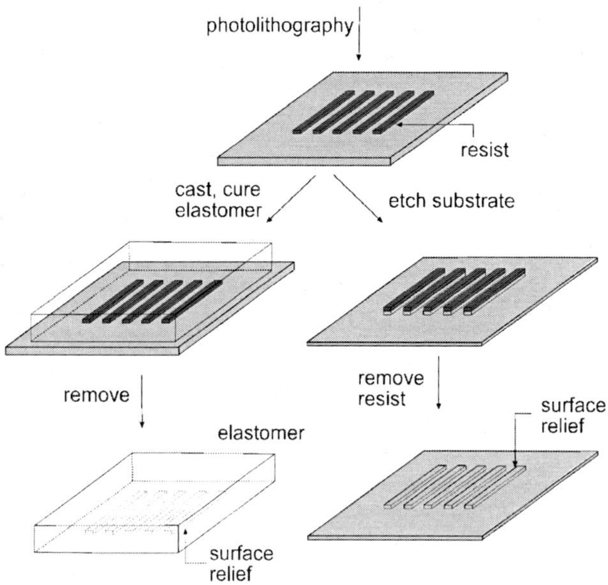

Fig. 10.10. Micro-contact printing process. Schematic illustration of two methods for producing high resolution stamps. The first step of both involves patterning a layer of resist on a flat substrate. This structure, known as the "master", is converted to a stamp either by etching or by molding. In the first case (right frames), the resist acts as a mask for etching the underlying substrate. Removing the resist yields a stamp. This structure can be used directly as a stamp to print patterns or to produce additional stamps. In the molding approach (left frames), a prepolymer is cast against the relief structure formed by the patterned resist on the substrate. Curing (thermally or optically) and then peeling the resulting polymer away from the substrate yields a stamp. In this approach, many stamps can be made with a single "master" and each stamp can be used many times.

10.5.2
Micro-contact Printing

Micro-contact printing (μCP) uses high resolution rubber stamps with "inks" (e.g. alkanethiols) that form self assembled monolayers (SAMs) on the surface (e.g. thin gold film) that is printed [13]. In one μCP approach, these SAMs act as resists for etching material in the unprinted areas, as illustrated in Fig. 10.11.

Figure 10.12 shows patterns of thin films of gold and silver generated in this manner. The typical edge resolution is ~50–100 nm; features with dimensions greater than ~200 nm can be generated easily, over large areas in a single patterning step. One of its advantages is that it is applicable not only to ultraflat surfaces of silicon wafers but also to curved objects and even to plastic substrates.

Figure 10.13 shows examples of high-resolution patterns formed on small scale cylindrical objects – optical fiber and microcapillary tubes [1]. These simple devices (integrated photomasks for optical fiber Bragg gratings and intravascular stents), require only one patterned layer.

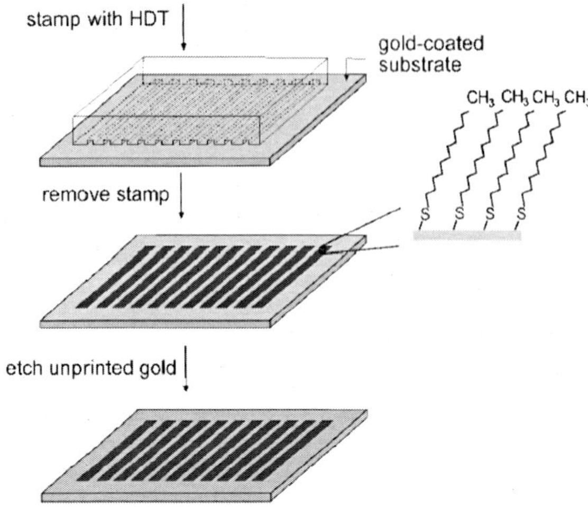

Fig. 10.11. Steps for micro-contact printing. The stamp is first inked with a solution of a material that is capable of forming a self assembled monolayer (SAM) on the substrate that will be printed. In the case illustrated here, the ink is a millimolar concentration of hexadecanethiol (HDT) in ethanol. The substrate consists of a thin layer of Au on a flat support. Removing the stamp after a few seconds of contact leaves a patterned SAM of HDT on the surface of the Au film. The printed SAM can act as a resist for the aqueous based wet etching of the exposed regions of the Au. The pattern of Au that results after etching and removal of the ink can be used to build devices of various types, including those for plastic electronics.

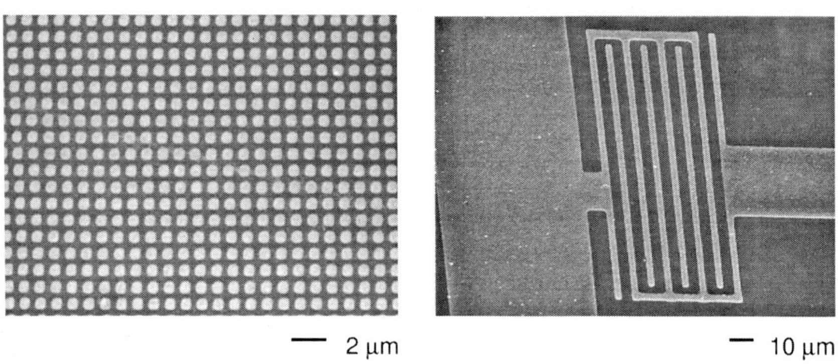

Fig. 10.12. Scanning electron micrographs of structures formed by micro-contact printing a self-assembled monolayer ink of hexadecanethiol on to a thin metal film of Au (left) and Ag (right) followed by wet etching of the unprinted areas. The left frame shows an array of Au (20 nm thick) dots with ~500 nm diameter. The right frame shows a printed structure of Ag (100 nm thick) in interdigitated geometry.

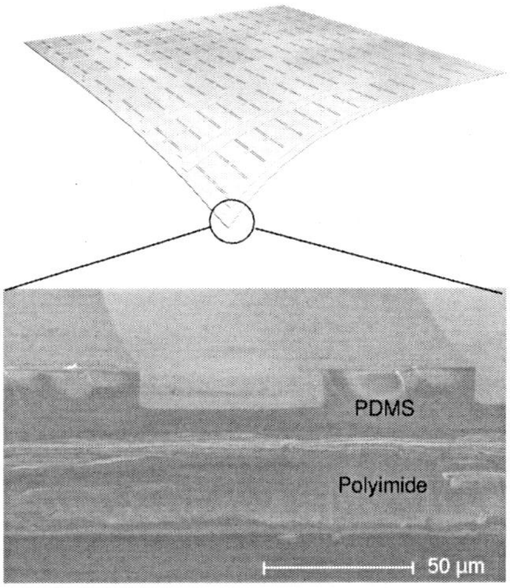

Fig. 10.13. Examples of structures micro-contact printed on to the curved surfaces of cylinders. The top frame shows 3 µm lines and spaces of Au printed on the outer surface of an optical fiber. The bottom frames show a free standing metallic structure in the form of an intravascular stent formed by printing, electroplating and then etching away the cylindrical support.

For plastic electronics and many other systems, multiple layers must be patterned on top of each other with good overlay registration. A challenge with µCP is that the elastomeric stamps tend to deform during the printing. This deformation can alter, in ways that are difficult to control, the precise alignment of features. There are at least two simple strategies to reduce these problems:

1. eliminate significant mechanical manipulation of the stamp during printing [40], and
2. implement composite stamps that use PDMS layers bonded to stiff supports [41].

The second approach has the advantage that it provides flexibility in the choice of means to perform the printing. Figure 10.14 shows a stamp with this construction, designed for plastic electronics applications [42]. It consists of a thin layer of PDMS on top of a sheet of polyimide. The relatively high in-plane modulus of the polyimide prevents distortions that can frustrate registration. Its small thickness enables the stamp to be bent in a manner that facilitates printing.

Figure 10.15 shows a vector diagram of distortions measured across a stamp designed for a relatively large area plastic active matrix backplane circuit. (These stamps show an overall isotropic shrinkage of ~0.025%, which can be incorporated into the design of the master. Misalignments associated with this shrinkage are not shown in the figure.) The histogram of displacements on the right indicates maxi-

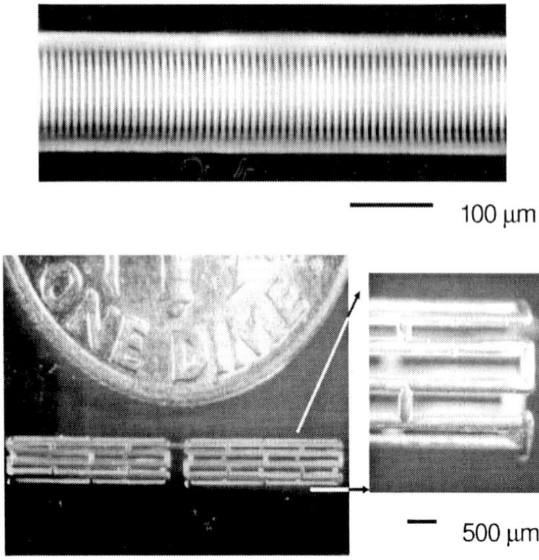

Fig. 10.14. High resolution stamp on a plastic substrate. Bending and resolution are illustrated.

mum distortions of less than a few microns. This value, which is more than a factor of ten smaller than that previously reported by us for this application [40], should satisfy many interesting applications in plastic electronics (e.g. flexible displays).

Micro-contact printing can be used in plastic electronics to form high-resolution source/drain electrodes with short channel lengths [14]. Depositing an organic semiconductor on top of these electrodes yields a transistor with a layout like that

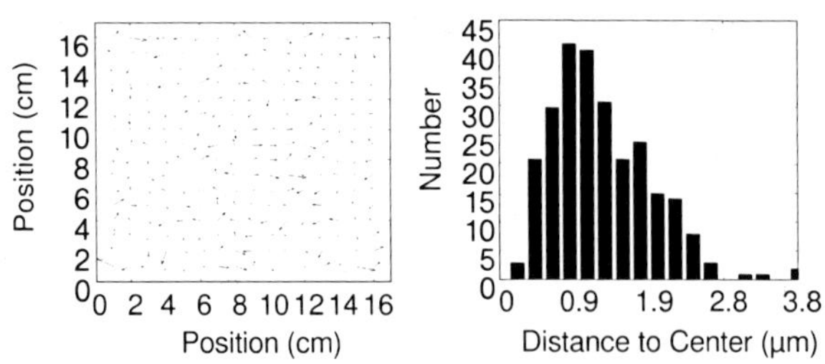

Fig. 10.15. Vector diagram (left frame) of feature misalignment in a composite stamp designed for an active matrix circuit for a display. The histogram in the right frame shows the distribution of alignment errors observed in this system. These micron-level distortions are small enough to satisfy the requirements of many applications in plastic electronics. A small overall isotropic shrinkage (~0.025%) in the stamp is not shown here; this dimensional change can be incorporated into the design of the master.

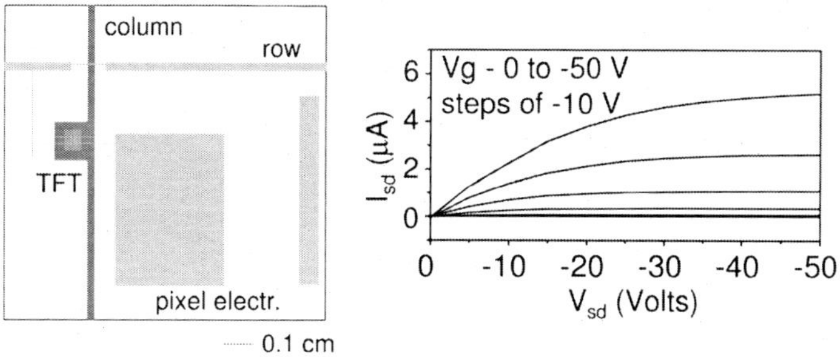

Fig. 10.16. Schematic illustration of the layout of a unit cell in an active matrix display circuit (left frame). It consists of a transistor connected to row and column electrodes and a pixel electrode pad. This circuit drives an overlying layer of electronic ink. The frame on the right shows the current–voltage characteristics of a typical transistor in a large active matrix circuit that contains several hundred unit cells like the one illustrated in the left frame. The structure consists of a substrate (PET), a gate electrode (ITO), a gate dielectric (GR), source and drain electrodes (20 nm Au and 1.5 nm Ti), and a layer of the organic semiconductor pentacene (25 nm).

shown in Fig. 10.1. This basic approach has been used to build large area plastic backplane circuits for active matrix electronic paperlike displays [40]. In that system, µCP with "inks" of hexadecanethiol ($\sim$1 mM in ethanol) defined source/drain electrodes and interconnections in thin films (20 nm) of Au. Glass resin films spin cast to thicknesses of $\sim$1 µm served as gate dielectrics [40, 43]. The substrate consisted of a 250 µm thick sheet of poly(ethylene terephthalate) (PET) with a thin (100 nm) coating of indium tin oxide (ITO) for the gate.

Figure 10.16 shows the layout of a unit cell (i.e. a single pixel driver) in this display and typical current–voltage characteristics of the drive transistor. The channel length in this case is $\sim$15 µm; the leads that connect the source/drain electrodes to the row and pixel electrodes are $\sim$15 µm wide. Images of the entire circuit and a display, built by laminating this circuit against an unpatterned sheet of electronic ink, appear in Fig. 10.17. The high resolution of µCP is important to achieving the necessary performance (transistor current output and switching speed) for this application. Resolution limits of existing forms of µCP prevent patterning of transistors with $L < \sim 0.5$ m. An alternative technique, described below, provides resolution for building devices with L deep into the nanometer regime.

10.5.3
Nanotransfer Printing

Nanotransfer printing (nTP) is a more recent high resolution printing technique. It uses surface chemistries as interfacial "glues" and "release" layers (rather than "inks") to control the transfer of solid material layers from relief features on a stamp to a substrate [10–12, 44]. This approach is purely additive (i.e. material is only deposited in locations where it is needed) and it can generate complex two or three-dimensional structures in single or multiple layers with nanometer resolu-

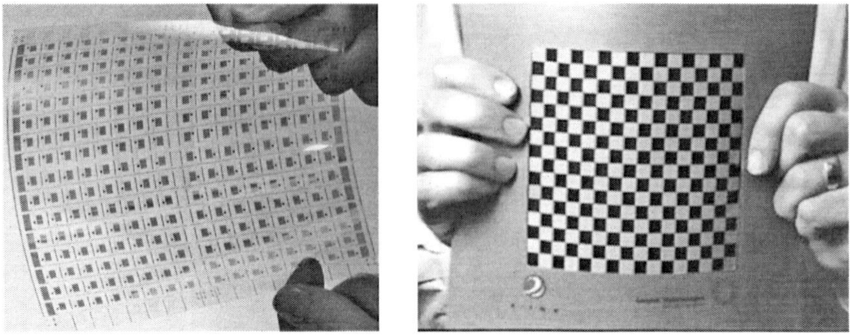

Fig. 10.17. Mechanically flexible plastic active matrix circuit (left frame) for an electronic paperlike display (right frame). Micro-contact printing defined the critical features in the circuit.

tion. It has been explored so far by our group and, in independent efforts, by groups at Princeton [45] and IBM/Zurich [46].

Figure 10.18 schematically illustrates the process. It uses elastomeric stamps similar those of μCP, or hard stamps formed using the etching procedures outlined in Fig. 10.10. Depositing a thin layer of solid material on to the raised features of the stamp constitutes the "inking" step. All of the examples described here use electron beam evaporation of a thin layer of Au (20–50 nm) or a bilayer of Au (20–50 nm)/Ti (2–5 nm).

Other material systems are also possible. Contacting this coated stamp with a substrate that supports a suitable surface chemistry results in covalent bonding between the coating and the substrate. If the surface of the stamp is treated such that the coating does not adhere well to it, then removing the stamp leaves a pattern with the geometry of the relief features. nTP can be used with SAM and other surface chemistries for printing on to flexible or rigid substrates with hard inorganic or soft polymer stamps. The stamps typically have depths of relief > 0.2 μm for patterning metal films with thicknesses < 50 nm.

Figure 10.19 shows an example of patterning on micron length scales using nTP. Fabrication of this pattern involved transfer of a 20 nm Au coating from a PDMS stamp formed by casting against photoresist patterned by contact mode photolithography. A vertical, collimated flux of Au and slightly re-entrant sidewalls on the relief features of the stamp ensured that Au was deposited only on the raised and recessed regions and not the side-walls. For the case of Fig. 10.19, a SAM formed by exposing a silicon wafer to a silane solution (3-mercaptopropyltrimethoxysilane) yields exposed thiol groups that enable transfer of Au from the stamp to the substrate [11]. Sulfur–gold bonds can form when this surface is brought into contact with an Au-coated (20 nm) PDMS stamp.

Figure 10.20 shows a printed pattern that has nanometer resolution [11]. In this case a hard stamp of GaAs, formed by electron beam lithography and etching, was used. The metal coating consisted of a bilayer of Au (20 nm) and Ti (5 nm). The

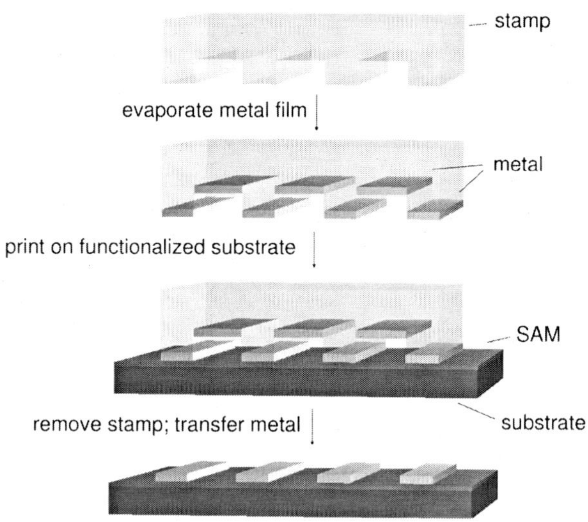

Fig. 10.18. Schematic illustration of steps for nanotransfer printing. Depositing a layer of metal using a collimated flux normal to the surface of the stamp yields a thin discontinuous coating on the raised and recessed regions of the stamp but not on its side-walls. Contacting this coated stamp to a substrate that supports suitable surface chemical groups leads to covalent bonding between the metal and the substrate. In one example, self assembled monolayers with exposed thiol groups enable printing of Au patterns on GaAs substrates. If the adhesion between the metal and the stamp is poor, then removing the stamp leaves a pattern in the geometry of the relief features.

gold adheres poorly to the surface of the GaAs. The Ti layer forms a ~3 nm oxide layer under ambient conditions. Exposing this oxide to an oxygen plasma breaks bridging oxygen bonds, thus creating defect sites where water molecules can adsorb. The result is a titanium oxide surface with some fractional coverage of hydroxyl (OH) groups (titanol) [47, 48]. The substrate is a thin film of PDMS (10–50 μm thick) cast on to a sheet of PET (175 μm thick). Exposing the PDMS to an oxygen plasma produces surface (OH) groups (silanol) [49]. Placing the plasma oxidized, Au/Ti-coated stamp on top of this substrates leads to intimate, conformal contact between the raised regions of the stamp and the substrate, without applying any pressure to the stamp.

The soft, conformable PDMS is important in this regard. It is likely that a dehydration reaction takes place at the (OH)-bearing interfaces during contact; this reaction results in permanent Ti–O–Si bonds that produce strong adhesion between the two surfaces. Peeling the substrate and stamp apart transfers the Au/Ti bilayer from the raised regions of stamp (to which the metal has extremely poor adhesion) to the substrate. The frames on the right show images of the metal-coated stamp before printing (top) and the transferred pattern (bottom). The resolution appears to be limited only by the resolution of the stamp itself, and perhaps by the grain size of the metal films. Although we have not yet quantified the accuracy in multilevel registration that can be achieved with nTP, we expect it to be similar to that of

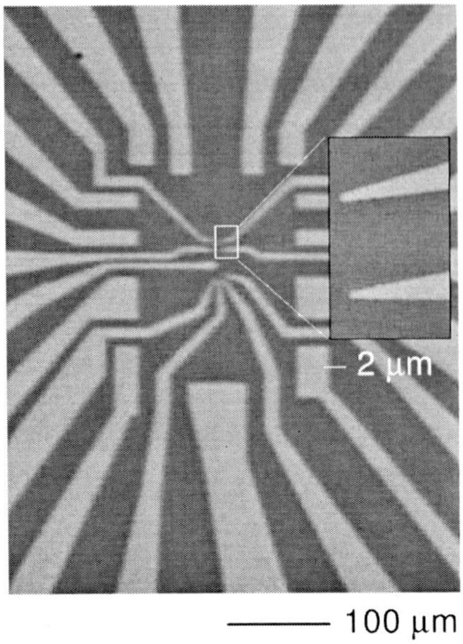

Fig. 10.19. Steps for nanotransfer printing a thin layer of Au on to a silicon wafer using a self-assembled monolayer (SAM) surface chemistry. Plasma oxidizing the surface of the wafer generates OH groups. Solution or vapor phase exposure of the wafer to 3-mercaptopropyltrimethoxysilane yields a SAM with exposed thiol groups. Contacting an Au-coated PDMS stamp to this surface produces thiol linkages that bond the gold to the substrate. Removing the stamp completes the transfer printing process. The resolution demonstrated here is limited only by the contact photolithography used to generate the master for the stamp.

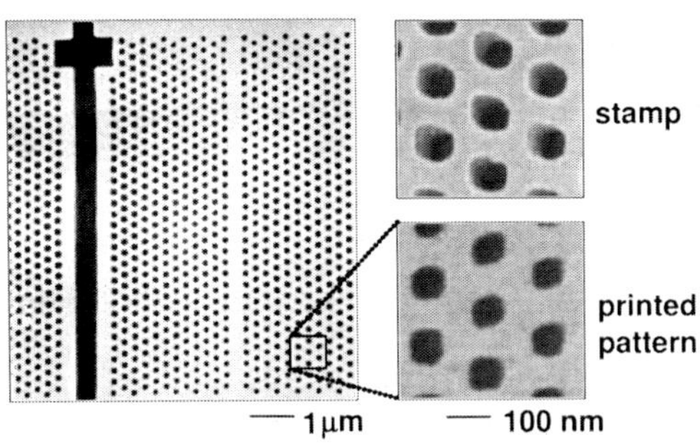

Fig. 10.20. Illustration of the usage of a PDMS stamp to print a high resolution Au/Ti layer onto a flexuble subtrate. Separating the stamp from the substrate results in the transfer of the Au/Ti layer from the rised regions onto the substrate.

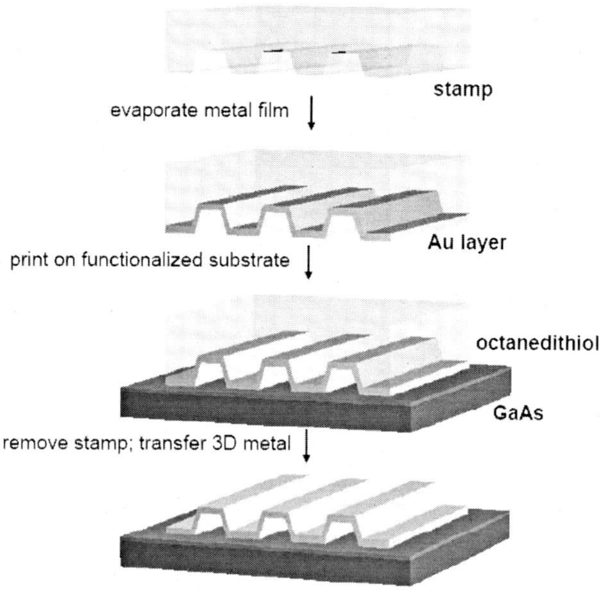

Fig. 10.21. Schematic illustration of steps for forming a type of three-dimensional structure by nTP. Printing with a stamp that has a conformal coating of metal produces sealed nanochannels on the substrate. This illustration depicts printing of a structure of Au on to a GaAs substrate by use of an octanedithiol monolayer.

embossing techniques that use rigid stamps or to that of μCP with composite stamps. In addition to two-dimensional (2D) patterns like those in Figs 10.19 and 10.20, it is possible to build certain classes of three-dimensional (3D) structures with nTP [12]. Figure 10.21 schematically illustrates one example. Here a conformal coating of Au on the stamp yields, after printing, arrays of closed nanochannels.

Figure 10.22 shows a scanning electron micrograph of a structure formed in this manner. In this case, a SAM formed from octanedithiol on a substrate of GaAs provides the surface chemistry for transferring the Au nanochannel structures (20 nm Au thickness) [44]. The PDMS stamps were generated by casting and curing against a pattern of relief on an InP wafer formed by interference lithography and etching. Because the stamp is soft and nTP is purely additive, it is easy to build multilayer structures by printing multiple times on a single substrate. Figure 10.22 also shows a multilayer nanochannel lattice structure formed in this way. Cold welding bonds one Au nanochannel structure to the next. We observed no degradation in the structural integrity that would prevent more than the 10 layers illustrated here.

Our experience indicates that the processing windows for the current versions of nTP are narrower than those of μCP. At least three processing aspects are important to achieving high fidelity with nTP:

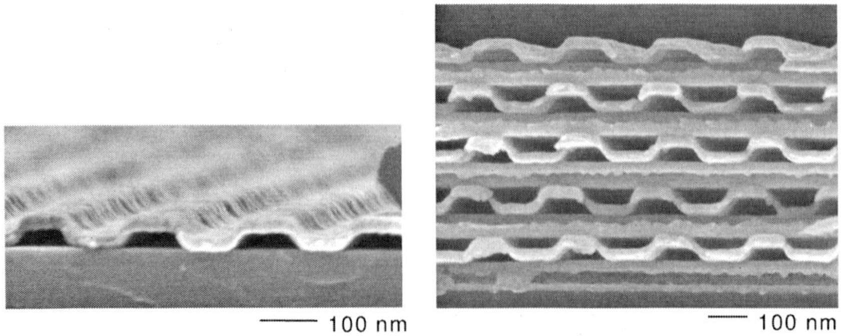

Fig. 10.22. Scanning electron micrograph of a printed nanochannel structure made of a thin layer (20 nm) of Au bonded to a GaAs wafer by an octanedithiol monolayer (top frame). The bottom frame shows a multilayer nanochannel structure formed by multiple printing steps. The nanochannels bond to one another by gold–gold cold welding. This structure consists of ten layers of crossed channels. Additional layers are possible.

1. the deposition procedures must be carefully controlled to avoid cracking or buckling of the metal films when PDMS stamps are used;
2. PDMS stamps must be handled carefully before and during the printing to avoid surface strains that can damage the metal coatings; and
3. the surface chemistry must provide a high density of bonding sites, and the surfaces of the stamps and substrates must come into uniform, intimate contact to enable efficient transfer.

For the first issue, high deposition rates and surface treatments that facilitate wetting of the metals on the surfaces of the stamps are important. For the second, composite stamp designs that use rigid backings and relatively thin PDMS layers are helpful. In the case of the third issue, suitable chemistries must be employed with substrates with low roughness. With well controlled procedures, patterns over large areas (several cm^2) that are free of defects or cracks (as observed by high resolution scanning electron and atomic-force microscopy) can be produced [42]. We note that in many cases, XPS analysis shows that a residual, ultrathin layer of organic material can transfer from the stamp with the metal film to the substrate when PDMS based stamps are used. This layer is on the order of 1–4 nm thick, depending on the processing conditions. It can be removed using various dry or wet etching procedures [42].

Like μCP, nTP can pattern electrodes for plastic electronic components. Figure 10.23 shows an optical micrograph (inset) and current–voltage characteristics (left frame) of a transistor that incorporates interdigitated source/drain electrodes of Au/Ti patterned by nTP on PDMS/PET using procedures described previously [10].

Laminating this structure against a substrate that supports an organic semiconductor, gate dielectric (GR) and gate electrode (ITO) forms a transistor with the source/drain electrodes on top of the semiconductor. The transistor shown here

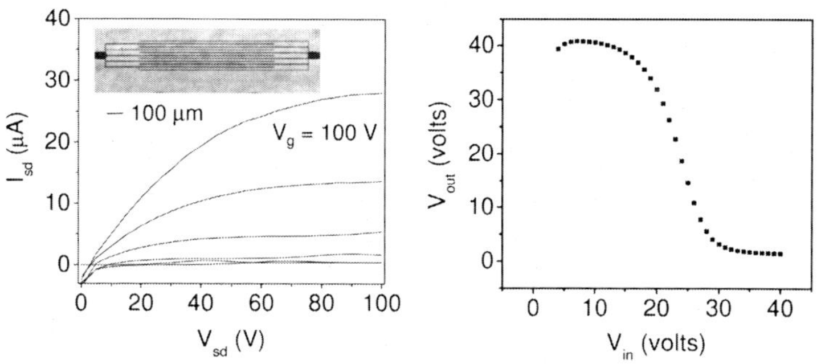

Fig. 10.23. The frame on the left shows current–voltage characteristics of an n-channel transistor formed with electrodes patterned by nanotransfer printing. Laminating these electrodes against a substrate (PET) that supports an organic semiconductor (FCuPc), a gate dielectric (GR) and a gate (ITO) completes the device. The inset shows an optical micrograph of the interdigitated electrodes. The frame on the right shows the transfer characteristics of a simple complementary logic gate that uses this device and a similar one for the p-channel (pentacene) transistor.

uses the n-type organic semiconductor copper hexadecafluorophthalocyanine (FCuPc). The right frame of Fig. 10.23 shows the transfer characteristics of a laminated complementary logic gate whose electrodes and connecting lines are defined by nTP. The p-channel transistor in this circuit uses pentacene for the semiconductor [10].

The high resolution of nTP and its multilayer capabilities are useful for plastic electronic systems. Figure 10.24 shows a scanning electron micrograph and

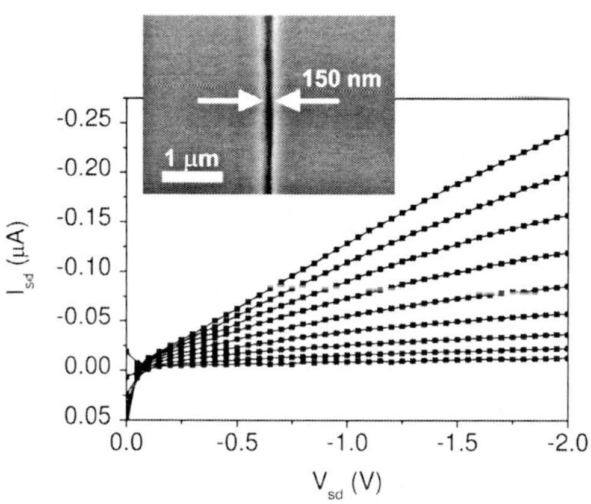

Fig. 10.24. Current–voltage characteristics from a nanoscale organic transistor formed by contact of a metal coated stamp against a thin film of the organic semiconductor pentacene. The gate voltage varies between +1.0 V and −2.2 V in steps of 0.4 V. The channel length is ~150 nm. The inset shows a scanning electron micrograph of the separation between the source and drain electrodes on the stamp.

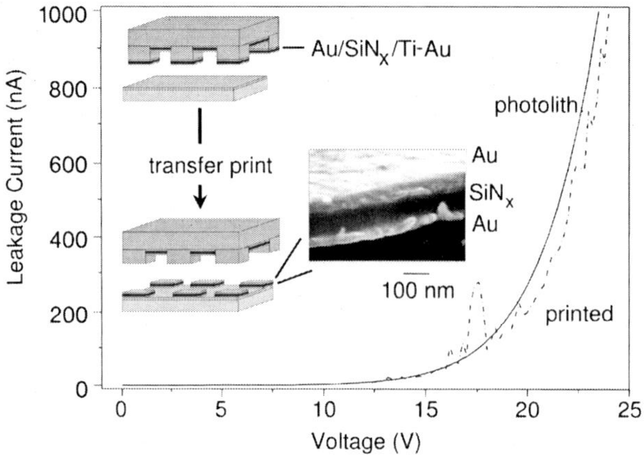

Fig. 10.25. Multilayer thin film capacitor structure printed in a single step on to a plastic substrate using the nanotransfer printing technique. A multilayer of Au/SiN_x/Ti/Au was first deposited on to a silicon stamp formed by photolithography and etching. Contacting this stamp to a substrate of Au/PDMS/PET forms a cold weld that bonds the exposed Au on the stamp to the Au-coating on the substrate. Removing the stamp produces arrays of square (250 μm × 250 μm) metal/insulator/metal capacitors on the plastic support. The dashed line shows the measured current–voltage characteristics of one of these printed capacitors. The solid line corresponds to a similar structure formed on a rigid glass substrate using conventional photolithographic procedures. The slightly higher level of noise in the printed devices results, at least partly, from the difficulty in making good electrical contacts to structures on the flexible plastic substrate.

current–voltage characteristics of a transistor with $L = 150$ nm. In this device, an Au-coated stamp is simply placed into contact with the semiconductor (pentacene) to form a top contact device [50]. nTP can also form complex multilayer devices with electrical functionality on plastic substrates [11].

Figure 10.25 shows a metal/insulator/metal (MIM) structure of Au (50 nm), SiN_x (100 nm; by plasma enhanced vapor deposition), Ti (5 nm) and Au (50 nm) formed by transfer printing with a silicon stamp that is sequentially coated with these layers. In this case, a short reactive ion etch (with CF_4) after the second Au deposition removes the SiN_x from the side-walls of the stamp. nTP transfers these layers in a patterned geometry to a substrate of Au(15 nm)/Ti(1 nm) on PDMS(50 μm)/PET. Interfacial cold welding between the Au on the surfaces of the stamp and substrate bonds the multilayers to the substrate. These MIM capacitors have performance similar to devices fabricated on silicon wafers by photolithography and lift-off [11]. This example illustrates the ability of nTP to print patterns of materials whose growth conditions (high temperature SiN_x by PECVD in this case) prevent their direct deposition or processing on the substrate of interest (PET in this case). The cold welding transfer approach has also been exploited in other ways for patterning components for plastic electronics [45, 51].

10.6 Large Area Stamps, Molds, and Photomasks for Soft Lithography

10.6.1 Micro-contact Printing: A Path to Reel-to-reel Electronics

Flexography and μCP have basic similarities and ample differences. They both use flexible elastomeric printing plates and are "inked" with a slightly raised image, which is transferred on to a substrate. Flexography [52] can be briefly described as a reel to reel, high-speed printing process. It uses fast-drying inks for printing large areas (>1 m^2) on plastic, foil, acetate film, brown paper, and other materials used in the packaging industry with modest resolution (30 μm). In contrast, μCP [13, 14] has a demonstrated ability to pattern sub-micrometer features of Au and other metals, typically over small areas with potential applicability to large areas. It is well suited for the fabrication of conducting layers in high performance electronic devices such as backplanes for electrophoretic displays [40]. Commercializing high performance flexible displays might be achieved in a reel-to-reel process, in which large area μCP plates are used in a *flexography-like* process. A starting point on such an ambitious path, and the heart of this work, is learning how to produce flexible plastic printing plates that offer high resolution over large areas at a reasonable cost.

In this work, it is demonstrated that 12-inch × 12-inch polydimethylsiloxane (PDMS) stamps with micrometer-size features can be fabricated via a simple procedure. The size was chosen to fit the plate size requirements of small commercial tag and label flexographic presses. Stamps were used to pattern micrometer-size Au lines on to 12-inch × 12-inch Mylar substrates using standard thiol chemistry for the inks. The large PDMS stamps were constructed via a straightforward and inexpensive process. A standard photoresist (PR) coating was used to create relief structure in a master. The PR was exposed through a glass photomask using a standard exposure unit, which is nearly collimated over a 30-inch × 40-inch area and commonly available in printing shops. A PR master with lines and spaces (L/S), 1 to 10 μm wide was then used to produce the 12-inch × 12-inch PDMS stamps. The stamps were inked using established thiol chemistries and micrometer size Au lines were then patterned on to a Mylar substrate. This simple, low-cost approach to photoresist exposure enables the fabrication of masters up to 30 inch × 40 inch in size with micrometer resolution without complex photolithography setups that are typically used in semiconductor manufacturing or research facilities. It is also shown that that these same stamps can be used as molds for replica molding of photopolymers and as phase masks for exposing photoresist layers [53].

10.6.2 Inexpensive Approaches to Large-area Printing

Because most anticipated applications of plastic electronics systems cannot bear high processing costs, the scale-up of μCP presented here is based on simple, low-

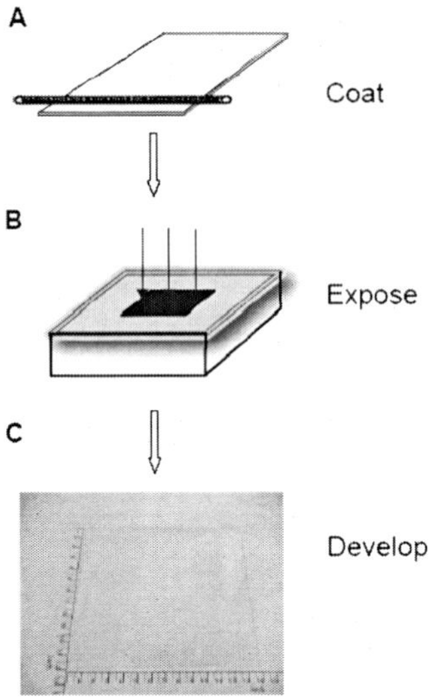

Fig. 10.26. Schematic illustration of steps for making 12-inch × 14-inch glass master. (a) A clean 12-inch × 14-inch × 0.7-mm glass sheet was coated with diluted photoresist using a Meyer to a 1.3 μm thickness and prebaked for 2 min at 115 °C. (b) Large-area UV exposure unit, contact between photomask (glass or plastic) and glass sheet is done by vacuum (c) completed glass master after development in MF-319 developer for 1 min and post bake at 115 °C for 2 min.

cost manufacturing steps. First, a large master for the PDMS stamp was fabricated using standard PR technology. A Shipley 1805 PR was coated on to a 12-inch × 14-inch clean glass sheet (Corning 1737) to 1 μm in thickness using a Meyer rod (Fig. 10.26a). A chrome glass photomask (g-PM, Advance Reproductions) was used to pattern this resist. The features in the g-PM included sets of source and drain electrodes with 1, 1.5, 2, 3, 5 and 10 μm channel lengths with lines measuring 500 μm long and 250 μm wide. Four imaging steps were performed using a small, inexpensive 5-inch × 5-inch g-PM to expose the 12-inch × 12-inch PR area. The mask was sequentially positioned at the four corners of the PR. At each position, the PR areas not in direct contact with the photomask were masked. Having established a vacuum contact between the g-PM and PR, the PR was exposed for 13 s. The PR was exposed using light from a ultraviolet (UV) source (L1261-OLEC lamp). This system (Douthitt) is commonly used in the printing industry for the exposure of large analog photopolymer plates (30 inch × 40 inch) used in color proofing (Fig. 10.26b). The bulb's spectral range and energy density over the exposure area are 360 to 385 nm and 3.8 ± 0.1 mW cm^{-2}, respectively. After having completed the exposures, the PR was post baked at 115 °C for 2 min before its development in MF-319 developer (Shipley) for 1 min (Fig. 10.26c). With this approach, large masters with 1 μm features can be achieved without an I-liner imaging.

The developed PR was then used as a master to fabricate the PDMS stamp

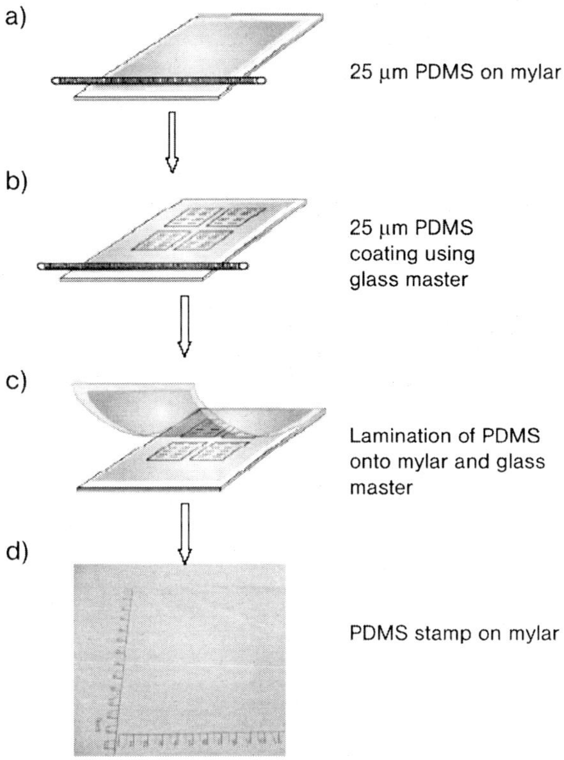

Fig. 10.27. Fabrication steps for a large area PDMS stamp.

(Fig. 10.27). A fluorinated silane self-assembled monolayer was first applied to the master to aid in the separation of master and stamp. The stamp was formed using one of the following procedures. For thinner stamps, PDMS (Sylgard 184) pre-polymer mixture (1:10) was poured over the PR master and a 250 μm thick, 13-inch × 13-inch Mylar film was placed on top of the liquid PDMS. The pre-polymer was evenly spread to less than 50 μm in thickness by rolling a Meyer rod over the surface of the Mylar backing film.

Pre-polymer of PDMS was cured with holding down the edges of the slightly oversized Mylar backing film using vacuum plate at 67 °C on hotplate for 18 h. Thicker plates were fabricated as follows. To ensure the PDMS stamp were absolutely flat and uniform, two PDMS pre-polymer mixture (1:5) layers were coated and put in contact via lamination. Thus, a 25 μm thick layer of PDMS pre-polymer was directly coated on the PR master surface and another 25 μm thick layer was coated on to the stamp's Mylar backing (Fig. 10.27a, b). The two pre-polymer layers were then laminated before the curing step.

Adhesion between the Mylar backing and the PDMS stamp was enhanced by pre-treating the Mylar surface with an adhesion promoter (1205, Dow Corning).

Curing laminated PDMS stamp at room temperature for 48 h followed by delaminating from glass master completed the stamp fabrication. The thickness of cured PDMS layer obtained by following the later procedure is ~50 μm (Fig. 10.27c, d). Atomic force microscopy (AFM) images of the developed PR are shown in Fig. 10.28a, b.

Because PDMS stamps are soft, a mold was fabricated (with Norland optical adhesive (NOA 73)) and used for the AFM analysis (Fig. 10.28c). A 25 μm NOA 73 photopolymer film was first coated on to a 12-inch × 14-inch glass sheet and laminated on to a PDMS stamp. The replication was completed by crosslinking the NOA 73 by UV exposure and removing the PDMS stamp. AFM images of the Au lines on a Mylar substrate fabricated using μCP from the stamp is shown in Fig. 10.28d.

To demonstrate the use of the resulting stamp, μCP was performed on a layer of Au on plastic. In particular, Ti/Au (2 nm/20 nm) film was deposited on to a 12-inch × 12-inch Mylar film (250 μm thick) using an e-beam evaporator. After the stamp had been inked with a 1 mM hexadecanethiol solution, the stamp was gently brought into contact with the Au-coated Mylar surface. The contact between the stamp and the Mylar was established in a gradual way to control the propagation of the wetting front (from top to bottom) and avoid the formation of trapped air bubbles. The thiol molecules were transferred from the raised regions of the stamp which come into contact with the Au surface. An aqueous ferro/ferricyanide etchant was then used to etch away the Au in the areas unprotected by the thiol molecules [13].

In addition, the stamps are phase masks that can be used to produce submicrometer features of photoresist via near field exposure [53]. In this case, a 12-inch × 12-inch plastic photomask on 250 μm thick Mylar sheet having 12 μm lines and spacings throughout was purchased from Advance Reproductions. A 12-inch × 12-inch PR layer was coated using a Meyer rod (measured thickness of 500 nm), followed by a prebake at 115 °C for 2 min. The developed PR was used to fabricate the PDMS stamp with 12 μm lines and spaces. The height of the lines relative to the trenches was 500 nm. The adhesion of photoresist to the 12-inch × 12-inch glass plate was improved by priming the glass with hexamethyldisilazane (HMDS). The HMDS was vacuum deposited on the glass and baked at 150 °C for 30 min before the PR coating. The patterned stamps were placed in contact with the unexposed PR-coated glass (thickness 500 nm) and exposed for 5 s in the Douthitt exposure system. After post-baking at 115 °C for 2 min, exposed regions were developed with MF-319 developer for 30 s (Fig. 10.28e, f).

Deformations in the stamps can cause difficulties in alignment and registration. To quantify these deformations, the exact positions of the features on the stamp were compared with those on the PR master [41]. Although absolute distortions are important, relative distortions between multiple prints of the same pattern are enough to prove the feasibility of using μCP technique for printing multilayer structures.

The measurement begins by placing the PDMS stamp on its master and then recording with an optical microscope the offsets between the positions of 36 fea-

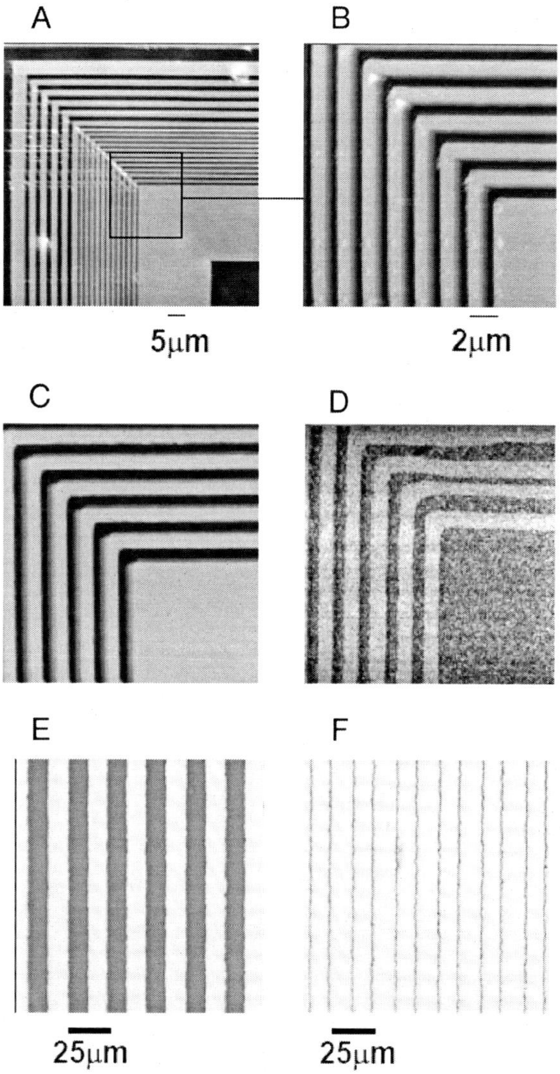

Fig. 10.28. (a) AFM of 1 and 2 μm line and spacing of developed photoresist on Kapton; (b) detail AFM of 1-μm lines and spaces of image (a); (c) AFM 1 μm line and spacing of replicated NOA 73 mold from PDMS stamp; (d) AFM 1 μm line and spacing of etched Au after thiol stamping using PDMS stamp; (e) optical micrograph of 12 μm line and spacing of photoresist on glass master; (f) optical micrograph submicron lines on glass master after phase shift lithography using PDMS photomask made from master (e).

tures of the master and of stamp in an array of 6 × 6 patterns equally spaced across the 12-inch × 12-inch area.

Figure 10.29 shows displacement offset vectors after subtracting the effects of rigid translations and rotations and isotropic shrinkage of the stamp (Root mean

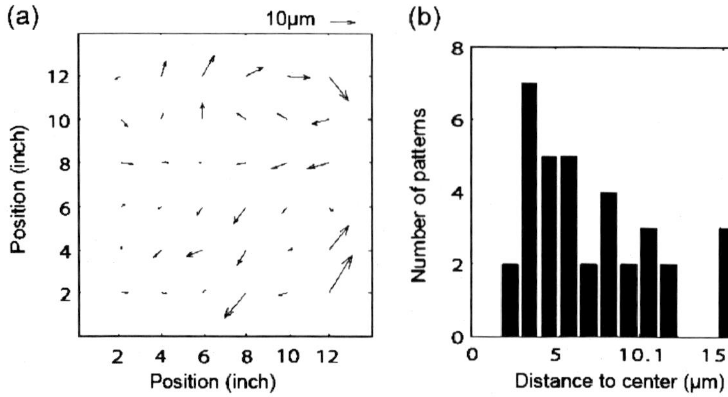

Fig. 10.29. Distortion measurements collected at 36 points equally spaced across a 12-inch × 12-inch stamp supported by a 250 μm thick Mylar backing. In this case a vacuum plate was used to hold down the Mylar backing on the glass master during the composite stamp curing step (67 °C on a hot plate for 18 h). (a) The top frame shows a vector diagram of misalignments between the stamp and its master. (overall translational, rotational misalignments, and isotropic shrinkage of the stamp are subtracted.) (b) The bottom frame shows a histogram plot of the lengths of the vectors illustrated in the top frame. The median distortion is approximately 6.5 μm.

square (RMS) value of residual distortion = 22 ppm (6.5 μm over 30 cm) for a stamp cured at 67 °C hotplate). The stamp used for this experiment consisted of a 50 μm thick PDMS layer supported by a 250 μm thick Mylar sheet. The algorithm used to perform these corrections was similar to that used by Menard et al. in the case of thin PDMS stamps backed with two Kapton layers [42]. The residual distortions have a median value of 6.5 μm. The approximate accuracy of the measurement method is estimated to be 1.5 μm. One of the factors that may facilitate the successful introduction of plastic electronics would be to provide a clear differentiation between any new approach being brought into the marketplace and the well established Si technology. Regarding plastic electronics as technologically capable of producing electronic devices over large areas, with high throughput, on to flexible substrates and perhaps with a reel-to-reel printing process would clearly emphasize the distinction. The aim of this work was to support this idea by proposing that μCP could perhaps be the starting point for the development of μ-flexography.

10.6.3
Registration Using the Lock-and-key Mechanism in Soft Imprinting

Soft imprinting using flexible stamp has ample potential to be applied for reel-to-reel process [54, 55]. To achieve fast process in reel-to-reel device fabrication, passive alignment process between flexible stamp and substrate is needed rather than active optical alignment procedure. As a passive alignment process, "lock and key" mechanism was studied between flexible stamp and substrate in soft imprinting.

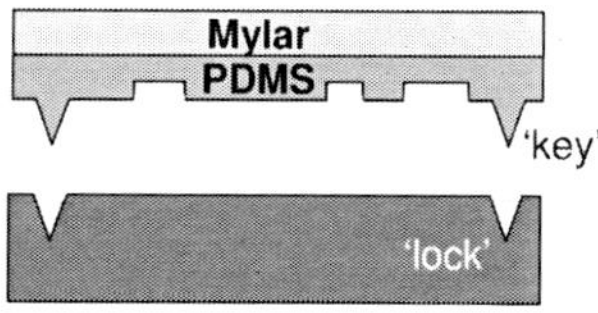

Fig. 10.30. Lock-and-key mechanism for registration in soft imprinting.

The lock and key mechanism is illustrated in Fig. 10.30. Key feature are fabricated on PDMS stamp that match the Lock feature fabricated on the stamp. Once Key feature on PDMS stamp matches in Lock feature on substrate, the desired features inside on stamp can be registered on to the substrate automatically. Originally, IBM suggested that this registration mechanism may be used for μCP in continuous process [56]. μCP, however, is difficult to use lock and key mechanism since sticky key feature of PDMS stamp is hard to fit into lock feature without help of lubricant. Also, self-assembled monolayer (SAM) ink on the relived surface on flexible stamp can be misprinted on to substrate during adjusting and matching lock and key, which results in low-resolution printing. On the other hand, soft imprinting is suitable for "lock and key" mechanism for high resolution patterning in micron scale, which use UV-curable resist. Before UV curing, the resist can act as a lubricant between stamp and substrate with helping matching lock and key.

Figure 10.31 shows fabrication procedure of PDMS/Mylar composite stamp having key feature and registration between the stamp and substrate having lock feature on wafer. Lock features made on Si (100) wafer using anisotropic etching in KOH solution. Si_3N_4 or SiO_2 layer acts as an etching resist of Si (100) during the anisotoropic etching of Si (100). Master for PDMS mold was made by photoresist patterning on etched Si (100) wafer with optical alignment between patterns and lock features so that photoresist patterns are positioned on center of lock features.

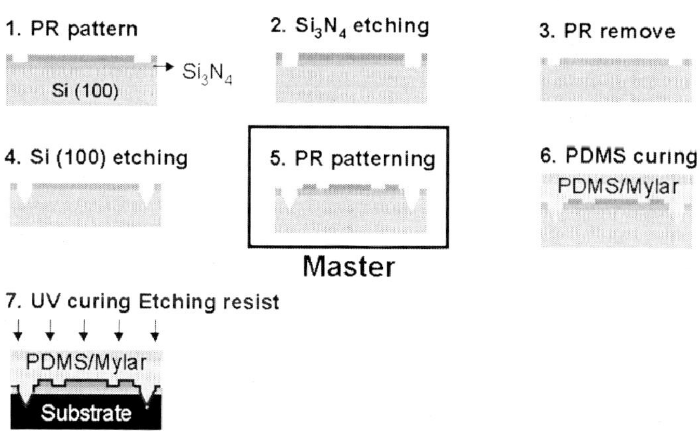

Fig. 10.31. Fabrication procedures for registration in soft imprinting.

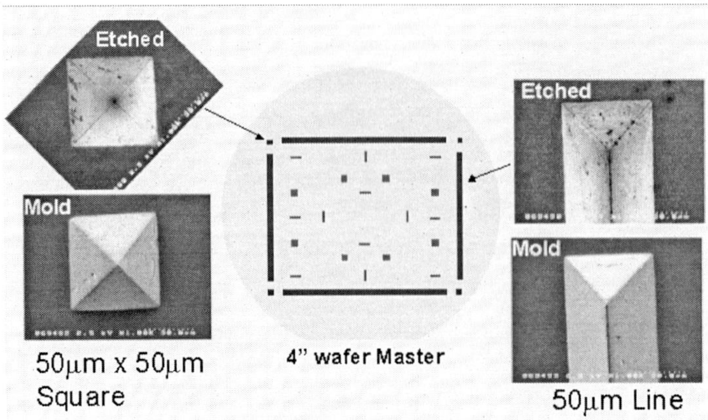

Fig. 10.32. SEM images of lock features on the etched master and key features on the PDMS/Mylar composite stamp molded from the master.

After molding PDMS stamp (Sylgard 184) with 250 μm thick Mylar backing film at room temperature so that shrinkage of PDMS be minimized, the composite stamp were placed on UV curable resist spin coated on an Si wafer substrate, which was made using same procedure with that of master. Offsets between photoresist patterns on substrate and relief patterns on PDMS stamp were investigated using optical microscope to study registration.

During matching between the key on stamp and lock on substrate, the prepolymer of etching resist acted as lubricant. Figure 10.32 shows SEM images of lock features on the etched 4 in wafer master and key features on the PDMS/Mylar composite stamp molded from the master. Blue features inside red lock lines are photoresist patterns to study registration. After matching lock and key, Fig. 10.33 shows that the offset values between 50 μm × 50 μm relief features on stamp and 50 μm × 50 μm photoresist patterns on substrate are all less than 5 μm on a 4-inch wafer. Currently, we are investigating registration between rubber stamp and plastic substrate using lock and key mechanism for multi layer device fabrication.

10.7 Conclusions

This chapter summarizes some of our recent work in printing techniques and plastic electronics. It also presents new data from printed transistors that use several different organic semiconductors in a variety of device geometries. In all cases, we observed good performance. μCP for the source/drain electrodes is attractive because it provides a simple and potentially low-cost route to high resolution (i.e. small channel lengths, L) structures that can be used to build transistors which

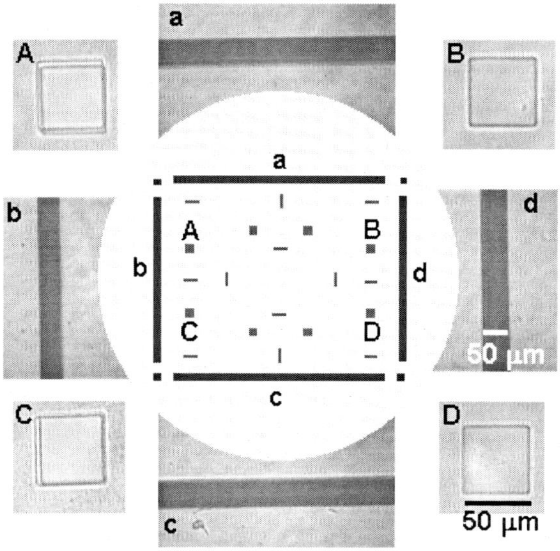

Fig. 10.33. Registered PDMS/Mylar stamp on Substrate using lock and key mechanism: A, B, C and D are optical microscope images of matched relief features of stamp on photoresist patterns of substrate; a, b, c, and d are optical microscope images of matched keys of stamp on to locks on substrate.

produce large "on" currents even with the moderate to low mobilities that are available from currently known organic semiconductors. It is also compatible with solution deposited metals (i.e. electrodeless silver). It is conceivable that it could be easily used for manufacturing in high volumes and large sizes, because it employs elastomeric materials of similar thickness and modulus to those used in standard flexography which currently exist in standard printing presses in reel-to-reel configurations. Its main disadvantage is that it is compatible with a narrow set of materials only. The most well developed forms of µCP use thiol inks, thin films of coinage metals and wet etching. Additional work will be required to invent methods for using this type of printing technique to pattern the dielectric and semiconductor layers with similar resolution. Techniques based on selective dewetting from printed hydrophilic/hydrophobic patterns may be promising.

Although the thermal transfer printing technique does not offer resolution as high as µCP, it does have enormous flexibility in the materials that can be patterned. It also benefits from its all "dry" and purely additive operation; it does not require etchants or solution routes for depositing or patterning the key materials. It is necessary, however, to balance the adhesion between the layers that are sequentially transfer such that interfaces have intimate physical and electrical contact. In addition, current commercial systems have pixel sizes of 5 µm × 2.2 µm, which limit the resolution. With the current DNSSA–PANI/SWNT formulation and a printer of this type, it is possible to print 10-µm channels and 25-µm lines.

For applications that demand sub-20 μm resolution, it may be possible to combine micro-contact and thermal transfer printing: high resolution source/drain electrodes printed with μCP could be used with other components that are patterned by thermal transfer. These and other strategies that combine and match different patterning techniques may provide an attractive means of building plastic circuits.

Acknowledgments

We wish to thank our collaborators at DuPont and Bell Labs, many of whom participated in some of the work described here.

References

1 Mirkin C A and Rogers J A, "Emerging Methods for Micro- and Nanofabrication", *MRS Bull*, 2001 **26** 506–507.

2 Sze S, *Semiconductor Devices: Physics and Technology*, New York, Wiley, 1985.

3 Drury C J, Mutsaers C M J, Hart C M, Matters M and de Leeuw D M, "Low-cost all-polymer integrated circuits", *Appl Phys Lett*, 1998 **73** 108–110.

4 Gelinck G H, Geuns T C T and de Leeuw D M, "High-performance all-polymer integrated circuits", *Appl Phys Lett*, 2000 **77** 1487–1489.

5 Sirringhaus H, Kawase T, Friend R H, Shimoda T, Inbasekaran M, Wu M and Woo E P, "High-resolution inkjet printing of all-polymer transistor circuits", *Science*, 2000 **290** 2123–2126.

6 Garnier F, Hajlaouir R, Yassar A and Srivastava P, "All-polymer field-effect transistor realized by printing techniques", *Science*, 1994 **265** 1684–1686.

7 Bao Z, Feng Y, Dodabalapur A, Raju V R and Lovinger A J, "High-performance plastic transistors fabricated by printing techniques", *Chem Mater*, 1997 **9** 1299–1301.

8 Rogers J A, Bao Z and Raju V R, "Non-photolithographic fabrication of organic transistors micron feature sizes", *Appl Phys Lett*, 1998 **72** 2716–2718.

9 Behl M, Seekamp J, Zankovych S, Torres C M S, Zentel R and Ahopelto J, "Towards plastic electronics: Patterning semiconducting polymers by nanoimprint lithography", *Adv Mater*, 2002 **14** 588–591.

10 Loo Y-L, Willett R W, Baldwin K and Rogers J A, "Additive, nanoscale patterning of metal films with a stamp and a surface chemistry mediated transfer process: Applications in plastic electronics", *Appl Phys Lett*, 2002 **81** 562–564.

11 Loo Y-L, Willett R W, Baldwin K and Rogers J A, "Interfacial chemistries for nanoscale transfer printing", *J Am Chem Soc*, 2002 **124** 7654–655.

12 Zaumseil J, Meitl M A, Hsu J W P, Acharya B, Baldwin K W, Loo Y-L and Rogers J A, "Three-dimensional and Multilayer Nanostructures Formed by Nanotransfer Printing", *Nano Lett*, 2003 **3** 1223–1227.

13 Kumar A and Whitesides G M, "Features of gold having micrometer to centimeter can be formed through a combination of stamping with an elastomeric stamp and an alkanethiol ink followed by chemical etching", *Appl Phys Lett*, 1993 **63** 2002–2004.

14 Rogers J A, Bao Z and Makhija A, "Printing Process suitable for reel-to-reel production of high performance organic transistors and circuits", *Adv Mater*, 1999 **11** 741–745.

15 Blanchet G, Loo Y-L, Rogers J A, Gao F and Fincher C, "Large area dry printing of organic transistors and circuits", *Appl Phys Lett*, 2003 **82** 463–465.

16 C. D. Dimitrakopoulos, D. J. Mascaro, "Organic thin-film transistors: A review of recent advances", IBM J. Res. & Dev V45 No. 1 January 2001.

17 G. B. Blanchet, G. Jaycox and C. Nuckolls, "Self-assembled three-dimensional conducting network of single-wall carbon nanotubes", Appl. Phys. Lett. **85**, 828–830 (2004).

18 G. B. Blanchet, M. Lefenfeld and J. Rogers, Appl. Phys. Lett. **84**, 296 (2004).

19 A. J. Heeger and A. G. MacDiarmid, Mol. Cryst. Liq. Cryst. 77, 1 (1981) and J. C. Chiang, A. G. MacDiarmid, Synth. Met. 13, 193 (1986).

20 Y. Yao, P. Smith and A. Heeger, Synth. Met. 48, 91 (1992).

21 M. Reghu, Y. Cao, D. Moses, and A. J. Heeger, Phys. Rev. B. 47, 1758 (1993) and A. J. Epstein and A. G. MacDiarmid, Synth. Met. 69, 179 (1995).

22 P. J. Kinlen, US Patent 5,863,465 (1999).

23 H. Kataura, Y. Kumazawa, Y. Maniwa, I. Umezu, S. Susuki, Y. Ohtsuka and Y. Aciba, Synth. Met. 103, 2555 (1999).

24 P. Petit, C. Mathis, C. Journet and P. Bernier, Chemical Physics letters, 305, 370 (1999).

25 A. B. Kaiser, G. Dusberg and S. Roth, Phys. Rev. B, 57, 1418 (1998).

26 C. Y. Yang, M. Reghu, A. j. Heeger, Y. Cao, Synth. Met. 79, 27 (1996).

27 J. E. Fisher, H. Dai, A. Thess, R. Lee, N. M. Hanjani, D. L. Dehaas and R. E. Smalley. Phys. Rev. B. 55, 4921 (1997).

28 R. S. Kohlman, A. Zibold, D. B. Tanner, G. G. Ihas, T. Ishigure, Y. G. Min, A. G. MacDiarmid and A. J. Epstein, Phys. Rev. Lett. 78, 3915 (1997).

29 A. G. MacDiarmid, A. J. Epstein, Synth. Met. 65, 103 (1994).

30 K. Lee and A. J. Heeger, Synth. Met. 84, 715 (1997).

31 J. P. Pouget, M. E. Jozefowicz, A. J. Epstein, X. Tang and A. G. MacDiarmid, Macromolecules, 24, 779 (1991).

32 Moreau, W M. In *Semiconductor Lithography: Principles and Materials*; Moreau, W. M., Ed.; Plenum: New York, 1988.

33 Gibson J M, "Reading and Writing with Electron Beams", *Phys Today*, 1997 **50** 56–61.

34 Matsui S and Ochiai Y, "Focused Ion Beam Applications to Solid State Devices", *Nanotechnology*, 1996 **7** 247–258.

35 Sohn L L and Willett R L, "Fabrication of Nanostructures Using Atomic-Force Microscope-Based Lithography", *Appl Phys Lett*, 1995 **67** 1552–1554.

36 Betzig E and Trautman K, "Near-Field Optics – Microscopy, Spectroscopy, and Surface Modification Beyond the Diffraction Limit", *Science*, 1992 **257** 189–195.

37 Bard A J, Denault G, Lee C, Mandler D and Wipf D O, "Scanning Electrochemical Microscopy: A New Technique for the Characterization and Modification of Surfaces", *Acc Chem Res*, 1992 **23** 357.

38 Stroscio J A and Eigler D M, "Atomic and Molecular Manipulation with the Scanning Tunneling Microscope", *Science*, 1991 **254** 1319–1326.

39 Nole J, "Holographic Lithography Needs No Mask", *Laser Focus World*, 1997 **33** 209–212.

40 J. A. Rogers, Z. Bao, K. Balwin, A. Dodabalapur, B. Crone, V. R. Raju, V. Kuck, H. Katz, K. Amundson and P. Drzaic, "Paper-like electronic displays: Large-area rubber-stamped plastic sheets of electronics and microencapsulated electrophoretic inks", PNAS, 98, 4835 (2001).

41 J. A. Rogers, K. E. Paul and G. Whitesides, "Quantifying distortions in soft lithography", *J. Vac. Sci. and Technol.* **B16**, 88 (1998).

42 E. Menard, L. Bilbaut, J. Zaumseil, and J. A. Rogers, "Improved chemistries, thin film deposition techniques and stamp design for nanotransfer printing", *Langmuir* **20**, 6871 (2004).

43 Bao Z, Kuck V, Rogers J A and Paczkowski M A, "Silsesquioxane resins as high performance solution processable dielectric materials for organic transistor applications", *Adv Func Mater*, 2002 **12** 526–531.

44 Loo Y-L, Hsu J W P, Willett R L, Baldwin K W, West K W and Rogers J A, "High-resolution transfer printing on GaAs surfaces using alkane dithiol self-assembled monolayers", *J Vac Sci Techn B*, 2002 **20** 2853–2856.

45 Kim C, Shtein M and Forrest S R, "Nanolithography based on patterned metal transfer and its application to organic electronic devices", *Appl Phys Lett*, 2002 **80** 4051–4053.

46 Schmid H, Wolf H, Allenspach R, Riel H, Karg S, Michel B and Delamarche E, "Preparation of metallic films on elastomeric stamps and their application for contact processing and contact printing," *Adv Func Mater*, 2003 **13** 145–153.

47 Nakamura M, Kato S, Aoki T, Sirghi L and Hatanaka Y, "Role of terminal OH groups on the electrical and hydrophilic properties of hydro-oxygenated amorphous TiOx:OH thin films", *J Appl Phys*, 2001 **90** 3391–3395.

48 Nakamura M, Aoki T and Hatanaka Y, "Hydrophilic characteristics of rf-sputtered amorphous TiO_2 film", *Vacuum*, 2000 **59** 506–513.

49 Schueller O J A, Duffy D C, Rogers J A, Brittain S T and Whitesides G M, "Reconfigurable diffraction gratings based on elastomeric microfluidic devices", *Sens Act. A*, 1999 **78** 149–159.

50 Zaumseil J, Someya T, Baldwin K, Bao Z, Loo Y-L and Rogers J A, "Nanoscale Organic Transistors Formed by Soft Contact Lamination and Source/Drain Electrodes Supported by High Resolution Rubber Stamps", *Appl Phys Lett*, 2003 **82** 793–795.

51 Kim C, Burrows P E and Forrest S R, "Micropatterning of organic electronic devices by cold-welding", *Science*, 2000 **288** 831–833.

52 Flexography: Principles and Practices, 5th edn. Flexographic Technology Association (FTA), 2000.

53 J. A. Rogers, K. Paul, R. Jackman and G. Whitesides, *Appl. Phys. Lett.* **70**, 2658 (1997).

54 U. Plachetka, M. Bender, A. Fuchs, B. Vratzov, T. Glinsner, F. Lindner, and H. Kurz, *Microelectron. Eng.* **73/74**, 167 (2004).

55 B. Michel, A. Bernard, A. Bietsch, E. Delamarche, M. Geissler, D. Juncker, H. Kind, J.-P. Renault, H. Rothuizen, H. Schmid, P. Schmidt-Winkel, R. Stutz, and H. Wolf, *IBM J. RES. & DEV.* **45**, 697 (2001).

56 H. A. Biebuyck, N. B. Larson, E. Delamarche, and B. Michel, *IBM J. RES. & DEV.* **41**, 159 (1997).

11
Thin-film Transistor Fabrication by Digital Lithography

William S. Wong, Jürgen H. Daniel, Michael L. Chabinyc, Ana Claudia Arias, Steven E. Ready, and René Lujan

11.1
Introduction

Conventional methods for materials processing, such as thin-film deposition and photolithographic patterning provide the foundation for device fabrication in the microelectronics industry. Although these methods are scaleable and cost-effective, as demonstrated in the silicon integrated-circuit (IC) industry, they are not always the most appropriate methods for applications where further miniaturization is not an advantage, such as large-area electronics. Active-matrix liquid-crystal displays (AMLCD) [1], an industry that has grown to a $30 billion dollar business and X-ray imaging [2], another rapidly emerging application for thin-film transistor (TFT) devices are two examples of large-area electronics. Both of these applications rely on the ability to pattern transistor structures over large substrates. Particularly for AMLCD, the cost and difficulty of conventional patterning larger areas has slowed down the introduction of very large-format products. Many new materials systems, based on organic and polymeric semiconductors, are also rapidly being developed to address the intrinsic difficulties of large-area processing. Ideally, these materials are solution-processable, which can enable novel deposition and patterning processes such as jet printing for fabrication of TFT devices.

While these polymeric materials hold promise for low-cost electronics and simplified device processing, they still suffer from lower performance than conventional inorganic semiconductors. As the development of these novel materials continues, an alternative technology incorporating patterning schemes such as inkjet printing and conventional vacuum deposited thin films, may provide the required overlap of performance with low-cost fabrication of microelectronic circuits on large-area platforms.

In flat-panel display applications, manufacturers reduce production costs by increasing the mother-glass plate size, thus yielding more displays on a single plate of glass using a batch process. Figure 11.1 shows the trend for substrate size as a function of time with the current Generation 7 plate size having an area of 1.8×2.2 m^2. As the plate size increases in the coming years, it is unclear how

Organic Electronics, Materials, Manufacturing and Applications. Edited by Hagen Klauk

ISBN: 3-527-31264-1

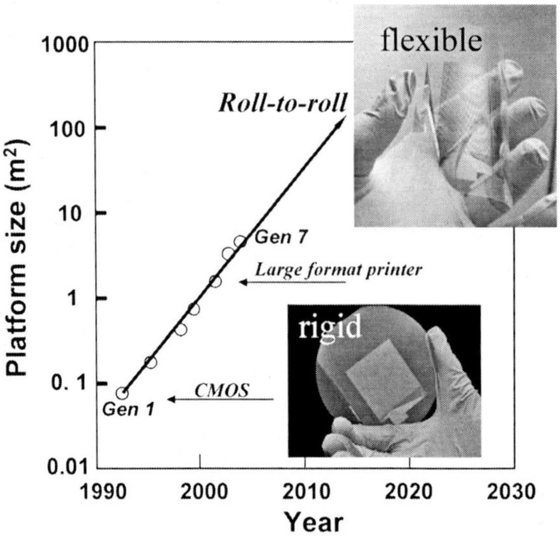

Fig. 11.1. AMLCD industry roadmap for mother-glass platform area from 1990 to present.

the manufacturers will approach large-area device processing. When AMLCD production goes past Generation 7, the display manufacturers may need to look beyond conventional fabrication methods. Large-area displays will have a larger pixel format that will not require high-resolution patterning techniques for the backplane. Low-resolution patterning processes, such as jet printing, may then be applicable.

Alternatives to conventional photolithography will enable both reduction in cost and complexity for large-area electronics fabrication and will also help enable a transition from rigid substrates to flexible platforms. The transition towards roll-to-roll processing of flexible electronics may be required to follow the roadmap for low-cost large-area electronics.

11.2
Jet-printed Patterning for Thin-film Transistor Processing

11.2.1
Introduction

A substantial fraction of TFT device processing cost lies in the photolithographic process; a large portion of this cost is from the initial investment of the lithography equipment. Because the fabrication of microelectronics on to large-area platforms does not necessarily require small features, the expense for high-resolution patterning equipment may be eliminated. Low-cost patterning techniques are more practical given the feature sizes required will be much larger than those for silicon ICs. For display applications, minimum features of 5–10 μm with 10–20 μm address busline widths are typical. In these applications, TFT arrays typically have a minimum feature size of approximately 10–50 μm, although the channel length is usually of the order of 1–5 μm. For the former dimension, many inexpensive and simpler techniques are available to define thin-film device features.

Contact printing is one promising process capable of achieving feature sizes comparable with photolithography [3]. The technique uses some form of master, fabricated separately and reused in a drum or flat-plate system, examples being: screen-printing, gravure, offset and micro-contact printing. The technique has been used to print etch masks of self-assembled monolayers (SAM) on gold for defining contacts on large-area (5-inch × 5-inch) electrophoretic ink displays [4]. The attainable feature sizes are comparable with conventional photolithography but layer-to-layer registration of only ~50 μm are typically demonstrated. Particle defects are a concern with contact lithography techniques and it is unclear whether a large-scale contact aligner is capable of patterning features with low defect density. In the case of patterned SAM etch masks, the highest quality results are typically achieved on Au films, which is incompatible with silicon processing, although patterning of Pd has recently been reported [5, 6].

Alternatively, for minimum feature sizes of 10–20 μm, many document printing techniques are capable of the patterning precision required for device definition. Direct-write techniques represent an emerging class of patterning processes that have already been demonstrated for fabricating solution-based organic transistors and optoelectronic devices [7–10]. The first demonstration of printed etch masks utilized laser printed toner for fabricating amorphous silicon TFTs [11–13]. These printed masks simplified the TFT fabrication process, however the minimum feature size of ~100 μm does not enable reasonable resolution in a display or an imager array. Inkjet printing, on the other hand, requires no master, is capable of fine features, and offers noncontact spatial control of small features. Features of ~20 μm is obtainable through ejection of small drop volumes of ~1.5 picoliters [14] and nozzleless inkjet print heads have reported drop diameters of the order of 1–2 μm [15].

11.2.2
Jet-printed Phase-change Etch Masks

The challenge for jet-print patterned TFT devices is achieving sufficient precision and reliability to fabricate electronic circuits having the desired feature size and performance. This section discusses the feasibility of using phase-change etch masks patterned by jet printing, in place of conventional photolithography.

The use of inkjetted liquids to directly write etch masks is a practical alternative to printed toner although jet printing also possesses inherent complexities. For example, when a resist that is dissolved in a solvent is put on a surface, the droplet size is largely determined by its wetting properties. Typically, small wetting angles are required to obtain good adhesion to a surface but this condition enables the liquid to spread and form relatively large features. On the other hand, if the liquid does not wet the surface because of its high surface energy then a large contact angle will form resulting in small drop features that have poor adhesion to the printed surface. Neither situation is desirable in semiconductor processing – the former increases the feature size and the latter gives unreliable patterning.

The use of a phase-change media circumvents many of these problems. A material slightly above its liquid/solid-phase transition temperature may be ejected from a jet-printing nozzle; the droplet solidifies quickly upon contact with a cooler surface. The feature size will then depend more on the cooling rate and less on the material's wetting properties, because a frozen droplet cannot spread. In this situation, the substrate temperature controls the printed feature size for materials having excellent wetting properties.

The phenomenon has been studied by Gao and Sonin [16] and it has been shown that the cross-sectional area and apparent contact angle of a printed bead is a function of the substrate temperature and the print-scan velocity. A quantitative analysis of jet-printed phase change materials was reported by Schiaffino and Sonin [17] for isothermal conditions. The results showed that the contact angles for the printed drops could be varied as a function of the substrate temperature. By adjusting the substrate temperature and the scanning speed, the feature size and print quality can be adjusted without having to modify the substrate surface energy.

Although drop-on-demand jet printing of low-temperature solders ($T_m < 300$ °C) have been reported [18], direct writing of many metals is unrealistic because of their high melting temperatures. For example, refractory metals such as Cr are commonly used as gate metals in TFT fabrication but the ejection of these metals in the liquid phase would be impractical in a jet-printing process. A more functional approach is to deposit the thin-film material by conventional techniques and subsequently use an etch-mask process, preferably patterned using a jet-printing technique, to define the device structures.

In addition to the small drop volumes and spatial placement of jet-printed drops, printing of phase-change waxes enables additional feature size and print quality management via the substrate temperature. Figure 11.2 shows the variation of printed line features as a function of the glass substrate temperature. At 30 °C, the rapid cooling of the phase-change material as it comes in contact with the glass is evident from the optical micrograph image showing jagged edges outlining the individual droplets (Fig. 11.2a). By raising the substrate temperature 10 °C (Fig. 11.2b), the wax cools more slowly enabling improved coalescence between subsequent droplets while minimizing the spreading of the line features. At 50 °C (Fig. 11.2c), the droplet spreading is more pronounced and print resolution is reduced because of a lower quench rate that results in significant feature spreading and contact between adjacent lines.

Once the printed feature size can be controlled, the quality of the etch features using the printed wax masks must be considered. This aspect was tested with both wet and dry etching schemes for metals, dielectric, and semiconductor etching. In all cases, the etched patterns matched the printed mask. Figure 11.3 shows a pat-

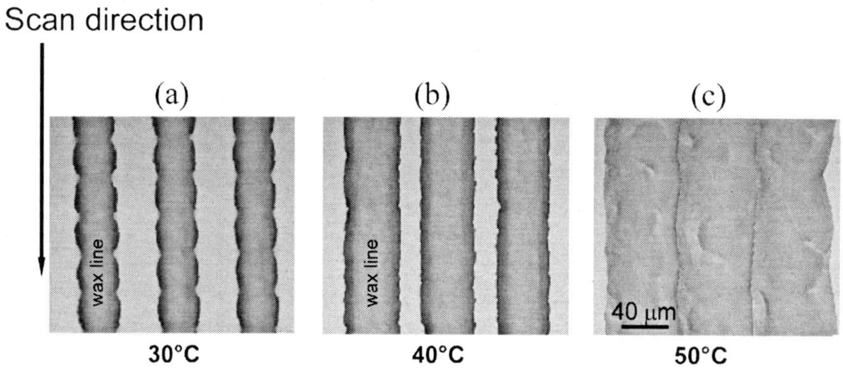

Fig. 11.2a–c. Variation of printed linewidths for phase-change etch masks. Temperatures were varied between 30 °C to 50 °C.

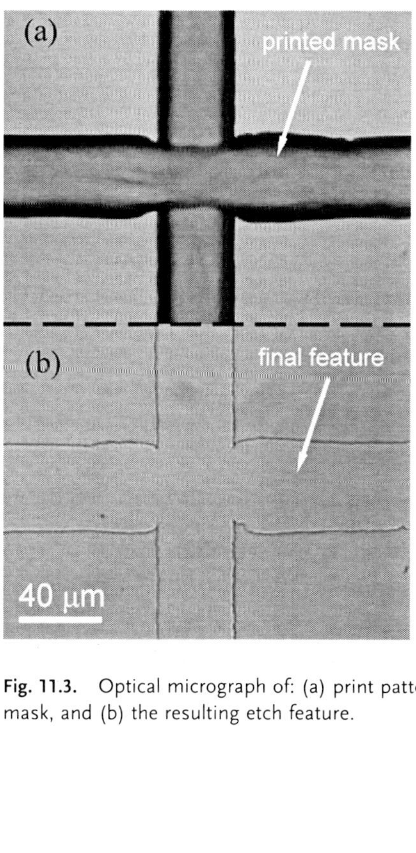

Fig. 11.3. Optical micrograph of: (a) print patterned wax etch mask, and (b) the resulting etch feature.

terned "cross" feature, printed by two orthogonal overlapping lines before etching (Fig. 11.3a) and after wax stripping (Fig. 11.3b). The final feature is well aligned with the printed pattern; no substantial under cut of the masked film is seen. The printed mask is comparable with photo resist patterned features and compatible with semiconductor device processing. The precise image transfer is an important aspect for the wax masking, particularly in the patterning of the source/drain contacts for the TFT and via contact holes in TFT arrays.

11.3
Digital Lithography

11.3.1
Digital Lithography for TFT Device Fabrication

Piezoelectric-based print heads are commercially available and have been used in document printing applications for many years. These print heads have minimum feature sizes as small as 40 μm and ejector densities as high as 1000 ejectors per print head. Other prototype technologies have been reported to have smaller drop size, such as nozzleless acoustic inkjet printing (AIP) [15, 19, 20], but the lack of commercial availability makes AIP impractical. Given its prevalence in the document-printing field, piezoelectric print heads by contrast, offer wide availability at low cost. These are desirable aspects if a tool, intended to replace the conventional photolithographic aligner and stepper along with the associated coater and developer, is to succeed.

By combining jet-printed lithography with digital imaging and processing, a novel patterning process, digital lithography, can be used to register virtual masks for TFT device patterning [21, 22]. The ability to correct the alignment of the mask layer through image processing before patterning is a significant advantage of the digital-lithographic process over other patterning methods. In this process, layer registration is performed by first positioning the process wafer roughly in the orientation of the previously defined layer. The coordinates of the alignment marks are then obtained through image capturing of the surface topography using a camera attached to a microscope objective. When the coordinates are obtained, the mask layer is then digitally processed, repositioned, and aligned to the process wafer before printing the mask pattern, eliminating the need to manipulate optics or mechanically adjust a mask aligner and process wafer. Optical micrographs of the patterning steps for the gate, island, and source/drain levels for a tri-layer amorphous silicon TFT stack are shown in Fig. 11.4. By removing the hardware and optical constraints of a conventional aligner, large-area coverage is no longer limited by the physical size of the mask or the field of view of the optics. Processing on flexible substrates is also simplified; the electronic imaging and alignment can adjust for global and localized distortions on a compliant substrate. Additional complexities such as mask maintenance and mask inspection are no longer required.

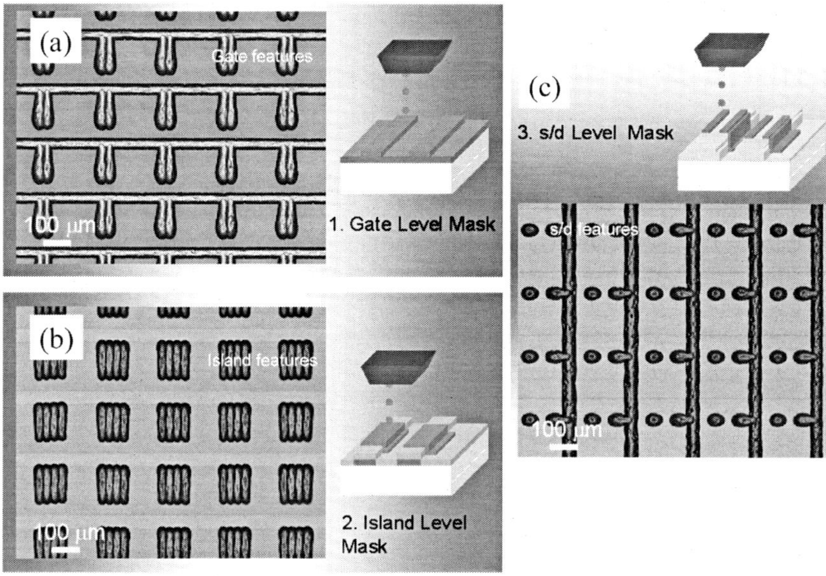

Fig. 11.4. Optical micrographs of patterned mask levels by digital lithography for (a) gate level, (b) island level, and (c) source/drain level.

11.3.2 Thin-film Transistor Device Structures

11.3.2.1 Amorphous Silicon TFTs

Hydrogenated amorphous silicon (a-Si:H) TFT devices were processed using digital lithography to test the printing technology using a well known semiconductor materials system. Both tri-layer TFTs and back-channel etch (BCE) TFTs were fabricated using digital lithography. For the tri-layer device, the channel length of the TFT is defined by the minimum droplet feature size that patterns the gate electrode. The process flow for this type of structure is shown in Fig. 11.5. With the current jet-printing technology, the channel length are ~40 µm and greater.

In contrast, the channel length of the BCE structure is defined by the source-to-drain contact spacing. While printed features may be large, the spatial control of the droplet placement is very precise, of the order of ±5 µm. Combining this control with the BCE TFT structure, a small feature gap can be defined to pattern the source/drain contacts to fabricate a shorter channel device. This structure is more appropriate for fine-feature print-patterned devices. Figure 11.6 shows a schematic comparison of the tri-layer and BCE structure.

As a first demonstration of the fabrication of TFT devices by digital lithography, tri-layer insulated-gate thin-film transistors were fabricated on four-inch glass substrates using a three-layer wax-mask process. Mask layers were used to define the

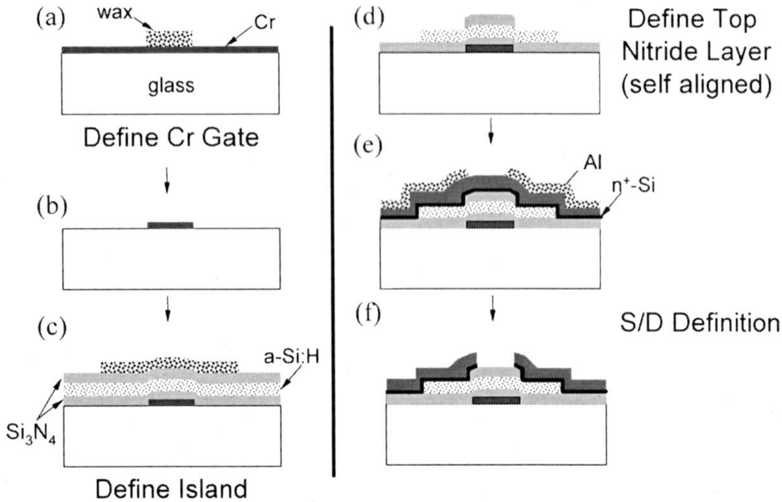

Fig. 11.5. Process flow for a bottom gate tri-layer a-Si:H TFT structure.

gate metal, active device island, and source-drain metal contacts. A bottom gate Cr electrode configuration was employed. First, a 100 nm thick Cr film was deposited on to glass that was then patterned using the Kemamide-based wax ejected from piezoelectric print head. Next, a 300 nm thick Si_3N_4 layer followed by a 50 nm a-Si layer and a 200 nm Si_3N_4 layer were deposited over the Cr gate creating the Si_3N_4/a-Si/Si_3N_4 (NSN) device stack.

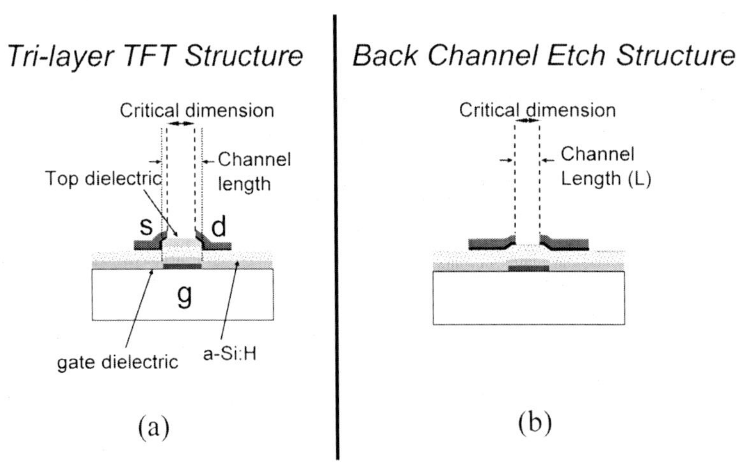

Fig. 11.6. Cross-sectional schematic for: (a) tri-layer TFT and (b) back-channel etch structure.

The TFT island level is patterned next using a two-step masking process. First an island mask layer is patterned around the gate electrode area. The active-device island feature was then formed by etching the NSN stack in a CF_4/O_2 plasma and the wax mask is then stripped in tetrahydrofuran solvent. Following the island definition, the top nitride layer was defined using a self-aligned process in which spin-on photoresist was applied on to the glass/Cr gate/NSN-island structures. In this step, the mask aligner is bypassed by using backside blanket ultra-violet exposure through the transparent glass substrate to define the top nitride feature, self-aligned to the Cr gate. The top nitride was then etched away and a 100 nm thick N^+–Si layer was deposited followed by the Al source-drain contact metal deposition. The source-drain metal was defined by printed-mask patterning and the device was completed by removing the exposed N^+–Si layer by plasma etching using the source-drain contacts as etch masks.

In a second demonstration of making shorter channel devices and taking advantage of the fine spatial control of the printing process, bottom gate BCE TFT were fabricated on four-inch glass substrates using a three-layer mask process. Each mask layer defined the gate metal, active device island, and source-drain metal contacts. First, a 100 nm thick Cr film was deposited on to glass that was then patterned using a Kemamide-based wax ejected from a multi-ejector piezoelectric print head. A 300 nm thick Si_3N_4 layer followed by a 50 nm a-Si:H layer and a 1000 nm N^+–Si layer were deposited over the Cr gate creating the Si_3N_4/a-Si:H/N^+–Si (NSN+) device stack. Registration of the subsequent layers was accomplished by alignment to the gate layer whose image and alignment mark location was captured by a camera mounted on a microscope objective. After deposition of the NSN+ layers the source/drain metal was deposited. The source-drain metal was then defined by digital lithography and the device was completed by removing the exposed N^+–Si layer by etching, using the top contact layer as an etch mask. The island features were then patterned over the channel region of the TFT followed by stripping of the wax mask with tetrahydrofuran to complete the device stack. This process was used to fabricate TFT arrays having 75 dpi resolution over a 128 × 128 pixel array. A portion of a completed array is shown in the micrograph of Fig. 11.7.

11.3.2.2 Polymeric TFTs by Digital Lithography

The replacement of conventional patterning processes with digital lithography is the first step in simplifying the fabrication of TFT devices. Solution-processable semiconductors and metals can be used to replace conventional vacuum deposition processes such as chemical and physical vapor deposition with jet printing. The move to these types of material will enable a transition from a conventional subtractive process for TFT fabrication to an all additive process. The number of fabrication steps for an all additive process using printing is lower compared with conventional deposition and photolithography, simply by replacing the masking and etching steps for defining the device structure. Additive printing limits the choice of materials to those that can be ejected as a liquid, however.

The combination of digital lithography with an additive printing process for de-

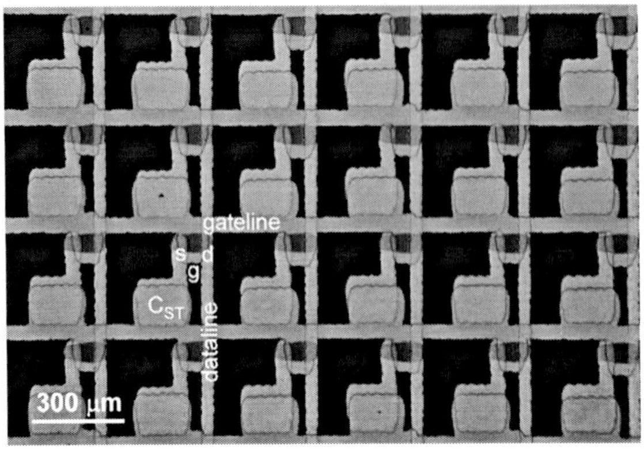

Fig. 11.7. Optical micrograph of a completed TFT array patterned by digital lithography.

position of polymer-based semiconductors is one approach to combine conventional materials (e.g. metals) with novel semiconductors and processing technologies. This hybrid approach to TFT fabrication uses the digital lithography process described above and printing of the polymer semiconductor is ejected from solution using a piezoelectric jet-printer [19, 20].

The TFT fabrication process on glass substrates starts with 100 nm of Cr for the gate metal, and is followed by a PECVD 200 nm thick Si_3N_4 dielectric with a 30 nm thick SiO_2 surface layer. The source drain metal is Cr/Au. Each of these layers is patterned using printed wax masks and chemical etching, steps a to d in Fig. 11.8. The surface is modified with a solution deposition of a self-assembled monolayer of octyltrichlorosilane (OTS-8) before inkjet printing deposition of the semiconductor. It has been shown that the OTS-8 layer affects the structural order of PQT-12 in thin films, improving the performance of the TFT [23]. Encapsulation and possibly other subsequent layers may be needed on the TFT, but these are not discussed here.

The printing process is designed to fabricate a bottom gate TFT with coplanar source and drain contacts (i.e. contacts deposited after the dielectric and before the semiconductor). From the point of view of TFT processing, the bottom gate coplanar structure has the advantage that the two metals and the dielectric layers can be completed before the semiconducting polymer is deposited, reducing the risk of damage to the polymer from subsequent processing. In some polymeric TFTs, the coplanar structure gives significant contact resistance effects compared with the staggered structure [24], but previous results and the data below show that the PQT-12 device does not have a significant contact resistance for the channel lengths that are used here (30–50 µm).

Figure 11.9 is an optical micrograph of the pixels within the printed 128 × 128 pixel array. The pixel size is 340 µm, corresponding to a resolution of 75 dots per inch (dpi). The metal line-width is about 50 µm, and is determined by the size of the wax drop from the jet-printer. However, the printing precision and good line

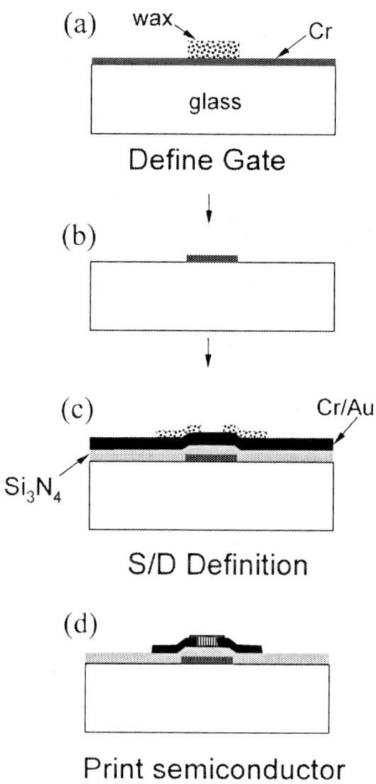

Fig. 11.8. Process flow for a bottom gate, co-planar polymeric TFT structure.

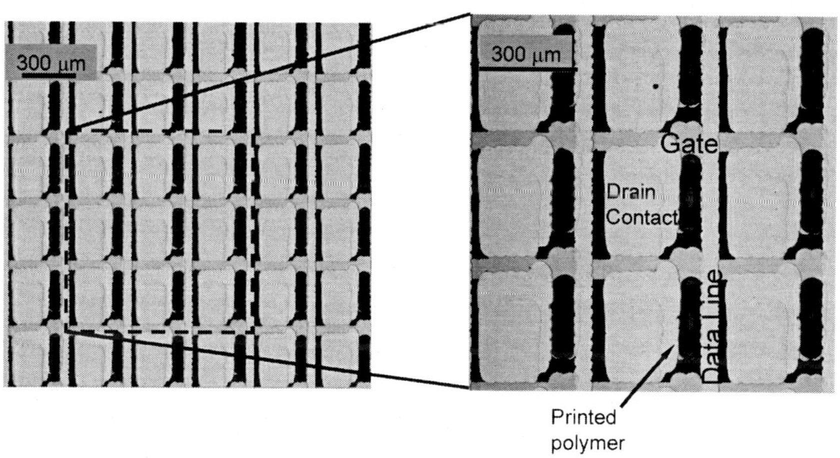

Fig. 11.9. Optical micrograph of a completed polymeric TFT array patterned by digital lithography and jet-printed PQT-12.

edge definition enable two lines to be placed closer together than the printed feature size, which is useful because both the TFT channel and the parasitic capacitance are defined by a gap rather than a line width, as discussed below. The printed polymer can be seen between the TFT contacts. Note that the polymer is confined to the region of the gate electrode. Extension of the polymer beyond the gate may cause higher leakage because the conduction in this region is not controlled by the gate. Also, the well defined area of jet-printed PQT-12 prevents the formation of a continuous layer of polymer between two TFTs connected by the same gate line, that would form a conduction path and causes significant cross-talk between pixels.

11.3.3 Thin-film Transistor Device Characteristics

11.3.3.1 a-Si:H TFTs

By using the tri-layer process flow described in Fig. 11.5, TFT devices were fabricated and tested in comparison to transistors made by conventional semiconductor device processing. The threshold voltage for the print patterned devices, having a width to length ratio (W/L) of 1 (with ~10 μm of gate to source/drain overlap) were measured to be ~2–3 V. Figure 11.10 shows the transfer characteristics for a typical discrete device having a measured field-effect mobility of 1.2 cm^2 V^{-1} s^{-1} and an on/off ratio of 10^8. The output characteristics are shown in the inset of Fig. 11.10. Gate-bias stress has been observed for the print-patterned TFT and has been found to be comparable with conventionally processed devices (Fig. 11.11).

Figure 11.12 shows the I–V transfer characteristics of a typical BCE device within the 128 × 128 pixel array having a measured threshold voltage of 2.8 V and a sub-threshold slope of 0.5 V $decade^{-1}$. The W/L ratio of these transistors is

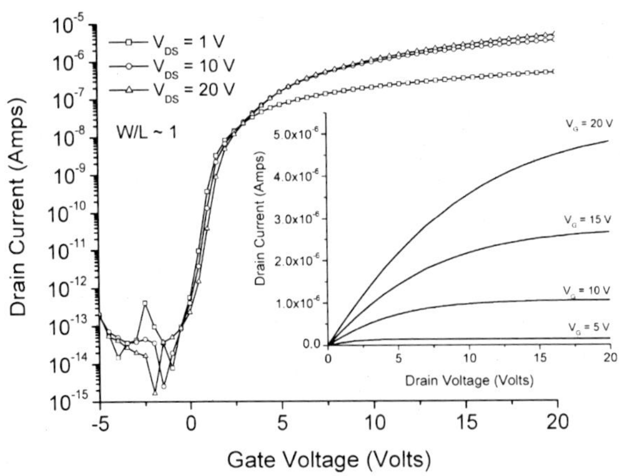

Fig. 11.10. I–V transfer and output curves for a typical tri-layer TFT device patterned by digital lithography.

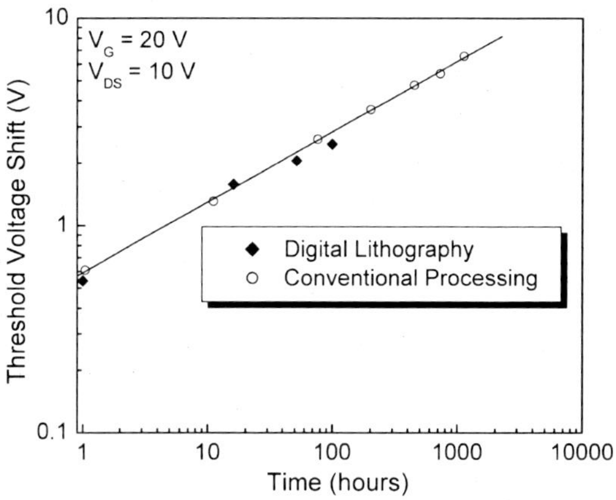

Fig. 11.11. Voltage threshold shift as a function of time for conventionally processed (open circles) and digital lithographically processed (closed diamonds) TFT. Conventionally processed TFT data from Ref. [1].

larger ($W/L = 1.5$) than that for the tri-layer devices, patterned using the same minimum printed feature size. A carrier mobility of 1 $cm^2\ V^{-1}\ s^{-1}$ was found for the linear and saturation regime. The output characteristics for a typical device in the array are shown in the inset of Fig. 11.12. One concern with the digital lithographic process was poor electrical contact at the metal–semiconductor interface because of wax residue after mask stripping. To verify the integrity of the contact,

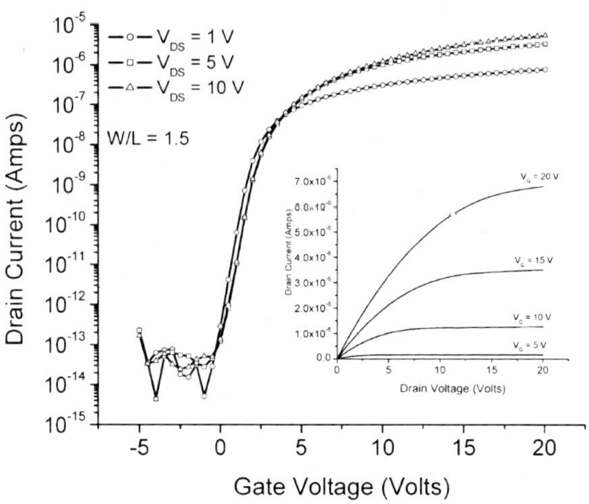

Fig. 11.12. *I–V* transfer and output curves for a typical BCE TFT device patterned by digital lithography.

the current–voltage (I–V) characteristics at low drain voltages shown in Fig. 11.12 (inset) were found to be linear, indicative of an ohmic contact. Variation of the current–voltage characteristics over the 40 mm × 40 mm array area was within 10%. Coupled with the measured I–V transfer curves, the TFT processed by digital lithography compare well with conventionally processed devices. The resulting device performance, with an on/off ratio of 10^8, is comparable with state-of-the-art a-Si:H TFT devices fabricated using conventional photolithographic techniques.

11.3.3.2 **Printed Polymeric TFTs**

Figure 11.13 shows typical transfer and output characteristics for a TFT taken from a printed array. The devices exhibit a mobility of 0.05–0.1 $cm^2\ V^{-1}\ s^{-1}$, and an off current of $\sim 10^{-12}$ A, giving an on/off ratio of $\sim 10^6$. The turn-on potential is close to 0 V and the sub-threshold slope is 2.5 V $decade^{-1}$. In addition, there is minimal bias stress effect. The output data shows good saturation and no sign of significant contact resistance. All these properties are similar to those obtained for spin-coated materials [23]. A set of 100 TFT were measured in a single array to determine the performance statistics. The probe test shows a high yield of working TFT (in this case it is 100%), and reasonable uniformity in both the on and off-current. In one series of measurements, we obtained an average mobility of 0.06 $cm^2\ V^{-1}\ s^{-1}$, with a standard deviation of 0.02 $cm^2\ V^{-1}\ s^{-1}$.

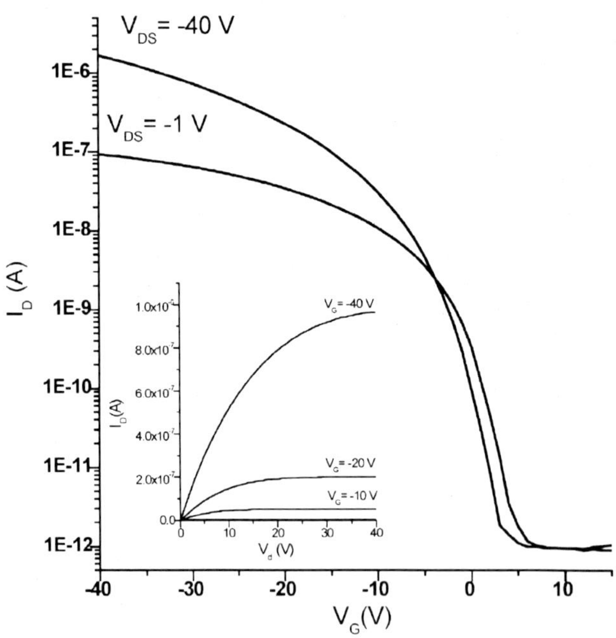

Fig. 11.13. I–V transfer and output curves for a typical polymeric TFT device patterned by digital lithography and jet-printed PQT-12.

11.4
TFTs on Flexible Substrates

11.4.1
Introduction

Future displays are projected to be flexible, light-weight, and of large area. Low power consumption, excellent daylight readability, and low cost are additional requirements, particularly for applications such as electric paper. Reflective displays based on electrophoretic inks are promising for flexible displays with paper-like visual appearance [25–28, 35]. Flexibility is desirable for ruggedness and for conformal or roll-up displays in addition to reducing processing cost and complexity in a roll-to-roll process.

While directly addressed (segmented) displays are sufficient for simple message signs, with active matrix addressing, arbitrary images can be displayed [27–30]. Large-area active matrix backplanes made by conventional lithography suffer from relatively high fixed (equipment-based) and variable (materials-based) cost, however. As the platform size increases, the need for flexible substrates and processing on these novel platforms will become important for cost reduction. Using the electronic imaging and registration capabilities of digital lithography, processing of flexible platforms will be simplified to correct for global and localized distortions compared with conventional rigid photolithographic masks.

The fabrication of large-area electronics, integrated on to polymeric substrates, would also require low-temperature processing of either organic or inorganic semiconductors. While the low-temperature aspects of the organic and polymeric semiconductor materials are described elsewhere, the use of inorganic materials, such as Si-based TFTs processed at low-temperature (<200 °C), is another approach to enable high-performance backplanes on flexible sheets. Combined with the electronic imaging and alignment of the digital lithographic process, high-performance and low-cost TFT device fabrication would be enabled. The processing developed in this type of approach could be used for large-scale printable electronic materials systems such as solution processable semiconductors for polymeric-based TFTs.

The challenge to this approach is combining digital lithography with low-temperature device processing to obtain high-performance devices fabricated using conventional thin-film vacuum deposition. Several groups have demonstrated the fabrication of low-temperature a-Si:H ($T < 150$ °C) although electrical stability was shown to be relatively poor or not reported. For the reported cases, a positive threshold voltage shift is typically observed because of charge trapping in the gate dielectric [31–34]. Improved device stability would help to enable high-performance TFT backplanes and, when combined with digital lithographic processing, would enable low-cost flexible electronics.

11.4.2
TFT Pixel Design Considerations

The previous sections have described how an a-Si:H and polymeric-based TFTs can be fabricated using the digital lithographic process. While a layer-to-layer registra-

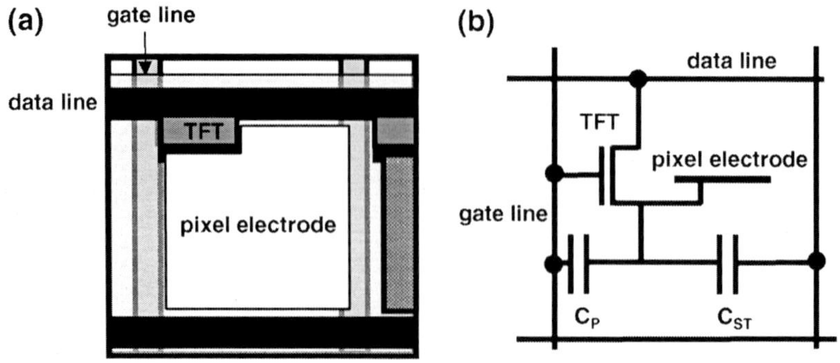

Fig. 11.14. Schematic diagram of a typical pixel circuit for an active-matrix array.

tion of ±5 μm is possible, this feature size is still relatively large when compared with conventional photolithographic processing. As a result, when devices fabricated through digital lithography are used in a backplane configuration, one must consider the design criteria for the TFT within a pixel.

For example, active–matrix backplanes for reflective displays using electrophoretic media is a potential application for large-area electronics. Electrophoretic media are usually charged colored particles suspended in an insulating fluid [35–37]. When the fluid is held in an electric field between two conductive plates, the particles will move between the plates creating a dark or a reflective state. Because the contrast in an electrophoretic display is because of reflection, no backlight is required and the display has relatively low-power consumption compared with a transmissive liquid-crystal display.

The design of active-matrix TFT (AM-TFT) backplanes for capacitive media, such as electrophoretic ink, has been extensively reviewed [1] and we discuss here some of the most relevant issues for printed backplanes. A simplified equivalent circuit for a pixel in an AM-TFT backplane is shown in Fig. 11.14.

To display an image, the data is written line by line to charge the pixel pad; this charge must be held at the pixel to switch the media between its on-state and its off-state while the other pixels are addressed. The RC time constant for charging the pixel is given by Eq. (1) where R_{on} is the resistance of the TFT in the pixel and C_{ST} is the capacitance of the storage capacitor (neglecting the contribution of the resistance and capacitance of the gate and data address lines).

$$T_{RC} = R_{on}C_{ST} \qquad (1)$$

One of the main parasitic elements in a pixel that is especially relevant for printing is the overlap of the gate and pixel electrode of the TFT. This overlap causes a parasitic capacitance, C_P, that reduces the potential at the pixel pad by an amount referred to as the feed-through potential V_{FT} (Eq. (2), where ΔV_G is the difference between the gate on/off voltage). If the V_{FT} is large relative to the data voltage, the voltage retained at the pixel may not be large enough to switch the media.

$$V_{FT} = \Delta V_G C_P/(C_P + C_{ST}) \qquad (2)$$

The design goals for the electrical behavior of display backplanes can be met using digital lithographic fabrication with a-Si:H and polymeric semiconductors. The on-current requirement can be met for a 75 dpi display with an a-Si:H TFT having a W/L of unity and also can be met, with appropriate scaling, by PQT-12. The features defined by digital lithography are well registered, but require the addition of a storage capacitor, C_{ST}, to compensate for C_P because of the metal overlap between features. Typical printed pixel structures achieve a T_{RC} of ~80 μs with a feed-through voltage of ~2 V; refresh rates of ~25 Hz (assuming the display has 500 lines) are achievable with these characteristics and are adequate for many electrophoretic display applications.

11.4.3
Digital Lithography for Flexible Backplanes

Low-temperature deposition of the a-Si:H-based thin-film transistor stack was accomplished using a similar process flow described earlier in Section 11.3.2. Here the deposition conditions were modified such that the highest process temperature did not exceed 150 °C. For the integration on to flexible platforms, a poly(ethylene naphthalate) (PEN) material provided by DuPont–Teijin Films, was used as a substrate [38]. During the processing, a localized runout of approximately ±4 μm was measured on the deposited multilayers on the PEN. Device features were aligned using the ±5 μm accuracy of the digital-imaging system of the digital lithographic system.

Digital lithographically processed TFT arrays having 128 × 128 pixel with 75 dpi resolution were fabricated directly on the PEN. Figure 11.15 shows a photograph of the completed flexible array. The I–V transfer and output curves for a typical TFT pixel ($W/L \approx 1$) within the array are shown in Fig. 11.16. The device has on/off ratios of $>10^8$, sub-threshold slopes of 0.5 V decade^{-1}, and field-effect mobility of

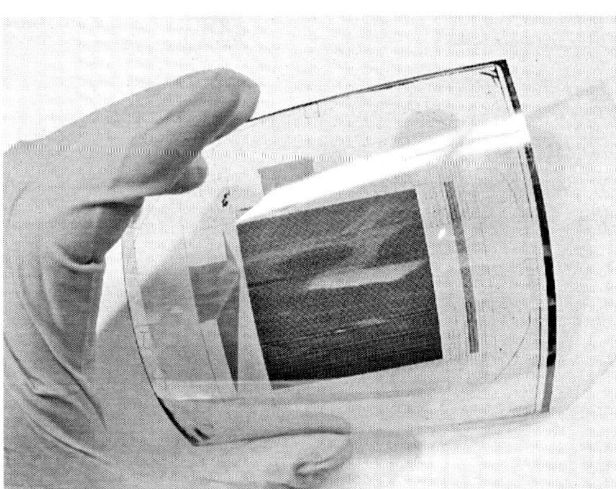

Fig. 11.15. Photograph of a 128 × 128 pixel backplane array on PEN.

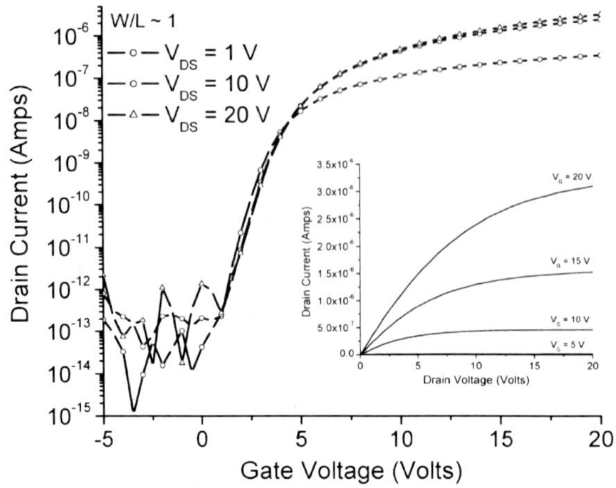

Fig. 11.16. *I–V* transfer and output curves for a typical TFT device patterned by digital lithography on PEN.

~1 $cm^2\ V^{-1}\ s^{-1}$, comparable with conventional high-temperature devices fabricated on glass substrates.

The electrical stability of the TFT was measured using dc bias followed by a pulsed recovery measurement. The dc bias conditions were applied with the gate potential set at 20 V and the source/drain bias at 1 V. The devices were stressed for 60 min followed by the pulsed recovery measurement having an on-time pulse of 25 ms. and an off time of 10 s. Figure 11.17 shows the measurement of the

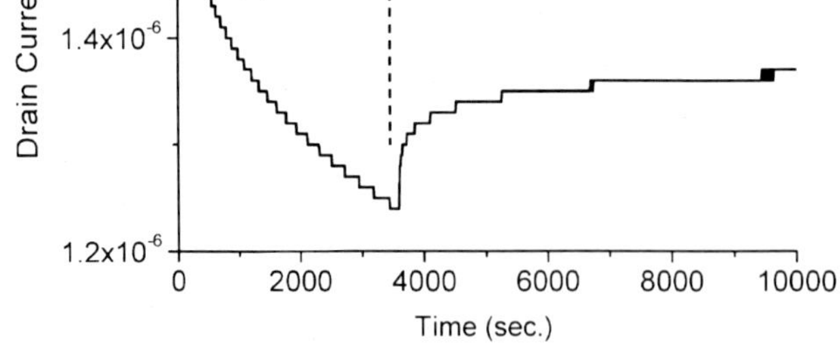

Fig. 11.17. Drain current as a function of time for dc and pulsed gate bias conditions. The TFT in this case was biased for 3600 s before pulsed recovery measurements were started.

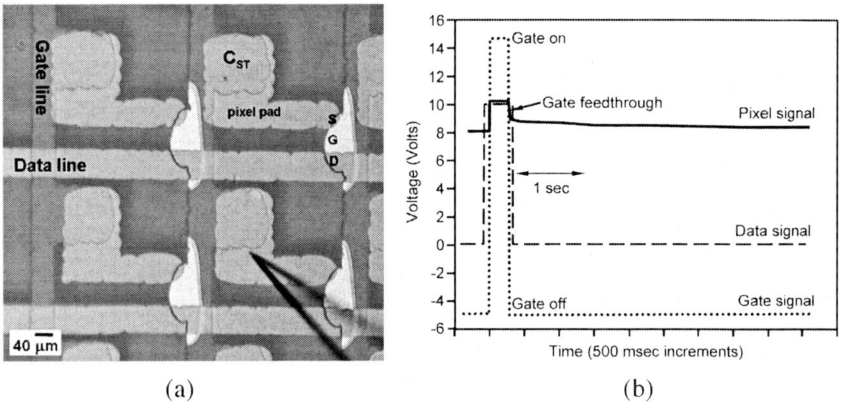

Fig. 11.18. (a) Optical micrograph of a completed pixel on PEN. (b) Picoprobe measurements showing the pixel response of the same TFT under scanning operation.

drain current as a function of time. The drop in current during the dc-bias measurement indicated the TFT threshold voltage is shifting positive, most probably because of trap filling in the gate dielectric [33]. The current recovery of the pulse measurement is relatively fast, suggesting charge trapping and de-trapping within the dielectric as the dominant current degradation mechanism. Although the devices showed some threshold voltage shift under dc bias, the measured low-temperature TFTs were relatively stable when compared with conventional TFTs, the former having a measured threshold potential shift of ~3 V after gate-bias stress. Under pulsed operations, such as in an electrophoretic display backplane, such devices should perform as well as conventional high-temperature TFT.

Figure 11.18a shows a close-up of a flexible 128 × 128 pixel backplane. The pixel layout is based on a storage-capacitor-on-gate design. All of the patterning was performed using wax printing without any photolithography steps. In this experimental backplane design, the pixel pad area was chosen to be relatively small to study the effect of the fill factor on the display media. The pixel response was measured with a Picoprobe and is illustrated in Fig. 11.18b. The measurement shows that the pixel gets charged to the level of the data line (10 V) when the gate is switched high ($V_{\text{gate-ON}} = 15$ V) and the charge remains substantially stored on the pixel pad for several seconds after turning off the gate ($V_{\text{gate-OFF}} = -5$ V). The feedthrough potential, because of the coupling of the pixel with the gate line, was typically 1–2 V at the chosen potentials in this pixel design.

11.5 Display Applications with Print-patterned Backplanes

The pixel layout was optimized with respect to the storage capacitance (C_{ST}), parasitic capacitance (C_{P}) and transistor dimensions. The goal was to achieve short

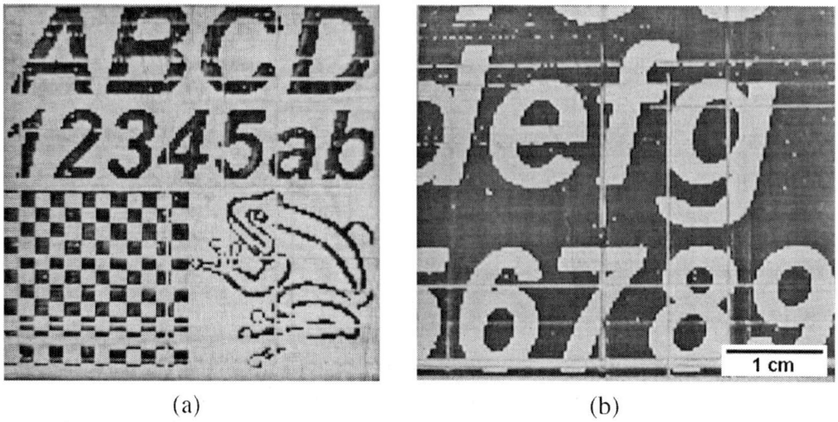

Fig. 11.19. Photograph of a electrophoretic media display integrated on to a digital lithographically processed a-Si:H backplane. The substrate in this example was glass.

pixel charging times, sufficient charge storage and an adequate fill factor for the pixels. Figure 11.19 shows a two-particle electrophoretic display driven by a 128×128 pixel array having 75 dpi resolution on a glass substrate. Integrating the arrays with electrophoretic ink tested the "printed" backplanes. In a simple setup, polymeric-based cell structures (200 μm pitch with 50 μm high walls) were patterned on to an indium-tin-oxide coated glass plate. This plate is then laid on to the backplane, entrapping the electrophoretic ink between the glass plate and the backplane. The surface of the active-matrix backplanes was coated with a black color filter material to shield the TFT from light and to provide a dark background for better contrast of the electrophoretic ink.

Bi-level driver chips drove the active matrix displays with data potential levels around 0 V (low) and 20 V (high). The counterplate potential was held constant. Figure 11.19a shows a photograph of an electrophoretic display driven by an amorphous silicon backplane that was patterned on a glass substrate using wax printing. In this display the electrophoretic ink was a black and white two-particle ink. Figure 11.19b shows the same display addressed with a different pattern.

11.6 Conclusions

Development of the digital lithographic process has resulted in the ability to fabricate high-quality thin-film transistor devices. The technology has the capability of replacing conventional photolithographic processes for applications in large-area electronics where small feature size is not a requirement. Printing technology has been shown to be capable of patterning TFT device and arrays with state-of-the-art performance on both conventional rigid substrates and novel flexible platforms. The ability to electronically image and align to existing patterned features gives the digital lithographic process a substantial advantage over conventional patterning technologies. By eliminating the conventional "hard" mask with an electronic image file, digital lithography enables unconstrained patterning of electronic devices over very large areas. In addition, the non-contact patterning of etch masks minimizes the density of particle defects found in contact-style patterning processes.

For fabrication of backplanes in the display industry, these results hold the promise of a low-cost, low-complexity alternative to the photolithographic process, although several aspects to the technology still require investigation. Can a jet-printing tool be made which has a throughput comparable with that of a conventional stepper/aligner processing Generation 7 mother glass? Will the cost of such a tool be sufficiently less than that of a conventional aligner to affect the fixed costs of building a new foundry? How small can the printed drop shrink to improve the patterning resolution? These questions still remain and are currently being investigated in the laboratory. If these questions can be positively resolved, digital lithography is a promising technology that combines the capability of multi-layer registration, digital imaging, electronic mask patterns, non-contact patterning, and large-area coverage into a compelling alternative for conventional photolithographic patterning.

Acknowledgments

The authors gratefully acknowledge A. Salleo, R.B. Apte, J.-P. Lu, S. Limb, B. Russo, M. Rosenthal, M. Young, and R.A. Street of PARC for assistance with the experiments and for helpful discussions. This work is partially supported by the Advanced Technology Program of the National Institute of Standards and Technology (contract numbers 70NANB7H3007 and 70NANB3H3029).

References

1 T. Tsukada, Active-Matrix Liquid-Crystal Displays in *Technology and Applications of Amorphous Silicon*, R.A. Street (Ed.), Springer Verlag, Heidelberg, **2000**.
2 R. A. Street, Large Area Image Sensor Arrays, in *Technology and Applications of Amorphous Silicon*, R.A. Street (Ed.), Springer, Heidelberg, **2000**.
3 Y. Xia and G.M. Whitesides, "Soft lithography," *Angew. Chem. Int. Ed.*, **1998**, *1998*, 550.
4 J.A. Rogers, Z. Bao, K. Baldwin, A. Dodabalapur, B. Crone, V.R. Raju, V. Kuck, H. Katz, K. Amundson, J. Ewing, and P. Drzaic, *Proc. Natl. Acad. Sci.*, **2001**, *98*, 4835.
5 J.C. Love, D.B. Wolfe, M.L. Chabynic, K.E. Paul, and G.M. Whitesides, "Self-assembled monolayers of alkanethiolates on palladium are good etch resists," *J. Am. Chem. Soc.*, **2002**, *124*, 1576.
6 D.B. Wolfe, J.C. Love, K.E. Paul, M.L. Chabinyc, and G.M.

WHITESIDES, "Fabrication of palladium-based microelectronic devices by microcontact printing," *Appl. Phys. Lett.*, **2002**, *80*, 2222.

7 H. SIRRINGHAUS, T. KAWASE, R.H. FRIEND, T. SHIMODA, M. INBASEKARAN, W. WU, E.P. WOO, "High-resolution inkjet printing of all-polymer transistor circuits," *Science*, **2000**, *290*, 2123.

8 S.-C. CHANG, J. BHARATHAN, Y. YANG, R. HELGESON, F. WUDL, M.B. RAMEY and J.R. REYNOLDS, *Appl. Phys. Lett.*, **1998**, *73*, 2561.

9 G. PERÇIN, T.S. LUNDGREN, and B.T. KHURI-YAKUB, *Appl. Phys. Lett.*, **1998**, *73*, 2375.

10 T.R. HEBNER, C.C. WU, D. MARCY, M.H. LU, and J.C. STURM, "Inkjet printing of doped polymers for organic light emitting devices," *Appl. Phys. Lett.*, **1998**, *72*, 519.

11 H. GLESKOVA, S. WAGNER, and D. SHEN, "Electrophotographic patterning of thin-film silicon on glass foil," *IEEE Electron Dev. Lett.*, **1995**, *16*, 418.

12 H. GLESKOVA, S. WAGNER, and D.S. SHEN, *J. Non-Cryst. Solids, pt. B*, **1998**, *227–230*, 1217.

13 E.Y. MA and S. WAGNER, "a-Si:H thin film transistors on rollable 25 μm thick steel foil," *Mat. Res. Soc. Proc.*, **1998**, *507*, 13.

14 B. HADIMIOGLU, S. ELROD, and R. SPRAGUE, *Proc. 2001 IEEE Ultrasonics Symposium*, **2001**, *1*, 627.

15 S.A. ELROD, B. HADIMIOGLU, B.T. KHURI-YAKUB, E.G. RAWSON, E. RICHLEY, C.F. QUATE, N.N. MANSOUR, and T.S. LUNDGREN, *J. Appl. Phys.*, **1989**, *65*, 3441.

16 F. GAO, A.A. SONIN, *Proceedings: Mathematical and Physical Sciences*, **1994**, *444*, 533.

17 S. SCHIAFFINO and A.A. SONIN, *J. Fluid Mech.*, **1997**, *343*, 95.

18 D.J. HAYES, W.R. COX, and M.E. GROVE, *J. Electron. Manuf.*, **1998**, *8*, 209.

19 K.E. PAUL, W.S. WONG, S.E. READY, and R.A. STREET, *Appl. Phys. Lett.*, **2003**, *83*, 2070.

20 A.C. ARIAS, S.E. READY, R. LUJAN, W.S. WONG, K.E. PAUL, A. SALLEO, M.L. CHABINYC, R. APTE, ROBERT A. STREET, Y. WU, P. LIU, and B. ONG, *Appl. Phys. Lett.*, **2004**, *85*, 3304.

21 W.S. WONG, S. READY, R. MATUSIAK, S.D. WHITE, J.-P. LU, J. HO, and R.A. STREET, *Appl. Phys. Lett.*, **2002**, *80*, 610.

22 W.S. WONG, S.E. READY, J.P. LU, and R.A. STREET, *Electron Dev. Lett.*, **2003**, *24*, 577.

23 A. SALLEO, M.L. CHABINYC, M.S. YANG, and R.A. STREET, *Appl. Phys. Lett.*, **2002**, *81*, 4383.

24 R.A. STREET and A. SALLEO, *Appl. Phys. Lett.*, **2002**, *81*, 2887.

25 E. KISHI, Y. MATSUDA, Y. UNO, A. OGAWA, T. GODEN, N. UKIGAYA, M. NAKANISHI, T. IKEDA, H. MATSUDA, K. EGUCHI, *Proc. S.I.D.* **2000**, *5.1*, 24.

26 R.C. LIANG, J. HOU, J. CHUNG, X. WANG, C. PEREIRA, Y. CHEN, *Proc. S.I.D.*, **2003**, *20.1*, 838.

27 S. INOUE, H. KAWAI, S. KANBE, T. SAEKI, T. SHIMODA, *IEEE Trans. Electr. Dev.*, **2002**, *49*, 1532.

28 J. DANIEL, B. KRUSOR, P. NYLEN, R. STREET, *IDRC 03*, **2003**, *P-32*, 263.

29 J.H. DANIEL, B.S. KRUSOR, N. CHOPRA, R.A. STREET, P.M. KAZMAIER, S.E. READY, J. HO, *Mat. Res. Soc. Symp. Proc.*, **2004**, *808*, A10.8.1.

30 P. KAZLAS, J. AU, K. GERAMITA, H. GATES, M. STEINER, C. HONEYMAN, P. DRZAIC, K. SCHLEUPEN, B. WISNIEFF, R. HORTON, R. JOHN, *Proc. S.I.D.*, **2001**, *12.1*, 152.

31 R. PLATZ, S. WAGNER, C. HOF, A. SHAH, S. WIEDER, B. RECH, *J. Appl. Phys.*, **1998**, *84*, 3949.

32 M.J. POWELL, *Appl. Phys. Lett.*, **1983**, 6, 597.

33 C.-S. YANG, L.L. SMITH, C.B. ARTHUR, and G.N. PARSONS, *J. Vac. Sci. Technol. B*, **2000**, *18*, 683.

34 A. SAZONOV, D. STRYAHILEV, A. NATHAN, L.D. BOGOMOLOVA, *J. Non-Crystalline Solids*, **2002**, *299–302*, 1360.

35 B. COMISKEY, J.D. ALBERT, H. YOSHIZAWA, and J. JACOBSON, *Nature*, **1998**, *394*, 253.

36 G.H. GELINCK, H.E.A. HUITEMA, E. VAN VEENENDAAL, E. CANTATORE, L. SCHRIJNEMAKERS, J.B.P.H. VAN DER PUTTEN, T.C.T. GEUNS, M.

Beenhakkers, J.B. Giesbers, B.-H. Huisman, E.J. Meijer, E.M. Benito, F.J. Touwslager, A.W. Marsman, B.J.E. van Rens, and D.M. de Leeuw, *Nature Mater.*, **2004**, *3*, 106.

37 I. Ota, M. Tsukamato, and T. Ohtsuka, *Proc. S.I.D.*, **1977**, *18*, 243.

38 K.R. Sarma, C. Chanley, S. Dodd, J. Roush, J. Schmidt, G. Srdanov, M. Stevenson, R. Wessel, J. Innocenzo, G. Yu, M. O'Regan, W.A. MacDonald, R. Eveson, K. Long, H. Gleskova, S. Wagner, J.C. Sturm, *Proc. SPIE – Int. Soc. Opt. Eng.*, **2004**, *5080*, 180.

12 Manufacturing of Organic Transistor Circuits by Solution-based Printing

Henning Sirringhaus, Christoph W. Sele, Timothy von Werne, and Catherine Ramsdale

12.1 Introduction to Printed Organic Thin Film Transistors

Solution or liquid-based printing techniques have been used in the graphic art printing industry for more than 500 years to replicate information on paper in high volumes and at very low cost. The development of functional materials with well defined structural, chemical, physical, or biological functionality that can be processed from solution or liquid phase has inspired many research groups in different scientific and technological disciplines to explore the potential of adapting graphic art printing techniques to the manufacture of functional devices and structures (for a review, see, for example, Ref. [1]). The vision of many of these approaches is to integrate advanced functionality other than graphic information on to large-area substrates, for example flexible paper, plastic or large-area rigid glass, and be able to do this at a manufacturing cost that is much lower than could be achieved with more conventional manufacturing techniques. A visionary example of such a printed functional system would be an ambient intelligent device comprising a power source, for example a printed battery or solar cell, an information input device, for example wireless link, photodetector, touch panel, or other sensing device, an integrated circuit of printed field-effect transistors (FETs) to provide information processing capability, and an information output device, for example a memory or display.

To a large extent the applications for such printed electronic systems are driven by the need to reduce the manufacturing cost of existing electronic devices, for example large-area active matrix liquid-crystal displays (AMLCDs) or radiofrequency identification (RFID) tags. The display industry has been driven by impressive cost reductions of approximately 20–25% for each doubling of the cumulative area of flat panel displays produced. It is expected that by 2010 large-area LCD panels for applications in LCD TV should be produced at a cost of approximately \$1 per square inch. To continue to be able to achieve these cost reductions despite the very high capital cost for Generation 6 and 7 display manufacturing fabrication

Organic Electronics, Materials, Manufacturing and Applications. Edited by Hagen Klauk

ISBN: 3-527-31264-1

plant, direct-write printing is considered a very attractive large-area manufacturing alternative to conventional vacuum deposition and photolithography. Similarly, the use of RFID tags is limited at present to applications such as access control, animal tagging, or pallet-level supply-chain tagging, in which the relatively high cost of conventional silicon RFID tags of more than \$0.2 can be tolerated. To enable ubiquitous replacement of optical barcodes by RFID tags, manufacturing solutions for integrated RFID tags which can be printed directly onto the packaging or an adhesive label at a cost of \$0.01 or less must be found.

In addition to such primarily cost-driven applications printing might also enable the use of electronic devices in modes, applications, and environments which are currently not accessible by more conventional electronics, for example electronic circuits on flexible substrates which can either be bent during usage or can be shaped once into a nonrigid form. One of the currently most attractive and near-term applications for flexible electronics are flexible electronic paper displays for applications in electronic books, newspapers, and shelf labels [2]. To achieve the necessary contrast and information content the characteristics of an electronic paper display medium require these to be driven by an active matrix array of thin film transistors (TFTs) on a flexible substrate. This is an emerging display application which is currently not served by conventional inorganic silicon technology, at least partly for technological reasons. More demanding, long-term applications include flexible organic electroluminescent displays driven by an active matrix of organic TFTs [3]. Flexible substrates, in particular low-cost, low-temperature plastic substrates, for example poly(ethylene terephthalate) (PET) or poly(ethylene naphthalate) (PEN), limit the window of process temperatures which is available to achieve optimum materials performance and stability to less than 150 °C, which is challenging for many inorganic materials, including amorphous silicon [4]. Solution-processible conjugated polymer semiconductors tend to exhibit good performance and stability when processed close to room temperature. Furthermore, plastic substrates exhibit significant substrate distortion when exposed to temperature/humidity variations or mechanical stress. If these occur during device processing in between two patterning steps, a first pattern on the substrate, for example a gate electrode pattern for the TFT, is distorted relative to a subsequent pattern, for example a source-drain pattern. Such distortions are difficult to compensate for in a conventional photolithographic process based on a set of photomasks, particularly if the distortion pattern changes from substrate to substrate because of variations of substrate history or process conditions. Some direct-write printing techniques including local materials deposition techniques such as inkjet printing are capable of compensating for such distortions locally and maintain accurate registration over a large substrate area by first measuring the distortion, and accordingly adjusting the position for the materials deposition unit.

There is a range of difficult technological and scientific hurdles that will need to be overcome before reliable products can be manufactured with the required performance at low cost by commercial printing techniques. A complete set of solution processible materials including not only conductors, semiconductors and dielectrics, but also planarization, isolation, passivation, and encapsulation layers meeting the required performance criteria must be developed. These must be formulated into functional inks which are compatible with the viscosity, surface tension, and solvent compatibility requirements of the specific printing techniques. They must be printed and processed in such a way that little or no degradation of device performance occurs compared with what is achievable with these materials in more conventional test structures.

The graphic arts printing equipment and printing processes must also be improved and adapted to meet the rather strict design rules for electronic circuits. Typical resolution capability of standard offset, gravure, screen, or inkjet printing is of the order of 50–100 μm. Clearly, these are not necessarily fundamental limitations. Higher resolution produces little benefit in how visual images are perceived by the human eye and has therefore not been a critical issue in commercial printing. Many applications of printed transistors require small critical feature size and linewidth, however, to achieve adequate circuit switching speed, to minimize parasitic capacitances and undesired cross talk, or to be able to achieve the required integration density. The switching frequency of a logic circuit scales with the channel length L and the mobility μ according to the following relationship:

$$v \propto \frac{\mu}{L^{\alpha}} \tag{1}$$

With current mobilities of solution-processed organic TFTs switching speeds are limited to 1–10 Hz if channel lengths and linewidths of 50–100 μm are used [5, 6]. This is more than 3–4 orders of magnitude lower than what is achievable with conventional lithographic processing [7, 8], and is not sufficient for most applications. Speed improvements can be achieved by mobility improvement, and by reduction of parasitics and channel length. α equals 1 if the gate linewidth is not scaled on reduction of L, i.e. the parasitic capacitance between the gate and the source/drain/channel is not changed as L is reduced. Even more significant improvement in switching speed ($\alpha = 2$) is achieved if the overlap capacitance is scaled down in the same way as L, for example by using a self-aligned gate configuration.

This article is aimed at providing a critical review of the current status of printing-based manufacturing of TFT devices and circuits based on solution-processible organic semiconductors. Section 12.2 gives a detailed overview of the various different printing approaches that have been explored in the literature. In Section 12.3 we will focus on a recently developed self-aligned printing approach which is capable of realizing very high resolution feature sizes, as an example of how innovative ways of patterning can overcome some of the traditional limitations of printing. Finally, in Section 12.4 we will discuss the current performance and reliability of solution-processed organic TFTs (OTFTs) with a particular emphasis on first-generation applications in active-matrix displays based on bistable electronic paper.

12.2 Overview of Printing-based Manufacturing Approaches for OTFTs

A broad range of commercial printing techniques can be considered for printing of functional materials [9]. Table 12.1 provides a comparison of the most important techniques in terms of some of their characteristics, main features and applications. Organic transistors have been demonstrated using several of these techniques. In most cases this has been at the level of individual layers of a discrete device being patterned by printing while other layers were processed in more conventional ways. A few groups have, however, developed integrated manufacturing processes for printed TFTs, and demonstrated small-scale integrated circuits and

Tab. 12.1. Comparison of different graphic art printing techniques.

Printing technique	Layer thickness (μm)	Feature size (μm)	Viscosity (mPas)	Throughput ($m^2\ s^{-1}$)	Registration (μm)	Features/ issues	Examples of graphic art applications
Letterpress	0.5–1.5	>50	50000–150000	1	<200		Books
Flexography	0.8–2.5	80	50–500	10	<200	Wide range of substrates, medium quality	Packaging; newspaper; labels
Gravure	0.8–8	75	50–200	60	>10	Large run length; high quality	Magazines; plastic film and metal foils; bank notes
Pad	1–2	20	>50	0.1	>10	Nonplanar objects	Toys; CDs; pens
Offset	0.5–1.5	10–50	40000–100000	5–30	>10	High quality; need for ink additives	Newspapers; magazines; books
Screen	30–100	20–100	500–50000	2–3	>25	Wide range of inks; medium quality	Textiles; PCBs; CDs; large posters
Inkjet	<0.5	20–50	1–30	0.01–0.5	5–20	Digital data; local registration	Desktop; variable data

devices. These are reviewed here and an attempt is made to assess the relative strengths and weaknesses of the respective printing approaches without taking into account their respective stage of development.

12.2.1
Screen Printing

This is a process in which ink is pushed through a screen comprising a fine mesh of plastic or metal fibers (Fig. 12.1E). The pattern is defined by filling certain openings of the mesh with a stencil material. The screen is coated with the ink, and us-

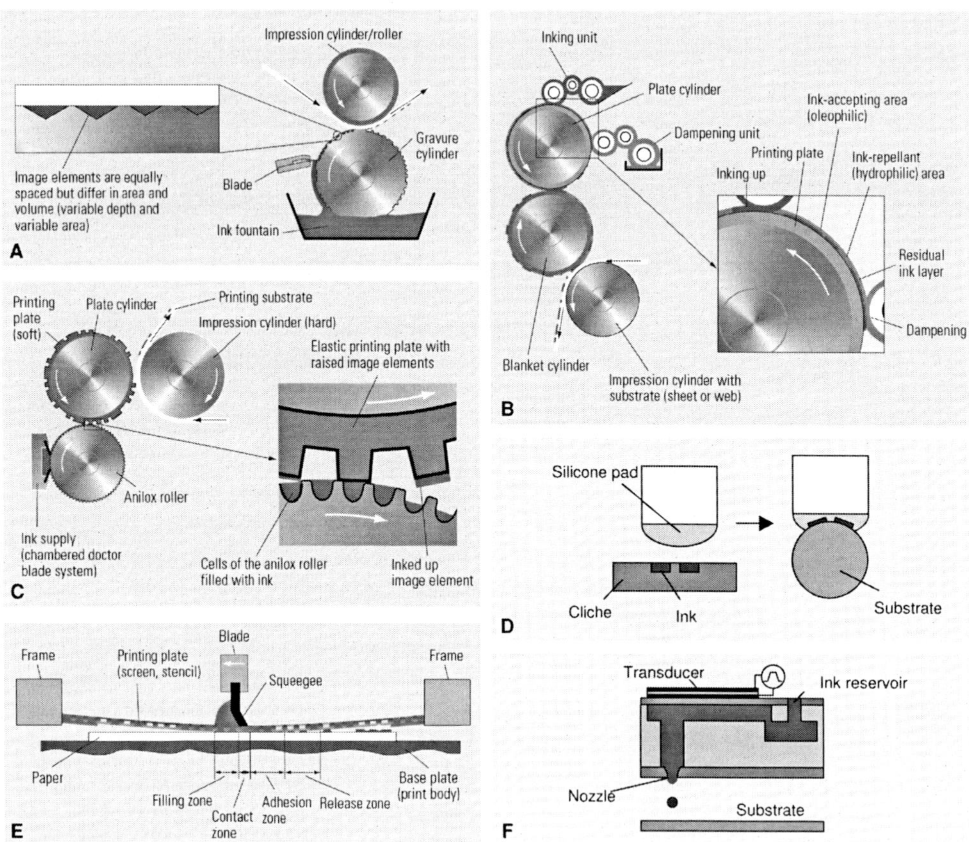

Fig. 12.1. Schematic diagram of operating principle of several graphic art printing techniques: (A) gravure, (B) offset, (C) flexography, (D) pad printing, (E) screen printing, (F) inkjet printing (A, B, C, D, E reprinted with permission from Ref. [9]. Copyright (2001) by Springer).

ing a so-called squeegee the mesh is brought into contact with the substrate, and the ink is pressed through the openings of the screen to define the desired pattern on the substrate. Screen printing requires a high ink viscosity, but is applicable to a large range of materials. It tends to produce relatively thick (>1 μm) films, because the wet film thickness is determined by the thickness of the screen. It is possible to print thinner films from lower viscosity inks, but this often results in degraded pattern definition, because of uncontrolled flow of ink on the substrate. The resolution is determined by the mesh size, and is of the order of 100 μm using standard screens, although screens with resolution of the order of 20 μm have become available. The most common screen-printed passive components are conductive traces and printed resistors, having found widespread use in electronic devices such as keyboards or membrane switches in hand-held or medical devices. The common material basis of printed passive components are thermoplastic or epoxy resins, filled with inorganic powders or particles, such as silver. For OTFT fabrication the technique has so far been limited to patterning of top-level interconnects or electrodes, for which the associated thickness profile does not interfere with the deposition of other critical layers. It can also be used for deposition of thick continuous or coarsely patterned layers such as thick dielectric or encapsulation layers. The first demonstration of a printed organic TFT [10] was made using screen-printed graphite source-drain (channel length 200 μm, thickness 10 μm) and gate electrodes and vacuum evaporated sexithiophene organic semiconductor. Screen printing has also been used for defining source-drain electrodes ($L = 100$ μm, thickness 10 μm) of bottom-gate FETs in both top-contact [11] and bottom contact configuration [12], and for patterning of gate electrodes and interconnects in top-gate FETs [13].

12.2.2
Offset Printing

Offset printing is the most widespread commercial printing technique. It is based on bringing the ink into contact with a printing plate comprising oleophilic/ink-accepting, and hydrophilic/ink-repellent surface areas. The ink is transferred selectively on to the oleophilic regions of the printing plate, and from there on to the substrate via an intermediate blanket cylinder (Fig. 12.1B). A water-based dampening solution is used to prevent the transfer of ink on to the nonprinting regions. Offset printing is a high-throughput printing technique, and therefore principally very attractive for electronic manufacturing. Finding the right compromise between meeting the requirements for ink viscosity, surface tension, evaporation rate, etc., to produce high-quality printed structures on a given substrate and simultaneously achieving optimum device performance can, however, make ink formulation a challenging task, and occasionally necessitates the use of additives. Offset printing has been used to print source-drain structures of PEDOT/PSS for top-gate OTFTs with channel lengths of 30–200 μm, and line widths of 100–200 μm [6].

12.2.3 Gravure Printing

Gravure printing is a very high-throughput printing technique based on a metal cylinder with an engraved or etched pattern of cells, which are filled with ink by doctor blading (Fig. 12.1A). The cells can be of variable size and depth, and are used to transfer the ink on to a flexible substrate when the latter is brought into contact with the rotating cylinder. In graphic arts printing gravure is limited to print runs above a million copies because of the significant cost of making the engraved cylinder. It is capable of printing lower viscosity inks from volatile solvents making it somewhat easier to achieve low film thickness without the use of additives than in the case of offset printing. To the best of our knowledge gravure printing has not yet been used for the fabrication of OTFTs.

A related noncontinuous, and lower-throughput technique is pad printing, where a flexible silicone stamp is used to pick up ink from a cliché to transfer it on to a substrate which can have a nonplanar shape (Fig. 12.1D). Pad printing has been used to print a pattern of etch resist on to a film of polyaniline to define source-drain electrode structures with channel length of 20–60 μm, and linewidths > 100 μm for top-gate FETs. A conducting carbon-based ink was also pad-printed to define bumps for via-hole interconnections, and gate electrodes. Using a combination of pad printing, and blade coating all-printed OTFT-based ring oscillators with a switching frequency of 0.8 Hz were demonstrated [5].

12.2.4 Flexography

The principle of flexography is similar to that of the well known letterpress technique. However, in flexography the printing plate is flexible. It comprises a pattern of protruding, ink-receptive regions separated by nonprinting wells, which is inked with the help of an anilox roller (Fig. 12.1C). The ink can be transferred from the protruding regions of the plate on to a broad range of absorbent and nonabsorbent substrates. Its main application in graphic arts printing is in packaging where its somewhat lower print quality compared to offset or gravure printing can be tolerated. To the best of our knowledge flexography has not yet been used for fabrication of OTFTs.

The related laboratory technique of soft lithography has, however, been employed in different modes to fabricate components of OTFTs. In its basic form soft lithography is based on a polydimethoxysilane (PDMS) stamp which is inked with a self-assembled monolayer (SAM) which can be transferred selectively to the substrate from the protruding regions of the stamp [14]. SAM patterns defined in this way have been used as etch masks for the etching of gold source-drain electrodes with high resolution [15–17]. Using PDMS stamps as phase shift masks in a photolithographic process gold source-drain electrodes have been defined with channel lengths of only 0.1 μm [18]. Selective solution deposition of polymer electrodes has been achieved by micromolding in capillaries (MIMIC). It has been proposed that the soft lithography technique could be used in a reel-to-reel mode [15]. Soft lithography has been used for definition of SAM surface energy patterns for dewetting of organic semiconductor layers deposited by continuous coating techniques [19]. Using PDMS stamps polymer-stabilized catalytic nanoparticles have been selectively transferred as a patterned seed layer for subsequent electroless plating [20]. Similarly, SAM patterns produced by soft lithography on PEN substrates have been used to induce selective electroless plating of nickel to occur only in the hydrophilic regions of the substrate [21]. The latter technique has the advantage that for patterning of narrow conducting lines it overcomes common problems associated with the collapse of the stamp occurring if large areas of the substrate are not supposed to come in contact with the stamp. Generally, both flexography and soft lithography tend to suffer from stamp distortions making accurate pattern registration over large areas challenging, although this can be overcome to some extent by mounting the stamp on a rigid carrier [22].

12.2.5 Inkjet Printing

Inkjet printing has become one of the most widespread graphic arts printing techniques in the home and office desktop printing market, for example for printing of full-color, high-resolution photographs. In the commercial printing arena inkjet technology is widely used for digital proofing, or short-run, wide-format digital printing such as posters for outdoor advertising. One of the traditional drawbacks of inkjet printing is its lower throughput capability. This is being improved for the graphic arts market to allow integration of inkjet printers with offset presses to print customized information in magazines. Generation 7 inkjet manufacturing equipment is being developed for applications in LCD manufacturing (printing of color filters and other components) and for polymer light-emitting diode displays, where inkjet patterning is one of the most promising technologies for patterning the red, green, and blue-emitting polymers [3]. The most common inkjet technology for printing electronic materials is drop-on demand (DOD) printing (Fig. 12.1F). In thermal DOD printing droplets are generated by heating the wall of the ink chamber causing formation of vapor bubbles and ejection of droplets though a nozzle orifice. In piezoelectric DOD printing a pressure wave in the ink chamber is generated by applying a voltage pulse to a piezoelectric stack or plate resulting in formation of droplets at the nozzles.

Piezoelectric DOD inkjet printing has been used for direct-write patterning of PEDOT/PSS conducting polymer source-drain electrodes [23] and of nanoparticle metal electrodes and interconnects [24, 25]. Linewidths of 20–100 μm are achievable with state-of-the-art inkjet printing equipment with droplet volumes of 1–30 pL. Micron-scale linewidths have been demonstrated using sub-picoliter droplet volumes, which are only achievable though with advanced droplet generation technology [26]. A metal patterning process based on printed etch masks for wet chemical etching of vacuum-deposited inorganic metal source-drain electrodes with $L = 40$–400 μm has also been developed [27, 28]. Active semiconducting layers have been inkjet patterned either by direct printing of polymer semiconductor inks [25, 27] or by patterning of mask patterns for SAM surface energy patterns to induce selective dewetting of the active semiconducting material [29]. A process for definition of via-hole interconnections by inkjet-printing of solvent for local etching/dissolution of continuous dielectric and semiconducting layers has been reported [30]. Its relatively high-resolution, and the relative ease of formulating inkjet printable inks for a broad range of functional materials has made inkjet one of the most promising techniques for printable electronics in spite of its throughput not being as high as that of offset or gravure.

12.2.6 Laser-based Dry-printing Techniques

In the graphic arts printing industry focused laser beams are used for fabrication of printing plates for offset printing, and for coding and marking. An array of focused laser beams of 5 μm spot size has been used to direct-write a pattern into a hydrophobic polymer to expose in certain regions the surface of the underlying hydrophilic substrate, and in this way define a surface energy pattern to enhance the resolution of inkjet printing. In this way source-drain electrodes for printed OTFTs were defined with a channel length of only 5 μm [25]. Laser printing has also been used in a thermal transfer mode, wherein a sacrificial substrate is coated with a continuous layer of a functional material together with a light absorbing layer that converts light into heat. Upon local laser irradiation using an array of focused laser beams the functional material is locally transferred from the sacrificial substrate on to the final substrate held in close proximity during the laser exposure. Optionally a release layer is used which vaporizes upon laser irradiation to facilitate the transfer of the functional material. This technique has been used for patterning of source-drain and gate electrodes of a polyaniline–carbon nanotube composite conductor, but is applicable in principle to the other layers of the device also [31]. The technique seems not to be suitable for patterning high conductivity interconnects, because thick metal films with good thermal conductivity are difficult to transfer.

12.2.7 Other Nonlithographic Manufacturing Approaches

In addition to the above mentioned techniques based on adaptation and application of established graphic arts printing techniques and equipment several groups have developed novel direct-write printing and patterning techniques for the deposition and patterning of OTFT materials. Examples include selective electropolymerization of a conducting polymer on a patterned conducting substrate followed by transfer of the source-drain pattern on to a flexible substrate [32], selective dewetting or lift-off of conducting polymer patterns using sacrificial laser printed toner patterns [33]. Vertical OFETs in which the critical channel length is not defined by a lateral patterning step, but by the thickness of a dielectric film in a multilayer stack have also been demonstrated using solid state embossing [34] and laser ablation [35]. Shtein [36] developed a vapor jet printing technique which uses a cavity in which an organic material is heated and sublimed, such that a collimated beam of vaporized molecules is emitted through a small nozzle. Nanotransfer printing is based on depositing a continuous metal layer on to a patterned PDMS stamp followed by transfer of the metal from the protruding regions of the stamp to the substrate under conditions where the metal adheres more strongly to the substrate than to the stamp [37, 38]. The related technique of soft contact lamination, in which the stamp remains in contact with the substrate has been shown to produce source-drain contacts with small contact resistance [39]. Transfer of metal electrodes from a rigid stamp has also been reported, but in this case a continuous strike layer on the substrate had to be used to achieve transfer of the metal by cold welding [40]. Submicrometer source-drain electrodes have been fabricated by underetching of a first metal pattern followed by evaporation of a second metal pattern using the overhanging resist from the first etching step to define the channel length [41].

One of the important general requirements for manufacturing of OTFTs by any technique is the ability to register accurately multiple patterned layers with respect to each other. This is crucial in order to achieve a high integration density, as well as to reduce parasitic capacitances, which can become prohibitively large if conducting patterns for source-drain and gate need to be made very large simply to ensure geometric overlap between them. Registration is similarly critical in conventional graphic arts printing in order to avoid visible image artifacts due to color-to-color misalignment, Moire effects etc. Sophisticated registration control is implemented on state-of-the-art offset presses involving precise register holes in the printing plate, in-line measurement of misalignment of registration marks, web tension control, and active control of lateral and circumferential cylinder position to compensate for the expansion of the web when it comes in contact with the dampening solution. In this way color registration accuracy of the order of tens of microns can be achieved. Similarly, in gravure printing the web shrinks when the ink is dried between printing steps, and this can be compensated for by re-dampening of the web when it leaves a drying unit. However, for both offset and gravure printing the registration capability is principally a global one, i.e., it is very challenging to compensate for local distortions of a previously defined pattern which might arise because of mechanical, thermal or chemically induced stresses in between processing steps. This is particularly problematic if the manufacturing process does not consist of a sequence of in-line printing steps involving a single printing technique. In the case of electronic circuits it is likely that a manufacturing process for electronic devices will comprise a sequence of in-line deposition and patterning steps involving different printing techniques, since it is unlikely that one particular printing technique could satisfy the requirements for all the different layers which need to be deposited. One of the attractive features of inkjet printing is that it is possible to compensate for local distortions through a combination of optical detection and distortion compensation on the software level.

12.3 High-resolution, Self-aligned Inkjet Printing

As stated above, a common challenge for most of the above printing-based manufacturing approaches are the limitations on critical feature size, linewidth and integration density which are imposed by the limited resolution capability of standard graphic arts printing techniques. The size of a printed feature depends on the amount of ink which is deposited onto the substrate, and the spreading characteristics of the ink on the substrate. We have previously demonstrated high-resolution inkjet printing based on a combination of printing with substrate surface energy prepatterning [23]. Before deposition of the functional ink a substrate surface energy pattern comprising lyophobic (ink repellant), and lyophilic surface regions is defined using a high resolution patterning technique such as photolithography [23], soft lithographic stamping [42] or direct-write laser patterning [25]. In the subsequent printing step the spreading of the ink on the substrate is confined by the surface energy pattern. This technique enables definition of channel lengths of several microns, and even submicrometer dimensions (Fig. 12.2) [43], and linewidths of the order of 20 μm using standard inkjet printing heads with a droplet volume of 2–30 pL. The concept of surface-energy assisted inkjet printing has also been extended to fabrication of self-aligned electrodes with small geometric overlap, and correspondingly small parasitic capacitance to other conducting patterns of the device. In one approach a self-aligned surface energy pattern for an upper layer was defined with the help of a topographic contrast defined during patterning of a lower layer of the device [34]. In another approach a surface energy pattern

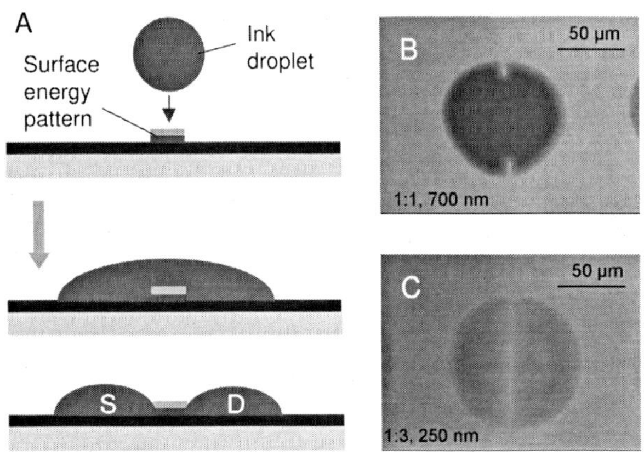

Fig. 12.2. (A) Schematic diagram of high-resolution inkjet printing based on surface-energy barrier induced ink dewetting; inkjet printed droplets of high (B) and low-viscosity (C) PEDOT/PSS droplets printed on top of a submicrometer hydrophobic rib on top of a hydrophilic substrate. The low viscosity ink dewets successfully off a 250 nm wide ink barrier (Reprinted with permission from Ref. [43]. Copyright (2005) by Nature Publishing Group).

was self-aligned with respect to a previously defined electrode pattern with the help of photopatterning of a self-assembled monolayer exposed through the back of the substrate [44].

For submicrometer channels the need to pre-define a surface energy pattern becomes prohibitively expensive and challenging, particularly for the low-cost, large-area applications at which OTFT technology is aimed. We have recently demonstrated a bottom-up, self-aligned inkjet printing technique, that removes the need for any substrate prepatterning, and is capable of defining sub-100 nm critical features with two simple additive printing steps using standard inkjet printing equipment without the need for any lithography or precise relative alignment [45]. The method comprises the steps of:

- inkjet printing a first conductive pattern on to the substrate,
- selectively modifying the surface of the first conductive pattern to be of low surface energy without modifying the surface of the substrate, and
- inkjet printing a second conductive pattern partially overlapping with the first conductive pattern, but not requiring precise relative alignment.

The droplets of the second pattern are repelled by and flow off the low-energy surface of the first pattern and dry with their contact line in close proximity to the edge of the first pattern, but forming a small self-aligned gap (Fig. 12.3A). The size of this gap is not defined by any top-down patterning step as in the surface-energy pattern assisted printing approach described above, but is entirely controlled by the dynamics of the contact line motion when the droplets flow off the surface of the first electrode. Self-aligned printing is a general micro and nanostructuring concept which can be implemented in a variety of different modes depending on how the surface modification and the second printing step are performed. The discussion here is limited to conducting polymer electrodes (PEDOT/PSS), but we have found that the technique is equally powerful with printable metals.

12.3.1 Self-aligned Printing by Selective Surface Treatment

The basis for the self-aligned printing technique is the creation of a hydrophobic layer on the surface of the first electrode. This can be achieved by a post-printing surface modification of the first printed pattern, for example by a Freon plasma treatment of the substrate after the deposition of the first printed pattern. The CF_4 treatment results in formation of a low-energy, fluorinated surface layer on the first conductive PEDOT electrode, but leaves the glass substrate in a high energy state. In fact, we have found that the CF_4 treatment beneficially further lowers the contact angle of the ink on the glass substrate, presumably due to etching of the SiO_2 surface, resulting in a high surface energy contrast between the first PEDOT pattern and the bare substrate. When printing the second electrode pattern at variable distances from the first pattern we observed a strong surface repulsion effect between the ink droplets of the second pattern, and the fluorinated surface of

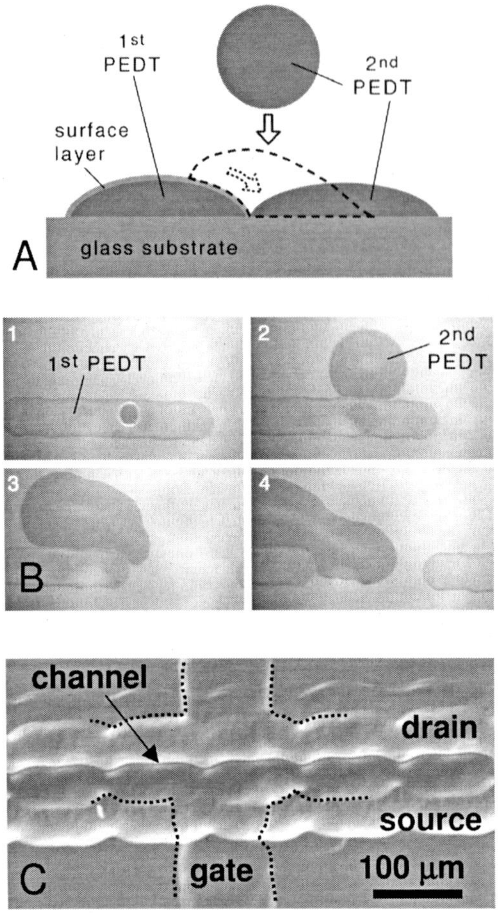

Fig. 12.3. (A) Schematic diagram of SAP process; (B) video sequence showing ink flowing off the surface of a first conductive pattern of CF_4 treated PEDOT/PSS; (C) optical micrograph of all-printed, sub-100 nm SAP TFT with top-gate electrode overlapping with SAP defined channel.

the first pattern. Owing to the high receding water contact angle complete dewetting was observed even if the second pattern was printed overlapping significantly with the first pattern (Fig. 12.3B). As discussed below the two conductive electrodes defined in this way are in very close proximity, but electrically isolated from each and can be used as the source and drain electrode of an FET when a gate electrode is printed overlapping with the SAP gap structure (Fig. 12.3C). A critical process condition is the receding contact angle, which if much lower than the advancing contact angle due to surface inhomogeneities manifests itself in pinning of the contact line of the receding liquid flowing off the first pattern. However, under optimized conditions receding contact angles on the surface-modified PEDOT/PSS are sufficiently high to allow the ink to flow off completely. Advancing and receding water contact angles were 104–119° and 85°, respectively, on CF_4 treated PEDOT/PSS.

12.3.2
Self-aligned Printing by Surface Segregation

In an alternative implementation of the SAP concept a component can be mixed into the ink that has a tendency to segregate to the surface during the drying of the ink on the substrate, for example a surfactant molecule with a polar headgroup, and a nonpolar tailgroup (Fig. 12.4A). Surfactants exhibit complex self-organization in solution, and are key components of many biological systems, such as membranes, and also have important applications in many fields of chemical engineering, such as detergents or foams. In polar liquids such as water the surfactant tail groups avoids contact with water by segregation to the surface of the liquid, and forming a low surface tension surface layer with the tail groups directed towards the interface with air and the polar head groups in contact with water [46].

Using surfactants we have produced self-aligned low-surface energy coatings on the surface of the first conductive ink pattern that are absent on the surface of the higher-surface-tension bare substrate, and can be used for the successful repulsion of a second conductive ink droplet. For nonoptimized surface conditions it was only possible to print the second pattern flowing against, but not overlapping with the first pattern. For overlapping printing conditions incomplete dewetting was ob-

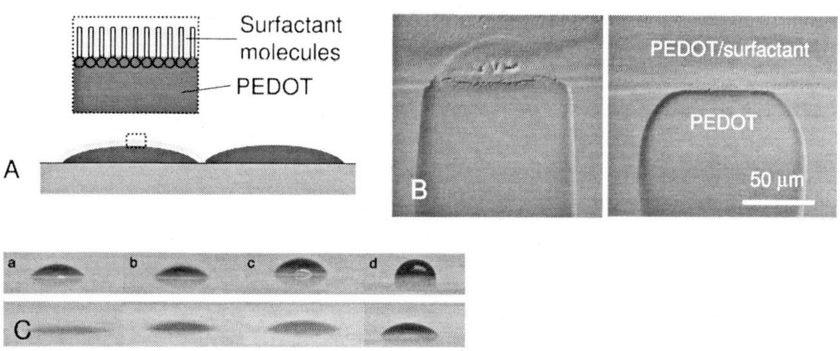

Fig. 12.4. (A) Schematic diagram of SAP mode based on surfactant segregration; (B) optical micrographs of PEDOT/PSS droplet dewetted off the surfactant modified first PEDOT electrode for nonoptimized (left) and optimized (right) conditions. (C) Static contact angle measurements of water-based PEDOT/PSS droplets on surfactant-modified PEDOT surface for different surface conditions (a: as deposited, b: annealed 150 °C, c: annealed 90 °C under high humidity; d: annealed 90 °C under high humidity + 150 °C in N_2) measured immediately after placing the droplet on the surface (top) and 30 s later (bottom).

served manifesting itself in pinning of the contact line of the liquid receding from the first pattern (Fig. 12.4B, left) and resulting in significant electrical leakage current between the two conductive electrodes. Again, a critical process parameter to achieve sufficient dewetting is the receding contact angle, which, if much lower than the advancing contact angle, leads to pinning of the contact line. It is also important to consider the stability of the contact angle in time, since the surfactant surface coating of the first electrode might redissolve, when coming in contact with the ink of the second pattern. Through optimization of the process conditions it has been possible to achieve high , stable contact angles for PEDOT droplets deposited on surfactant-segregated PEDOT surfaces (Fig. 12.4C) leading to complete dewetting of PEDOT ink droplets off the surface of a first printed PEDOT line even when overprinted (Fig. 12.4B, right). Cationic surfactants, such as benzalkonium chloride (BC), hexadecyltrimethylammonium bromide (CTAB), and di-*n*-decyldimethylammonium chloride (DD50) were found to interact more favorably with the PEDOT/PSS polyelectrolyte, compared to anionic surfactants such as sodium dodecylsulfate (SDS), and provided greater surface repulsion. This is possibly related to the tendency of negatively charged PSS to segregate to the surface of the PEDOT providing favorable electrostatic interactions with cationic surfactants segregating to the surface.

12.3.3 Self-aligned Printing by Autophobing

As shown below the two SAP modes described above are capable of defining very high resolution, sub-100 nm features. However, for some applications it might be desirable to have the ability to define self-aligned gaps of micrometer dimensions leading to potentially more reliable feature definition and higher patterning yield. This is also possible using the SAP concept in an autophobing mode, in which surfactant molecules from the first ink droplet adsorb on to the surface region surrounding the droplet and causing the ink to be repelled by the surface region surrounding it. When the ink recedes as a result this creates a self-aligned hydrophobic surface energy barrier of surfactant molecules with well defined length around the edge of the first printed pattern (Fig. 12.5A). Evidence for autophobing of surfactant containing solutions of PEDOT was found when printing successive droplets of PEDOT/DD50 into the same location of the glass substrate such that the volume of liquid ink on the substrate increases in time. The large droplet created on the surface in this way exhibited stick-jump motion, i.e. the contact line was moving in a discontinuous fashion [47]. The apolar hydrocarbon chains of the surfactant molecules, extending away from the solid surface, force the droplet contact line to recede, and exposing a narrow hydrophobic region. The low surface energy surfactant layer stops the contact line movement (stick), until the supply of additional liquid sufficiently increases the contact angle to overcome (jump) the hydrophobic barrier.

This phenomenon can be exploited by printing the second droplets at a well defined distance away from the first pattern (without overlapping) such that the ink

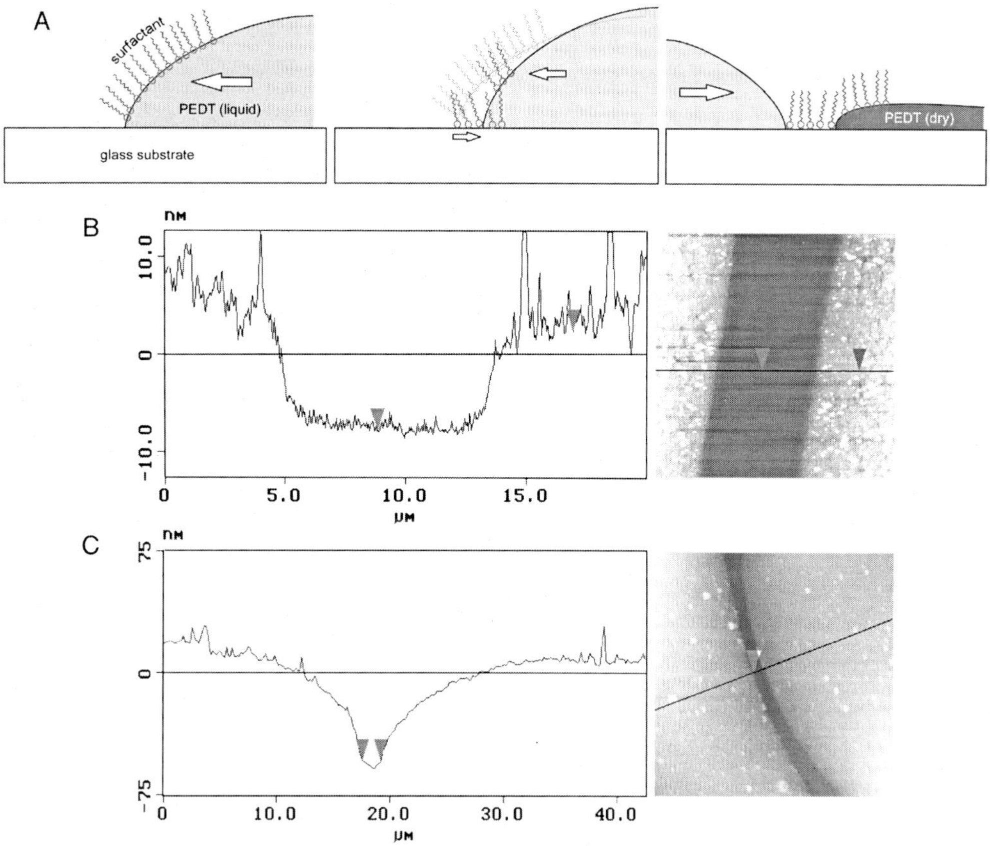

Fig. 12.5. (A) Schematic diagram of SAP mode based on surfactant-induced autophobing; AFM topographs and cross sections of PEDOT SAP gaps produced in the autophobing mode with DD50 concentrations of 1 g L^{-1} (B) and 0.5 g L^{-1} (C).

flows towards the region around the first pattern modified by the surfactant. The surfactant barrier is capable of stopping the flow of the second ink droplets towards the edge of the first pattern, and defining a gap between the two patterns which is defined by the width of the surfactant modified region surrounding the first pattern. One of the key requirements is that the surfactant is capable of adsorbing to the bare substrate surface. On a bare glass/SiO_2 surface which tends to be negatively charged cationic surfactants were found to provide better repulsion than anionic surfactants, while on glass surfaces modified with NH_3^+ terminated self-assembled monolayers the opposite behavior was observed. Using the autophobing mode of SAP we have been able to define SAP gaps of both micrometer and submicrometer dimension with the width of the gap being controllable by the surfactant concentration (Fig. 12.5C).

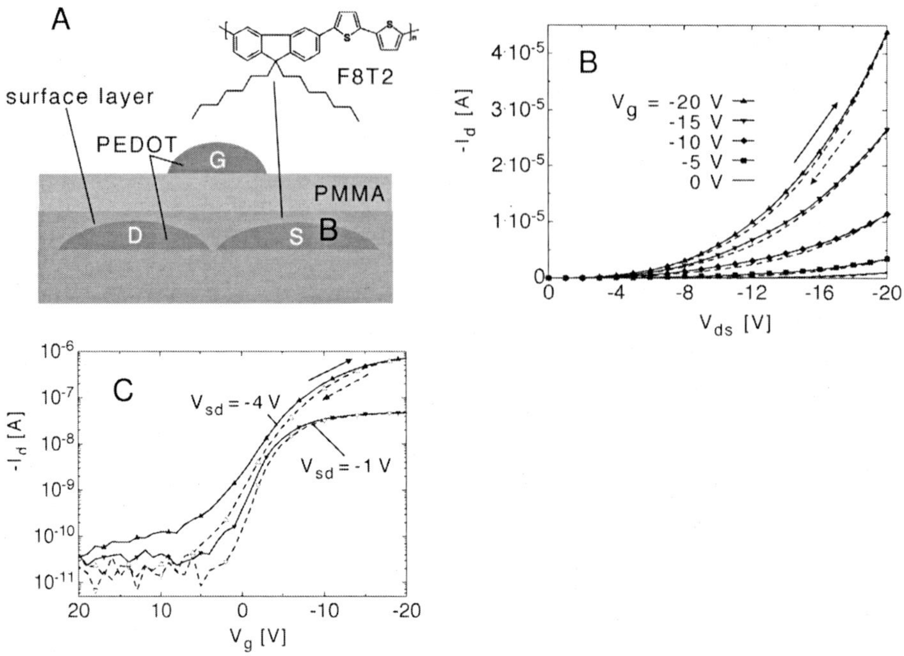

Fig. 12.6. (A) Schematic diagram of top-gate SAP TFT; Output (B) and transfer (C) characteristics of SAP TFT with F8T2 semiconducting layer. (Reprinted with permission from Reference [45]. Copyright (2005) by Wiley).

In the following we discuss the electrical characteristics of field-effect devices with SAP defined source and drain electrodes. The discussion is focused on devices fabricated using post-deposition CF_4 plasma treatment, as this technique has to date been found to be the most reliable in the case of PEDOT electrodes. Top-gate FET devices were fabricated with SAP PEDOT source and drain electrodes (Fig. 12.6A), and inkjet-printed PEDOT gate electrodes. As semiconductor we have used poly(9,9-dioctylfluorene-co-bithiophene) (F8T2), which is a well understood model polymer with good environmental stability and moderately high mobility of 2×10^{-3} cm^2 V^{-1} s^{-1} in an unaligned configuration. FETs with SAP fabricated source-drain electrodes exhibit characteristic short-channel behavior. No saturation of the drain current is observed when the device is operated in the saturation regime ($|V_d| > |V_g - V_T|$) (Fig. 12.6B). In spite of the short channel length ON–OFF ratios greater than 10^4 are achieved as along the source-drain potential is kept below 5 V (Fig. 12.6C). When higher source-drain voltages were applied, the ON–OFF current ratio was observed to decrease. This behavior is also characteristic of short-channel FETs.

The channel length defined by the SAP technique is not defined by a top-down patterning step, i.e. the gap dimensions are not known a priori. We have used a range of metrology tools to resolve and determine the channel dimensions defined

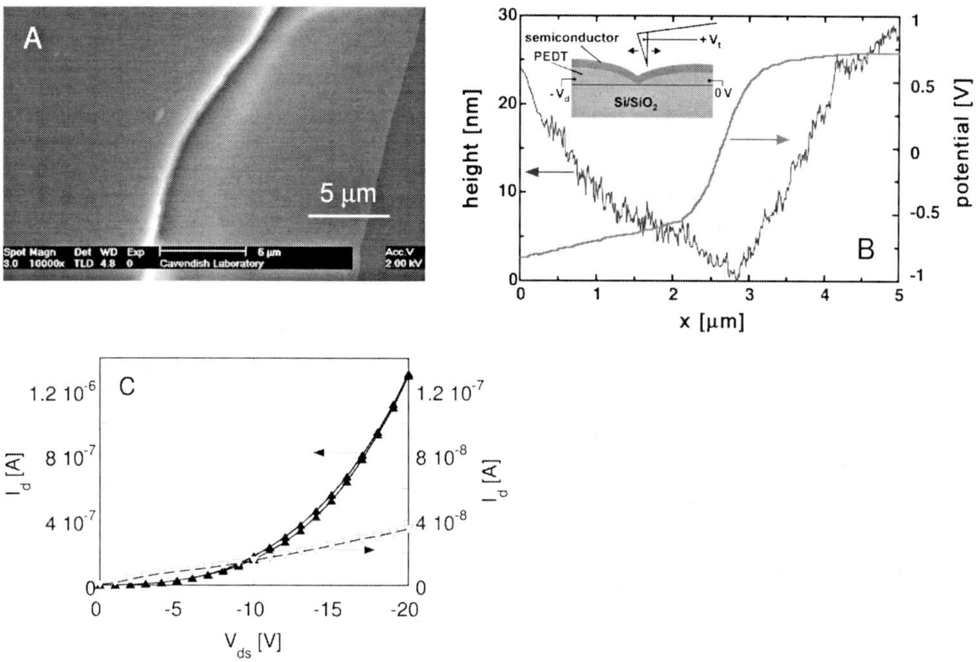

Fig. 12.7. Assessment of channel length of SAP-defined source-drain electrodes: (A) top-view environmental scanning electron microscopy (ESEM) image of channel region; (B) Scanning Kelvin probe microscopy of bottom-gate SAP FET; (C) Comparison of $I_d(V_{sd})$ characteristics at $V_g = -20$ V for SAP FET (open symbols) and a $L = 500$ nm FET (solid symbols) fabricated by dewetting.

by the SAP technique. In Atomic Force Microscopy (AFM) the gap between the two PEDOT/PSS electrodes is not clearly discernible (see topography cross section in Fig. 12.7B). The thickness profile of the two electrodes is thinning towards the gap, but the two electrodes appear to be touching each other, although no leakage current (<10 pA) could be detected between the two electrodes even at high source-drain potentials of 20 V. Similarly, in plan-view environmental scanning electron microscopy (ESEM) (Fig. 12.7A) the gap cannot be clearly resolved, but appears to be of submicrometer dimension. Scanning Kelvin Probe Microscopy (SKPM) measurements were performed in air using a tapping-mode AFM to measure directly the potential profile of a bottom-gate SAP FET (Fig. 12.7B). Most of the electrical potential drops across the SAP gap across a distance less than 0.5 µm, proving unambiguously that the SAP channel is of submicrometer dimension. However, the true dimension of the channel is possibly much lower, and cannot be deduced from these SKPM measurements since the resolution is limited by the experimental setup. Higher-resolution SKPM measurements can be performed using noncontact SKPM measurements in vacuum [48], and will be reported elsewhere.

To obtain a more accurate estimate of the channel length of the SAP structures a channel length scaling analysis has been performed. The ON current of the SAP devices is higher by a factor of about 400 than that of long-channel devices printed with two spatially separate source-drain electrodes ($L = 25$ μm), but otherwise identical device configurations. We have also compared the SAP characteristics to those of submicrometer FETs with known channel dimension ($L = 500$ nm) fabricated by the dewetting technique, again with otherwise identical device configuration (see discussion above). The SAP device has about an order of magnitude higher ON current than the $L = 500$ nm device (Fig. 12.7C), which gives a high level of confidence that the SAP gap is less than 100 nm, possibly as short as 50–60 nm.

We would like to emphasize that the SAP gap is not determined by the thickness of the hydrophobic surface layer formed on the surface of the first electrode. The process does not rely on the surface layer being electrically insulating. In fact, this would be undesirable because it is necessary that the surface modification of the first electrode is performed in such a way that the electrode still allows efficient injection of charges from both electrodes into the semiconducting layer. The process is also not based on the specific surface chemistry of PEDOT, since we have observed similar behavior when printing conductive inks against photolithographically defined gold electrodes modified by self-assembled monolayers. We believe that the SAP gap dimensions are determined by the balance of surface forces which repel the ink from the surface of the first electrode, and the pressure exerted onto the contact line of the second droplet when it is spreading back towards the first electrode after having flown off its surface.

Relatively high gate voltages of 10–20 V are required to switch the FETs, owing to the large thickness ($d = 120$–130 nm) of the dielectric layer required to prevent electrical shorts. This relatively thick dielectric layer does not fulfill basic scaling laws which require the dielectric thickness d to be significantly smaller than the channel length L. An approximate scaling rule for long-channel operation of TFTs is $d/L < 0.1$. In our current configuration this requirement is not fulfilled, and this is responsible for the short-channel effects in the device characteristics, in particular the lack of saturation in the output characteristics of the TFT, and the decrease of ON–OFF current ratio with increasing source-drain voltage. In the future it will be important to combine the SAP technique with much thinner dielectrics, such as ultrathin polymer dielectrics [49] or self-assembled monolayer dielectrics [50].

One of the critical questions with any advanced process technology providing submicrometer feature definition is an assessment of the achievable uniformity and yield of gap formation. To investigate this active-matrix arrays of 4800 transistors have been fabricated with SAP defined source-drain electrode. The leakage current between source and drain electrodes were measured prior to deposition of the semiconducting active layer. Most remarkably among 350 tested FETs only a single electrical short could be detected, in only three devices was the leakage current above the noise level of the measurement system (Fig. 12.8A). Similarly, after completing the device fabrication measurement of the ON and OFF currents of 150

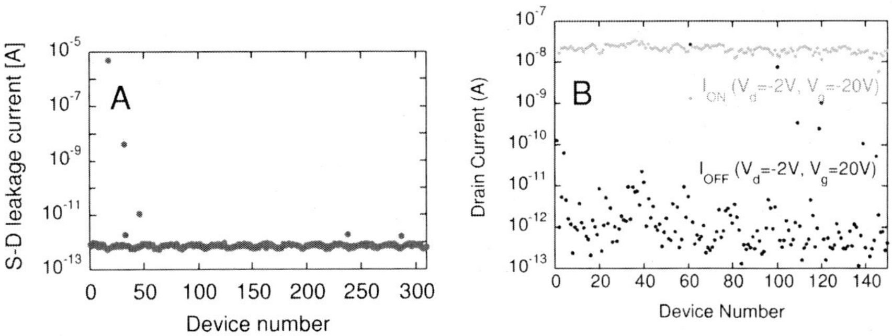

Fig. 12.8. (A) Leakage current measurements at $V_{sd} = 20$ V for 320 SAP PEDOT source-drain electrode structures prior to deposition of the semiconducting layer; (B) Variations of the ON and OFF current of an active matrix array of 150 SAP FETs. (Reprinted with permission from Ref. [45]. Copyright (2005) by Wiley).

SAP fabricated FETs showed that in spite of the submicrometer channel length the drain current in the ON state of $2.1 \pm 0.07 \times 10^{-8}$ A is encouragingly uniform and none of the devices shorts electrically between source and drain electrodes (Fig. 12.8B). Less than 10% of the devices show OFF currents above the noise level of the measuring setup ($>10^{-11}$ A). Preliminary optical microscopy investigations suggest that the few devices with higher leakage currents contain particle defects in the SAP gap. These initial uniformity and yield results are quite remarkable given the sub-100 nm dimension of the channel and indicate that the SAP approach with its bottom-up patterning mechanism might provide a scalable approach to nanopatterning over large areas with high yield.

As shown above the SAP devices have ON currents that are about 2–3 orders of magnitude higher than that of devices with channel length $L > 20$ μm which can be fabricated readily by straight inkjet or other printing techniques without the risk of electrical shorts. This increased ON current results in an associated improvement in circuit switching speed. By incorporating SAP FETs into simple logic circuits such as inverters it has been shown that an increase in speed by about 2 orders of magnitude is indeed achievable [45]. We emphasize that this speed enhancement is achievable even without corresponding scaling of the other device dimensions such as the linewidth of the gate electrode which determines the parasitic capacitance of the next logic stage which needs to be charged with the ON current of the FET. If the gate linewidth could be reduced in the same way as the channel length an even more significant speed improvement ($1/L^2$ scaling as opposed to mere $1/L$ scaling) could be achieved.

The self-aligned printing technique is a novel lithography-free, bottom-up nanoscale printing technique, which uses the unique flow properties of small liquid droplets to define with high yield and uniformity critical features that are more than two orders of magnitude smaller than previously achievable with straight, additive printing, and comparable to that of sophisticated state-of-the art deep-UV or electron-beam lithography. The technique is potentially applicable to other liquid-based, graphic-art printing techniques. It might provide an approach to overcome the limitations in device performance imposed by the limited resolution of commercial printing techniques without sacrificing the other attractive attributes of printing. By using polymer or molecular semiconductors with mobilities exceeding 10^{-1} cm^2 V^{-1} s^{-1}, and thinner polymer dielectrics switching frequencies in the MHz range will be achievable with SAP printed organic circuits.

12.4 Performance and Reliability of Solution-processed OTFTs for Applications in Flexible Displays

In this section we review the device performance and reliability which is achievable with state-of-the-art solution-processed OTFTs with a particular emphasis on comparing these to the requirements for electronic paper display applications. The performance of solution-processible OTFTs which is generally benchmarked against that of amorphous silicon (a-Si) thin film transistors (TFTs) with field-effect mobilities of 0.5–1 cm^2 V^{-1} s^{-1} and ON–OFF current ratio of 10^6–10^8 has improved significantly over the last 5 years. The current record mobility value for polymer OFETs is 0.6 cm^2 V^{-1} s^{-1} for a solution-processed polythiophene derivative based on dithienothiophene [51], while for vacuum-sublimed pentacene OFETs mobility values of up to 5 cm^2 V^{-1} s^{-1} have been achieved [52]. These are headline mobility values, which are not necessarily yet achievable in a stable, high-volume production process, but as discussed below there are now several classes of polymers which allow achieving consistent mobilities in the 10^{-2}–10^{-1} cm^2 V^{-1} s^{-1} range and ON–OFF current ratios of 10^6–10^8 in a manufacturing environment.

Several groups have reported that amorphous polymers based on triarylamine similar to those used in xerographic applications allow achieving high field-effect mobilities of 10^{-3}–10^{-2} cm^2 V^{-1} s^{-1}, combined with good operating, environmental and photostability. Veres et al. have reported high-performance FETs with field-effect mobilities of up to 6×10^{-3} cm^2 V^{-1} s^{-1}, low threshold voltage, and good device stability based on a range of polytriarylamine (PTAA) derivatives [53, 54]. These are used in combination with apolar, low-*k* polymer dielectrics. With structurally related (9,9-dialkylfluorene-alt-triarylamine) (TFB) in contact with benzocyclobutene dielectric very stable device operation during continuous switching at 120 °C without device degradation was demonstrated [49]. A prototype microcrystalline polymer with high field-effect mobilities of 0.1–0.3 cm^2 V^{-1} s^{-1} is regioregular poly(3-hexylthiophene) (P3HT) [55, 56]. However, P3HT has poor environmental stability, particularly during exposure to light and moisture. The oxidative stability of P3HT can be improved by increasing the ionization potential of the polythiophene backbone by either disrupting its ability to adopt a fully planar conformation through the side chain substitution pattern [57], or by incorporating partially conjugated co-monomers into the main chain [58]. These materials maintain the beneficial microcrystalline, lamellar self-organization motive of the parent

P3HT polymer, and as a result exhibit similar field-effect mobilities, but have significantly improved environmental and operating stability.

An alternative route to solution-processible organic semiconductors is to use precursors to small molecule semiconductors, such as pentacene [59] or tetrabenzoporphyrin [60], which can be converted into their fully conjugated, insoluble form by thermal or irradiative [61] treatment on the substrate. Pentacene precursors have been shown to yield field-effect mobilities of 0.01–0.1 $cm^2\ V^{-1}\ s^{-1}$ [62], and 0.1–0.8 $cm^2\ V^{-1}\ s^{-1}$ [63] after thermal conversion at 150–200 °C. Small molecule organic semiconductors can also be rendered solution processible by attachment of flexible side chains [64–66]. Due to the relatively low solubility of these molecules the growth of uniform thin films of these molecules remains challenging, however.

Many organic semiconductors show p-type conduction only. Electron field-effect conduction has been reported in a few, relatively high electron affinity ($EA > 3.5$ eV) small molecule organic semiconductors deposited from vacuum phase, and solution-processed organic semiconductors (for a review see Ref. [67]). It has been shown recently [68] that electron conduction is in fact a generic feature of a much broader class of organic semiconductors including those with normal electron affinities of 2.5–3.5 eV if the right dielectric is used which avoids trapping of electrons at the interface. This research opens up new ways to realize n-type devices and complementary logic circuit configurations.

It is generally well appreciated now that the choice of the right dielectric is crucial for achieving optimum field-effect mobility, device stability and reliability. In the same way as silicon MOS technology owes much to the quality of the $Si-SiO_2$ interface, dielectrics for organic FETs have recently received significant attention. For a solution-processed active interface, in which either the gate dielectric material is deposited from solution on to a solution-processible semiconducting material or vice versa, it is critical to avoid dissolution or swelling effects during deposition of the upper layer, which can lead to interfacial mixing and increased interface roughness. The preferred approach to achieve this is to choose orthogonal solvents for the deposition of the multilayer structure [23].

In principle, for a given thickness of dielectric a high k dielectric should be preferable to a low-k dielectric for an FET application which requires the FET to exhibit a high drive current at low drive voltage. Various solution-processible high-k dielectrics for low-voltage OFETs have been used in the literature such as anodized Al_2O_3 [69] ($\varepsilon = 8–10$), or TiO_2 [70] ($\varepsilon = 20–41$), (for a review see [54]). Many polar, high-k polymer dielectrics, such as polyvinylphenol ($\varepsilon = 4.5$), or cyanoethylpullulan ($\varepsilon = 12$) are hygroscopic, and susceptible to drift of ionic impurities during device operation, and cannot be used for ordinary TFT applications [71]. Veres et al. [53] have shown that the benefit of using a high-k dielectric can be partly or completely eliminated by the field-effect mobilities being higher in contact with low-k dielectrics with $\varepsilon < 3$ than dielectrics with higher k. The latter usually contain polar functional groups which are randomly oriented near the active interface, which is believed to increase the energetic disorder at the interface beyond what naturally occurs due to the structural disorder in the organic semiconductor film itself resulting in a lowering of the field-effect mobility. Low-k dielectrics also have the ad-

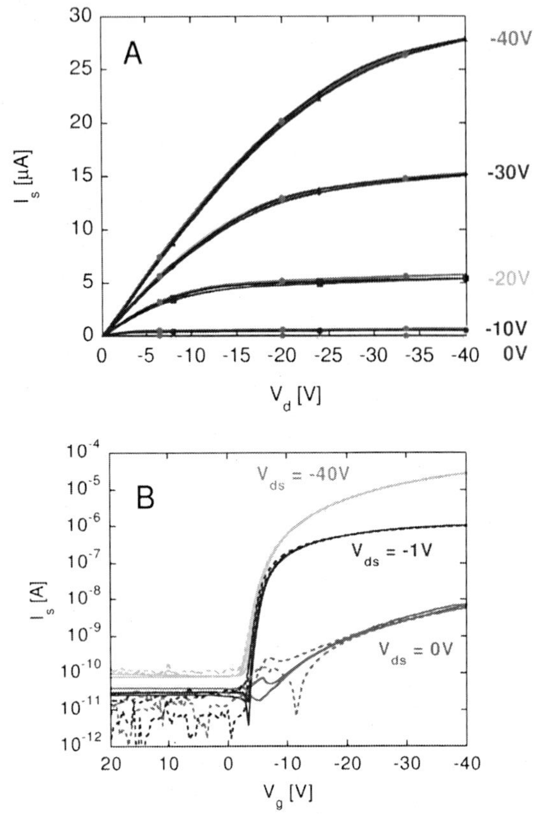

Fig. 12.9. Output (A) and transfer characteristics (B) of a state-of-the-art OTFT ($L = 10$ μm) measured in air and light without encapsulation. In (A) measurements directly after device manufacture and several weeks later after storage under unprotected conditions are compared. (Reprinted with permission from Ref. [72]).

vantage of being less susceptible to ionic impurities, which can drift under the influence of the gate field causing device instabilities.

Figure 12.9 shows the output and transfer characteristics of a state-of-the-art, polymer FET fabricated using the Plastic Logic direct-write manufacturing process ($L = 10$ μm). No encapsulation of the TFT is present other than what is naturally provided by the presence of the PET substrate on the bottom and an inkjet printed silver gate electrode on the top. The device exhibits a field-effect mobility of 0.04 $cm^2\ V^{-1}\ s^{-1}$ and ON–OFF current ratio of 5×10^5. The threshold potential is $V_T = 5$–6 V. These basic TFT performance values can be achieved consistently in a manufacturing environment, and are sufficient to drive a 100 dpi electronic paper display with A5 size.

Figure 12.9A also compares characteristics measured in ambient air and light directly after device manufacture and several weeks later after the device had been exposed to air, light, and varying levels of humidity in an experiment meant to investigate the shelf life stability under typical environmental conditions without special protection of the device [72]. No evidence for device degradation is observed. These recent improvements in the shelf as well as operational life of OTFTs have been achieved as a result of using organic semiconductors with better inherent stability, better understanding of the requirements for gate dielectrics, and by more controlled manufacturing processes.

The operational stability of these devices has been investigated for typical drive conditions during display operation by switching the devices ON and OFF continuously for 10^7 switches at a refresh frequency of 50 Hz corresponding to continuous operation for 54 h. This measurement was performed on an unencapsulated TFT exposed to air and light. The device is formed in a top-gate configuration on a PET substrate with a polyfluorene-based polymer semiconductor (SC), and a spin-coated polymer dielectric. Figure 12.10A shows that there is no measurable degradation in the ON and OFF current during this lifetime test. We note that for applications in bistable, electronic paper displays the TFTs only need to be operated when the information on the display needs to be changed, and therefore 10^7 switches corresponds already to a realistic number of image updates which might occur during the total lifetime of the display.

One of the key requirements for active matrix display addressing is the ability to achieve good TFT uniformity across the active matrix array. Figure 12.10B shows the typical variation of TFT ON and OFF currents of 100 TFTs on a PET substrate measured across an array containing 4800 devices in total. The typical variation of TFT ON current is less than 10%, and the average value of the threshold voltage of these devices is $V_T = -3.9$ V with small standard deviation of only 0.3 V. The yield of devices that exhibit an ON–OFF current ratio exceeding 10^5 is 95–100% varying from substrate to substrate, and is limited by defects associated with the manual handling of the substrate. Figure 12.11 shows a photograph of a 10 inch diagonal black-and white electronic paper display (10 ppi) fabricated by laminating an active

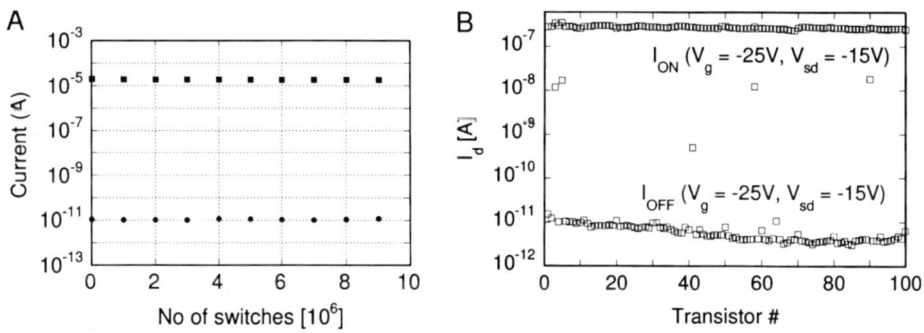

Fig. 12.10. (A) Measurement of operational stability of OTFT ON and OFF current during continuous switching at 50 Hz for 10^7 cycles under typical display driving conditions. (B) Variations of the ON and OFF current of an active matrix array of 100 OTFTs on a PET substrate. (Reprinted with permission from Ref. [72]).

Fig. 12.11. Photograph of a 10″ diagonal OTFT driven active matrix electronic paper display on a PET substrate.

matrix array of these OTFTs on a PET substrate with an electrophoretic E Ink Imaging Film. The displays exhibit excellent, paper-like contrast and very good bistability. They can be bent repeatedly without degradation of contrast to a radius of curvature of 5 mm. Plastic Logic has announced that based on these TFTs it is planning to develop A4 size flexible electronic paper displays with resolution up to 150 ppi in its recently opened prototype manufacturing line.

The investigation and further improvement of device reliability and stability remains of course a central topic for the development of OTFTs, and significant efforts are required to prove that OTFTs meet stringent lifetime requirements for real products. Also, from a scientific perspective, much better understanding of the mechanisms for residual device degradation is still required. However, the recent advances in improving device reliability reviewed here show that even without device encapsulation state-of-the-art OTFTs can exhibit similar operational stability as inorganic amorphous silicon TFTs. Lifetime issues are much less a concern for OTFTs than in the related field of organic LEDs where commercialization has been held back for several years by remaining lifetime and stability problems.

12.5 Conclusions

In this article we have reviewed the current status of the application of graphic arts printing techniques to manufacturing of organic thin-film transistor circuits. Recent progress on the materials and device technology has led to very significant performance and reliability improvements of solution-processed OTFTs which now meet the requirements for first-generation applications, such as active matrix electronic paper displays on flexible substrates, or very simple, low-frequency intelligent labels. However, even the first-generation applications require tightly controlled manufacturing processes which are able to define high-resolution, critical features and linewidths, maintain accurate layer-to-layer registration, and deposit and pattern all layers without process-induced degradation in materials performance. To achieve this with commercial printing techniques requires development of improved and customized printing equipment, optimized materials formulations, as well as development of stable manufacturing processes. In terms of more demanding applications, such as high-resolution LCD or OLED displays, or standard compatible RFID tags there is still a significant gap between the application requirements and what can be achieved with fully printed OTFTs. Whether this gap can be filled by future improvements of materials performance and printing equipment in combination with good process engineering remains to be seen. It appears likely that for some applications, in particular for performance logic, new printing approaches capable of producing much smaller features than what appears realistically achievable with current printing technology will be needed. The self-aligned printing technique reviewed here is potentially such an approach, which provides a scalable route to significant performance enhancements of printed OTFTs by device miniaturization. In spite of the significant challenges ahead the vision of printable electronics remains a very appealing one, and we can expect that its potential will start to be realized in real products over the next 3–5 years.

Acknowledgments

It is a pleasure to acknowledge many wonderful students, postdoctoral research fellows and colleagues who have made possible the work reported in this review, in particular Drs Takeo Kawase, and Jizheng Wang for their work on surface-energy assisted printing, Dr Tatsuya Shimoda, and Professor Sir Richard Friend for many interesting discussions, and the entire team of engineers and scientists at Plastic Logic for making organic transistors a commercial reality.

References

1 Sirringhaus, H. and Shimoda, T., *Mrs Bulletin*, **2003**, *28*, 802–803.
2 Granmar, M. and Cho, A., *Science*, **2005**, *308*, 785–786.
3 Halls, J., *Information Display*, **2005**, *21*, 10–15.
4 Chikama, Y., Nishiki, N., Nakatani, Y., Watanabe, N., Gotoh, H., Umeda, H. and Oka, W., Development of a transmissive a-Si TFT LCD with transparent flexible substrate, 11th International Display Workshops IDW'04, Niigata, Japan, 2004.
5 Knobloch, A., Manuelli, A., Bernds, A. and Clemens, W., *Journal of Applied Physics*, **2004**, 96, 2286–2291.
6 Brandt, N., Fischer, T., Fügmann, U., Hahn, U., Hübler, A. and Zielke, D., Offset Printed Functional Polymer Structures for Transistors, Technologies for Polymer Electronics – TPE 04 (TITK). Internationales Symposium, Rudolstadt, 2004.
7 Gelinck, G.H., Geuns, T.C.T. and Leeuw, D.M.d., *Appl. Phys. Lett.*, **2000**, *77*, 1487–1489.
8 Fix, W., Ullmann, A., Ficker, J. and Clemens, W., *Applied Physics Letters*, **2002**, *81*, 1735–1737.
9 Kipphan, H., Handbook of Print Media, Springer, Berlin, 2001.
10 Garnier, F., Hajlaoui, R., Yassar, A.

and Srivastava, P., *Science*, **1994**, *265*, 1684–1686.

11 Bao, Z.N., Feng, Y., Dodabalapur, A., Raju, V.R. and Lovinger, A.J., *Chemistry of Materials*, **1997**, *9*, 1299.

12 Brandon, E.J., West, W. and Wesseling, E., *Applied Physics Letters*, **2003**, *83*, 3945–3947.

13 Bock, K.H., Reel to reel manufacturing of printed electronics and systems, Printed Electronics Europe 2005, Cambridge, 2005.

14 Xia, Y.N. and Whitesides, G.M., *Annual Review of Materials Science*, **1998**, *28*, 153–184.

15 Rogers, J.A., Bao, Z., Makhija, A. and Braun, P., *Adv. Mat.*, **1999**, *11*, 741–745.

16 Tate, J., Rogers, J.A., Jones, C.D.W., Vyas, B., Murphy, D.W., Li, W.J., Bao, Z.A., Slusher, R.E., Dodabalapur, A. and Katz, H.E., *Langmuir*, **2000**, *16*, 6054–6060.

17 Leufgen, M., Lebib, A., Muck, T., Bass, U., Wagner, V., Borzenko, T., Schmidt, G., Geurts, J. and Molenkamp, L.W., *Applied Physics Letters*, **2004**, *84*, 1582–1584.

18 Rogers, J.A., Paul, K.E., Jackman, R.J. and Whitesides, G.M., *Applied Physics Letters*, **1997**, *70*, 2658–2660.

19 Kagan, C.R., Breen, T.L. and Kosbar, L.L., *Applied Physics Letters*, **2001**, *79*, 3536–3538.

20 Ng, W.K., Wu, L. and Moran, P.M., *Applied Physics Letters*, **2002**, *81*, 3097–3099.

21 Zschieschang, U., Klauk, H., Halik, M., Schmid, G. and Dehm, C., *Advanced Materials*, **2003**, *15*, 1147.

22 Burgin, T., Choong, V.E. and Maracas, G., *Langmuir*, **2000**, *16*, 5371–5375.

23 Sirringhaus, H., Kawase, T., Friend, R.H., Shimoda, T., Inbasekaran, M., Wu, W. and Woo, E.P., *Science*, **2000**, *290*, 2123–2126.

24 Huang, D., Liao, F., Molesa, S., Redinger, D. and Subramanian, V., *Journal of the Electrochemical Society*, **2003**, *150*, G412–G417.

25 Burns, S.E., Cain, P., Mills, J., Wang, J.Z. and Sirringhaus, H., *Mrs Bulletin*, **2003**, *28*, 829–834.

26 Murata, K., http://unit.aist.go.jp/nanotech/research/inkjet.html.

27 Paul, K.E., Wong, W.S., Ready, S.E. and Street, R.A., *Applied Physics Letters*, **2003**, *83*, 2070–2072.

28 Arias, A.C., Ready, S.E., Lujan, R., Wong, W.S., Paul, K.E., Salleo, A., Chabinyc, M.L., Apte, R., Street, R.A., Wu, Y., Liu, P. and Ong, B., *Applied Physics Letters*, **2004**, *85*, 3304–3306.

29 Chabinyc, M.L., Wong, W.S., Salleo, A., Paul, K.E. and Street, R.A., *Applied Physics Letters*, **2002**, *81*, 4260–4262.

30 Kawase, T., Sirringhaus, H., Friend, R.H. and Shimoda, T., *Adv. Mat.*, **2001**, *13*, 1601–1605.

31 Blanchet, G.B., Loo, Y.L., Rogers, J.A., Gao, F. and Fincher, C.R., *Applied Physics Letters*, **2003**, *82*, 463–465.

32 Becker, E., Parashkov, R., Ginev, G., Schneider, D., Hartmann, S., Brunetti, F., Dobbertin, T., Metzdorf, D., Riedl, T., Johannes, H.H. and Kowalsky, W., *Applied Physics Letters*, **2003**, *83*, 4044–4046.

33 Okuzaki, H., Ishihara, M. and Ashizawa, S., *Synthetic Metals*, **2003**, *137*, 947–948.

34 Stutzmann, N., Friend, R.H. and Sirringhaus, H., *Science*, **2003**, *299*, 1881.

35 Parashkov, R., Becker, E., Ginev, G., Riedl, T., Brandes, M., Johannes, H.H. and Kowalsky, W., *Applied Physics Letters*, **2004**, *85*, 5751–5753.

36 Shtein, M., Peumans, P., Benziger, J.B. and Forrest, S.R., *Advanced Materials*, **2004**, *16*, 1615.

37 Loo, Y.L., Willett, R.L., Baldwin, K.W. and Rogers, J.A., *Applied Physics Letters*, **2002**, *81*, 562–564.

38 Hur, S.H., Khang, D.Y., Kocabas, C. and Rogers, J.A., *Applied Physics Letters*, **2004**, *85*, 5730–5732.

39 Zaumseil, J., Someya, T., Bao, Z.N., Loo, Y.L., Cirelli, R. and Rogers, J.A., *Applied Physics Letters*, **2003**, *82*, 793–795.

40 Kim, C., Shtein, M. and Forrest, S.R., *Applied Physics Letters*, **2002**, *80*, 4051–4053.

41 Scheinert, S., Doll, T., Scherer, A., Paasch, G. and Horselmann, I., *Applied Physics Letters*, **2004**, *84*, 4427–4429.

42 Sirringhaus, H., Buergi, L., Kawase, T. and Friend, R.H., *Polymer Transistor Circuits fabricated by Solution Processing and Direct Printing*. In: *Thin Film Transistors*, C.R. Kagan and P. Andry (Eds.), Marcel Dekker Inc, New York, 2003, pp. 427.

43 Wang, J.Z., Zheng, Z.H., Li, H.W., Huck, W.T.S. and Sirringhaus, H., *Nature Materials*, **2004**, *3*, 171–176.

44 Ando, M., Kawasaki, M., Imazeki, S., Sasaki, H. and Kamata, T., *Applied Physics Letters*, **2004**, *85*, 1849–1851.

45 Sele, C.W., von Werne, T., Friend, R.H. and Sirringhaus, H., *Advanced Materials*, **2005**, *17*, 997.

46 Porter, M.R., Handbook of surfactants, Blackie Academic & Professional, London, 1994.

47 Lee, S.W. and Laibinis, P.E., *Journal of the American Chemical Society*, **2000**, *122*, 5395–5396.

48 Burgi, L., Sirringhaus, H. and Friend, R.H., *Applied Physics Letters*, **2002**, *80*, 2913–2915.

49 Chua, L.L., Ho, P.K.H., Sirringhaus, H. and Friend, R.H., *Applied Physics Letters*, **2004**, *84*, 3400–3402.

50 Halik, M., Klauk, H., Zschieschang, U., Schmid, G., Dehm, C., Schutz, M., Maisch, S., Effenberger, F., Brunnbauer, M. and Stellacci, F., *Nature*, **2004**, *431*, 963–966.

51 McCulloch, I., Annual Meeting of the American Chemical Society, San Diego, USA, 2005.

52 Baude, P.F., Ender, D.A., Haase, M.A., Kelley, T.W., Muyres, D.V. and Theiss, S.D., *Applied Physics Letters*, **2003**, *82*, 3964–3966.

53 Veres, J., Ogier, S.D., Leeming, S.W., Cupertino, D.C. and Khaffaf, S.M., *Advanced Functional Materials*, **2003**, *13*, 199–204.

54 Veres, J., Ogier, S., Lloyd, G. and de Leeuw, D., *Chemistry of Materials*, **2004**, *16*, 4543–4555.

55 Bao, Z., Dodabalapur, A. and Lovinger, A.J., *Appl. Phys. Lett.*, **1996**, *69*, 4108–4110.

56 Sirringhaus, H., Tessler, N. and Friend, R.H., *Science*, **1998**, *280*, 1741–1744.

57 Ong, B.S., Wu, Y.L., Liu, P. and Gardner, S., *Journal of the American Chemical Society*, **2004**, *126*, 3378–3379.

58 Heeney, M., Bailey, C., Genevicius, K., Shkunov, M., Sparrowe, D., Tierney, S. and McCulloch, I., *Journal of the American Chemical Society*, **2005**, *127*, 1078–1079.

59 Herwig, P.T. and Mullen, K., *Advanced Materials*, **1999**, *11*, 480.

60 Aramaki, S., Sakai, Y. and Ono, N., *Applied Physics Letters*, **2004**, *84*, 2085–2087.

61 Afzali, A., Dimitrakopoulos, C.D. and Graham, T.O., *Advanced Materials*, **2003**, *15*, 2066.

62 Gelinck, G.H., Geuns, T.C.T. and de Leeuw, D.M., *Applied Physics Letters*, **2000**, *77*, 1487–1489.

63 Afzali, A., Dimitrakopoulos, C.D. and Breen, T.L., *Journal of the American Chemical Society*, **2002**, *124*, 8812–8813.

64 Laquindanum, J.G., Katz, H.E. and Lovinger, A.J., *Journal of the American Chemical Society*, **1998**, *120*, 664–672.

65 Katz, H.E., *Chemistry of Materials*, **2004**, *16*, 4748–4756.

66 Mushrush, M., Facchetti, A., Lefenfeld, M., Katz, H.E. and Marks, T.J., *Journal of the American Chemical Society*, **2003**, *125*, 9414–9423.

67 Dimitrakopoulos, C.D. and Malenfant, P.R.L., *Advanced Materials*, **2002**, *14*, 99.

68 Chua, L.L., Zaumseil, J., Chang, J.F., Ou, E.C.W., Ho, P.K.H., Sirringhaus, H. and Friend, R.H., *Nature*, **2005**, *434*, 194–199.

69 Majewski, L.A., Grell, M., Ogier, S.D. and Veres, J., *Organic Electronics*, **2003**, *4*, 27–32.

70 Majewski, L.A., Schroeder, R. and Grell, M., *Advanced Materials*, **2005**, *17*, 192.

71 Sandberg, H.G.O., Backlund, T.G., Osterbacka, R. and Stubb, H., *Advanced Materials*, **2004**, *16*, 1112.

72 Reynolds, K., Burns, S., Banach, M., Brown, T., Chalmers, K., Cousins, N., Creswell, L., Etchells, M., Hayton, C., Jacobs, K., Menon, A., Siddique, S., Ramsdale, C., Reeves, W., Watts, J., Werne, T.v., Mills, J., Curling, C., Sirringhaus, H., Amundson, K. and McCreary, M.D., *Proceeding of the International Display Workshop (IDW)*, **2004**.

IV
Devices, Applications, and Products

13
From Transistors to Large-scale Integrated Circuits

Gerwin H. Gelinck, Erik van Veenendaal, Eduard J. Meijer, Eugenio Cantatore, H. Edzer A. Huitema, Pieter J.G. van Lieshout, Fred J. Touwslager, Alwin W. Marsman, and Dago M. de Leeuw

13.1
Introduction

Microelectronics has been the enabling technology to develop modern electronic systems, especially in the area of data and signal manipulation. Silicon technology has been the dominant force in integrated device manufacturing and is likely to retain this position in the foreseeable future. On the other hand, 'plastic transistors', i.e. transistors in which an organic material is used as the active semiconductor [1], may form the basis of a new low-cost microelectronic technology on flexible substrates [2]. The use of organic materials has a number of important advantages over conventional techniques using mainly inorganic materials, like amorphous silicon. The low process temperature, typically less than 150 °C, creates the possibility to use a wide range of plastic substrates instead of glass. Furthermore, the (thermo)mechanical properties of organic semiconductors are compatible with plastic substrates. The possibility to use flexible plastic substrates to obtain thin, flexible organic electronics has been shown by several groups [3–7].

In recent years charge carrier mobilities comparable to that of amorphous silicon have been reported using a variety of organic semiconductors [1]. One of the key goals now is the integration of these organic transistors using a cost-effective fabrication process. A useful fabrication classification can be made according to the way the organic semiconductor is applied: by vapor deposition or by solution processing. A vapor deposited organic semiconductor generally yields higher thin-film transistor performance. Solution processing is however preferred for high-volume, low-cost production. It simplifies the manufacturing process compared to the conventional chemical vapor deposition techniques, especially for large areas. Preferably all layers, and not only the semiconductor, are processed from solution [8–12].

Next to the deposition, solution processing opens a plethora of alternative patterning technologies such as printing [13, 14], stamping [15], selective dewetting [16], inkjet printing [11]. Here, the ultimate goal is a roll-to-roll solution-based process leading to ultra-low cost electronics. For an early market entry, on the other

Organic Electronics, Materials, Manufacturing and Applications. Edited by Hagen Klauk

ISBN: 3-527-31264-1

hand, we argue that it is best to use standard photolithography techniques for patterning of the layers. The mature photolithography and spincoating equipment used in the established IC and AMLCD industry then allows the use of a mature knowledge base to quickly move up the learning curve towards a fully industrialized process. This approach was followed in the work described here.

The chapter is organized as follows: An analysis of organic transistor operation will be given in the next section. Both current–voltage (Section 13.2.2) and capacitance–voltage characteristics will be discussed (Section 13.2.3). A model for the transistor in the accumulation regime will be introduced that can explain the basic features of the organic transistor. Section 13.3 is devoted to the manufacturing and characterization of integrated circuits. In Section 13.3.1, we describe in more detail the technology used in our group to manufacture (flexible) organic transistors and integrated circuits. A brief analysis of transistor modeling will follow in Section 13.3.2. As an example of how this model can be used to understand and optimize circuit performance, we evaluate the link between transistor behavior and inverter performance with reference to some basic digital building blocks, and the effect of dimension downscaling on switching speed in Section 13.3.3. In Section 13.3.4 we will present a simple analysis of integrated circuits, again focusing on the relation between device and circuit performance.

13.2 Discrete Devices

13.2.1 Basic Device Operation of Organic Transistor

An organic field-effect transistor (FET) consists of a number of discrete layers, formed from insulating, semiconductor or electrically conducting material (Fig. 13.1). It can roughly be considered as a parallel plate capacitor, where one conducting electrode, the gate, is electrically insulated from the other electrode, which is

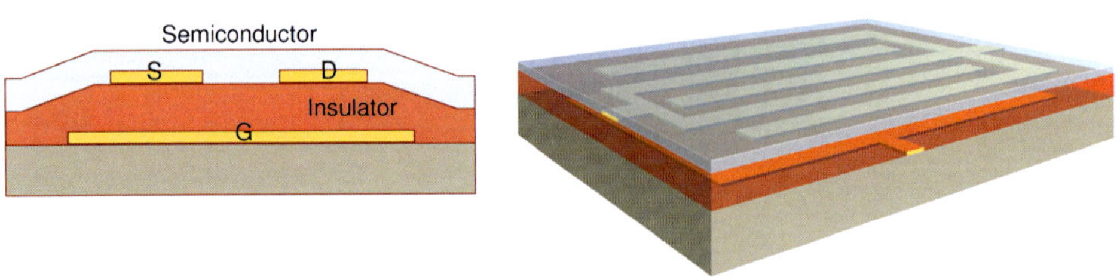

Fig. 13.1. Schematic structure of the organic thin-film transistor. Starting from the bottom, the transistor consists of the following layers: substrate (brown), gate electrode G, (yellow), gate insulator layer (red), the source, S, and drain, D, electrodes (yellow), semiconductor (grey).

the semiconductor layer. The basic function of the organic FET is the same as that of conventional metal–insulator–semiconductor field-effect transistors. By applying a voltage to the gate electrode, it is possible to modulate the current between two other terminals, the source and drain electrode. We note that organic FETs typically operate in accumulation, rather than in inversion [17, 18] and the majority carriers are typically holes (p-type semiconductor). Here, we will discuss p-type organic semiconductors only. If the source and drain electrodes are held at ground and a negative voltage is applied to the gate, positive charge is induced near the insulator–semiconductor interface, and a conductive channel of accumulated holes is established between the source and the drain terminals. The transistor is in its 'ON' state. If a positive gate voltage is applied to the gate relative to the source and drain electrodes, the semiconductor becomes depleted of mobile charge carriers, resulting in the condition known as depletion, the 'OFF' state. Instead of the threshold voltage defined in silicon metal–insulator–semiconductor FETs [19] as the onset of strong inversion, we use the flatband voltage to describe the onset of charge carrier accumulation in organic accumulation transistors [18]. The source voltage is always fixed to ground, $V_S = 0$ V. This implies that the drain voltage is always negative, $V_D \leq 0$ V.

13.2.2
Current–Voltage Characteristics

For the description of the current-voltage characteristics it is instructive to look at the current contributions in the different operating regimes. In the accumulation mode, the charge carriers induced by the field-effect predominantly carry the current. At the flatband voltage V_{FB}, only charge carriers in the bulk of the semiconductor carry the current. This is shown schematically in Fig. 13.2. With increasingly positive V_G an increasingly thick depletion layer is formed in the semiconductor until finally the semiconductor is depleted of mobile charges over its whole thickness. The current below V_{FB} is therefore determined by both the bulk conductivity of the semiconductor and the ease at which the depletion layer thickness increases with positive gate voltage. The latter in turn is related to the dopant density in the semiconductor [20].

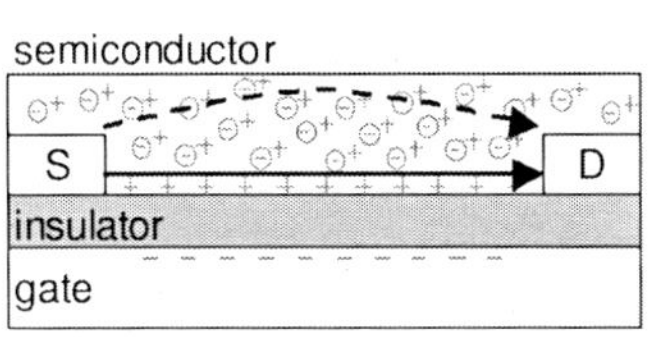

Fig. 13.2. Schematic of an organic accumulation-mode FET, showing a p-doped semiconductor: + indicates a positive charge in the semiconductor, ⊖ indicates a negatively charged counter-ion. In accumulation the transistor current is composed of the field-effect current (solid arrow) and a bulk current (dashed arrow).

The field effect mobility (μ_{FE}) has been conventionally defined as a term relating the thin-film conductance along the channel to the field-induced charge per unit area, or:

$$\mu_{FE} = \frac{g}{Q_{ind}} \tag{1}$$

where g is the channel conductance (Ω^{-1}), and Q_{ind} is the induced charge (C cm^{-2}).

Using Eq. (1) as the basic definition for field-effect mobility, it was found that in organic FETs, as in a-Si FETs, the mobility depends on the induced charge [1]. The Q_{ind} dependence is experimentally observed as a gate-bias dependent mobility, because:

$$Q_{ind} = C_i(V_G - V_{FB}) \tag{2}$$

where C_i the insulator capacitance per unit area. Two models have been proposed to explain the field-dependent mobility in organic FETs. The trapping–release model, as proposed originally for a-Si transistors [21], was found to give a reasonable description of the characteristics of organic transistors [22, 23]. In this model the assumption is made that most of the charge carriers are trapped in localized states. Then the amount of (temporarily) released charge carriers to an extended-state transport level (the valence band for classical p-type semiconductors) depends on the energy level of the localized states, the temperature, and the gate voltage. However, while extended-state transport may occur in highly ordered materials, we do not expect it to play a role in disordered organic films, where the low-mobility charge carriers are strongly localized. An alternative model was proposed by Vissenberg and Matters [24]. In this model, the motion of charge carriers in organic semiconductors is typically described by hopping transport, which is a phonon-assisted tunneling mechanism from site to site. Combining the hopping model with the Gaussian disorder representation, as proposed by Bässler [25], has resulted in a useful description of organic transistor operation [24, 26, 27]. Because of the hopping of carriers between localized states, which are distributed in energy, the experimentally determined charge carrier mobility has a thermally activated behavior and depends on the charge carrier density, as is illustrated in Fig. 13.3 for a pentacene transistor. Also shown in Fig. 13.3 are the characteristics of a solution-processed pentacene FET [28] in the accumulation regime as a function of temperature. The solid lines depict the results of the model. The model gives a very reasonable description of the organic FET operation in accumulation.

13.2.3
Capacitance–Voltage Characteristics

The capacitance, C, of the transistor can be changed by depleting or accumulating charge in the semiconductor at the interface with the insulator. In depletion, the

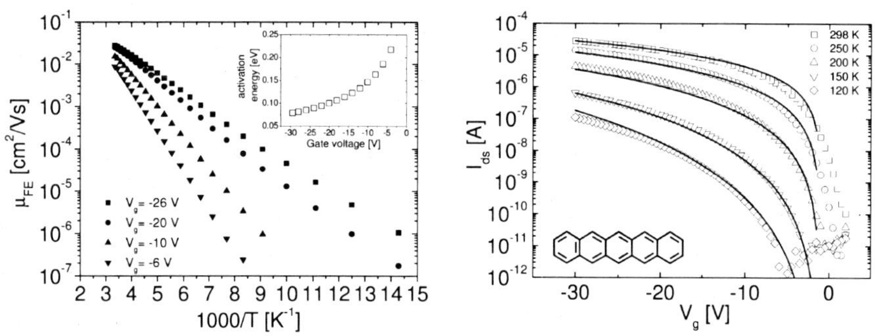

Fig. 13.3. Temperature dependence of the field-effect mobility (left) and the I_{ds}-V_g characteristics (right) of a pentacene FET. The inset in the right panel shows the dependence of the activation energy on the gate voltage. The solid lines are modelled as outlined in [24]. The transistor width is 2 cm and length is 10 μm, drain voltage −2 V. The samples are measured under vacuum (10^{-7} mbar).

capacitance corresponds to the geometrical capacitance of the electrodes, i.e. the spatial overlap between gate with source and drain multiplied by the capacitance of the insulating layer per unit area. When the device goes from depletion into accumulation, C increases as a result of the formation of an accumulation channel. At low frequency the charges residing in the channel can respond to changes in applied voltage. At sufficiently high frequencies this is not longer the case, and the increase in C becomes smaller. The frequency behavior can therefore provide information on the time response, and hence charge transporting properties, of the channel charges.

Using impedance spectroscopy we measured transistors with solution processed pentacene as the semiconductor. Fabrication details are discussed in Section 13.3.1. We have made use of a ring-type transistor, in which the source electrode forms a closed ring around the transistor channel and the drain electrode, at which the current is monitored. Using this geometry, the measurements are insensitive to parasitic currents that may flow outside the transistor area [29].

We calculate the total device capacitance from the modulus of the impedance, Z, and its phase angle, Θ, using: $C = -\sin\Theta/\omega|Z|$, where $\omega = 2\pi \times f_{mod}$, where f_{mod} is the a.c. modulation frequency. Typical capacitance versus gate bias (C–V) curves are given in Fig. 13.4 for different modulation frequencies. In depletion, i.e. for positive V_G, the capacitance, C_{depl}, of ~1.4 pF is equal to the geometrical capacitances of the source-gate and drain-gate electrodes. The slight frequency dependency mainly reflects the dispersion in the dielectric constant of the gate insulator. In accumulation, the capacitance depends on both V_G and frequency. At low frequency in accumulation, the capacitance is 2.6 pF. The abrupt change in capacitance at $V_G \approx 0$ V is close to the onset voltage of accumulation as determined by I–V measurements on this device. The difference between the capacitance in accumulation and depletion corresponds to the geometrical channel capacitance, C_{ch}, and thus equals $L \times W/C_i$, where C_i is the insulator capacitance per unit area. At

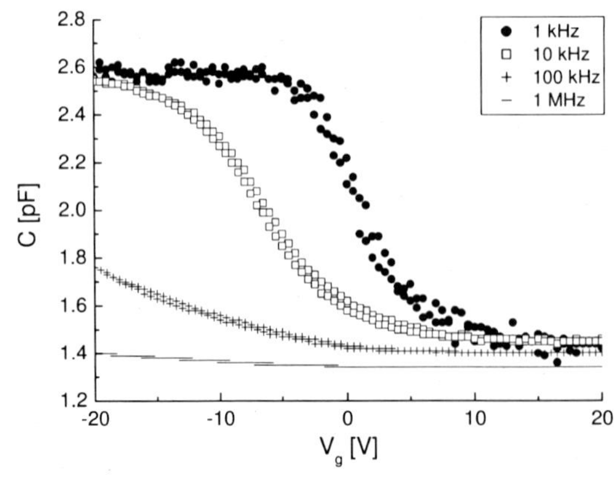

Fig. 13.4. Capacitance versus gate voltage of a pentacene FET for several frequencies. ($L = 20\ \mu m$, $W = 500\ \mu m$).

higher frequencies the measured capacitance decreases (but is equal to, or higher than, C_{depl}) and depends on V_G.

The frequency dependency is easily explained if we realize that charges only contribute to the measured value of C if they can respond in the time or frequency of the measurement. Hence, the observed CV characteristics in accumulation reflect a change in response time. For low frequencies and negative V_G the mobile channel charge has sufficient time to respond to changes in applied voltage. At higher frequencies this is no longer the case, and C decreases. Because the mobility of the charges depends on V_G in the semiconductor layer, C depends on V_G.

13.3
Fabrication and Characterization of Integrated Circuits

The technology to fabricate integrated circuits requires combination of active transistors with suitable interconnect at a pitch dense enough for the application envisioned, on large area and with sufficient yield. It is crucial however to demonstrate the manufacturability of circuits with a complexity around 1000 transistors, as all applications of organic electronics will need this kind of complexity. Good results have been booked in this area with digital circuits comprising 650 [30], 864 [31], and 1888 [35] transistors, and the work is definitely ongoing.

With an increase in integration scale it is becoming increasingly necessary to develop tools for the realistic simulation and optimization of the circuitry. An enabling factor is to improve the understanding of device and circuit operation and the awareness of the link between the two. This can be achieved via in-depth transistor modeling, circuit simulation and clever qualitative analysis, as will be discussed after we have introduced our technology.

13.3.1 Fabrication

A 25 μm thin freestanding foil, as used as a substrate material for our transistors, is too flexible to be processed directly. So, the first process step is to laminate the foil onto a rigid support. This enables the use of standard off-the-shelf patterning and deposition equipment, like spincoaters, photolithography tools etc. After processing the support is removed without influencing the properties of the transistors and can be reused. The demand on the registration of the different layers with respect to each other determines the choice of substrate material. Polyethylene naphthalate, PEN, has been found very suitable. The material is very well resistant against the chemicals used, and can be obtained in a highly crystalline form with a melting point of 265 °C. This high crystallinity makes that the material can withstand the highest process temperature of 170 °C whilst its glass transition temperature is only 121 °C [32]. A typical registration better than 2.5 μm over a 150 mm wafer is achieved and integration of transistors over large areas with a relatively low overlap of 5 μm is possible.

A bottom gate transistor is made. The transistors are combined into integrated circuits using a three or four mask process. The electrodes and interconnect lines are made by patternwise exposure of either highly conducting polyaniline (PANI) (9,10) or poly(3,4-ethylenedioxythiophene) (PEDOT) [12, 33] thin films. The deep-UV positive PANI process is based on photochemical reduction, while the water-borne I-line PEDOT lithography is a negative process based on photo-crosslinking. The minimum feature size in both cases is 2.5 μm lines and 1 μm spacings. The sheet resistance amounts to 1–2 kOhm square^{-1}. For some applications this resistance is too high [34]. Therefore, we also made devices with gold electrodes in which the gold was patterned either by standard photolithography [34] or by micro-contact printing [36]. The gate dielectric is a photoresist in which contact holes are defined. Upon spincoating the top electrode layer, the holes are filled and form the vias. The contact resistance is negligible, far less than 1 kΩ. Devices are finished upon spincoating the polymeric semiconductor. Stack integrity is maintained in the whole process by crosslinking previously deposited films and by the use of proper solvents. For all technology options mentioned above we could scale up the processing technology to 150-mm laminated foils. A detail of a polymeric IC is shown in Fig. 13.5, and a picture of a finished 150-mm wafer is shown in Fig. 13.6.

The bottom gate approach makes it possible to integrate discrete devices into larger functional logic without additional processing steps. Because the semiconductor is applied in the last step a large freedom in the choice of semiconductors is obtained [10, 37]. Here, we present only results obtained with solution-processed pentacene: a 100 nm thick precursor pentacene film is spincoated, and subsequently converted to pentacene [28]. By patterning the semiconductor layer using a subtractive photolithography process [38] parasitic lateral leakage currents are minimized, and as a result the off-current of the transistors is decreased by at least

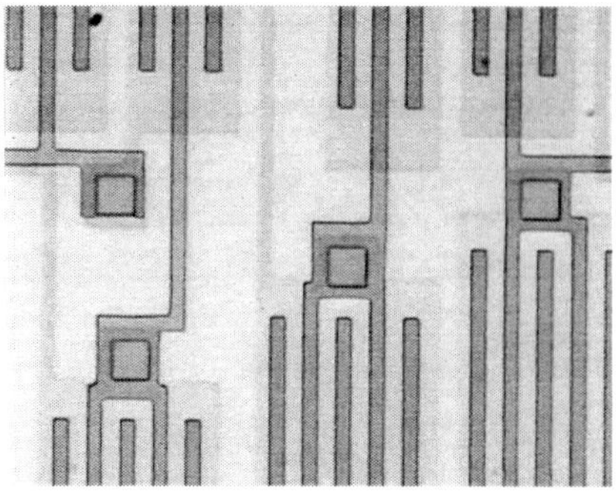

Fig. 13.5. Top-view differential interference contrast microphotograph of a bottom gate polyaniline-insulator-polyaniline structure. The channel length is 2.5 μm. The squares are the vias of 5×5 μm^2.

two orders of magnitude without sacrificing mobility and without large shifts in threshold voltage.

Typical transfer and output characteristics of a solution-processed pentacene FET with channel length $L = 20$ μm and $W = 1000$ μm are shown in Fig. 13.7. The characteristics were obtained under ambient conditions. The field effect mobility of the FET is 0.01 $cm^2\ V^{-1}\ s^{-1}$ at $V_G = -20$ V. On less typical wafers mobilities as high as 0.25 $cm^2\ V^{-1}\ s^{-1}$ have been observed. The drain current modulation $I_D(V_D = -1\ V, V_G = -20\ V)/I_D(V_D = -1\ V, V_G = 10\ V)$ of 10^5 is routinely obtained. By improving the uniformity and by reducing the parameter spread the integration level could be increased to about 10^3 transistors.

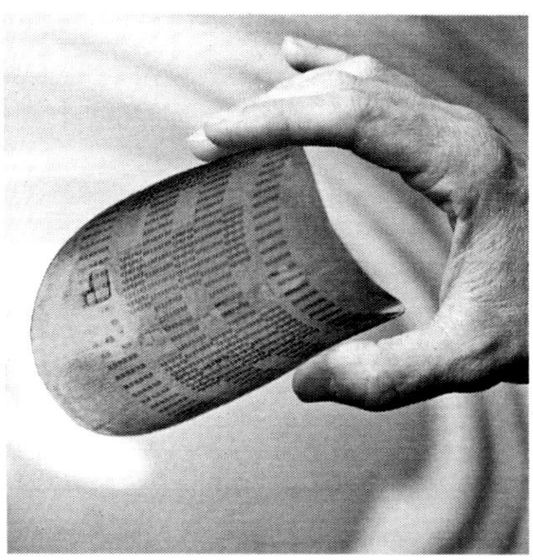

Fig. 13.6. Fully processed 150-mm wafer foil containing all-polymer transistors and integrated circuits.

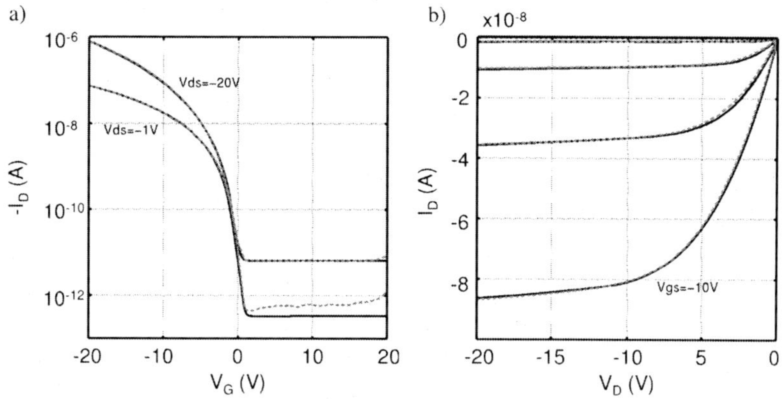

Fig. 13.7. Transfer characteristics (a) and output characteristics (b) of a typical pentacene FET with $L = 20$ μm and $W = 1000$ μm. Dotted lines are the measured curves, full lines are the modelled curves.

An important point of concern of organic transistors is their lifetime, both on the shelf and under operation. In a nitrogen atmosphere the pentacene devices do not degrade over a period of months, hysteresis is negligible, and the shifts of the threshold voltage during operation are minimal. Without encapsulation the characteristics of these devices show a decrease in mobility of typically 30% after 3 months when stored in a climate chamber at 25 °C and a relative humidity of 50%. With PEDOT and PANI electrodes, there is a significant decrease of the mobility in time, and during operation considerable shifts of the threshold voltage occur in the direction opposite to the applied bias [39]. These threshold shifts are significantly smaller (typically less than 1.5 V after biasing the device for one hour at a gate bias of +30 or −30 V) when gold electrodes are used, in line with the hypothesis that mobile ions in the gate dielectric, stemming from the electrodes, are co-responsible for the threshold shifts. This is a major improvement over previously reported values of threshold voltage shift under similar stress conditions [39–41]. Lifetime can be improved further by encapsulation, similar to vapor-deposited pentacene [42].

13.3.2 Modeling

In Fig. 13.7 the experimental drain currents are compared with modeled characteristics. Below, the simple nine-term OFET model that is used here will be outlined.

Three regimes of operation are distinguished:

1. the accumulation regime;
2. the sub-threshold regime; and
3. the depletion regime.

For gate voltages more negative than a certain onset voltage, referred to as the threshold voltage, an accumulation channel of holes is formed. For gate voltages slightly positive to the threshold voltage, the drain current changes exponentially with the gate-source voltage. This is the subthreshold regime. For more positive gate voltages the channel is depleted of mobile holes and only a small off-current is observed.

Charge transport in the accumulation channel is described by the percolation model [24] based on thermally activated tunneling of holes between localized states in an exponential density of states, described in Section 13.2.2. In the accumulation regime this Variable Range Hopping (VRH) model yields a gate-voltage dependent field-effect mobility of the form:

$$\mu_{\mathrm{FE}} = \mu_0[[-V_{\mathrm{G}} + V_{\mathrm{T}}]]^{\gamma} \tag{3}$$

where V_{T} is the threshold voltage, μ_0 is the mobility prefactor, γ is the mobility exponent and $[[x]] = 0.5x + 0.5|x|$. In the VRH model γ can be written as $2(T_0 - T)/T$, where T_0 is a measure of the width of the exponential density of states and T the absolute temperature.

Using the gradual channel approximation – the lateral electric field does not significantly influence the perpendicular electric field – the drain current in the accumulation regime reads:

$$I_{\mathrm{D}}^{\mathrm{acc},1} = -\frac{\mu_0 W C_{\mathrm{i}}}{L(2+\gamma)}[[[-V_{\mathrm{G}} + V_{\mathrm{T}}]]^{2+\gamma} - [[-V_{\mathrm{G}} + V_{\mathrm{T}} + V_{\mathrm{D}}]]^{2+\gamma}] \tag{4}$$

Although the gradual channel approximation implies that this expression is only valid in the linear regime ($|V_{\mathrm{D}}| \ll |V_{\mathrm{G}} - V_{\mathrm{T}}|$), it is also found to be a reasonable description of the drain current close to saturation ($V_{\mathrm{D}} = V_{\mathrm{G}} - V_{\mathrm{T}}$).

To obtain a correct description of the drain current close to saturation a linear correction parameter κ is introduced:

$$I_{\mathrm{D}}^{\mathrm{acc},2} = I_{\mathrm{D}}^{\mathrm{acc},1} \cdot \min(1 + \kappa[[V_{\mathrm{D}}]], 1 + \kappa[[-V_{\mathrm{G}} + V_{\mathrm{T}}]]) \tag{5}$$

This expression is independent of the drain voltage in the saturation regime ($V_{\mathrm{D}} < V_{\mathrm{G}} - V_{\mathrm{T}}$), reflecting the fact that for these drain voltages a depletion region will develop around the drain that causes the drain current to saturate. However, this is only true in first approximation. Channel shortening of the saturated transistor gives rise to a gently increasing drain current in the saturation regime. The depletion region around the drain extends an amount ΔL into the channel, resulting in an effective channel length of $L - \Delta L$. In the depletion region around the drain the electric field is high and charge transport is space charge limited (SCL) [43, 44]. In the model we use a simple expression for the SCL current:

$$I_{\mathrm{D}}^{\mathrm{SCL}} \propto W\frac{[[V_{\mathrm{G}} - V_{\mathrm{T}} - V_{\mathrm{D}}]]^2}{\Delta L^3} \tag{6}$$

An accurate description of the SCL current requires two-dimensional analysis. Current continuity requires that the SCL current equals the current in the shortened saturated transistor, i.e. the saturation current of Eq. (3) multiplied with the factor $L/(L - \Delta L)$. Solving the equations for ΔL, the drain current reads:

$$I_\mathrm{D}^{\mathrm{acc},3} = I_\mathrm{D}^{\mathrm{acc},2} + \lambda \frac{W^{1/3}}{L} ([[V_\mathrm{G} - V_\mathrm{T} - V_\mathrm{D}]] \cdot |I_\mathrm{D}^{\mathrm{acc},2}|_{V_\mathrm{D}=V_\mathrm{G}-V_\mathrm{T}}|)^{2/3} \tag{7}$$

where λ is the channel-shortening term. This equation describes the on-current in the accumulation regime well.

The sub-threshold behavior for gate voltages slightly positive to V_T, is described using the expression:

$$I_\mathrm{D}^{\mathrm{sub}} = I_0 |V_\mathrm{D}|^{\sigma} \exp\left[\frac{\ln 10}{S}[[V_\mathrm{G} - V_\mathrm{T}]]\right] \tag{8}$$

where S is the inverse subthreshold slope and I_0 and σ are fitting terms. This equation is inspired by Si MOS modeling [21], but lacks physical justification.

Finally, the off-current in the depletion regime is described as a simple ohmic current:

$$I_\mathrm{D}^{\mathrm{off}} = \frac{V_\mathrm{D}}{R} \tag{9}$$

where R is the off-resistance.

The total drain-source current then reads:

$$I_\mathrm{D}^{\mathrm{total}} = I_\mathrm{D}^{\mathrm{acc},3} + I_\mathrm{D}^{\mathrm{sub}} + I_\mathrm{D}^{\mathrm{off}} \tag{10}$$

The model contains a total of nine terms $-\mu_0$, γ, V_T, κ, λ, S, σ, I_0 and R. These terms are either extracted from the characteristics or some of them are fitted together to the characteristics in the appropriate regime.

In Fig. 13.7 a FET with channel length $L = 20$ µm is described accurately using $\gamma = 1.03$, $V_\mathrm{T} = 0.02$ V, $S = -0.44$ V decade^{-1}. In Fig. 13.8 the experimental and modeled transfer characteristic for $V_\mathrm{D} = -1$ V is plotted together with the corresponding characteristics of a series of FETs on the same process evaluation module with progressively smaller channel length. This scaling series can be modeled with the same γ and V_T as for the $L = 20$ µm device. The variation in μ_0 is within 10% of the average value. The saturation correction parameter κ is inversely proportional to L. The channel shortening parameter λ should be constant, but is slightly increasing with $1/L$. The subthreshold slope becomes less steep for smaller channel lengths. The parameters σ and I_0 behave as true fitting parameters. The off-resistance R, finally, is constant, which seems to imply it is not an intrinsic parameter of the FET channel.

The nine-term model description of the current equations is very useful in cir-

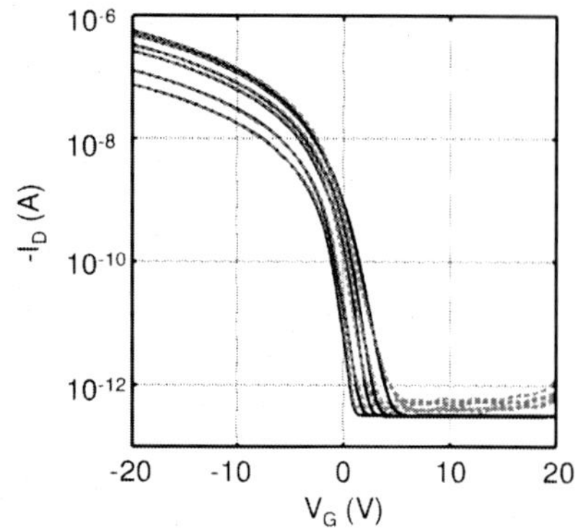

Fig. 13.8. Transfer characteristics ($V_D = -1$ V) of a series of neighbouring pentacene FETs with different channel lengths, $L = 20, 10, 5, 4, 3, 2.5$ μm and $W = 1000$ μm. Dotted lines are the measured curves, full lines are the modelled curves.

cuit simulations for circuit design. We will exemplify this using the threshold voltage as example. Figure 13.9 shows that on a typical wafer the extracted threshold voltage is normally distributed in a small range of less than 2 V centered around 0 V. As will become clear in the next sections, both spread and mean value impacts the performance of certain types of logic gates.

13.3.3
Analysis of Inverters

The basic building block of integrated circuits is the inverter gate. In this section the basic performance of two widely used inverter types will be analyzed, focusing on the link between device characteristics and circuit performance.

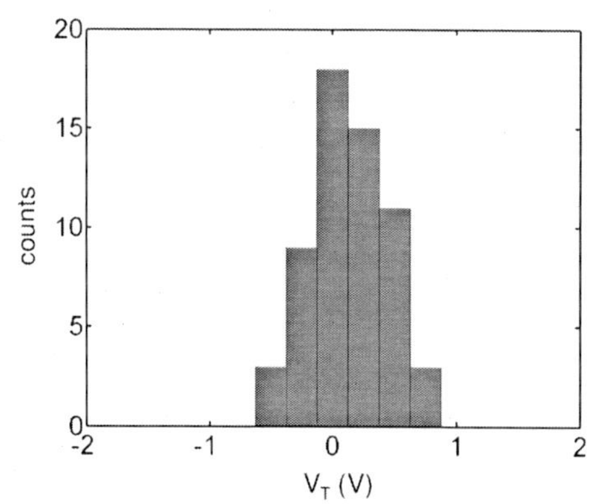

Fig. 13.9. Distribution of the threshold voltage V_T for 60 pentacene FETs with $L = 20$ μm spread uniformly over a 150-mm wafer (one FET per process evaluation module).

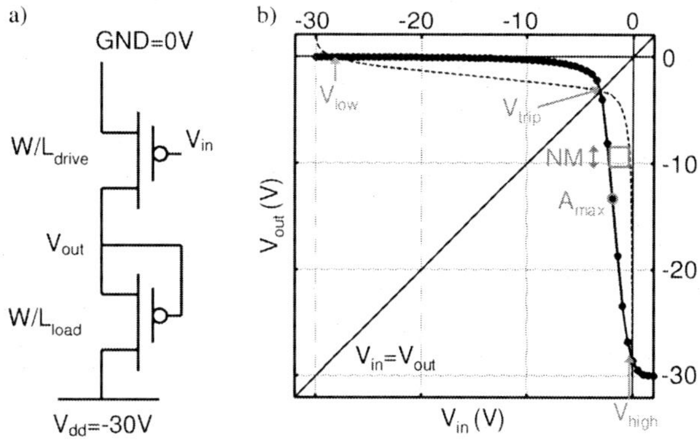

Fig. 13.10. Schematic of a "$V_{GS} = 0$" inverter (a) and transfer characteristics of a typical "$V_{GS} = 0$" inverter (b). The trip voltage ($V_{trip} = -3.2$ V), the characteristic switching input and output voltages ($V_{high} = -0.1$ V and $V_{low} = -27.9$ V), the maximum gain ($A_{max} = 10.7$) and the noise margin (NM = 1.9 V) are also shown. The noise margin is defined as the side of the largest square that can be inscribed between the transfer characteristics and the characteristics obtained by exchanging input and output axes (thin dashed line).

The first logic design we consider will be called here "$V_{GS} = 0$" logic, as the load (or pull-down) transistor has the source connected to the gate (Fig. 13.10A). The basic operation of this logic gate can be outlined if we keep in mind that the pull-down transistor is chosen to be wider than the driver transistor. When the input V_{in} is high (V_{in} = GND), the V_G of the driver and of the load transistor are both zero. The load transistor, however, is wider and can supply a higher current than the driver. As the current in the two transistors must remain the same, the V_D of the load transistor will have to decrease, and so the output node will be pulled down towards V_{dd}, obtaining a low output. When the input is low ($V_{in} = V_{dd}$), the driver transistor is strongly switched-on and can easily pull up the output node against the small current provided by the load. When the inverter displays voltage amplification, i.e. the gain $A = |dV_{out}/dV_{in}|$ is larger than unity, this inverter can be used to drive subsequent inverters without losing logic integrity: the output voltage will always be close to either GND or V_{dd}. Cascaded inverters with no voltage amplification yield an ambiguous logic state output somewhere between V_{dd} and GND.

Some important terms in the static input–output characteristic of an inverter are the separation of logic high (V_{high}) and low (V_{low}) level, the maximum gain (A_{max}), the position of the trip point (V_{trip}) and the noise margin (NM). Figure 13.10B shows the transfer characteristics of a typical "$V_{GS} = 0$" inverter, in which the meaning of all these parameters is illustrated. V_{high} is always very close to GND. The value of V_{low} is determined by the voltage divider formed by the resistance of the driver transistor with $V_{GS} = 0$ and the resistance of the load transistor with $V_{GS} = 0$. Thus, the separation of logic levels depends mainly on the ratio between the width of the load and of the driver transistor, if they have the same length. This ratio (rw) should be chosen large enough ($rw \approx 10$) to ensure that the distance between the logic high and low level is close to the bias voltage V_{dd}. In an ideal inverter the trip point V_{trip}, where the input and the output voltage are the same, should coincide with the point of maximum gain and be at the center of the supply range. In Fig. 13.10B V_{trip} is reached already for small negative input voltages. The asymmetric position of the trip point strongly reduces the noise margin of the "$V_{GS} = 0$" logic, i.e. to a small fraction of V_{dd}. This asymmetry is due to the fact that the driver transistor (with small positive threshold) rapidly switches on as the input voltage goes low, and easily pulls up the weak load. One would then prefer for the driver transistor a negative threshold. On the other hand a positive threshold is needed for the load, if we want to operate it at zero V_G.

A key performance indicator of an inverter is the switching speed: the sum of the pull-down delay and the pull-up delay. In "$V_{GS} = 0$" logic the switching speed is inherently low. The switching speed is determined by pull-down delay as the low pull-down current provided by the load is much smaller than the transient pull-up current provided by the driver.

One way to increase the operating frequency is to use the so-called diode load logic [45] (Fig. 13.11A). In this case, when the input is low ($V_{in} = V_{dd}$), the V_G of both the driver and the load transistors are initially equal to V_{dd}. If the driver transistor has to be able to pull-up the output node, the driver must be wider than the load transistor by an appropriate factor *rw*. When, on the other hand, the input is high (V_{in} = GND), the load transistor will be at first heavily on and will easily pull-down the output node against the driver, which is almost off. This type of logic preferably requires normally off transistors whereas p-type transistors are typically normally on. Figure 13.12A illustrates transfer characteristics of diode-connected load inverters constructed from inverters that are slightly normally on ($V_T \approx 0.5$ V). The inverters have again an asymmetric input–output characteristic because the driver transistor (with positive threshold and larger than the load) switches on very rapidly when the input goes low. Consequently the noise margin is very low.

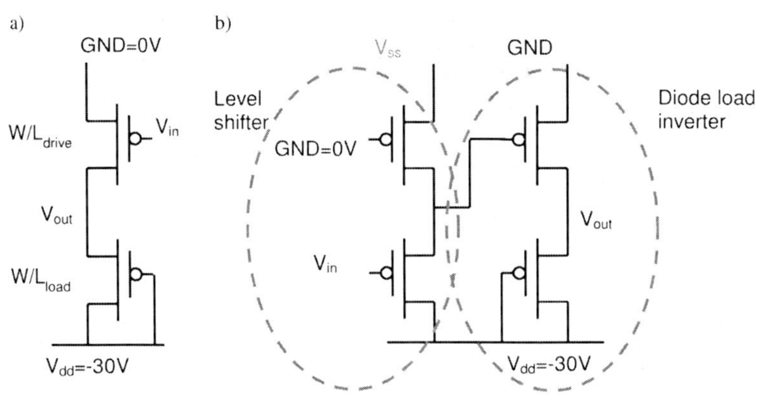

Fig. 13.11. Schematic of a diode load inverter (a) and of a diode load inverter with level shifter (b).

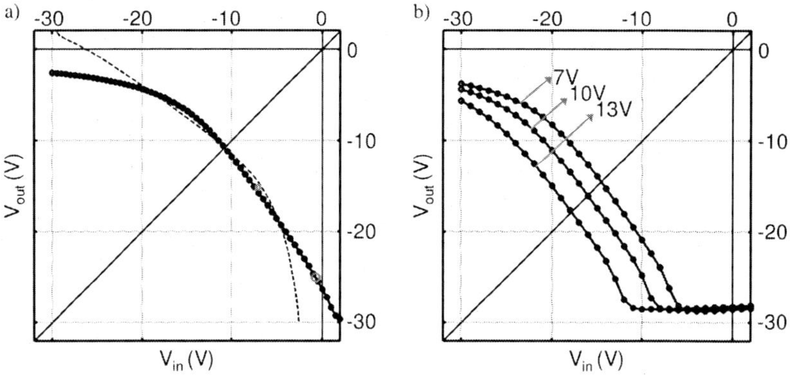

Fig. 13.12. Transfer characteristics of a typical diode load inverter (a) [$V_{trip} = -10.8$ V, $A_{max} = 2.1$, NM $= 0.5$ V] and of a typical diode load inverter with level shifter for different values of the shifting voltage V_{ss} (b) [7 V: $V_{trip} = -14.5$ V, $A_{max} = 2.3$, NM $= 1.6$ V; 10 V: $V_{trip} = -16.0$ V, $A_{max} = 2.5$, NM $= 2.4$ V; 13 V: $V_{trip} = -17.9$ V, $A_{max} = 2.4$, NM $= 2.6$ V].

The switching speed is increased considerably. As the pull-down transistor is diode-connected, a transistor that is fully on provides at the beginning of the transient the pull-down current. The pull-up delay in diode-connected logic is also faster than in "$V_{GS} = 0$" logic.

The critical dependence of noise margin on the position of the switch-on voltage illustrates the Achilles heel of both types of logic inverter. We note that in fact a threshold voltage close to 0 V is expected for transistors in which the semiconductor is not highly doped, has nearly-ohmic source-drain contacts and negligible interface states [19]. This makes this dilemma fundamental [46]. Moreover, when the inverter is used with a high duty cycle, the driver and load transistor are stressed in a different way. In the "$V_{GS} = 0$" inverter, for instance, the driver transistor is stressed whereas the load transistor, which is always off, is stressed to a smaller extent. The negative gate stress will result in a more positive threshold voltage for the driver transistor [39], erasing the already small noise margin of the inverter altogether.

The inverter characteristics can be made more symmetrical by using a so-called level shifter [4, 35] (Fig. 13.11B). The additional stage of two identical FETs has the same effect as adjusting the threshold voltage of the drive transistor to negative values. By setting the shifting voltage V_{ss} to a value above GND, ideally the inverter characteristics shift by an amount V_{ss} – GND. The noise margin increases significantly with increasing V_{ss}. Experimental results for the diode-logic inverter with level shifter are shown in Fig. 13.12B. This inverter is more stable than the "$V_{GS} = 0$" inverter: the driver and the load transistor are stressed equally during operation at a high duty cycle and V_{ss} can compensate for threshold voltage shifts. The drawback of using diode load inverters with level shifter is that it doubles the transistor count (potential yield problem, see below) and it requires a second voltage rail.

13.3.4
Analysis of Integrated Circuits

The switching speed of organic integrated circuits can be estimated from the performance of the individual transistors and is roughly proportional to μ_{FE}/L^2 [19] where L is the channel length of the transistor, and μ_{FE} is the field-effect mobility. To reach higher switching speeds, the search for higher mobility materials is therefore important, but it is also of great interest to downsize the transistor geometries. This is illustrated in Fig. 13.13 in which the clock frequency of seven-stage ring oscillators is plotted vs. channel length. Since contact resistances are negligible in these devices, the switching speed is roughly proportional with μ/L^2 for "$V_{GS} = 0$ V" logic. By using diode-connected logic the frequency increases a factor of ten, culminating in a clock frequency of 13 kHz for an $L = 0.75$ μm diode-connected logic ring oscillator.

To test our technology and study the manufacturability of circuits of growing complexity, a set of 4 different sequential code generators were designed in the same mask set. The length of the code as well as the transistor count increases from one circuit to the other. The smallest circuit is a finite state machine that evolves through a fixed sequence of states. The other circuits are based on a sequencer that serializes bits contained in a write once memory. Fixed start and stop sequences are interleaved with the programmable pattern. Combinatorial logic based on "$V_{GS} = 0$" logic inverters and so-called NAND gates have been used to build the code generators. The largest circuit produces a 64-bit code, with 48 programmable bits. It includes about 700 transistors and is laid-out with 2.5 μm design rules. Its area is 3.2×2 mm^2. The output sequence measured on one of the 48 bit code generators is reported in Fig. 13.14.

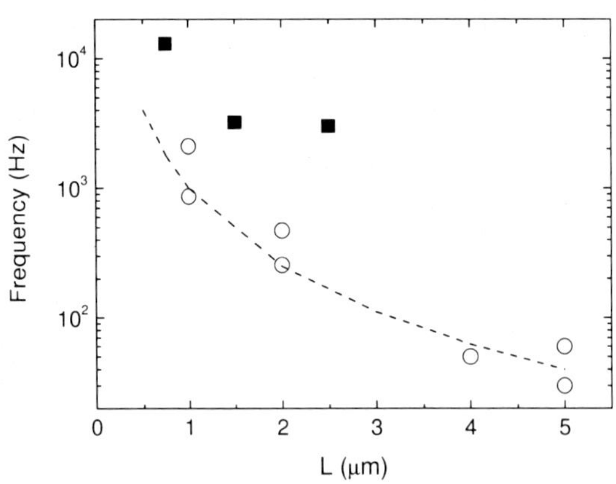

Fig. 13.13. Switching frequency of seven-stage ring oscillators as a function of channel length, L. Open circles: "$V_{GS} = 0$ V" logic. Closed squares: diode-connected logic. The dashed line shows the proportionality to μ/L^2.

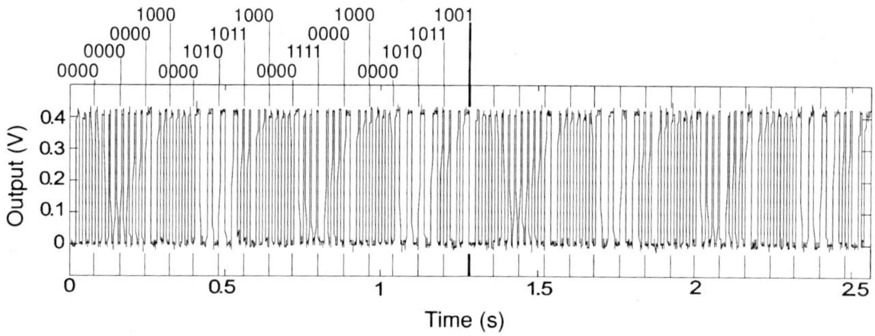

Fig. 13.14. The output pattern of a 48-bit code generator. An external clock operating at 50 Hz is applied. The graph shows two full periods of 64 bits, 48 information bits plus 16 start bits. The code is indicated above.

Programming of the code occurred either through mask design [47] or after manufacturing by mechanically punching through appropriate intact capacitors to make them conducting [9]. Both concepts are undesirable from an application point of view. Therefore we developed an electrically programmable read-only element that could be integrated in a code generator without additional processing steps [48]. The operation of the read-only memory element resembles that of interruption of an electrical safety fuse, since it is based on irreversible conductivity breakdown of circuit interconnect lines made of polyaniline (PANI) doped with camphorsulfonic acid.

Yield measurements on the different code generators typically show a strong decrease with increasing complexity [30]. It is important to find the root cause of this. Yield loss of digital circuits can have a number of possible causes: the process defectivity is too high; the transistor performance is only marginally adequate to guarantee correct circuit functionality; or the circuit design does not provide enough margin to accommodate for device parameter variation.

On the same wafer a number of test structures for the different technology steps, discrete transistors as well as simple logic gates are included. This makes it is possible to correlate circuit performance with basic information on the quality of the technology and on the performance of the devices. Analysis of discrete transistors and logic gates showed that the noise margin of our circuits is limited by the very asymmetric position of the trip point, and that the spread in onset voltage, possibly in combination with gate-field induced shifts in the onset voltage during operation, can partly explain the decreasing yield of the circuits with more than 100 transistors. On the other hand, recent measurements show a correlation between yield loss and total transistor area, suggesting that hard faults also contribute to yield loss [49]. A way to improve yield would then be to use a logic style that allows a significant reduction in terms of transistor count per function. This can be realized using a suitable kind of dynamic logic. We applied this concept to build large dynamic shift registers, containing up to 4000 transistors. These developments are addressed in Chapter 14 of this book.

Acknowledgments

The authors would like to thank the other members of the Polymer Vision team for their help in realizing the results described. DuPont Teijin Films is thanked for the supply of backplane foils.

References

1 C.D. Dimitrakopoulos, P.R.L. Malenfant, *Adv. Mater.* **2002**, *14*, 99–117.
2 D. Voss, *Nature* **2000**, *407*, 442–444.
3 J.A. Rogers, Z. Bao, K. Baldwin, A. Dodabalapur, B. Crone, H.E. Katz, V. Kuck, K. Amundson, P. Drzaic, *Proc. Of National Academy of Sciences* **2001**, *98*, 4835–4840.
4 H. Klauk, M. Halik, U. Zschieschang, G. Schmid, W. Radlik, *IEDM Technical Digest* **2002**, 557–560.
5 C.D. Sheraw, L. Zhou, J.R. Huang, D.J. Gundlach, T.N. Jackson, M.G. Kane, I.G. Hill, M.S. Hammond, J. Campi, B.K. Greening, J. Francl, J. West, *Appl. Phys. Lett.* **2002**, *80*, 1088.
6 M.G. Kane, J. Campi, M.S. Hammond, F.P. Cuomo, B. Greening, C.D. Sheraw, J.A. Nichols, D.J. Gundlach, J.R. Huang, C.C. Kuo, L. Jia, H. Klauk, T.N. Jackson, *IEEE Electron Device Letters* **2000**, *21*, 534–536.
7 J. Krumm, E. Eckert, W.H. Glauert, A. Ullmann, W. Fix, W. Clemens, *IEEE Electron Device Letters* **2004**, *25*, 309–401.
8 F. Garnier, R. Hajlaoui, A. Yassar, P. Srivastava, *Science* **1994**, *265*, 1684–1686.
9 C.J. Drury, C.M.J. Mutsaers, C.M. Hart, M. Matters, D.M. de Leeuw, *Appl. Phys. Lett.* **1998**, *73*, 108–110.
10 G.H. Gelinck, T.C.T. Geuns, D.M. de Leeuw, *Appl. Phys. Lett.* **2000**, *77*, 1489–1491.
11 H. Sirringhaus, T. Kawase, R.H. Friend, T. Shimoda, M. Binbasekaran, W. Wu, E.P. Woo, *Science*, **2000**, *290*, 2123–2126.
12 F.J. Touwslager, N.P. Willard, D.M. de Leeuw, *Appl. Phys. Lett.* **2002**, *81*, 4556–4558.
13 G. Blanchet, J.A. Rogers, *J. Inf. Sci. Technol.*, **2003**, *47*, 296–303.
14 A. Hübler, U. Hahn, W. Beier, N. Lasch, T. Fischer, *IEEE Polytronic Conference* **2002**, 172–176.
15 S.K. Park, Y.H. Kim, J.I. Han, D.G. Moon, W.K. Kim, *IEEE Trans. Elec. Dev.* **2002**, *49*, 2008–2015.
16 M.L. Chabinyc, W.S. Wong, A. Salleo, K.E. Paul, R.A. Street, *Appl. Phys. Lett.* **2002**, *81*, 4260–4262.
17 G. Horowitz, R. Hajlaoui, H. Bouchriha, R. Bourguiga, M. Hajlaoui, *Adv. Mater.* **1998**, *10*, 923–927.
18 E.J. Meijer, C. Tanase, P.W.M. Blom, E. van Veenendaal, B.-H. Huisman, D.M. de Leeuw, T.M. Klapwijk, *Appl. Phys. Lett.* **2002**, *80*, 3838–3840.
19 S.M. Sze, *Physics of Semiconductor Devices*, 2nd edn. Wiley, New York, **1981**.
20 E.J. Meijer, C. Detcheverry, P.J. Baesjou, E. van Veenendaal, D.M. de Leeuw, T.M. Klapwijk. *J. Appl. Phys.* **2003**, *93*, 4831.
21 K. Lee, M. Shur, T.A. Fjeldly, T. Ytterdal, *Semiconductor Device Modeling for VLSI*, Prentice–Hall, Englewood Cliffs, NJ, **1993**, p. 500–504.
22 P.V. Necliudov, M.S. Shur, D.J. Gundlach, T.N. Jackson, *J. Appl. Phys.* **2000**, *88*, 6594–6597.
23 G. Horowitz, R. Hajlaoui, P. Delannoy, *J. Phys. III* **1995**, *5*, 355–371.
24 M.C.J.M. Vissenberg and M. Matters, *Phys. Rev. B* **1998**, *57*, 12964–12967.

25 H. Bässler, *Phys. Stat. Sol. B* **1993**, *175*, 14–55.

26 C. Detcheverry, M. Matters, *Proc. ESSDERC 2000*, **2000**, 328–331.

27 E. Calvetti, L. Colalongo, Zs.M. Kovasc-Vanja, Solid State Electronics **2005**, *49*, 567–577.

28 P.T. Herwig, K. Müllen, *Adv. Mater.* **1999**, *11*, 480–483.

29 H. Klauk, D.J. Gundlach, J.A. Nichols, T.N. Jackson, *IEEE Trans. Elec. Dev.* **1999**, *46*, 1258–1263.

30 E. Cantatore, C.M. Hart, M. Digioia, G.H. Gelinck, T.C.T. Geuns, H.E.A. Huitema, L.R.R. Schrijnemakers, E. van Veenendaal, D.M. de Leeuw, *ISSCC 2003 Dig. Tech. Papers*, **2003**, 382–383.

31 B. Crone, A. Dodabalapur, Y.Y. Lin, R.W. Filas, Z. Bao, A. LaDuca, R. Sarpeshkar, H.E. Katz, W. Li, *Nature* **2000**, *403*, 521–523.

32 W.A. MacDonald, *J. Mater. Chem.* **2004**, *14*, 4.

33 F.J. Touwslager, N.P. Willard and D.M. de Leeuw, *Synthetic Metals* **2003**, *53*, 135–136.

34 H.E.A. Huitema, G.H. Gelinck, K.E. Kuijk, C.M. Hart, E. Cantatore, D.M. de Leeuw *J. SID (Soc. Information Display)* **2002**, *10*, 195–202.

35 G.H. Gelinck, H.E.A. Huitema, E. van Veenendaal, E. Cantatore, L. Schrijnemakers, J.B.P.H. van der Putten, T.C.T. Geuns, M. Beenhakkers, J.B. Giesbers, B.-H. Huisman, E.J. Meijer, E. Mena Benito, F.J. Touwslager, A.W. Marsman, B.J.E. van Rens, D.M. de Leeuw, *Nature Materials* **2004**, *3*, 106–110.

36 J. Schellekens, D. Burdinski, M. Saalmink, M. Beenhakkers, G.H. Gelinck, M.M.J. Decré, *Mat. Res. Soc. Symp. Proc. Vol. EXS-2* **2004** M2.9.

37 N. Stutzmann, in preparation

38 I. Kymissis, S.D. Dimtrakopolous, S. Purushothaman, *J. Vac. Sci. Technol. B* **2002**, *20*, 956–959.

39 S.J. Zilker, C. Detcheverry, E. Cantatore, D.M. de Leeuw, *Appl. Phys. Lett.* **2001**, *79*, 1124–1126.

40 D. Knipp, R.A. Street, A. Völkel, J. Ho, *J. Appl. Phys.* **2003**, *93*, 347–355.

41 A. Salleo, R.A. Street, *J. Appl. Phys.* **2003**, *94*, 471–479.

42 Y. Qiu, Y. Hu, G. Dong, L. Wang, J. Xie, and Y. Ma, *Appl. Phys. Lett.* **2003**, *83*, 1644–1646.

43 M. Shur, M. Hack, J.G. Shaw, *J. Appl. Phys.* **1989**, *66*, 3371–3380.

44 L. Torsi, A. Dodabalapur, H.E. Katz, *J. Appl. Phys.* **1995**, *78*, 1088–1093.

45 W. Fix, A. Ullmann, J. Ficker, D. Brennan, D. Welsh, J. O'Brien, H. Rost, W. Clemens, *Proc. 23rd ESSDERC* **2002**, 527–529.

46 A possible way to solve this technologically is by introducing a monolayer of molecular dipoles at the gate insulator–semiconductor interface, thereby giving the option to control the interface states and hence switch-on voltage. See for instance Kobayashi et al, *Nature Materials*, **2004**, *3*, 317–322.

47 D.M. de Leeuw, G.H. Gelinck, T.C.T. Geuns, E. van Veenendaal, E. Cantatore, B.H. Huisman, *IEDM Technical Digest* **2002**, 293–296.

48 A.W. Marsman, C.M. Hart, G.H. Gelinck, T.C.T. Geuns, D.M. de Leeuw, J. Mater. Research, **2004**, *19*, 2057–2060.

49 E. Cantatore, E. van Veenendaal, manuscript in preparation.

14
Roll-up Active-matrix Displays

H. Edzer A. Huitema, Gerwin H. Gelinck, Erik van Veenendaal, Fred J. Touwslager, and Pieter J. G. van Lieshout

14.1
Introduction

Currently, rigid, glass-based displays are the only flat panel displays on the market. Non-rigid displays are under research and development and will enter the market in small volumes in the coming years. It is important to distinguish the different types of non-rigid display from each other. In general, three categories of non-rigid display can be defined – conformable displays, flexible displays, and roll-up or rollable displays. The proposed definition of each of the three categories is shown in Table 14.1.

Conformable displays can typically replace rigid displays everywhere design freedom is important, for example advertisement, car dashboards, etc. If the conformable display is thin enough it can also be used to replace glass-based displays in the mobile market (e.g. in pdas, cell phones, etc). The mechanical properties of a conformable display are less demanding than those of flexible and rollable displays, because it needs to be bent only once.

The distinction between flexible and rollable displays is the roll radius. In general, the roll radius is a function of the display thickness. According to a crude rule of thumb the roll radius is 50 times the display thickness, although more detailed analysis shows that adhesion between the layers in a flexible or rollable display is also crucial. It is important to distinguish rollable from flexible displays, because rollable displays enable small devices with a large display whereas flexible displays cannot be stored in a device by rolling, because the roll radius is too large. Therefore rollable displays enable a much wider range of applications than flexible displays.

Rollable displays are particularly attractive for mobile applications, because of their low weight, small form factor, and robustness when rolled into the device, as will be discussed in Section 14.7 on roll-up display device concepts. The availability of rollable display devices will greatly stimulate advances in electronic books, newspapers and magazines, and new services offered by (third generation) mobile net-

Organic Electronics, Materials, Manufacturing and Applications. Edited by Hagen Klauk

ISBN: 3-527-31264-1

Tab. 14.1. The three categories of non-rigid flat panel displays and their characteristics.

Category	Bending radius (cm)	Bendability	Advantages over rigid displays
Conformable		Bend once	Can be fit to an uneven or curved surface
Flexible	>2	Bend many times	Thin, bendable, lightweight, rugged
Rollable	<2	Bend many times	Ultra thin, lightweight, rugged, rollable into a small volume when not in use

work operators. These applications currently depend on the more fragile, heavy, and bulky laptops and pdas or on smaller mobile-phone displays.

Rollable displays are currently under development at Polymer Vision (2003). In this chapter state-of-the art in rollable active-matrix displays is discussed. First, the active-matrix technology and design will be presented, followed by the electronic ink film and integration of this film with the rollable active-matrix backplane. This is followed by sections on functional rollable displays and the product concepts that are enabled by these displays. In the last section the status of the research and development toward a rollable system on plastic is presented. This is the long-term goal of organic electronics, in which the first steps have been taken by producing a functional shift register circuit using the same organic electronics technology as is used for producing the rollable displays.

14.1.1 Non-rigid Display Research and Development Overview

Displays with flexible plastic substrates have already been demonstrated, but these are mainly direct-drive and passive-matrix displays based on liquid crystal (Park 2000, Lueder 1992) and polymer light-emitting diodes (Gustafsson 1992), and electrophoretic ink (Liang 2003). Direct drive displays can be used in simple applications, for example signage, price tagging, or smart cards. Passive matrix displays have the additional capability of displaying graphics and text. The limitations of passive matrix displays are maximum resolution of a few hundred rows, the long image update time when used in combination with certain display effects (e.g. electrophoretics), and low image quality, because of crosstalk.

Active-matrix displays differ from the aforementioned displays in that they have a switch incorporated into each pixel (Tsukada 2000). This removes the limitations encountered in passive matrix displays but requires more sophisticated processing equipment to be used. The dominant pixel switch technology is the amorphous silicon thin-film transistor (TFT) on glass (Tsukada 2000), although other technology has also been used (Brody 1996, Kuijk 1991). Currently, rigid active-matrix displays on glass are already widespread although flexible active-matrix displays on plastic substrates are still in the research and development phase. Plastic active-matrix back planes have been made with amorphous silicon (Gleskova 1999, Polach 2000) and with polycrystalline silicon as semiconductor (Young 1997). Complete plastic active-matrix LC displays using amorphous silicon diodes (Young 1997), amorphous silicon TFTs (Okamoto 2002), and a substrate-transfer process with polycrystalline silicon TFTs (Inoue 2002) have also been demonstrated. Alternatively, the amorphous silicon process has been applied to pliable steel foil, resulting in pliable electrophoretic displays (Chen 2003).

In the nineteen-nineties organic electronics, i.e. using an organic material as semiconductor, emerged as a promising thin-film technology for large-area integrated circuits (Voss 2000). The use of organic materials has several important advantages over conventional techniques using mainly inorganic materials, for example amorphous silicon. The low process temperature, less than 200 °C (Gelinck 2000), creates the possibility of using a wide range of plastic substrates instead of glass and the mechanical properties of organic materials are also compatible with plastic substrates. This technology therefore opens the way to plastic displays that are truly rollable.

In recent years field-effect mobilities comparable with those of amorphous silicon have been reported using a variety of organic semiconductors (Dimitrakopoulos 2002). One key goal is the integration of these organic transistors into functional electronic circuits. Small logic building blocks, for example ring oscillators (Brown 1995) and even more complex circuits such as code generators (Drury 1998), have been made.

The first active-matrix display with an organic semiconductor was reported in 2000 (Philips 2000). The display (Huitema 2001), shown in Fig. 14.1, was processed on glass, used a solution-processed semiconductor, contained 4096 pixels and was able to show monochrome images. This was rapidly followed by the first active-matrix displays with organic semiconductors on plastic substrates (Mach 2001, Sheraw 2002, Fujizaki 2003, Gelinck 2004). More recently a flexible QVGA active-matrix display with an organic semiconductor has been reported (Huitema 2003b). With a total of 76800 pixels this is currently the largest flexible display with organic transistors.

14.2
Rollable Active-matrix Backplane Technology

Organic semiconductors can be applied in two fundamentally different ways – by vapor deposition and by solution processing. Vapor deposition of an organic semiconductor usually yields higher thin-film transistor performance. Solution processing is, however, preferred for high-volume, low-cost production. The deposition of soluble organic materials, for example by spin-coating, simplifies the display manufacturing process, especially for large areas. Solution processing also opens a

Fig. 14.1. The first reported active-matrix display with an organic semiconductor. The display contains 64 by 64 pixels and is driven by 4096 polymer TFT, with solution processed PTV as the semiconductor. An image containing 256 gray levels is shown. The display is refreshed at 50 Hz.

plethora of alternative patterning technology, for example printing (Blanchet 2003, Huebler 2002), stamping (Park 2002), selective dewetting (Chabinyc 2002), and inkjet printing (Sirringhaus 2000). Here, the ultimate goal would be a low-cost roll-to-roll solution-based process. To fully exploit the mature production equipment used in the established AMLCD industry, however, it is best to use standard photolithography techniques for patterning of the layers. This enables use of a mature knowledge base for rapid movement up the learning curve toward a fully industrialized process.

Our TFT technology is based on a bottom-gate device architecture (Fig. 14.2). This geometry is comparable with the inverted staggered-electrode structure com-

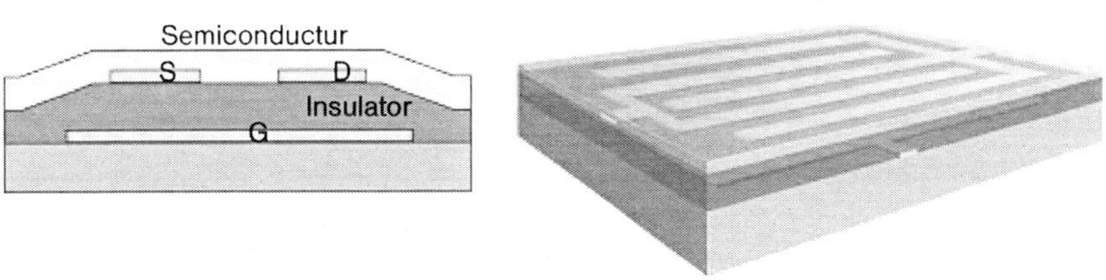

Fig. 14.2. Cross-section of a TFT.

monly used for amorphous silicon TFTs (Tsukada 2000). The thin-film transistors are processed on 150 mm diameter, 25 μm thick poly(ethylene naphthalate) (PEN) films laminated on a removable support. The organic semiconductor and gate dielectric are processed from solution. Because of the specifications of sheet conductivity of the column and row lines in the display, metallic electrodes are used (Huitema 2003a). Gate electrodes and first-level interconnect lines are made by patterning gold using photolithography techniques. The gate dielectric is a 350 nm thick photo-imageable polymer (polyvinylphenol) that is spincoated and subsequently exposed to UV light to define contact holes. Source-drain electrodes, pixel pad, and second level interconnect lines are defined in the second gold layer. On top of this stack, a 100 nm thick precursor pentacene film is spincoated. Synthesis of this precursor and conversion procedure to pentacene are described elsewhere (Herwig 1999). Finally, the semiconductor is patterned using a subtractive photolithography process (Kymissis 2002).

The key features of rollable active-matrix technology are that, first, a thin flexible foil is glued on to a rigid support; the functional layer stack is then processed and, finally, the foil containing the microelectronic devices is delaminated from its support without degradation of the devices. A rigid support can be re-used. This enables the use of standard off-the-shelf patterning and deposition equipment. Typically, registration better than 2.5 μm over a 150 mm wafer for a four-mask process is achieved. Integration of transistors over large areas with a relatively low overlap of 5 μm is possible. This enables the production of transistors with sufficiently small parasitic stray capacitances for active-matrix displays and row driver circuits.

The pixel TFT performance has been evaluated on a QVGA active matrix backplane. In total, 380 devices in a number of random columns have been measured. The transfer and output characteristics of a typical organic pixel TFT are shown in Fig. 14.3. To extract data from these characteristics the model presented in

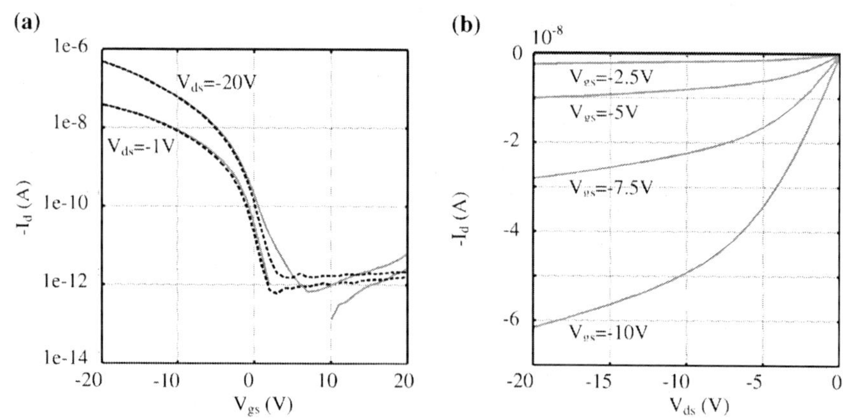

Fig. 14.3. Characteristics of a pixel TFT with typical mobility 0.01 $cm^2\ V^{-1}\ s^{-1}$: (a) transfer characteristics; (b) output characteristics.

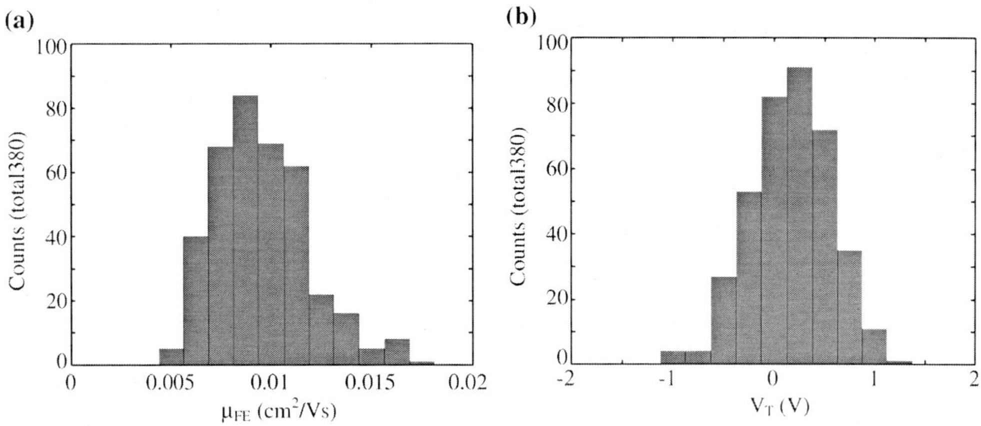

Fig. 14.4. Impression of the uniformity of a QVGA active-matrix backplane based on measurements of 380 pixel TFT. (a) Distribution of the mobility, extracted from the transfer characteristics for $V_{gs} = -20$ V and $V_{ds} = -1$ V. (b) Distribution of the threshold voltage, obtained by fitting Eq. (1) to the transfer characteristics for $V_{ds} = -1$ V.

Eqs. (1) and (2) of Chapter 13 of this book are used. The field-effect mobility is 0.010 $cm^2\ V^{-1}\ s^{-1}$ (linear mobility extracted from the transfer characteristics for $V_{gs} = -20$ V and $V_{ds} = -1$ V), and the leakage current is a few picoamps.

An impression of the uniformity of the active-matrix backplane is given in Fig. 14.4. In Fig. 14.4a the distribution of the mobility of the pixel TFT is shown. The average field-effect mobility is 0.010 $cm^2\ V^{-1}\ s^{-1}$ with a standard deviation of 0.0024 $cm^2\ V^{-1}\ s^{-1}$. The average on-current is -1.0 μA, measured for $V_{gs} = -25$ V and $V_{ds} = -15$ V (typical for charging a pixel to -15 V). The average off-current is -5.7 pA, measured for $V_{gs} = +10$ V and $V_{ds} = -15$ V (typical for leakage of a pixel charged to $+15$ V). Hence our pixel TFTs have an on/off ratio exceeding 10^5. The distribution of the threshold voltage of the pixel TFT is shown in Fig. 14.4b. The threshold voltage of a TFT is obtained by fitting Eq. (1) to the transfer characteristics for $V_{ds} = -1$ V. The threshold voltage has a Gaussian-like distribution around 0.2 V with a small standard deviation of 0.4 V. The average value of the power term γ obtained in the same analysis is 1.1 (standard deviation 0.1).

14.3
Roll-up Active-matrix Backplane Design

The display contains 240 rows and 320 columns. The pixel size is $300 \times 300\ \mu m^2$ resulting in a display diagonal of 4.7 inches. An overview of the display design is shown in Fig. 14.5. All contacts to the display are situated on the left. The row electrodes are arranged horizontally and the column electrodes are arranged vertically. The three small blue rectangles are the contacts to the common electrode.

Fig. 14.5. Overview of the QVGA active-matrix rollable display design.

A field shielded pixel structure is used. The cross-section of the active-matrix stack is shown in Fig. 14.6. The first four layers, defining the TFT, are identical with the stack presented in Section 14.2. The rows of the display are processed on the first metal level whereas the columns are processed on the second metal level. In the field-shielded pixel design, the pixel electrode is defined in a third metal level of gold, resulting in a six-mask process. The pixel pad overlaps the storage capacitor, TFT, and column lines with a ~6 μm thick polyvinylphenol layer acting as inter-layer dielectric. The optical aperture thereby increases to over 95%. The TFT channel length (L) and width (W) are 5 μm and 140 μm, respectively.

The equivalent circuit of one pixel is shown in Fig. 14.7. The display is driven row-by-row. During one frame all the rows are sequentially selected by applying a potential that switches the TFT from the nonconducting (positive voltage) to the conducting state (negative voltage). During the row selection period the pixel ca-

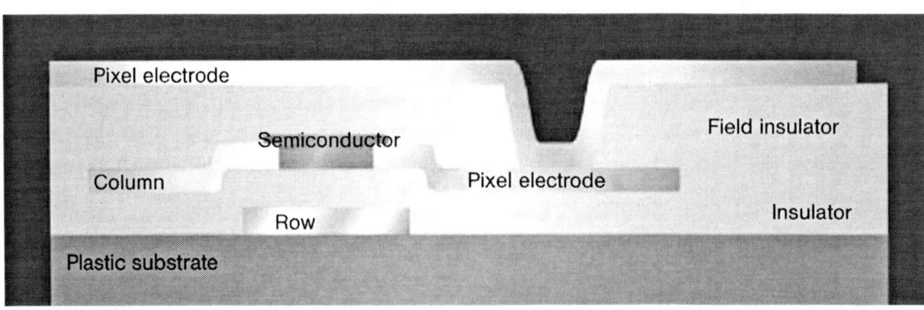

Fig. 14.6. Cross-section of the organic-electronics rollable active-matrix layer stack. The stack is built using six photolithography steps. A third metal layer is used to shield the electric field of the TFT and the electrodes from the electronic ink film.

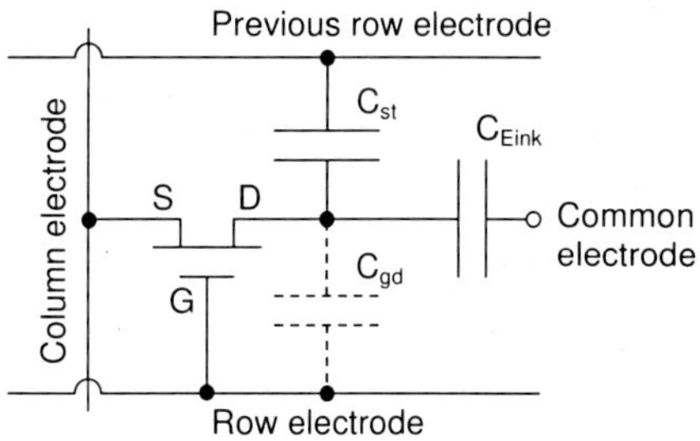

Fig. 14.7. Equivalent circuit of one pixel. The source, drain, and gate of the TFTs are indicated by *S*, *D*, and *G*, respectively. The storage capacitor is indicated by C_{st}, the pixel capacitor by C_{Eink}, and the parasitic gate-drain capacitance by C_{gd}.

pacitors of the selected row are charged to the potential applied to the column electrodes. The pixel capacitance consists of the sum of the capacitors on the drain side of the TFTs in Fig. 14.7. During the remaining frame time (i.e. the hold time) the other rows are addressed. The TFTs are then in their nonconducting state and the charge on the pixel capacitors must be retained. Typical potentials used on the gate lines are −25 V during row selection and +25 V during the hold time. Typical column voltages are −15 V, 0 V, and +15 V. It is important to note is that the electronic ink is addressed by DC driving – there is no inversion of the voltage polarity on the pixels every frame, as is commonly used to drive LC displays (Section 14.6).

To estimate the transistor requirements for our display, we make use of a transistor model (Detcheverry 2000) that includes a gate-voltage-dependent field-effect mobility:

$$\mu_{FE} = \mu_0(-V_{gs} + V_T)^{\gamma} \tag{1}$$

where V_T is the threshold voltage, γ is a fitting term that has a value close to unity for the organic semiconductor, and μ_0 is equal to the mobility at $-V_{gs} + V_T = 1$ V. By combining Eq. (1) with the classical analytical Shockley expression for the drain current in a MOS transistor, Eq. (2) is derived for the drain current in the linear regime:

$$I_d = \frac{W}{L}\frac{\mu_0 C_i}{(\gamma+2)}[(-V_{gs} + V_T)^{\gamma+2} - (-V_{gs} + V_T + V_{ds})^{\gamma+2}] \tag{2}$$

where C_i is the gate insulator capacitance per unit area.

14.3.1
Field-effect Mobility Effects

The required field-effect mobility for our display can be determined by performing circuit simulations with the circuit shown in Fig. 14.7. The term μ_0 in Eq. (1) is

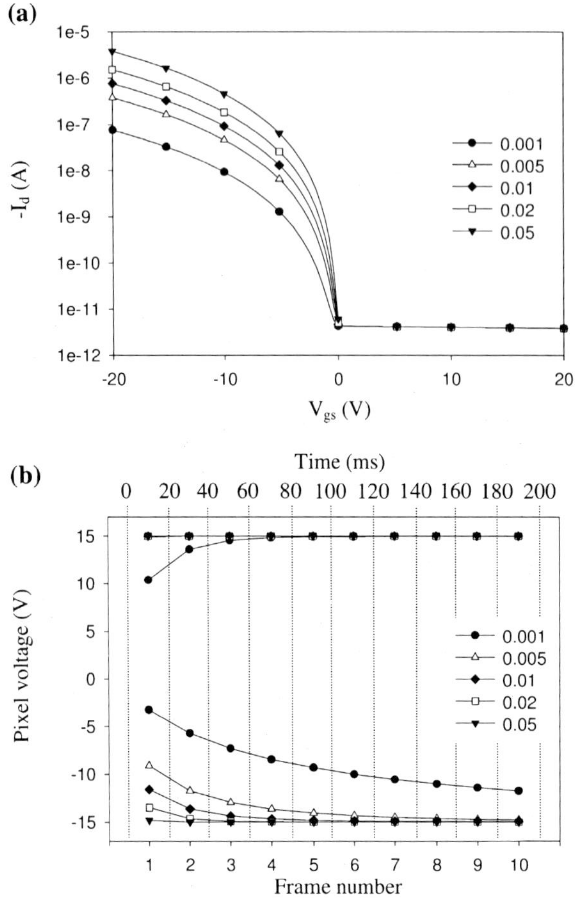

Fig. 14.8. (a) Simulated transfer characteristics for five different values of μ_0 in Eq. (1) for $V_{sd} = 20$ V. The resulting field-effect mobility at $V_{sg} - V_T = 25$ V and $\gamma = 1.1$ is indicated in the legend. The leakage current was set to 3 pA. (b) Simulated charging behavior of a pixel with the same five values for the mobility. The vertical axis shows the pixel voltage, the horizontal axis the frame number. The simulated frame rate was 50 Hz.

varied between simulations. The resulting mobility is determined at $V_{gs} = -20$ V, with a constant value for γ of 1.1, because this is a fixed value for our technology. The transfer characteristics of the simulated transistor are shown in Fig. 14.8a. The on-current increases with increasing field-effect mobility. The off-current is taken sufficiently small in these simulations to suppress any effects of leakage currents on pixel voltage. The resulting charging behavior of the pixel is shown in Fig. 14.8b. The vertical axis shows the average pixel voltage during one frame; the horizontal axis shows the frame number. Charging toward a voltage of -15 V and charging to a voltage of +15 V is simulated, because these voltages are the minimum and maximum voltages used during display addressing, as will be discussed in Section 14.6. The gate voltage is −25 V during the line selection time and +25 V during the hold time.

Charging to −15 V is slower than charging to +15 V, because of the smaller value of the source-gate voltage. At a mobility of 0.001 $cm^2\ V^{-1}\ s^{-1}$ charging to +15 V is incomplete during the first three frames whereas charging to −15 V is incomplete for more than ten frames. Incomplete charging of the pixels can reduce the optical performance of our display in at least two different ways. The first is reduced switching speed, because of the decreasing drift velocity of the electronic ink particles with decreasing pixel voltage. The second is nonuniformity between pixels, because the sensitivity of the pixel voltage toward small differences in the mobility increases when pixel charging is incomplete.

At a mobility of 0.01 $cm^2\ V^{-1}\ s^{-1}$ charging to +15 V is complete whereas charging to −15 V results in a pixel voltage of −12 V during the first frame and −14.8 V after five frames. The average pixel voltage during the first five frames is −14 V, which is a charging ratio of 93%. As the smallest pulses supplied to the pixels are five frames or longer when generating images with four gray levels at a refresh rate of 50 Hz, the effect of incomplete charging during the first frame is small. By comparing the average pixel voltage with the switching curves (not shown here) it was verified that it is possible to generate at least four gray levels with sufficient uniformity at a mobility of 0.01 $cm^2\ V^{-1}\ s^{-1}$.

14.3.2
Leakage Current Effects

The transfer characteristics of simulated transistors with different leakage currents are shown in Fig. 14.9a. The field-effect mobility is taken sufficiently high during the simulations to suppress any effects from incomplete pixel charging. The column voltage is set to 0 V during the hold time and +15 V and −15 V during the line selection period. The simulated pixel voltage is shown in Fig. 14.9b. The vertical axis shows the average pixel voltage during one frame and the horizontal axis the leakage current. At a leakage current of 1 nA the average pixel voltage is only 60% of the voltage supplied on the column electrode. Optically this large leakage current negatively affects the switching speed of the electronic ink particles and the image uniformity, because of vertical cross talk, i.e. the current leakage between the column electrode and the pixel during the hold time. When the leakage current is smaller than 50 pA, more than 97% of the charge on the pixel is retained during the hold time. This results in a voltage drop smaller than 0.5 V. By comparing the switching speed at 15 V with the switching speeds at 14.5 V (not shown here) it was verified that it is possible to generate at least four gray levels with sufficient uniformity at a leakage current of 50 pA. An even smaller leakage current further reduces the voltage drop and enables more gray levels.

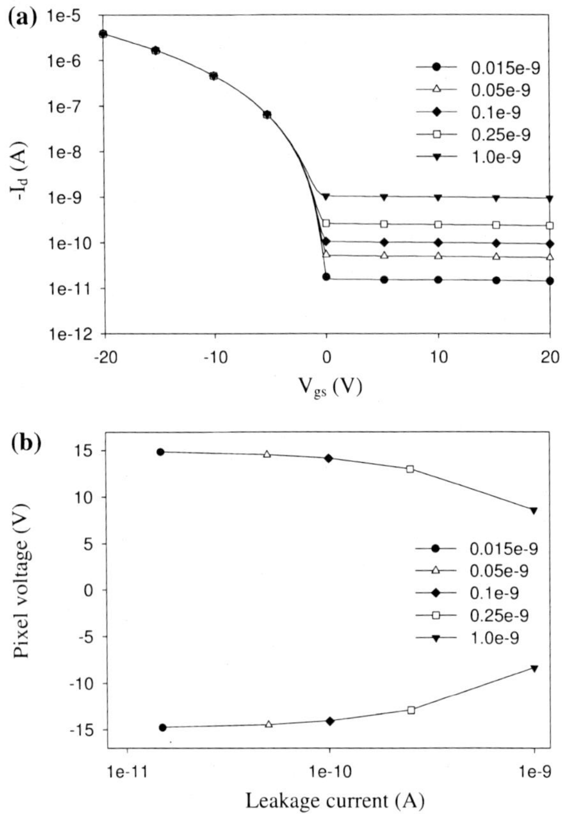

Fig. 14.9. (a) Simulated transfer characteristics for $V_{sd} = 20$ V at five different values of the leakage current, indicated in the legend. The field-effect mobility was set to 0.05 cm^2 V^{-1} s^{-1}. (b) Simulated charging behavior of a pixel with the same five values for the leakage current. The vertical axis shows the pixel voltage, the horizontal axis the leakage current. The simulated frame rate was 50 Hz.

14.4 The Electronic Ink Film

The electronic ink is supplied by E ink corporation (Comiskey 1998). The film consists of electrophoretic microcapsules in a polymer binder, coated on to a 25 µm polyester/indium tin oxide sheet (Fig. 14.10). Optical contrast is achieved by moving black and white sub-micron particles with opposite charge in a transparent fluid within a microcapsule. Depending on which sub-micron particles are closest to the viewer, light is scattered back (white state) or absorbed (black state). The electrophoretic effect is multi-stable – without any electric field the microcapsules keep their switching state. This greatly reduces the power consumption of the display (Ritter 2001).

The maximum white state reflectivity that can be achieved in active-matrix dis-

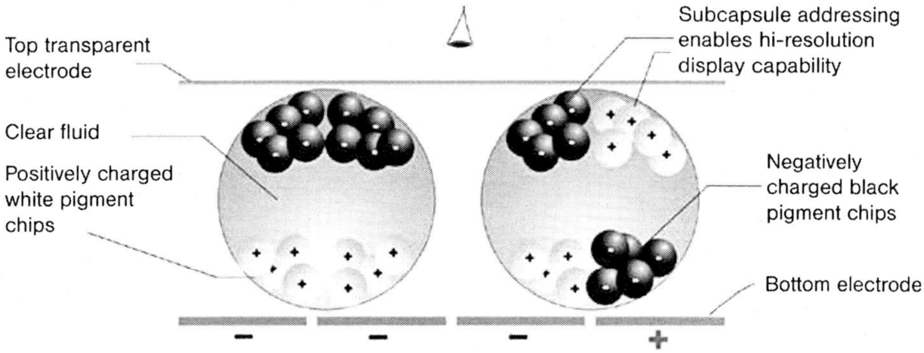

Fig. 14.10. Schematic cross-section of electronic ink as supplied by E Ink Corporation.

plays is currently 35% of perfect white. This is in between the currently achievable reflection of LC displays (~15%) and that of paper (~80%). In addition, the electronic ink is a diffuse reflector, just like paper. This adds to the paper-like viewing experience and omits the need for polarizers and diffusors normally found on LC panels, thus reducing panel thickness and cost.

14.5 Roll-up Display Integration

Integration of the active-matrix backplane and the electronic ink frontplane is a simple lamination process at a low temperature. After lamination the final stage of display processing is removal of the display from its rigid support. This support can then be re-used in the fabrication process. The cross-section of the complete stack of the rollable active-matrix display is shown in Fig. 14.11. The complete display is only 100 μm thick, which is the same thickness as a normal piece of paper.

14.6 Functional Active-matrix Roll-up Displays

The electronic ink is multi-stable. The display is therefore operated using differential driving – during an image update pixel addressing takes into account the initial switching state and the final switching state of that pixel (Zhou 2004). For example, when a pixel is in the low-reflectance state and remains in the same state in the next image ideally no pulse should be applied to that pixel, whereas a full length pulse is required when the pixel is changed from the low reflectance to the high reflectance state. The application of shorter pulses (pulse-width modulation) or a lower pixel voltage (amplitude modulation) results in intermediate reflectance

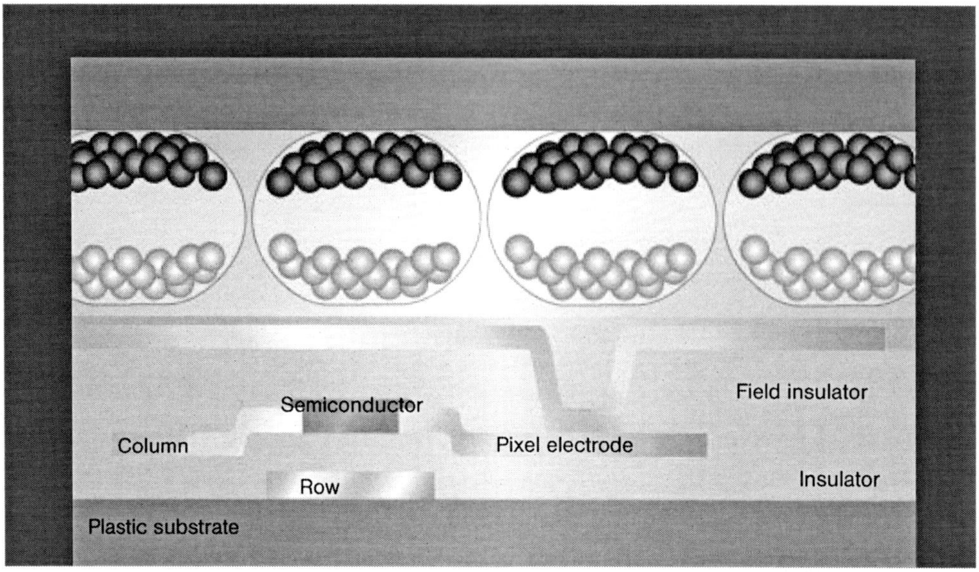

Fig. 14.11. Cross-section of the rollable active-matrix display. Total display thickness is 100 μm.

levels (i.e. gray levels). Another important difference from driving of LC displays is that the electronic ink is sensitive to the polarity of the applied voltage – a positive pixel voltage drives the black ink particles toward the viewer whereas a negative pixel voltage drives the white ink particles towards the viewer.

We use pulse width modulation to drive the display, because this enables use of simple three-level column drivers. The display is refreshed at 50 Hz, resulting in a minimum gray-to-gray pulse-width of 20 ms. The column voltage levels are −15 V, 0 V and +15 V whereas the rows are operated at −25 V during row selection and +25 V during the rest of the frame time. The common electrode voltage is set to 4 V to compensate for the kickback voltage.

Pictures of a functional display are shown in Fig. 14.12. Images containing four gray levels are displayed. The black to white switching time is 600 ms and a complete image update takes 1.5 s. A summary of the display specifications is shown in Table 14.2.

Display lifetime is a crucial aspect of market acceptance. Shelf and operational lifetime are general measures of the lifetime of displays For rollable displays the bending lifetime is an another consideration. The current operational lifetime is 168 h when operating the display at a 10% duty cycle (every 10 s the display is addressed for 1 s) at room temperature. The operational lifetime test must be performed with a duty cycle, because the display is multi-stable. The current shelf lifetime at room temperature is 6 months. Bending lifetime is excellent – the display has been rolled to a radius of 7.5 mm more than 10 000 times without any degradation.

Fig. 14.12. Two pictures of the rollable QVGA active-matrix electronic ink display. Images with four gray levels are displayed.

Tab. 14.2. Specifications of the rollable QVGA display.

Property	Value
Panel size	4.7 inch (72 mm × 96 mm)
Resolution	320 × 240 (85 ppi)
Flexibility	A bending radius of 7.5 mm
Display weight	1.6 g
Display type	Electrophoretic (E Ink)
Maximum reflection	35%
Reflective contrast	10
Number of gray levels	4
Image update time	1.5 s
Driving	Pulse width modulation
Frame rate	50 Hz
Display thickness	100 μm
Aperture ratio	97.5%
Number of pixels	76 800
Pixel size	300 μm × 300 μm
Row voltages	±25 V
Column voltages	−15 V/0 V/+15 V
Common electrode voltage	4 V
TFT channel length	5 μm
TFT channel width	140 μm
Measured mobility	0.01 $cm^2\ V^{-1}\ s^{-1}$
Measured on-current	1.0 μA
Measured off-current	6 pA
Measured threshold voltage	0.2 V

14.7
Roll-up Display Device Concepts

The most exciting feature of rollable display devices is that the display area will be larger than the device area itself. Even for the QVGA display presented in the previous paragraphs integration into a device yields a display area that is more than 150% of the device area. For larger display sizes this advantage becomes even greater. This is a huge step forward compared with current mobile devices, for example a mobile phone or a PDA, for which the display area is 30 to 70% of the device area.

Several rollable display device concepts are shown in Fig. 14.13. These devices can be used as a mobile e-reader (a)–(c), an e-book (d), a wearable GPS device (e) or a compact mobile phone with video conferencing capability (f). All the concepts have the advantage of a large display for a small device.

The e-reader will probably be on the market in two years. The mobile phone concept shows a display with full color and video capabilities that will not be on the market very soon – it still requires approximately 5 years development time.

14.8
Towards a System-on-plastic: Driver Integration

With the availability of an OTFT as basic building block it is possible to realize more functionality on the backplane than just the active matrix. This enables shifting functional blocks, for example drivers, from outside the backplane on to the backplane, which is advantageous for the complexity and cost of the total system. When drivers are integrated on to the backplane, the amount of silicon is reduced, as also is the number of connections needed from the driving electronics to the backplane.

Because the electrophoretic display effect that is used is multi-stable, the row drivers are only operational during an image update. This is different from the commonly used LC display effects that need to be driven continuously during use of the display. In view of bias stress effects, this makes the electrophoretic display effect ideal for integration of row drivers in organic electronics.

14.8.1
Row Driver Integration

Shift registers, which can be used as row drivers, only need clock and data signals and supply lines. The number of connections between the shift register and the external addressing hardware is therefore only approximately ten in total. The number of contacts to a complete QVGA display with integrated row drivers will be approximately 350, resulting in higher contact reliability. As the width of the circuitry is only a few millimeters, the footprint of the display will become smaller. Figure 14.14a shows the layout of one of our QVGA designs. Figure 14.14b shows the

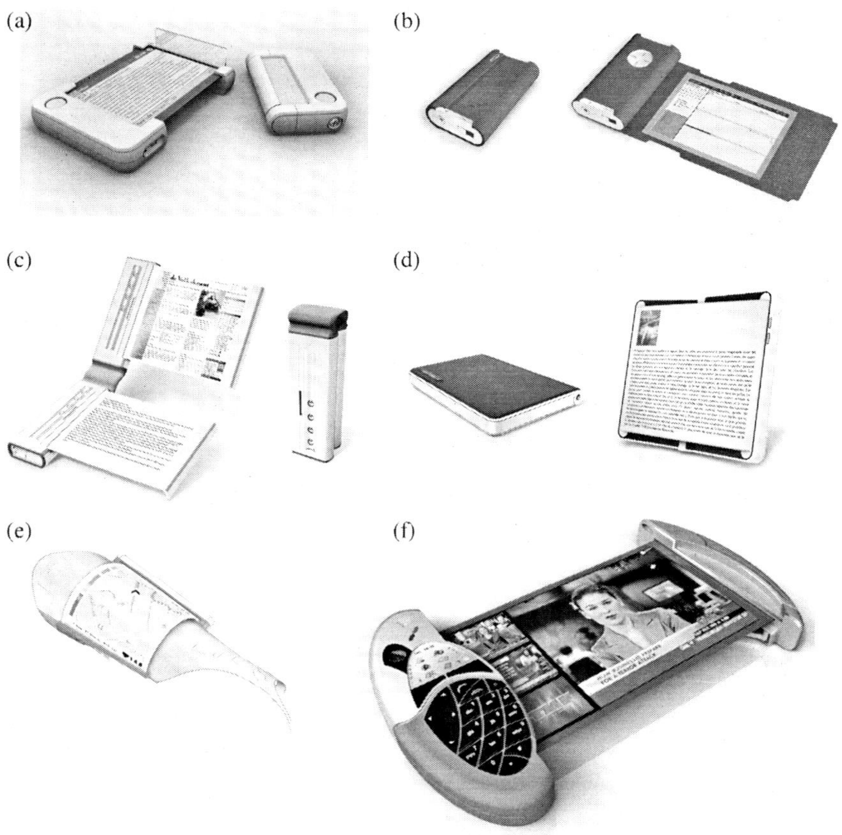

Fig. 14.13. Product concepts with rollable displays. (a) Rollable display e-reader device where the display is rolled out of the device on use. (b) Rollable display e-reader in which the display is wrapped around the device body. (c) Rollable display e-reading device using two displays. (d) Rollable e-book device in which the display rolls out of the device when the book is opened. (e) Wearable rollable display in which the display can be rolled over the arm upon use. (f) Future full color rollable display device showing the latest news broadcast.

same active matrix, but now combined with a shift register. Both the reduced number of contacts and the reduced footprint are obvious from this figure.

Another advantage of driver integration is that the number of signals going to the internal drivers is decoupled from the actual number of rows present in the display. This simplifies the design of the external driving electronics and increases the flexibility of the driver platform.

In a shift register, a selection pulse is shifted from one stage to the next at every clock pulse. The frame rate and the number of rows in the display determine the requirements for the clock frequency. As the image update time of the display is 600 ms, a frame rate of approximately 10 Hz is sufficient for black and white images. This enables e-reading applications. The corresponding minimum clock fre-

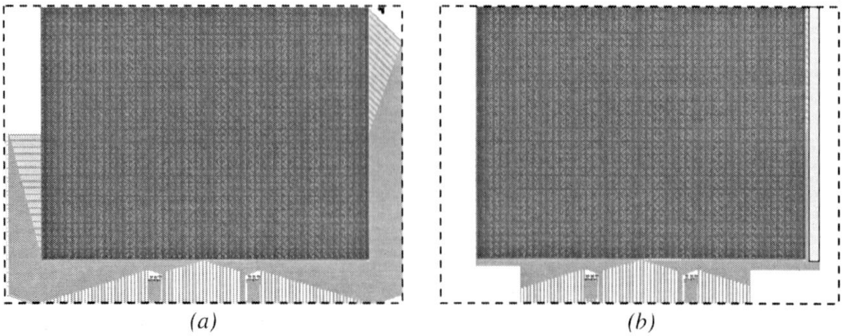

Fig. 14.14. (a) Layout of the display without integrated drivers. (b) Artist's impression of the display with integrated row driver.

quency of the shift register for our display is 2.4 kHz. For accurate rendering of gray levels, the frame rate, and thus the clock frequency of the row shift register, should be higher.

14.8.2 Stand-alone Shift Registers

In recent years several types and size of stand-alone shift register have been studied. Static topologies with 32 stages have been designed and realized. Results for these shift registers were published elsewhere (Gelinck 2004). Later, dynamic shift registers, ranging in size from 32 to 240 stages, have also been studied. Figure 14.15 shows a photograph of a foil with a display and a cell with several of the static and dynamic 32-stage shift register variants.

As displays and shift registers are fabricated using the same technology, it is expected that results from shelf-life and bending tests on displays are also valid for shift registers.

The dynamic topology contains 18 TFTs per stage. The complete dynamic 32-stage shift register contains 576 TFTs. The area per stage is 0.45 mm^2. This shift register functions at a clock frequency of up to 11 kHz, corresponding to a maximum frame rate of 45 Hz for the QVGA display. Approximately 80% of the shift registers are fully functional up to the 32nd stage. Measurements, of which results are depicted in Fig. 14.16, have shown an operational lifetime exceeding 4 h at a clock frequency of 2.8 kHz and a supply voltage of 20 V.

Shift registers have recently been made with 120 stages, half the number of rows of the QVGA display. The topology of the 120-stage shift register is similar to that of the 32-stage version. Only TFT sizes were optimized. Figure 14.17 shows measurement results for the new shift register. The top graph shows the clock signal and the lower graphs show the output waveform of every 20th output. The shift register functions at a clock frequency of up to 2 kHz, which corresponds to a frame rate of 8 Hz for our QVGA display.

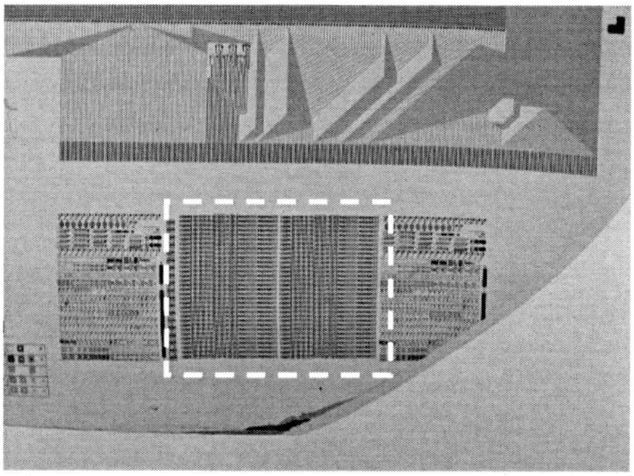

Fig. 14.15. Photograph of a flexible foil, containing both a QVGA display (top of photograph) and shift registers (inside the dashed rectangle).

The yield for fully functioning 120-stage shift registers, with a total of 2160 TFTs, is approximately 30%, which is in line with the increment in area compared with the 32-stage shift registers. Extensive operational life tests will be performed in the near future.

Table 14.3 summarizes some key properties of 120-stage shift registers. As is apparent from this table, the vertical cell pitch of the shift register has been matched to the row pitch of the QVGA display for easy integration in the near future.

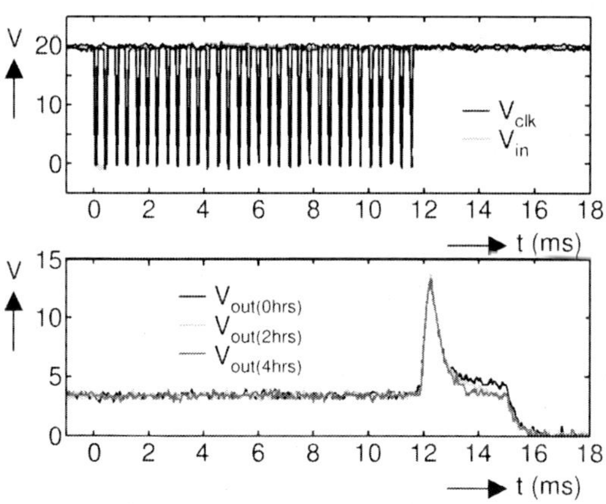

Fig. 14.16. Measurement results of the 32-stage shift register. The top graph shows the clock pulses and the start pulse; the bottom graph shows the output pulse of the 32nd stage after 0, 2, and 4 h respectively.

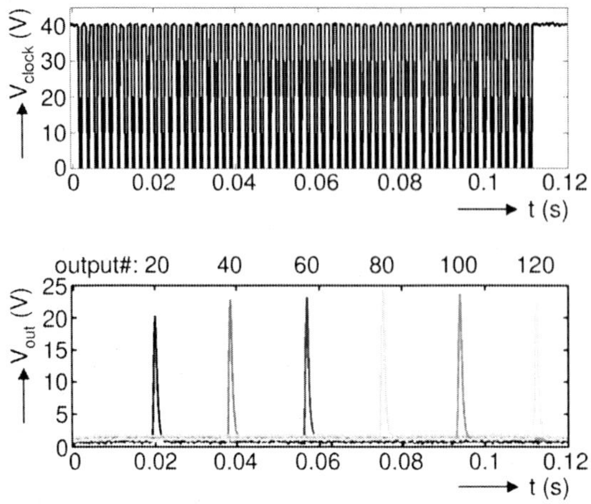

Fig. 14.17. Measurement results of the 120-stage shift register, showing the clock pulses and the output pulses for every 20th stage.

Tab. 14.3. Specifications of the 120-stage shift register.

Property	Value
Size per stage (w × h)	1800 μm × 300 μm
Area per stage	0.54 mm^2
Total area	65 mm^2
Max. clock frequency	2 kHz
Max. QVGA frame rate	8 Hz
Number of stages	120
TFTs per stage	18
Total number of TFTs	2160
Circuit yield	30%
Max. supply voltage	40 V
TFT channel length	5 μm

14.8.3
Integrated Shift Registers

The next step is to integrate a shift register with an active matrix to obtain a functional display with a reduced number of contacts. On the basis of results for the 120-stage shift register, a 240-stage version was developed. Table 14.4 lists the specifications of the integrated shift register. At the output, a voltage swing of

Tab. 14.4. Specifications of the integrated shift register.

Characteristic	Value
Select level	−25 V
Unselect level	25 V
Output load	80 pF
Clock frequency	1.2 kHz
QVGA frame rate	5 Hz
Rise/fall time	80 μs
Spike amplitude	5 V

−25...25 V is needed. The main challenge was to design an output stage capable of driving the capacitive load caused by a display row with this voltage swing within a small fraction (10%) of the row time.

A new dynamic output stage was developed to meet this specification. Figure 14.18 shows measurements of a single selection pulse under two load conditions (80 pF and 180 pF) at supply voltages of ±30 V. Rise and fall times are approximately 100 μs, which is a little over the specified value.

In view of yield, a shift register with 240 stages was divided in four 60-stage blocks, which could either be chained to one long shift register or operated separately. This way, defects do not immediately result in a completely inoperable shift register. Figure 14.19 shows photographs of both E Ink and SiPix displays with an integrated shift register of which part of the lower 60-stage blocks are functional.

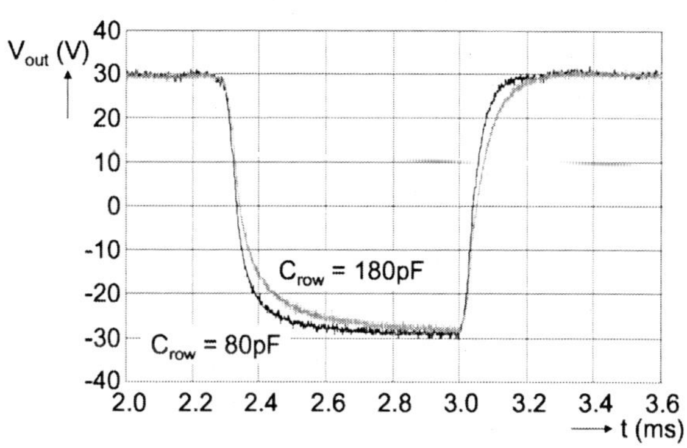

Fig. 14.18. Measurement results of the dynamic output buffer of the integrated shift register under two load conditions (80 pF and 180 pF).

(a) (b)

Fig. 14.19. rollable QVGA active-matrix electronic ink displays with a partly functional integrated shift register with electronic ink from E Ink Corporation (a) and SiPix Imaging (b).

Acknowledgment

The authors would like to thank the other members of the Polymer Vision team for their help in realizing the results described in this paper. Dago de Leeuw and Eugenio Cantatore are acknowledged for their numerous contributions to the polymer electronics technology that was used to produce the results described. Mark Hage is acknowledged for help with display measurements and Siebe-Jan van der Hoef and Albert Geven for realizing the display addressing hardware and software. E Ink Corporation and SiPix Imaging Inc. are thanked for supplying electrophoretic material and DuPont Teijin Films for the supply of plastic substrates.

References

Blanchet, G. and Rogers, J.A. (2003) Printing techniques for plastic electronics. *J. Inf. Sci. Technol.*, **47**, 296–303.

Brody, T.P. (1996) The birth and early childhood of active-matrix – A personal memoir. *J. SID*, **4**, 113–117.

Brown, A.R. Pomp, A. Hart, C.M. and de Leeuw, D.M. (1995) Logic gates made from polymer transistors and their use in ring oscillators. Logic Gates Made form Polymer Transistors and Their Use in Ring Oscillators. *Science* **270**, 972–974.

Chabinyc, M.L. Wong, W.S. Salleo, A. Paul, K.E. and Street, R.A. (2002) Organic polymeric thin-film transistors fabricated by selective dewetting. *Appl. Phys. Lett.*, **81**, 4260–4262.

Chen, Y. Kazlas, J.A.P. Ritenour, A. Gates, H. and McCreary, M. (2003) Flexible active-ematrix electronic ink display. *Nature*, **423**, 136.

Comiskey, B. Albert, J.D. Yoshizawa, H. and Jacobson, J. (1998) An electrophoretic ink for all-printed reflective electronic displays. *Nature*, **394**, 253–255.

Detcheverry, C. and Matters, M. (2000) Device simulation of all-polymer thin-film transistors. *Proc. ESSDERC*, 328–331.

Dimitrakopoulos, C.D. and Malenfant, P.R.L. (2002) Organic thin film transistors for large area electronics. *Adv. Mater.*, **14**, 99–117.

DRURY, C.J. MUTSAERS, C.M.J. HART, C.M. MATTERS, M. and DE LEEUW, D.M. (1998) Low-cost all-polymer integrated circuits. *Appl. Phys. Lett.*, **73**, 108–110.

FUJISAKI, Y. INOUE, Y. SATO, H. KURITA, T. TOKITO, S. and FUJIKAKE, H. (2003) Organic TFT-driven Liquid Crystal Cell with Anodic-oxidised Gate Insulator and Double Protection Layer. *IDW Digest*, 291–294.

GELINCK, G.H. GEUNS, T.C.T. and DE LEEUW, D.M. (2000) High-performance all-polymer integrated circuits. *Appl. Phys. Lett.*, **77**, 1487–1489.

GELINCK, G.H. HUITEMA, H.E.A. VAN VEENENDAAL, E. CANTATORE, E. SCHRIJNEMAKERS, L. VAN DER PUTTEN, J.B.P.H. GEUNS, T.C.T. BEENHAKKERS, M. GIESBERS, B. HUISMAN, B.-H. MENA BENITO, E. TOUWSLAGER, F.J. MARSMAN, A. VAN RENS B. and DE LEEUW, D.M. (2004) Flexible active-matrix displays and shift registers based on solution-processed organic transistors. *Nature Mater.*, **3**, 106–110.

GLESKOVA, H. WAGNER, S. and SUO, Z. (1999) Failure resistance of amorphous silicon transistors under extreme in-plane strain. *Appl. Phys. Lett.*, **75**, 3011–3013.

GUSTAFSSON, G. CAO, Y. TREACY, G.M. KLAVETTER, F. COLANERI, N. and HEEGER, A.J. (1992) Flexible light-emitting diodes made from soluble conducting polymers. *Nature*, **357**, 477–479.

HERWIG, P.T. and MÜLLEN, K.A. (1999) A Soluble Pentacene Precursor: Synthesis, Solid-State Conversion into Pentacene and Application in a Field-Effect Transistor. *Adv. Mater.*, **11**, 480–483.

HUEBLER, A. HAHN, U. BEIER, W. LASCH, N. and FISCHER, T. (2002) High volume printing technologies for the production of polymer electronic structures. *IEEE Polytronic Conference*, 172–176.

HUITEMA, H.E.A. GELINCK, G.H. VAN DER PUTTEN, J.B.P.H. KUIJK, K.E. HART, C.M. CANTATORE, E. HERWIG, P.T. VAN BREEMEN, A.J.J.M. and DE LEEUW, D.M. (2001) Plastic transistors in active-matrix displays. *Nature*, **414**, 599.

HUITEMA, E. GELINCK, G. VAN DER PUTTEN, B. CANTATORE, E. VAN VEENENDAAL, E. SCHRIJNEMAKERS, L. HUISMAN, B.-H. and DE LEEUW, D. (2003a) Plastic Transistors in Active-Matrix Displays. *ISSCC Digest*, 380–381.

HUITEMA, H.E.A. GELINCK, G.H. VAN VEENENDAAL, E. CANTATORE, E. TOUWSLAGER, F.J. SCHRIJNEMAKERS, L.R.R. VAN DER PUTTEN, J.B.P.H. GEUNS, T.C.T. BEENHAKKERS, M.J. VAN LIESHOUT, P.J.G. LAFARRE, R.W. DE LEEUW D.M. and VAN RENS, B.J.E. (2003b) A Flexible QVGA Display With Organic Transistors, *IDW Digest*, 1663–1664.

INOUE, S. UTSUNOMIYA, S. SAEKI, T. and SHIMODA, T. (2002) Surface-Free Technology by Laser Annealing (SUFTLA) and Its Application to Poly-Si TFT LCDs on Plastic Film With Integrated Drivers. *IEEE Trans. Electron Devices*, **49**, 1353–1360.

KUIJK, K.E. (1991) A Novel Diode Matrix for LC-TV Displays, *SID Digest*, 401–408.

KYMISSIS, I. DIMTRAKOPOLOUS, C.D. and PURUSHOTHAMAN, S. (2002) Patterning pentacene organic thin film transistors. *J. Vac. Sci. Technol. B*, **20**, 956.

LIANG, R.C. HOU, J. ZANG, H.M. and CHUNG, J. (2003) Passive Matrix Microcup Electrophoretic Displays. *IDMC Digest*, 351–354.

LUEDER, E. (1999) Passive and MIM-driven LCDs with plastic substrates. *IDW Digest*, 215–218.

MACH, P. RODRIQUEZ, S.J. NORTRUP, R. WILTZIUS, P. and ROGERS, J.A. (2001) Monolithically integrated flexible display of polymer-dispersed liquid crystal driven by rubber-stamped organic thin-film transistors. *Appl. Phys. Lett.*, **78**, 3592–3594.

OKAMOTO, M. OKADA, Y. BAN, A. OKA, W. MATSUDA, Y. and SHIBAHARA, S. (2002) Development of a Color Reflective TFT-LCD Using Plastic Substrates. *IDW Digest*, 315–318.

PARK, S.K. HAN, J.I. KIM, W.K. and KWAK, M.K. (2000) Development of 2-inch Plastic-Film STN-LCD with Uniform Cell Gap. *SID Digest*, 514–517.

PARK, S.K. KIM, Y.H. HAN, J.I. MOON, D.G. and KIM, W.K. (2002) High performance polymer TFT printed on a plastic substrate. *IEEE Trans. Elec. Dev.*, **49**, 2008–2015.

Philips press release, "Philips Research develops the world's first display using polymeric semiconductors as pixel drivers",

September 2000, (http://www.polymervision.com/).

Polach, S. Randler, M. Bahnmueller, F. and Lueder, E. (2000) A Transmissive TN display Addressed by a-Si:H TFT on a Plastic Substrate. *IDW Digest*, 203–206.

Polymer Vision is a venture inside the Philips Technology Incubator since 2003.

Ritter, J. (2001) Active-Matrix Electrophoretic Displays. Asia Display/IDW *Digest*, 343–346.

Sandoe, J.N. (1998) AMLCD on Plastic Substrates. *SID Digest*, 293–296.

Sheraw, C.D. Zhou, L. Juang, J.R. Gundlach, D.J. Jackson, T.N. Kane, M.G. Hill, I.G. Hammond, M.S. Campi, J. Greening, B.K. Francl, J. and West, J. (2002) Organic thin-film transistor-driven polymer-dispersed liquid crystal displays on flexible polymeric substrates. *Appl. Phys. Lett.*, **80**, 1088–1090.

Sirringhaus, H. Kawase, T. Friend, R.H. Shimoda, T. Binbasekaran, M. Wu, W. and Woo, E.P. (2000) High-resolution inkjet printing of all-polymer transistor circuits. *Science*, **290**, 2123–2126.

Tsukada, T. (2000) Active-Matrix Liquid-Crystal Displays. *Springer Series Mater. Sci.*, **37**, 7–93.

Voss, D. (2000) Cheap and cheerful circuits. *Nature*, **407**, 442–444.

Young, N.D. Bunn, R.M. Wilks, R.W. McCulloch, D.J. Deane, S.C. Edwards, M.J. Harkin, G. and Pearson, A.D. (1997) Thin-film-transistor- and diode-addressed AMLCDs on polymer substrates. *J. SID*, **5**, 275–281.

Zhou, G. Johnson, M. Cortie, R. Zehner, R. Amundson, K. Henzen, A. and van de Kamer, J. (2004) Addressing an Active Matrix Electrophoretic Display, IDW *Digest*, 1729–1732.

15
Active-matrix Light-emitting Displays

Shelby F. Nelson and Lisong Zhou

15.1
Introduction

An important potential commercial application of organic thin-film transistors is in backplanes for displays. The large demand for portable electronic equipment, for example cell phones, laptop computers, personal digital assistants, and digital cameras, has helped to make flat-panel-display manufacturing a $50 billion a year industry with approximately a 25% annual growth [1]. Although the flat panel industry is currently dominated by liquid-crystal technology, there has been a tremendous amount of interest in, and research on, organic light-emitting diodes (OLEDs). OLED displays have several advantages over liquid-crystal displays. Unlike liquid crystals, OLEDs are emissive, bright, and have a wider intrinsic viewing angle, and thus do not require backlights or filters. In principal this enables use of thinner, lighter, and potentially more power-efficient displays. OLEDs are used in a number of commercial products, and the OLED display market is expected to be enormous in the future.

An OLED is an electroluminescent device that consists of a stack of organic layers sandwiched between two conductive electrodes. When a bias is applied, current flows through the device and electrons and holes are injected into opposite sides of the organic layers. When they meet, their recombination results in emission of light characteristic of the band gap of organic material. Invented in 1985 [2] in devices made of small organic molecules, the field has grown to include polymeric light-emitting diodes (often referred to as PLEDs). Although there are differences between polymeric and small-molecule emitters in processing, stack design, and the particulars of stability issues, for the purposes of this article they are similar enough to include under the name "OLED".

For high information-content displays, active-matrix (AM) pixel addressing provides improved display performance and reduced power consumption. In active matrix addressing each individual pixel is controlled by one or more thin-film transistors (TFTs). To date, most AM OLED displays have used polysilicon TFTs as the active elements, because they can provide sufficient current at low voltages and acceptable device dimensions, and they are capable of integrated drive electronics

Organic Electronics, Materials, Manufacturing and Applications. Edited by Hagen Klauk

ISBN: 3-527-31264-1

[3–5]. For example, Sanyo and Kodak, together, commercially manufactured the 2.2-inch OLED displays for one of the Kodak EasyShare digital cameras. Improvements in the efficiency of OLEDs also enables lower-mobility TFTs, for example those based on a-Si:H and organic semiconductors, to be used as OLED drive devices [6, 7]. Compared with polysilicon TFTs, a-Si:H is a mature technology, with established infrastructure, and is more suitable for large displays. Organic TFTs with performance similar to that of a-Si:H [8, 9], although at an early stage of development, have the potential for low-temperature and low-cost manufacturing. Organic TFTs could potentially be manufactured on polymeric substrates in a roll-to-roll process, which would be a revolutionary change from current batch-process manufacture. Displays fabricated on polymeric substrates could be flexible, lightweight, and rugged, all of which are very attractive properties for portable electronics. The largest concern with using organic TFTs in AM OLED displays is whether sufficient steady current can be supplied at low voltages. In addition, several critical issues, for example dielectric integrity, device passivation and integration technique, and TFT uniformity, must still be addressed. Nonetheless, organic TFTs may be an option for large-area and low-cost displays.

OLEDs are current-driven devices, and the brightness is proportional to the current. The simplest design for an active-matrix OLED pixel has a select transistor, a drive transistor, a storage capacitor, and the emissive OLED. It is instructive to consider the constraints on these two fundamental transistors. The most significant requirement for the drive TFT is to provide enough drive current to achieve appropriate brightness. With the demonstration of an a-Si:H-driven active-matrix OLED display (for example, Samsung and Philips, 2005 SID) it is obvious this is not a problem for TFTs with a mobility of approximately 1 $cm^2\ V^{-1}\ s^{-1}$. Organic TFTs using pentacene for the semiconductor have this sort of device performance.

At low currents an OLED is non-emissive, and therefore the off-current requirement on the drive TFT is loose. In an active-matrix OLED display, however, the gate voltage on the drive TFT, held with the help of storage capacitor, needs to be held constant during the complete frame time. (The storage capacitor has been charged through the select TFT during the line-scan time; as for the LCD case, this requires sufficiently low off-current for the select TFT not to discharge the pixel capacitor during a frame time.) The size of the select TFT is such that it enables 99% of the data voltage to be charged on the storage capacitor. With a storage capacitor of comparable size for both LCD and OLED displays, a device with a small width over length (W/L) suffices for VGA-sized displays at 60 Hz.

Drive TFTs operate in a saturation regime ($V_{DS} > V_G - V_T$) whereas select TFTs are mostly in the linear regime. Because most pentacene TFTs for display applications use a bottom-contact structure for ease of patterning the metal contacts, the contact resistance is relative high – especially in the linear regime. As a result, most pentacene TFTs have lower linear mobility than saturation mobility. The higher on-resistance of select transistors will therefore increase (dis)charging times for the storage capacitor; this must be taken into consideration when determining the select-TFT size.

As OLED technology moves to longer lifetimes, the drive-TFT stability becomes more important. Given that the drive TFT is under a single polarity stress when turned on, a threshold voltage shift can occur, which will diminish the pixel brightness. Although threshold voltage shift is not a problem for poly-Si TFTs, it is a critical issue in both a-Si:H and pentacene TFTs. With current OLED technology advances, for example to higher efficiencies, lower turn-on voltages, and smaller drive currents for the same light outputs, some of these issues diminish, and pentacene TFTs still look plausible for some applications.

15.2 OLED Pixel Differences from LCDs

It is useful to understand some differences between requirements for LCD and OLED pixels. Although an OLED pixel needs more transistors than that of an LCD, the requirement of select transistors in OLED pixels is almost the same as in LCDs. They share the common goal of (dis)charging the storage capacitors that, for the most part, will have similar size, given that the LC capacitor is normally much smaller than the storage capacitor and can be neglected. So, usually a small transistor with $W/L \approx 3$ will suffice for VGA-sized displays. The drive transistor in an OLED pixel, however, will usually be much larger than the select transistor if it is to provide enough OLED current. For a-Si:H and pentacene TFTs, it will take much more area than poly-silicon, and can limit the pixel aperture ratio for bottom-emitting displays.

The TFTs in an LCD panel are operated under stress conditions of both positive and negative gate biases. Positive voltages are applied to the gate electrode in a pulsed mode. From one pulse to the next, the gate is biased to negative voltages. Therefore, it is expected that these positive and negative shifts tend to cancel each other out. Thus, we normally see much better stability of a-Si:H TFTs in an LCD panel than in a dc-driven device. It is different for TFTs on OLED panels, because the drive TFTs are under constant bias stress. As a result, usually a complex pixel design is necessary to complement device instability.

The TFTs in LCDs act as pure switches. As long as their gate biases are well above the threshold and data voltage, they pass the data voltage to the LC capacitor without distortion. There is no stringent requirement on the uniformity of transistors. The drive transistors in an OLED display determine the OLED current, however. A small change in threshold voltage will directly affect the current, and that, in turn, affects the brightness. As a result, OLED displays require extremely careful control of TFT uniformity.

15.3 Complex Pixel Design

The combination of tight requirements, and device uniformity and instability issues, for both organic and a-Si transistors, have given rise to more complex pixel designs for driving OLEDs. Many proposals have been published in the literature for both poly-Si and a-Si:H TFTs, most of which use a four-transistor (4-T) structure; we will describe a few of these as examples. Dawson et al. proposed a 4-T structure that uses the two additional poly-Si transistors to autozero the threshold voltage of the current-drive transistor [5]. This technique can eliminate the threshold voltage variations, but cannot deal with mobility variations. Another 4-T structure, based on a-Si:H, was proposed by Yi He, who used a current-programming method to compensate for both threshold and mobility variations [10]; this has the disadvantage that charging time for low data current is long. To overcome this problem, a 4-T current-mirror structure was proposed by Yumoto to scale the data current and reduce charging time at low data current [11]. As remarked on by Fish et al. [12], however, changes in both the transistors and the OLED must be accommodated. In their design the net effect of the different degradation mechanisms on light output is directly measured by means of an in-pixel photodetector, so an optical feedback loop ensures constant pixel brightness when integrated over a frame time. Such creative approaches will clearly be needed for high-quality displays.

15.4 Practical Design

To illustrate both the design considerations and the practical issues that arise in trying to integrate an organic TFT backplane with an organic LED frontplane, we will describe a pentacene-TFT-driven active-matrix OLED display built and tested at Penn State University in collaboration with Kodak. As Fig. 15.1 shows, the simplest two-TFT pixel layout was used, consisting of a select and drive TFT, storage capacitor, and the emissive OLED. During display operation, a pulsed bias is applied to the gate of the select TFT, turning the TFT on and enabling a simultaneous

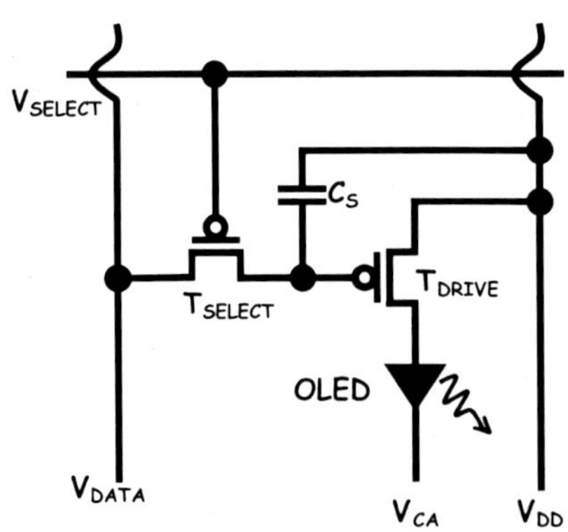

Fig. 15.1. Two TFT OLED pixel for p-type transistors.

data pulse to be applied to the gate of the drive TFT. The drive TFT is turned on and current can flow through the OLED. The storage capacitor retains the gate bias on the drive TFT until the pixel is again addressed, one complete frame time later.

A first rough calculation gives some basic data for the pixel. Although top emission is possible and may provide benefits, for example higher aperture ratio, and enable the use of opaque substrates, we chose the more stable and mature technology of a bottom-emission structure. The average current necessary to produce a bright display (100 cd m^{-2}) is approximately 10 mA cm^{-2} [2]. If we choose a pixel size of 500 μm × 500 μm and an aperture ratio of 50% on a bottom-emission display, a steady current of $I_{sat} = 12.5$ μA is necessary. Assuming a TFT operating in saturation, the required mobility can be calculated by use of Eq. (1).

$$\mu = \frac{I_{SAT}}{\frac{W}{2L} C_{OX} (V_{GS} - V_T)^2} \tag{1}$$

where C_{OX} is the capacitance per unit area of the dielectric, V_{GS} is the gate source voltage, and V_T is the threshold voltage.

To achieve the proper current levels for the drive TFT using a convenient device geometry ($W = 200$ μm, $L = 20$ μm), a silicon dioxide gate dielectric thickness of 300 nm, a threshold voltage of +10 V, and gate source voltage of −30 V, a mobility of 0.14 $cm^2\ V^{-1}\ s^{-1}$ is required, according to Eq. (1). For a well-fabricated pentacene TFT mobility is normally much higher than this requirement.

Because the select TFT is used only for (dis)charging the storage capacitor (the gate capacitor of the drive TFT is much smaller than the storage capacitor), we can use a small device geometry, $W = 20$ μm, $L = 20$ μm. This transistor is basically working in the linear region, and the on-resistance of the select TFT can be calculated by use of Eq. (2):

$$R_{ON} = \frac{V_{DS}}{I_D} = \frac{1}{\mu C_{OX} (V_{GS} - V_T) \frac{W}{L}} \tag{2}$$

Taking a modest mobility of 0.5 $cm^2\ V^{-1}\ s^{-1}$, and V_{GS} and V_T of −30 V and +10 V, respectively, the calculated on-resistance is only 2 MΩ. Assuming a storage capacitor of 1.2 pF, which conveniently corresponds to an area of 200 μm × 50 μm, the (dis)charging time is only 2.5 μs, much less than the required row time for a VGA-sized display at 60 Hz refresh rate (~30 μs).

During each frame time, the storage capacitor has to hold the charge so that the drive TFT can be maintained at a constant gate voltage, thus providing a constant current through the drive TFT and OLED. Assuming the select TFT has 10 pA off-current (although the off-current is normally less), the frame time is 16 ms (corresponding to 60 Hz refresh rate), and the storage capacitor is 1.2 pF. Therefore, the voltage decay, according to Eq. (3), is only approximately 0.13 V, which is relatively small compared with the normal operation potential range of 40 V. This results in a better than 99% retention rate for the display.

$$\Delta V = \frac{\Delta Q}{C_S} = \frac{I_{OFF} t_{FRAME}}{C_S} \tag{3}$$

All these calculations show that pentacene TFTs are feasible for active-matrix OLED display backplanes. For the Penn State/Kodak practical realization mentioned above, a 48 × 48 pixel bottom-emission display panel was designed on a 64 mm × 64 mm glass substrate. To obtain good yield, a design rule of 10 μm was used for the minimum feature size (line width or separation) for most structures on the test panel. The coarseness of this design rule is not related to the use of organic compounds, but rather to the simplicity of photolithographic processes.

15.5
AIM–SPICE Simulation of Pentacene TFT-driven OLEDs

Using these designs, we chose to simulate the behavior of the pentacene TFTs, the OLED, and the TFT/OLED active pixel characteristic in detail, using AIM-SPICE [13]. Automatic integrated circuit modeling–SPICE (AIM–SPICE) is a SPICE package designed to work under Windows. For assurance that variations in processing will not cause the backplane to fail, it is most useful to model the drive TFT from a pentacene TFT with a mediocre performance, for example a mobility of 0.6 $cm^2\ V^{-1}\ s^{-1}$ and a threshold voltage of 10 V. TFT and OLED device models were formed using measured device data and dimensions consistent with the pixel design above – the TFT has a channel width of 200 μm and channel length of 20 μm, and the OLED has an area of 1.25×10^{-3} cm^2.

Figure 15.2a, below, shows the drain current as a function of gate-source voltage. As expected, it has an off-current of less than 10 pA, a subthreshold slope of 2 V $decade^{-1}$, and the expected mobility and threshold voltage. Figure 15.2b shows the drain current as a function of drain voltage. Similarly, the simulation of a small molecule OLED with mediocre performance (or, in this example, an aged device) is shown in Fig. 15.3. Figure 15.4 shows the TFT-driven OLED current at different data voltages; clearly it responds to the data voltage very well. The on-current is well above the design target of 12.5 μA, as expected. The TFT/OLED on-current is lower than for discrete TFT because the potential drop over the OLED will partially consume the drain voltage the TFT, and the working point in the pentacene TFT itself is no longer in deep saturation. The OLED current is still not very different from the TFT current at lower drain-source bias. When designing a bottom-emission OLED panel using a-Si:H TFTs, the ITO anode is on the source terminal of the TFT, because a-Si:H is an n-type material. The data voltage applied

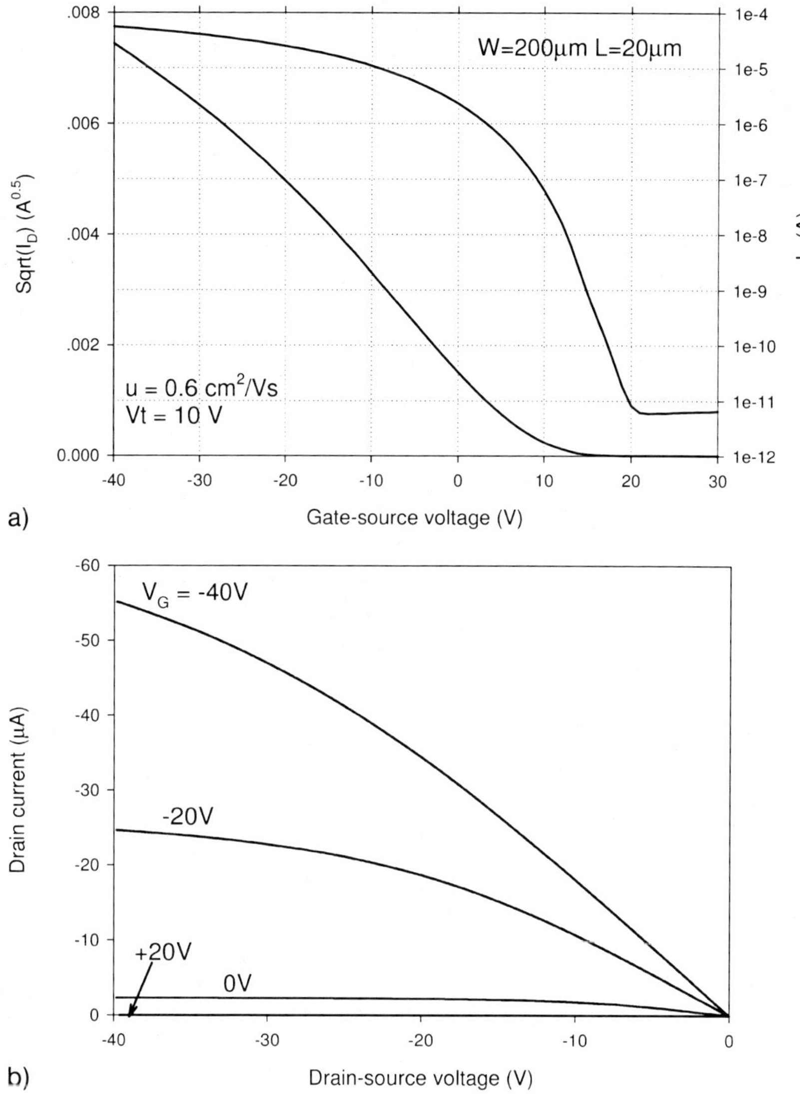

Fig. 15.2. AIM–SPICE simulation of a mediocre pentacene thin film transistor. A. Drain current as a function of gate-source voltage. B. Drain current as a function of drain voltage.

on the gate is shared between the TFT gate-source and OLED voltages, so the data voltage must be increased to accommodate it. On the other hand, pentacene TFTs are p-type devices, and the OLED is connected to the drain terminal, so the data voltage we apply is purely the gate-source voltage.

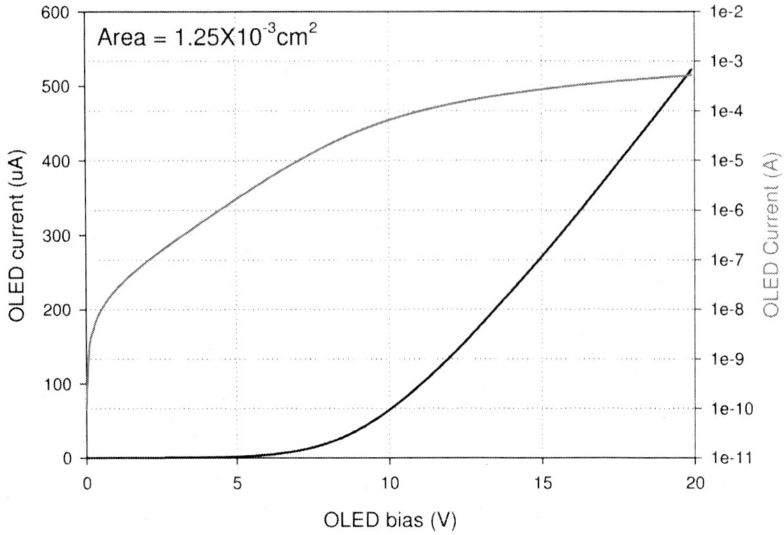

Fig. 15.3. AIM–SPICE OLED simulation, based on an aged device. Good agreement with data is achieved.

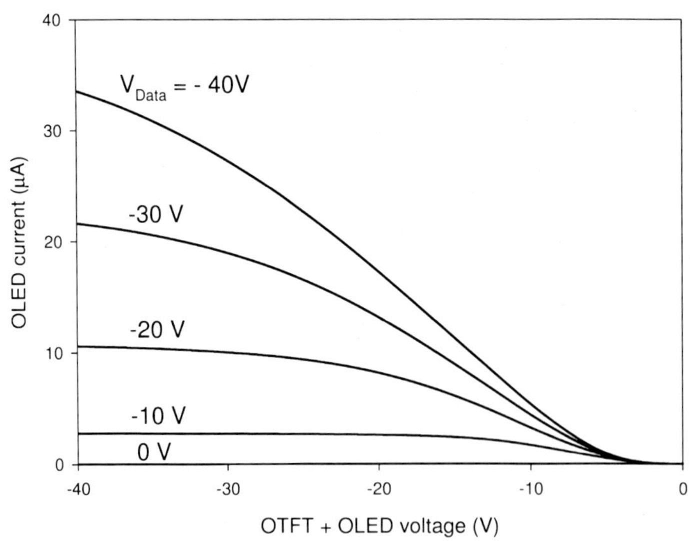

Fig. 15.4. AIM–SPICE OTFT + OLED simulation, showing good current response.

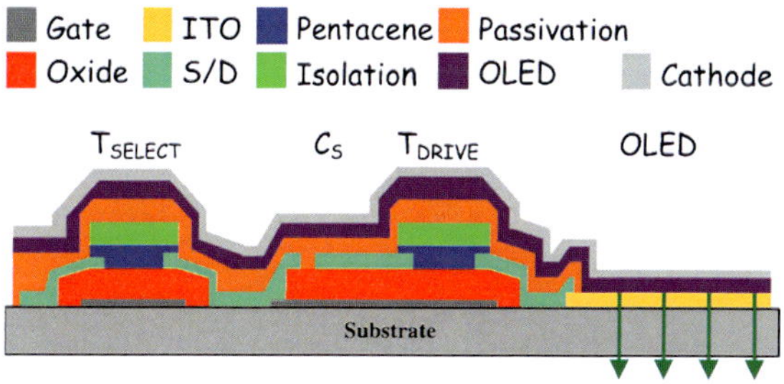

Fig. 15.5. Cross section of OLED bottom emitter pixel.

15.6 Fabrication Process

As with other backplane technology, we first fabricated the pentacene TFT backplane then deposited the OLED; this was followed by final display encapsulation. Figure 15.5 shows a cross section of the pixel, including the select TFT, storage capacitor, drive TFT, and OLED. The substrate is 0.7 mm glass coated with 85 nm ITO film. The ITO was patterned by photolithography and wet etching. Next, a 50-nm chrome gate layer was deposited by sputtering and patterned by wet etching. A 300 nm thick ion mill-sputtered SiO_2 was used as the gate dielectric and patterned with lift-off [9]. The ion mill-sputtered SiO_2 was chosen for two reasons: first it tends to have a smoother dielectric surface; second, and more important, it establishes a surface on which to grow octadecyltrichlorosilane (OTS). A 100 nm platinum source–drain layer was deposited by sputtering and patterned by lift-off. A self-assembled monolayer of OTS was formed on the sample by vapor prime in a vacuum oven for 3 h, which plays a key role in improving the pentacene TFT performance by reducing the surface energy of dielectric [8, 14]. Next, a 50 nm pentacene layer was thermally evaporated at 60 °C at a rate of 0.1–0.5 Å s^{-1}. Given that our pentacene thin films are normally in a light accumulation state on the field region, an isolation step was necessary to minimize the leakage current between the devices and conduction lines. As will be discussed in more detail in the next section, our best process was to first apply a layer of polyvinyl alcohol, pattern, apply a layer of parylene, and follow with more patterning. The OLED stack was deposited on to the TFT backplane through a shadow mask by thermal evaporation in an OLED deposition system. The organic layers were: 75 nm 4,4′-bis[*N*-(1-naphthyl)-*N*-phenylamino]biphenyl (α-NPD) as the hole transport layer and 75 nm tris(8-hydroxyquinoline)aluminum (Alq_3) as the electron-transport and emission layer. Next, a 220 nm magnesium/silver cathode was deposited through the shadow mask. Finally, the OLED display area was encapsulated by a glass slide and sealed with UV-cured epoxy.

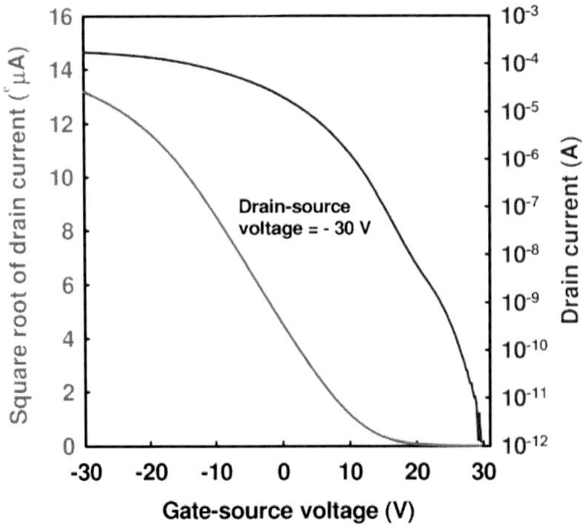

Fig. 15.6. Characteristic of a PVA-patterned pentacene TFT ($W = 240$ μm, $L = 10$ μm), showing relatively poor subthreshold slope, but relatively good on/off ratio in the current.

15.7 Device Passivation

Passivation is needed to insulate the backplane from the OLED stacks everywhere except the ITO and bonding contact areas. Unlike poly-Si and a-Si:H backplanes, on which both organic and inorganic passivation layers can easily work, the device passivation technique needs extra consideration for pentacene TFTs. We explored several different materials for passivation of pentacene TFTs, including poly(vinyl alcohol) (PVA), room temperature plasma-enhanced chemical vapor deposition silicon nitride (RT PECVD SiN), and vapor-deposited parylene.

Usually, pentacene TFTs suffer badly from organic solvents, which cause the pentacene thin film to become discontinuous by cracking or peeling off [15]. Instead, PVA was used because it is water soluble, photo-patternable, and easily developed in water. After PVA lithography, the field pentacene is etched away by oxygen-plasma dry etching. Normally, in devices patterned this way the mobility and threshold voltage are preserved and the off-current is very low, but the subthreshold slope degrades to more than 2 V decade^{-1}. Figure 15.6 shows an example of the characteristic of a PVA-patterned pentacene TFT.

A concern with PVA layers, however, is that the moisture trapped in the PVA could be detrimental to OLED lifetime. Some dry passivation methods are attracting more interest for this application.

RT PECVD SiN has been tried as the passivation layer for pentacene TFTs. The device was measured and compared before and after the deposition. Figure 15.7 shows TFT drain current as a function of gate-source voltage before and after passivation. The TFT performance is greatly degraded from the SiN deposition – the

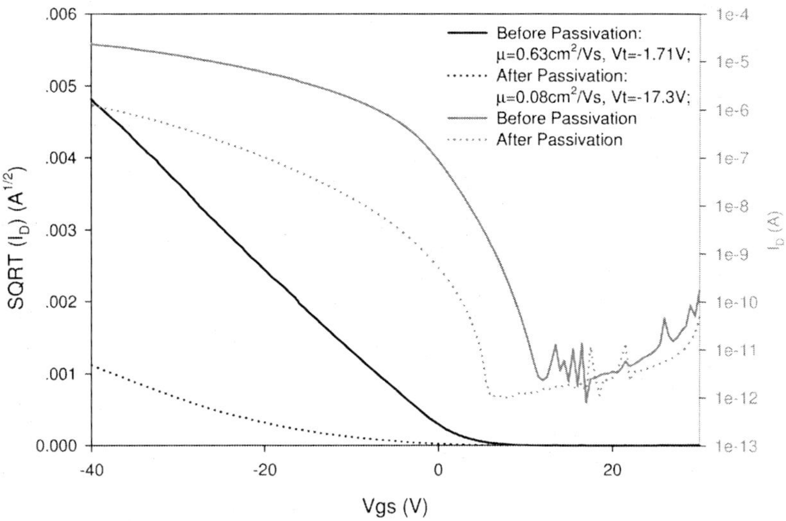

Fig. 15.7. TFT characteristics – drain current as a function of gate-source voltage before and after the SiN passivation, showing considerable degradation in both mobility and threshold potential as a result of the process.

field-effect mobility dropped from the original 0.6 $cm^2\ V^{-1}\ s^{-1}$ to 0.08 $cm^2\ V^{-1}\ s^{-1}$, and the threshold voltage shifted from an original -1.7 V to -17.3 V. It has been indirectly proved by exposure of pentacene TFTs to UV that the degradation was primarily because of the exposure to short wavelengths during the deposition [16].

Parylene is the generic name for the poly-*para*-xylylene class of polymers known for their ability to polymerize on surfaces from an active monomer gas with no liquid phase involved. Parylene-based polymers have been commercially available for over 25 years and have found widespread use as environmental and electrical isolation coatings for electronic circuits, automotive sensors, and medical substrates [17] owing to their ability to form conformal coatings. Parylene has some important advantages: virtually room-temperature deposition is used; it forms a uniform pinhole-free film with outstanding barrier properties with very low permeability to moisture and gases; and it has extreme chemical resistance against most common organic solvents and most acids and bases.

It also has superior electrical properties, for example extremely high dielectric strength and very high resistivity. For this work, parylene C was used because of its excellent combination of physical and electrical properties. Table 15.1 lists its important properties.

The parylene coating process is a three-step procedure that includes vaporization, pyrolysis, and polymerization. The parylene C coating process, shown schematically in Fig. 15.8, begins with the vaporization of the precursor dimer (di-*para*-xylylene), a granular white powder, at 150 °C and a pressure of 1 Torr. This

Tab. 15.1. Properties of Parylene C.

Dielectric strength (V cm^{-1})	2.7×10^6
Dielectric constant at 1 kHz	3.1
Moisture absorption (%, *w/w*)	0.06
Minimum process temperature (°C)	27
Volume resistivity (Ω m)	6×10^{16}
Glass transition temperature (°C)	80
Decomposition temperature (°C)	290
Refractive index	1.64
Deposition method	Vapor deposited
Patterning method	Dry etch
Solvent resistance	Good

results in a dimer gas that moves toward the pyrolysis chamber where, at 680 °C and 0.5 Torr, the dimer is cracked, yielding the monomer (a diradical *para*-xylylene). The monomer enters the room-temperature deposition chamber at 100 milliTorr where it is adsorbed by and polymerizes on all surfaces resulting in a

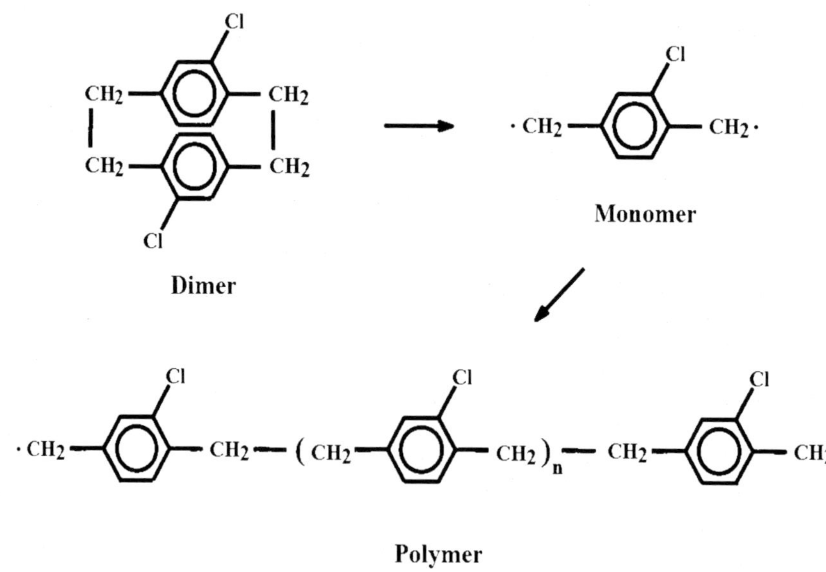

Fig. 15.8. Schematic diagram of the parylene C polymerization process.

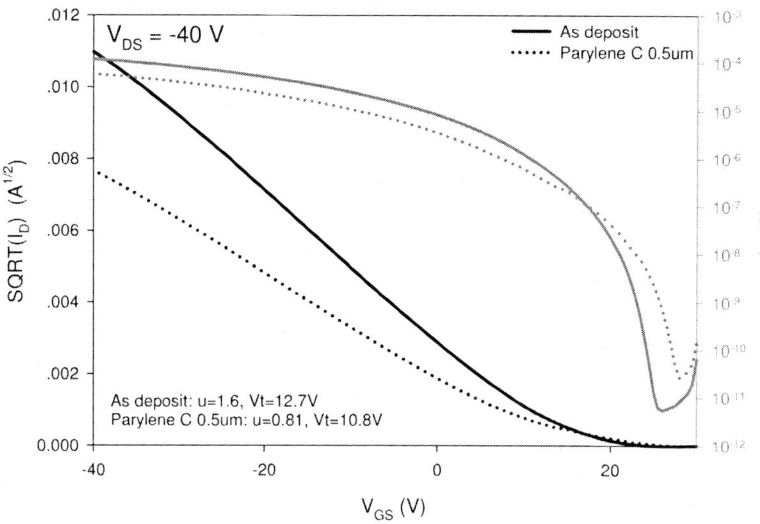

Fig. 15.9. Bottom-contact pentacene TFT characteristics before and after parylene passivation. The device has $W/L = 400/80$ μm. The passivation degrades the mobility whereas the threshold potential is relatively steady.

high molecular weight polymer (polychloro-*p*-xylylene). In our process, the parylene deposition was performed by means of a PDS 2010 Labcoater.

Although there are claims that parylene can passivate pentacene TFTs without degradation [18], this has only been shown where initial mobility before the passivation was less than 0.1 cm^2 V^{-1} s^{-1}. On our devices, for which starting mobility is uniformly much higher, the results reveal the degradation effect of parylene on good pentacene TFTs.

A 0.5-μm film of parylene-C was deposited on pentacene TFTs at room temperature. TFT characteristics were measured and compared before and after the deposition without patterning of the parylene. The bottom-contact TFT ($W/L =$ 400/80 μm) characteristics before and after parylene passivation are compared in Fig. 15.9. Field-effect mobility decreased by 50% from 1.6 cm^2 V^{-1} s^{-1} to 0.80 cm^2 V^{-1} s^{-1} whereas the threshold voltage stays the same.

To segregate our possible contact effects, the same experiment was repeated with gold top-contact-structure pentacene TFTs also. A 0.5-μm film of parylene-C was deposited and the TFTs were measured without a patterning of the parylene. The TFT ($W/L = 400/80$ μm) characteristics before and after parylene passivation are compared in Fig. 15.10. Field-effect mobility has a 40% decrease from 2.3 cm^2 V^{-1} s^{-1} to 1.4 cm^2 V^{-1} s^{-1}, while the threshold voltage stayed the same.

These results suggest that pentacene–parylene interaction usually degrades the pentacene active layer, although the degradation mechanism is not fully under-

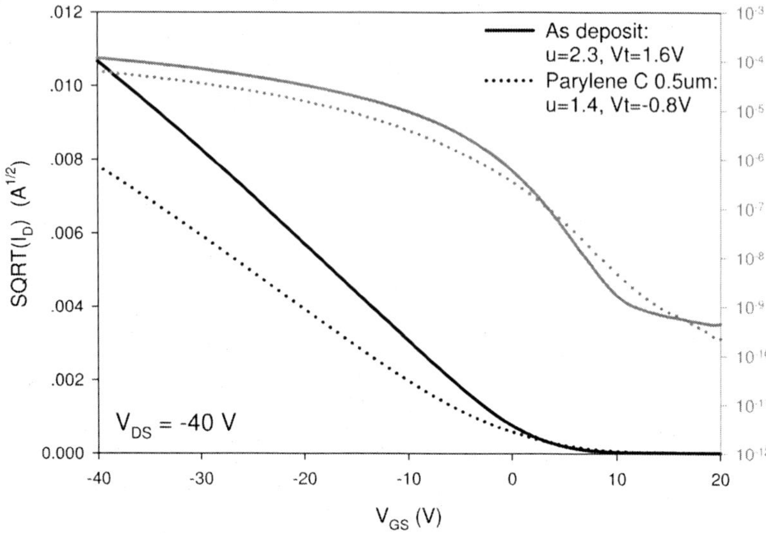

Fig. 15.10. Au top-contact pentacene TFT characteristics before and after parylene passivation. This device ($W/L = 400/80$ μm) has very similar characteristics to the bottom contact device in Fig. 15.9, suggesting degradation is not contact-related.

stood. The stress in this parylene film was measured to be relatively low, about 1×10^7 Pa, so we believe the degradation is more probably attributable to chemical reaction between pentacene and parylene-C film. Because the curves in Figs 15.9 and 15.10 overlap closely at the off and subthreshold regions, but have more differences in the saturation region, it is as if the devices lost some channel width. Pentacene usually grows in a 3D morphology with peaks and valleys in the film; perhaps, therefore, the parylene could reach down to, and react with, the interface pentacene molecular layers that conduct the channel current. The reaction, microscopically, would reduce the pathway of current conduction in the interface layer, and degrade the TFT mobility macroscopically.

15.8
PVA and Parylene

A direct solution to this problem would be a buffer layer between the pentacene channel layer and parylene passivation layer. Given that poly(vinyl alcohol) (PVA) has been used as a photo-patternable etch mask for pentacene, it can also work as this buffer layer. After PVA patterning of the pentacene TFT, 1 μm parylene was vapor deposited and patterned by standard photolithography and RIE dry etching. The TFT characteristics were measured before and after the parylene passivation.

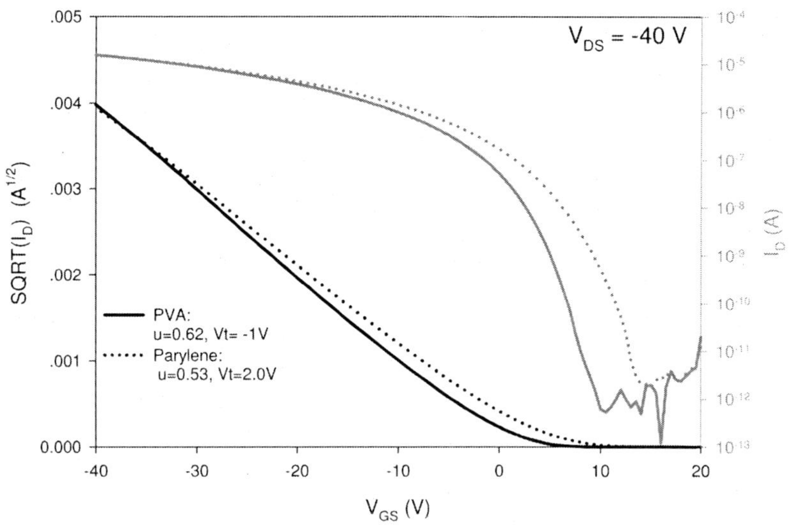

Fig. 15.11. Pentacene TFT characteristic change after parylene passivation with PVA as a buffer layer. The small change in mobility (less than 14%) suggests the PVA layer acts as a buffer between pentacene and the reactive parylene.

Figure 15.11 shows the plot of I_D against V_{GS}. The mobility has changed slightly, from 0.62 $cm^2\ V^{-1}\ s^{-1}$ to 0.53 $cm^2\ V^{-1}\ s^{-1}$, and the threshold potential has shifted from -1 V to 2 V. This result suggests that the etch-mask PVA can protect the pentacene TFT from parylene passivation by acting as a buffer layer.

15.9
Pentacene TFT Uniformity

Pentacene TFTs were fabricated as described above and tested. Figure 15.12 shows the typical characteristics of the drive TFT. Figure 15.12a is a plot of drain current against gate-source voltage at a drain source voltage of -40 V and Fig 15.12b is a plot of drain current against drain-source voltage for a range of gate-source voltage. The drive TFT has a saturation current of 70 μA at -40 V drain-source voltage, which clearly satisfies our design requirement. TFT characteristics can be extracted from Fig. 15.12a; the saturation field-effect mobility is 0.6 $cm^2\ V^{-1}\ s^{-1}$, the threshold voltage is 12.6 V, the subthreshold slope is 2.5 V $decade^{-1}$, and the on/off ratio is greater than 10^7.

A uniformity test array consisting of 105 TFTs ($W/L = 200/20$ μm) was included on the edge of the backplane design; the most important properties, field-effect mobilities and threshold voltages, are plotted in Fig. 15.13. The statistics on threshold voltage show the mean value was 13.7 V, the standard deviation 0.78 V, and

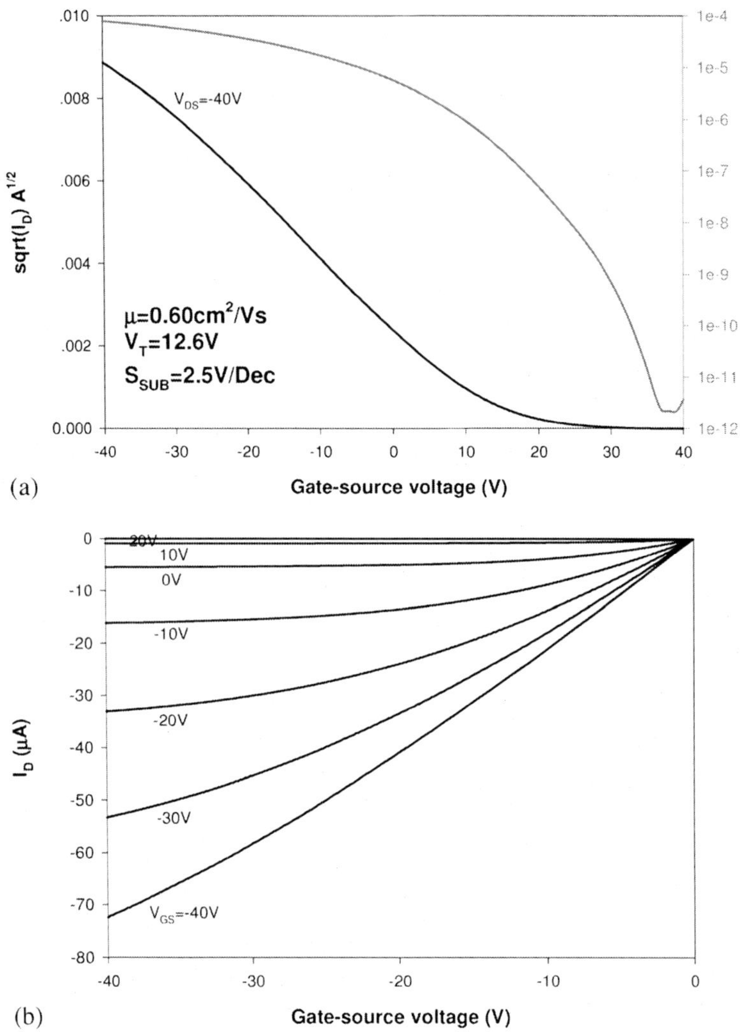

Fig. 15.12. Typical drive-TFT characteristics, as fabricated and passivated for a backplane. (a) I_d against V_{GS}, (b) I_d against V_{DS}.

the standard error 0.078; field-effect mobility statistics were mean value 0.584 $cm^2\ V^{-1}\ s^{-1}$, standard deviation 0.017 $cm^2\ V^{-1}\ s^{-1}$, standard error 1.7×10^{-3}.

Although the data here show we can achieve relatively uniform TFT performance, it is still not good enough for a display application, so a compensation circuit would be needed in the pixel design. Threshold voltage stability and uniformity, together, form perhaps the most risky feature of organic TFT backplanes for OLEDs.

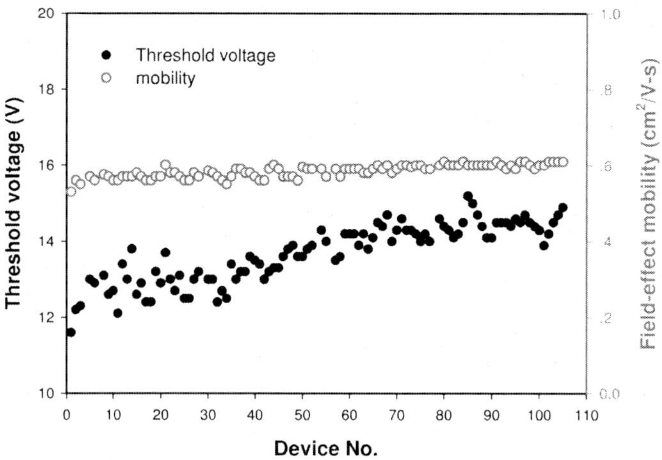

Fig. 15.13. Uniformity test of 105 drive TFTs. Mobility uniformity is excellent. Threshold voltage has an undesirably large standard deviation of 0.78 V.

15.10 Stability

Understanding and controlling stability is a critical issue for display applications. A simple bias stress test ($V_G = -40$ V, $V_{DS} = -40$ V) was conducted on pentacene TFTs for different times. Figure 15.14 shows the TFT characteristics before and after the bias stress. A few volts shift is observed in the threshold potential after a 20-min bias stress.

Organic TFTs stability has been studied in some detail by several different groups [9, 19–24], and the proposed mechanisms for threshold potential shift seem to be material-dependent and to vary for different dielectric/semiconductor systems. For example, in the regioregular polythiophene known as PQT-12 an irreversible shift in the threshold potential is attributed to charge trapping in disordered regions of the films [21]. In contrast, for an all-organic thin-film transistor made from solution-processed pentacene on a commercially available photoresist dielectric, the shifts in threshold are connected to mobile ions (associated with water) in the devices [25]. In yet another approach, bias stress-induced threshold voltage shifts were suppressed in evaporated pentacene devices with polyimide gate insulators by encapsulation and annealing, without a specification of the causal mechanisms [22].

In general, stability problems can be mitigated by using a lower driving voltage on the TFTs, making use of improvements in OLED turn-on voltage and current efficiency, or solved by designing a more complex pixel-driving circuit to compensate. More work remains, however, both in the fundamental device, in materials studies, and in engineering solutions, to make organic transistors a reliable technology.

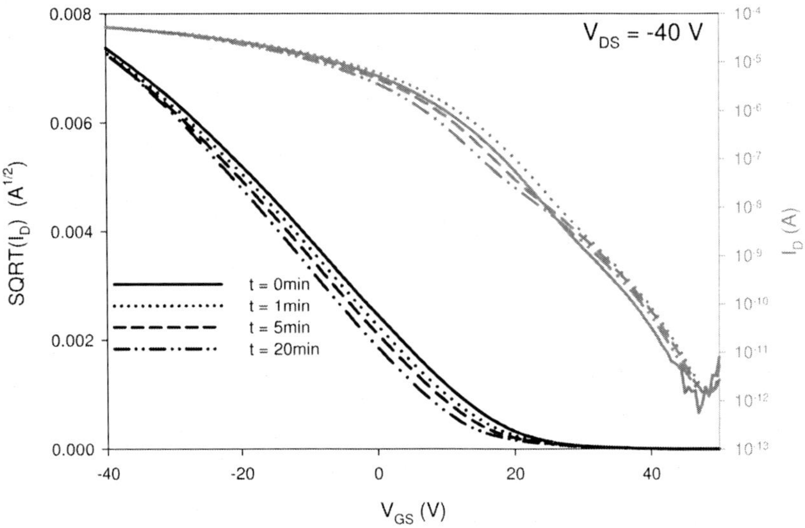

Fig. 15.14. Bias stress test of a drive TFT. A constant bias of $V_G = V_{DS} = -40$ V was applied for different times. Threshold potential shifts of a few volts are seen after 20 min continuous bias.

15.11 Integration of TFT and OLEDs

A passive OLED pixel was included on our active-matrix pentacene backplane for testing after OLED deposition and encapsulation. Figure 15.15 shows the I–V characteristics of a passive OLED pixel with an area of 2×2 mm^2. An isolated TFT-driven OLED pixel was also tested. Figure 15.16 shows the circuit schematic and OLED current for different data voltages. Clearly, OLEDs can be easily driven by a pentacene TFT.

Figure 15.17 shows a microscopic view of a pentacene TFT-driven OLED pixel on the display.

Figure 15.18 shows a 48×48 array driven with all of the pixels in the on or off state. With reference to Fig. 15.1, the pixel schematic, the display is driven under the conditions $V_{DD} = 20$ V and $V_{CA} = -10$ V. Under the "all on" condition $V_{SELECT} = 0$ V and $V_{DATA} = 0$ V whereas under the "all off" condition $V_{SELECT} = 0$ V and $V_{DATA} = 30$ V. There are many noticeable defects in the display because the processing was not conducted in a clean room. Several pixels were on when they are supposed to be off, and off when they are supposed to be on. The large number of defects demonstrates how important it is to process in a clean environment to yield large-area electronics. Attempts at minimizing the number of defects were made by more thoroughly cleaning the substrates during the process and trying

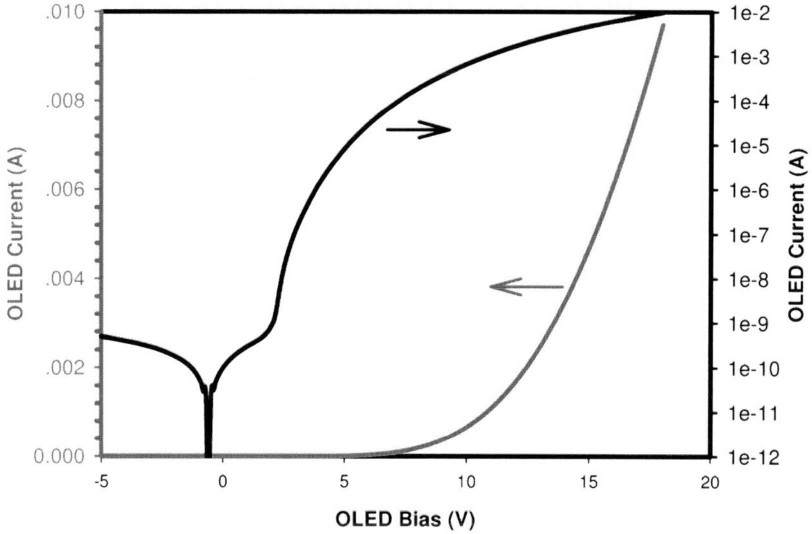

Fig. 15.15. Passive OLED test device current.

to reduce the number of particles. Although it was difficult to demonstrate a defect-free display, these results demonstrate that pentacene TFT backplanes are viable candidates for active-matrix OLED displays. Practical issues must still be resolved, however, for example TFT encapsulation and TFT uniformity and stability improvement.

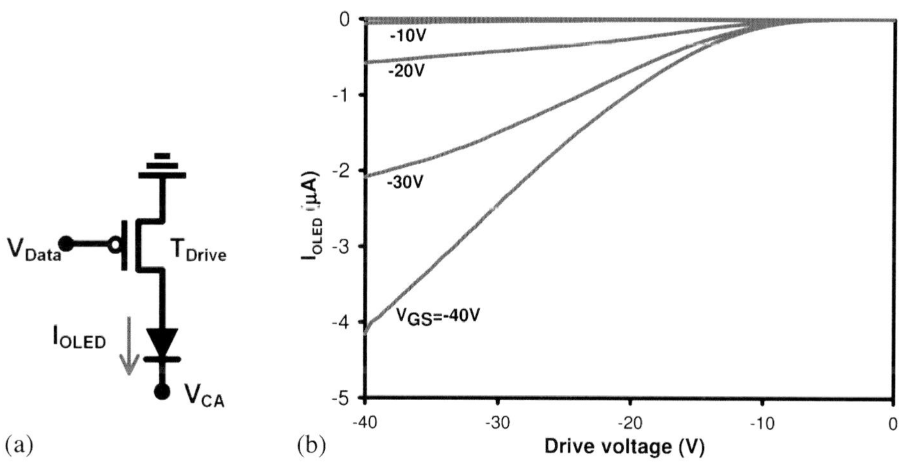

Fig. 15.16. (a) OLED active test pixel schematic, and (b) OLED current for different data voltages.

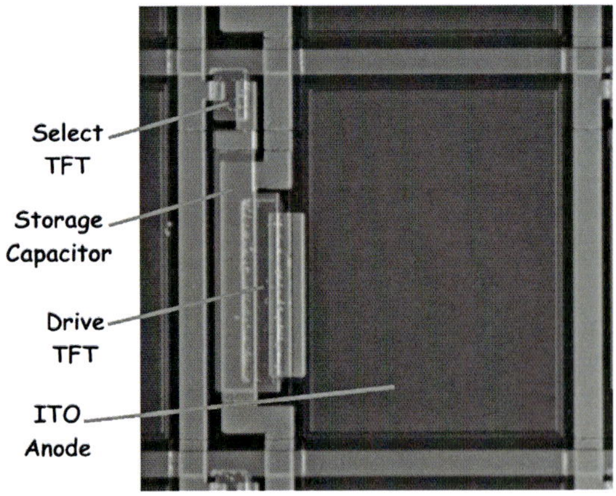

Fig. 15.17. Optical micrograph of a pentacene TFT-driven OLED pixel.

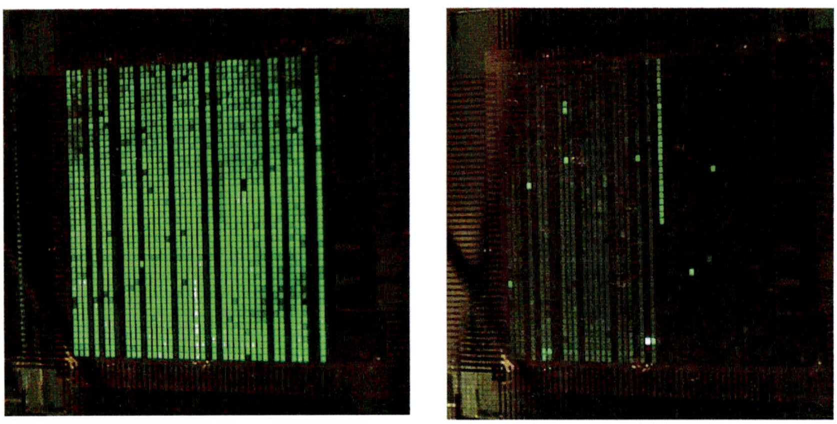

(a) (b)

Fig. 15.18. A 48 × 48 OTFT-driven OLED display on glass, driven with the pixels in (a) the all-on state and (b) the all-off state.

15.12 Flexible OLED Display

Although there have been several examples of flexible OLED displays on plastics, including passive-matrix OLED displays on poly(ethylene terephthalate) (PET) substrates from Pioneer and Universal Display Corporation [26, 27], and a-Si:H TFT-driven monochrome active-matrix OLED displays on poly(ethylene naphthalate) (PEN) from Honeywell [28], there have been no demonstrations of organic TFT-driven active-matrix OLED displays fabricated on polymeric substrates. Displays on polymeric substrates have advantages in both function and cost. With the advantages of low processing temperature, it is possible to use low-cost manufacturing methods, for example roll-to-roll processing, and produce inexpensive displays. In addition, flexible substrates make the display more attractive because it is flexible, lightweight, and rugged. In some applications the displays may have less stringent requirements on performance. For example, a display with less than 100% pixel yield, low resolution, or a short lifetime may be acceptable. Here, we report preliminary results on pentacene OTFT-driven active-matrix OLED displays on a flexible polymeric substrates.

An important part of this study was to recognize and address difficulties resulting from processing on polymeric substrates. Problems encountered included substrate handling, stability of the substrate, surface smoothness of the substrates, and effects of chemical exposure. Effective encapsulation of an OLED on a flexible polymeric substrate is a challenge but not one we will address in detail here.

15.13 Substrate Selection and Mounting

An excellent introduction to issues relating to substrates can be found in Chapter 7. We chose PET films as the substrate for our preliminary demonstration of a flexible active-matrix OLED display. The PET films had a nominal thickness of 125 μm. They were first cleaned by rinsing in acetone and IPA then heated at 150 °C for several hours under vacuum to reduce the level of moisture and volatile contaminants and to pre-shrink the material for improved thermal dimensional stability before processing. The films were laminated to glass carriers by use of a removable pressure-sensitive silicone gel adhesive, as shown in Fig. 15.19, to improve thermal coupling, maintain a flat surface, and provide rigid support during processing [29].

15.14 Thermal Dimensional Stability

Thermal and dimensional stability are critical in enabling a film to withstand the process temperatures, to ensure precision registration of the different layers in the

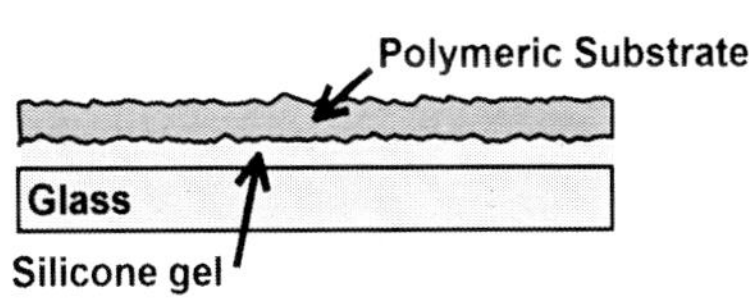

Fig. 15.19. Cross section of polymeric substrate mounted on a glass carrier.

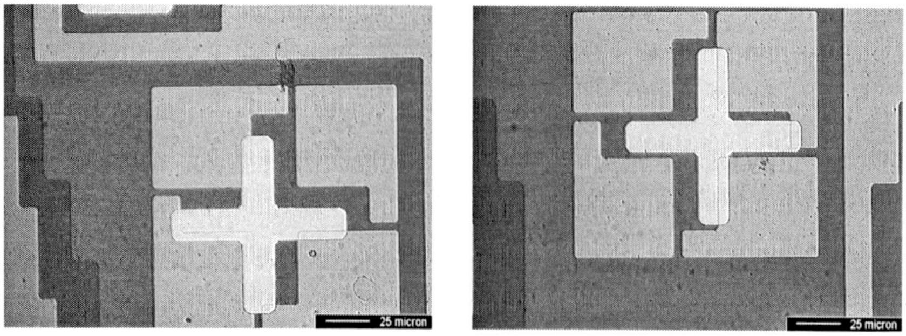

Fig. 15.20. Registration marks on PEN after a full OTFT process. Shrinkage can be determined by measuring the offsets.

final device, and for the multilayer device to be able to withstand thermal cycling. Plastic films undergo a variable and undesirable change in dimensions at the glass transition temperature (*Tg*) as a result of both molecular relaxation events associated with the increased mobility of the polymer chains and also because of "shrinkage" or "expansion" associated with the relaxation of residual strain within the oriented parts of the film structure. These changes in the plastic film can directly affect the integrity of the thin films on top. For example, we have observed that chromium thin films buckled after thermal evaporation on PET substrates.

As Fig. 15.20 shows, substrate shrinkage can be measured from misalignment of the registration marks after a standard pentacene TFT process. It was observed that shrinkage of ~500–1000-ppm (depending on directions and processes) can be observed for a 5-cm PET substrate whereas the range for a 5-cm PEN substrate is ~280–400 ppm. Large shrinkage not only requires large tolerance in mask design but also harms the integrity of the films deposited on the substrates.

Although pre-annealing the substrate can greatly reduce shrinkage, and careful process design can reduce the swelling effects of solvents and water, additional measures are still required for the dimensional stability.

15.15 Surface Quality

The substrate surface smoothness is critical to TFT performance. Device fabrication processes basically duplicate and/or worsen the surface roughness, which leads to smaller pentacene grains and results in deterioration of pentacene channel mobility. Atomic-force microscopy was used to characterize the surface roughness. Figure 15.21 shows an AFM image of our PET substrate surface before any process. The mean-square roughness and peak-to-valley roughness are 10 Å and 90 Å,

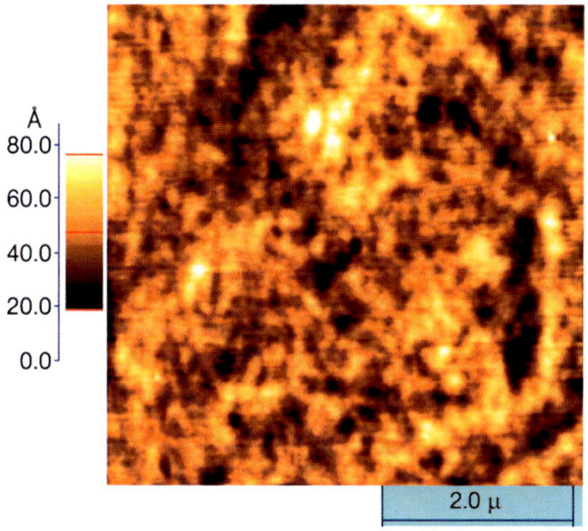

Fig. 15.21. AFM image of PET substrate surface. While visually boring, the surface roughness is sufficient to significantly reduce pentacene mobility if it is not in some way planarized.

respectively. The surface roughness measurement suggests that pentacene TFTs should at least have mediocre performance on PET substrates.

Cleanliness of the flexible substrates is also essential to ensure the integrity of subsequent layers such as conductive coatings and dielectric layers. Although the PET films have microscopically smooth surfaces, they also have surface scratches, spikes, and defects. In the absence of the harsh cleaning procedures possible on inorganic substrates, particulate matter on the polymeric substrates surfaces may also be an issue affecting the yield of display. In addition to improved manufacturing of the base film, the film surface must be engineered by adding a coating to improve planarization and scratch resistance [30].

15.16 Chemical Resistance

A wide range of solvents and chemicals is used when processing the different layers in the display and the effect of water and solvent absorption has a very detrimental effect on dimensional stability. This becomes increasingly important as processing moves toward roll to roll manufacture.

An inorganic layer, 500 nm RT PECVD SiN, was used to passivate the plastic substrates, protect the substrates from the fabrication process, and help maintain substrate dimensional stability by minimizing solvent or moisture uptake. Given that the Young's modulus of this inorganic layer is much higher than that of the plastic film, even the thin film helps maintain substrate dimensional stability.

15.17 Fabrication Process

The fabrication process is similar to that on glass substrates. The only difference is that a bilayer of gate dielectrics was used instead of single layer of SiO_2. A 300 nm thick RT PECVD SiN dielectric layer was first deposited using RT PECVD and patterned by CF_4 plasma dry etching. A 100 nm SiO_2 layer was then deposited by sputtering and patterned with liftoff. The first SiN layer was used to improve gate dielectric integrity and display yield. Our RT PECVD SiN layer usually has a rough surface, however, and devices built directly on this surface normally have very low mobility, less than 0.1 $cm^2\ V^{-1}\ s^{-1}$. SiO_2 was therefore used to improve dielectric surface smoothness and to provide a treatable surface from the self-assembly monolayer of octadecyltrichlorosilane [31].

15.18 Display Results

A 48 × 48 active-matrix OLED display was completed on flexible PET substrates. The completed display is shown bent, after wire bonding, in Fig. 15.22. Figure 15.23 shows the display brightness under different data voltages, and Fig. 15.24 demonstrates its flexible functionality during operation. Despite the many noticeable defects on the display, which are expected for displays fabricated outside a clean room, the display was characterized by good drive performance and uniformity.

Fig. 15.22. A 48 × 48 OTFT-driven OLED display on PET, shown after wire bonding.

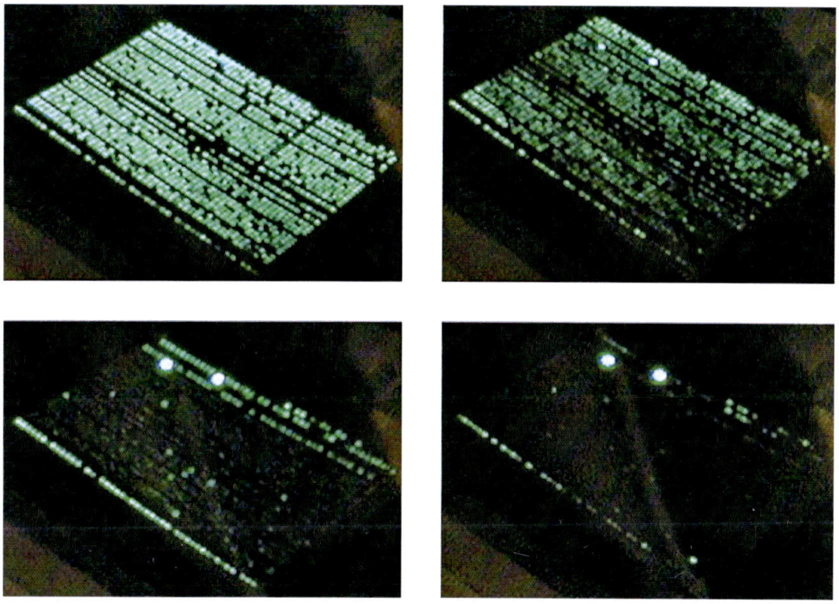

Fig. 15.23. The flexible OLED display brightness under different data voltages.

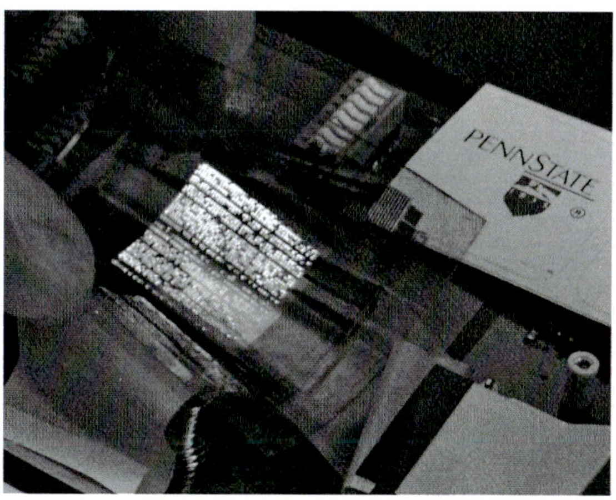

Fig. 15.24. The flexible OLED display working while bent.

15.19
Conclusion

We have demonstrated that OTFTs can be integrated with OLEDs to make prototype displays both on glass and on flexible substrates. Although an instructive exercise, we recognize that this process has not addressed some of the issues that arise when trying to make OTFT backplane technology. If low cost is a primary driver, patterning techniques and materials must be developed that are compatible with genuinely low-cost manufacture (probably quite unlike the standard lithography and vacuum-based processes used in these examples). The origins of the principle defects, and the remedies required to reduce them, are other important areas of research we have not addressed. Indeed, it can be difficult to understand defects effectively until the yield numbers from larger numbers of backplanes (processed under more ideal conditions) become available.

Despite the relative immaturity of this field, however, the ability of organic thin-film transistors to form a usable backplane is indubitable. The question of who will succeed in tackling the remaining issues, and with what sort of creative solutions, remains to be seen.

References

1 M. Fihn, "Flat panel display and equipment market outlook," presented at USDC and Needham and Company Boston Display Investor's Conference, **2001**.

2 C. W. Tang, S. A. Van Slyke, "Organic electroluminescent diodes," *Appl. Phys. Lett.* **1987**, *51*, 913–915.

3 R. M. A. Dawson, M. G. Kane, "Pursuit of Active Matrix Organic Light Emitting Diode Displays," *SID Symposium Digest of Technical Papers*, **2001**, *32*, 372–375.

4 R. M. A. Dawson, Z. Shen, D. A. Furst, S. Connor, J. Hsu, M. G. Kane, R. G. Stewart, A. Ipri, C. N. King, P. J. Green, R. T. Flegal, S. Pearson, W. A. Barrow, E. Dickey, K. Ping, S. Robinson, C. W. Tang, S. A. Van Slyke, C. H. Chen, J. Shi, M. H. Lu, M. Moskewicz, J. C. Sturm, "A Polysilicon Active Matrix Organic Light Emitting Diode Display with Integrated Drivers," *SID Symposium Digest of Technical Papers* **1999**, *30*, 438–441.

5 R. M. A. Dawson, Z. Shen, D. A. Furst, S. Connor, J. Hsu, M. G. Kane, R. G. Stewart, A. Ipri, C. N. King, P. J. Green, R. Y. Flegal, S. Pearson, W. A. Barrow, E. Dickey, K. Ping, C. W. Tang, S. A. Van Slyke, F. Chen, J. Shi, J. C. Sturm, and M. H. Lu, "Design of an Improved Pixel for a Polysilicon Active-Matrix Organic LED Display," *SID Symposium Digest of Technical Papers* **1998**, *29*, 11–14.

6 J.-J. Lih, C.-F. Sung, "Full-color active-matrix OLED based on a-Si TFT technology," *J. SID* **2003**, *11*, 617–620.

7 T. Chuman, S. Ohta, S. Miyaguchi, H. Satoh, T. Tanabe, Y. Okuda, M. Tsuchida, "Active Matrix Organic Light Emitting Diode Panel Using Organic Thin-Film Transistors," presented at SID, Seattle, Washington, **2004**.

8 D. J. Gundlach, C.-C. Kuo, S. F. Nelson, T. N. Jackson, "Organic thin film transistors with field-effect mobility > 2 $cm^2\ V^{-1}\ s^{-1}$," presented at 57th Device Research Conference Digest, **1999**.

9 H. Klauk, D. J. Gundlach, J. A. Nichols, T. N. Jackson, "Pentacene organic thin-film transistors for circuit and display applications," *IEEE T. Electron Dev.* **1999**, *46*, 1258–1263.

10 Y. He, R. Hattori, J. Kanicki, "Current-source a-Si:H thin-film transistor circuit for active-matrix organic light-emitting displays", *IEEE Electr. Dev. L.* **2000**, 21, 590–592.

11 A. Yumoto, M. Asano, H. Jasegawa, M. Sekiya, "Pixel-driving methods for large-sized poly-Si AMOLED displays,"

presented at International Display Workshop, **2000**.

12 D. Fish, S. Deane, J. Shannon, A. Steer, N. Young, "35.2: Improved Optical Feedback for AMOLED Display Differential Ageing Compensation," *SID Symposium Digest of Technical Papers* **2004**, *35*, 1120–1123.

13 T. A. Fjeldly, T. Ytterdal, M. Shur, *Introduction to Device Modeling and Circuit Simulation*, John Wiley & Sons, New York, **1998**.

14 D. J. Gundlach, C.-C. Kuo, C. D. Sheraw, T. N. Jackson, "Improved organic thin film transistor performance using chemically-modified gate dielectrics," presented at the International Society for Optical Engineering's 46th Annual Symposium on Optical Science and Technology, **2001**.

15 D. J. Gundlach, T. N. Jackson, D. G. Schlom, S. F. Nelson, "Solvent-induced phase transition in thermally evaporated pentacene films," *Appl. Phys. Lett.* **1999**, *74*, 3302.

16 L. Zhou, "a-Si:H and Pentacene TFT for Flexible Electronic Applications," in *Electrical Engineering Department*. University Park, Pennsylvania State University, **2005**.

17 Y. S. Yeh, W. James, H. Yasuda, "Polymerization of para-xylylene derivatives, VI. morphology of parylene N an parylene C films investigated by gas transport characteristics," *J. Polym. Sci. Part B* **1990**, *28*, 545.

18 I. Kymissis, C. D. Dimitrakopoulos, S. Purushothaman, "Patterning pentacene organic thin film transistors," *J. Vac. Sci. Technol. B* **2002**, *20*, 956–959.

19 H. L. Gomes, P. Stallinga, F. Dinelli, M. Murgia, F. Biscarini, D. M. de Leeuw, M. Muccini, K. Müllen, "Electrical characterization of organic based transistors: stability issues," *Polym. Adv. Technol.* **2005**, *16*, 227–231.

20 D. Knipp, R. A. Street, A. Volkel, J. Ho, "Pentacene thin film transistors on inorganic dielectrics: Morphology, structural properties, and electronic transport," *J. Appl. Phys.* **2003**, *93*, 347–355.

21 A. Salleo, F. Endicott, R. A. Street, "Reversible and irreversible trapping at room temperature in poly(thiophene) thin-film transistors," *App. Phys. Lett.* **2005**, *86*, 263505.

22 T. Sekitani, S. Iba, Y. Kato, Y. Noguchi, T. Someya, T. Sakurai, "Suppression of DC bias stress-induced degradation of organic field-effect transistors using postannealing effects," *Appl. Phys. Lett.* **2005**, *87*, 073505.

23 A. Salleo, R. A. Street, "Light-induced bias stress reversal in polyfluorene thin-film transistors," *J. Appl. Phys.* **2003**, 94, 471–479.

24 R. A. Street, A. Salleo, M. L. Chabinyc, "Bipolaron mechanism for bias-stress effects in polymer transistors," *Phys. Rev. B* **2003**, *68*, 085316.

25 S. J. Zilker, C. Detcheverry, E. Cantatore, D. M. d. Leeuw, "Bias stress in organic thin-film transistors and logic gates," *Appl. Phys. Lett.* **2001**, *79*, 1124–1126.

26 A. Yoshida, S. Fujimura, T. Miyake, T. Yoshizawa, H. Ochi, A. Sugimoto, H. Kubota, T. Miyadera, S. Ishizuka, M. Tsuchida, H. Nakada, "211: Invited Paper: 3-inch Full-color OLED Display using a Plastic Substrate," *SID Symposium Digest of Technical Papers* **2003**, *34*, 856–859.

27 M. S. Weaver, L. A. Michalski, K. Rajan, M. A. Rothman, J. A. Silvernail, J. J. Brown, P. E. Burrows, G. L. Graff, M. E. Gross, P. M. Martin, M. Hall, E. Mast, C. Bonham, W. Bennett, M. Zumhoff, "Organic light-emitting devices with extended operating lifetimes on plastic substrates," *Appl. Phys. Lett.* **2002**, *81*, 2929–2931.

28 K. R. Sarma, J. Schmidt, J. Roush, C. Chanley, S. Dodd, "AMOLED using a-Si TFT backplane on flexible plastic substrate," presented at Proc SPIE, Defense, Security, and Cockpit Displays, Orlando, Florida, USA, **2004**.

29 J. R. Huang, Y. Lee, T. N. Jackson,

C. R. Wronski, C. M. C. Toporow, G. J. Vendura, "Improved Thermal Control for a-Si:H Photovoltaic Cells Fabricated on Polymeric Substrates," presented at 26th IEEE Photovoltaic Specialists Conference, Los Angeles, CA, USA, **1997**.

30 W. A. MacDonald, "Engineered films for display technologies," *J. Mat. Chem.* **2004**, *14*, 4–10.

31 Y.-Y. Lin, D. J. Gundlach, S. F. Nelson, T. N. Jackson, "Stacked pentacene layer organic thin-film transistors with improved characteristics," IEEE *Electr. Device L.* **1997**, *18*, 606–608.

16
Large-area Detectors and Sensors

Takao Someya and Takayasu Sakurai

16.1
Introduction

Organic field-effect transistors (FETs) are of great interest, because of numerous advantages over inorganic semiconductors such as silicon and GaAs. First, organic semiconductor devices such as transistors can be manufactured on plastic films at ambient temperature and therefore are lightweight, thin, and mechanically flexible. Second, it is expected that organic transistor-based large-area circuits would be manufactured easily using ultralow cost processes such as printing technology and/or roll-to-roll techniques.

It has been reported that the mobility and on/off ratio of organic transistors with pentacene channel layers exceeds 1 $cm^2\ V^{-1}\ s^{-1}$ and 10^7, respectively [1–5], numbers comparable with those reported for amorphous silicon. Although thin-film, low-molecular-weight organic semiconductors such as pentacene are usually deposited under vacuum, many efforts have been made to fabricate plastic transistors by manufacturing processes compatible with printing technology [5, 6]. Organic transistors fabricated by inkjet [5, 6], micro-contact printing [7], and other printing methods have been developed and reported.

The two major driving applications for development of organic transistors are radio-frequency (rf) identification tags [8, 9] and flexible displays [7, 10] including paper-like screens or e-paper. In contrast, our group has proposed large-area electronics as another area with promising applications for organic transistors [11–16]. We have recently developed an electronic artificial skin for robots by integrating an organic transistor active matrix with rubber pressure sensors. In this new type of application high speed is not required and many advantages of organic transistors, for example large-area processability, flexibility, and low-cost, are benefits.

In this chapter we review recent progress of organic transistors for sensor applications. Emphasis is put on large-area, flexible pressure sensors suitable for electronic artificial skin and for photodetectors suitable for sheet image scanners. We also describe future prospects of large-area sensors and the other issues.

Organic Electronics, Materials, Manufacturing and Applications. Edited by Hagen Klauk

ISBN: 3-527-31264-1

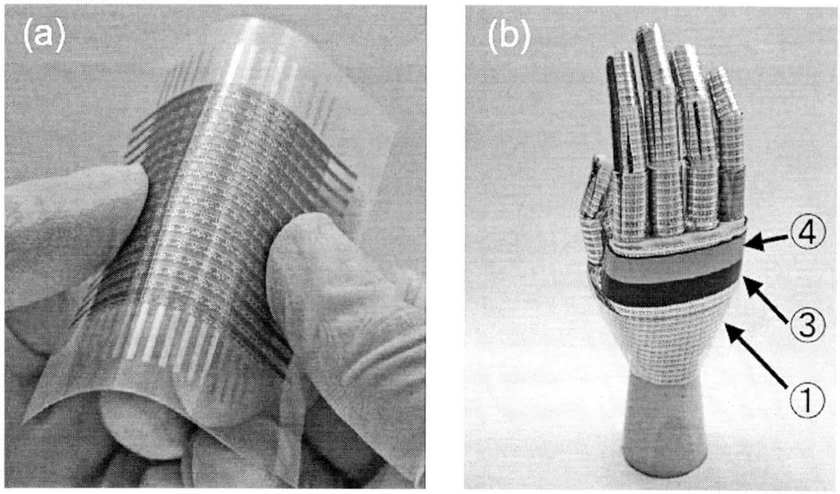

Fig. 16.1. A flexible, large-area pressure sensor. (a) Organic transistor active matrix is formed on a plastic film and integrated with pressure-sensitive rubber. (b) An image of electronic artificial skin attached to the robot surface. A plastic film with organic transistors (1), a pressure sensitive rubber sheet (3), a plastic film with top electrodes (4) are laminated together to form a large-area pressure sensor. An intermediate layer (2) is not seen.

16.2
Large-area Pressure Sensors

Large-area pressure sensor sheets are mechanically flexible, as shown in Fig. 16.1, and can therefore be wrapped around fine cylindrical bars, for example robot fingers. A sense of touch for humanoid robots is far behind the senses of sight and hearing. This is mainly because a flexible, large-area pressure sensor matrix has not been manufactured at reasonable cost. Flexible pressure sensors have been made from polymers or rubber. With increasing number of sensors in the matrix, however, problems associated with wiring cannot be overlooked; this makes it impossible to increase the density or total number of sensors to that comparable with human skin.

In this scheme we have overcome the above problem by introducing organic transistor integrated circuits as a flexible active matrix to read out pressure images, or distribution of pressure. As a result, we have successfully developed large-area, flexible pressure sensors with the number of pressure sensors exceeding 1,000. As shown in Fig. 16.2, the device is manufactured by laminating four different functional films:

1. a base plastic film with organic transistors,
2. a film with interconnection layer,
3. a pressure-sensitive rubber sheet, and
4. a film suspending copper electrodes for power supply.

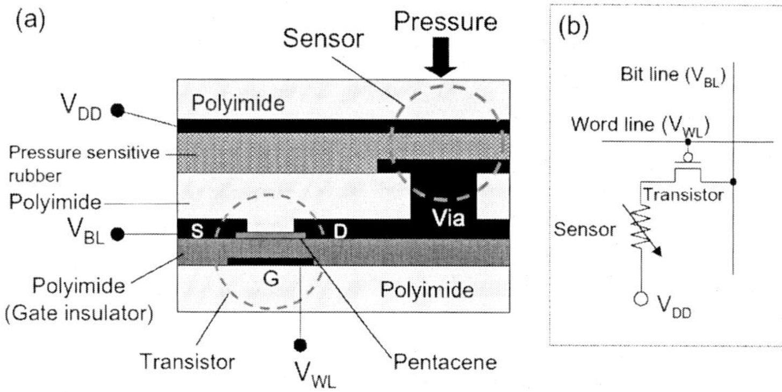

Fig. 16.2. Integration of organic transistors with pressure-sensitive rubber sheets. (a) A cross-sectional view of the device structure and (b) a circuit diagram of one sensor cell. The device is formed by laminating four different functional films. The pressure sensors are made by sandwiching a pressure-sensitive rubber sheet between two electrodes. One of the two electrodes is connected to the transistors through via holes. The source, drain, and gate electrodes and supply voltage are denoted S, D, G, and V_{DD}, respectively.

One sensor cell (sensel) consists of one organic transistor and one pressure sensor. We have made 16 × 16 or 32 × 32 sensel matrixes. The periodicity is 2.54 mm, which corresponds to 10 dots per inch (dpi).

Here we describe the manufacturing process flow. The base film is poly(ethylene naphthalate) (PEN) or polyimide, thickness 50–125 μm (Teijin Dupont Films, Q83 and Q65). First, gate electrodes consisting of 5 nm thick chromium and 100 nm thick gold are deposited in the vacuum system. Polyimide precursors (Kyocera Chemical, Kemittite CT4112A) are then spin-coated on the base film with gate electrodes and cured at 180 °C. Although polyimide is usually cured at temperature above 300 °C, the current material can be cured at 180 °C [17]. This curing temperature is compatible with the process with PEN films, which are low in cost and have quite low moisture absorption and water vapor permeability.

The channel layer is 50-nm-thick pentacene deposited in the vacuum sublimation system at ambient substrate temperature with fine metal masks. After deposition of pentacene thin film, 30-nm-thick gold layers are deposited as source and drain though fine metal masks.

The channel length L and width W of the transistors are 100 μm and 1.9 mm, respectively. The bare transistors without sensors have a mobility of 0.5 $cm^2\ V^{-1}\ s^{-1}$ and an on/off ratio of 10^6.

The plastic film with interconnection layers, denoted (2) in Fig. 16.1, can be made by a process similar to that used to manufacture flexible circuit boards. First, plastic films coated with copper foil are processed by a numerically controlled (NC) drilling machine to make via holes. Plating is then used to make interconnections between top and bottom sides though via holes. Finally, the copper layers are patterned by conventional photolithography and etching. Gold plating is occasionally employed to improve electronic interconnections.

The pressure sensitive rubber, denoted (3) in Fig. 16.1, is a 0.5 mm thick silicone rubber sheet containing graphite particles as conductors. On application of pressure to the rubber sheet, as indicated by the arrow in Fig. 16.2, spacing between graphite particles changes, resulting in big changes of resistance in a wide range from a few kΩ to complete insulator.

The copper electrodes for the power supply are suspended by means of a polyimide film, denoted (4) in Fig. 16.1. Those copper electrodes are not patterned.

As shown in Fig. 16.2, the pressure sensors are formed as pressure-sensitive rubber (3) sandwiched between two electrodes and connected to the transistors though via holes (2). Voltage bias V_{DD} is connected to the transistor when pressure is applied to the sensors, enabling detection of the distribution of pressure.

16.3 Organic Transistor-based Integrated Circuits

To read out pressure images from the sensor matrix, integrated circuits such as decoders and selectors are needed. We have also manufactured those integrated circuits with organic transistors and characterized the electronic performance of the system.

Figure 16.3 shows a circuit diagram of an electronic artificial skin system. Integrated circuits are formed by organic transistors with a pentacene channel layer; this has p-type conduction and consists of sensor matrix, column selector, and row decoder. The manufacturing process flow of the sensor matrix has been described above; other circuits are processed similarly.

The manufacturing process of selectors and decoders requires transistor–transistor interconnections with wiring gate electrodes and source/drain electrodes. To realize those interconnections, some spots of polyimide gate dielectric layers, uniformly coated on the base film, are removed by a CO_2 laser-drilling machine. The diameter of laser via holes is 90 μm. Although laser processes are widely used in the manufacture of flexible printed circuits, the current study is, to the best of our knowledge, the first demonstration to exploit this laser via technique for organic transistor integrated circuits.

Compared with conventional photolithography, laser drilling is a dry process and keeps the surface of polyimide gate dielectric layer away from water, etching solution, or other solvent, which often degrade the polyimide surface. We have confirmed that the electronic performance of transistors with laser via holes is identical with that without laser via holes.

The sensor matrix, column selector, and row decoder are first manufactured on separate films and then assembled with silver pastes and connecting tapes made of PEN films with gold stripes with 0.1 inch spacing. In this way, the circuits are manufactured physically by a "cut and paste" procedure. The sensor matrix consisting of 16 × 16 sensor cells has assembling electrodes with 0.1 inch spacing to glue to connecting tapes. Figure 16.3 shows a 16 × 16 sensor matrix. When a smaller sensor matrix, for example a 4 × 4 matrix, is needed, we can make it by

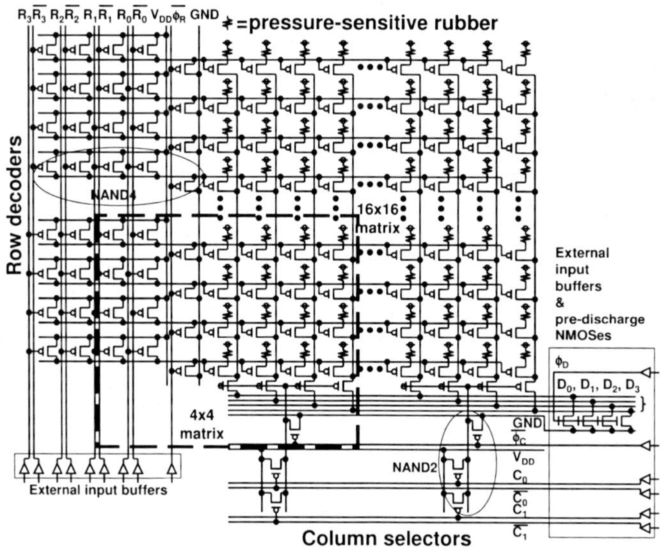

Fig. 16.3. Circuit diagram of electronic artificial skin consisting of a 16 × 16 access transistor matrix, column selector, and row decoder. The manufactured transistor with pentacene channel layer has p-type conduction. R0–R3 are row addresses, C0–C1 are column addresses, the bar indicates the reverse signal, ϕR-bar and ϕC-bar are activation signals of row decoder and column selector, respectively. D0–D3 are bit out. 0–4 and C–F are column addresses. GND is the ground, and V_{DD} is the power supply.

physically cutting the large matrix. Furthermore, if a sensor of nonrectangular shape is needed, we can also make it by cutting if the shape is convex. Such a mask-less process scheme should make it possible to reduce manufacturing costs substantially.

Figure 16.4 shows the waveforms of artificial skins. When pressure is applied to some areas of the sensor matrix, the pressure rubber of those parts becomes conductive and sensor cells pull bit lines up to the supply potential. If the supply potential is 40 V, the delay from activation to bit-out is 23 ms, from which the total time needed to scan the whole 16 × 16 matrix takes approximately 1 s. The delay for read out depends on supply voltage: If the voltage is 100 V, the delay is estimated to be a half. It is also shown by the simulation that the scan speed can be enhanced by one order of magnitude easily if the parasitic capacitance is suppressed by reducing the channel length and/or width of wiring of decoders and selectors.

If we consider the reasonable form of the practical artificial skin system, it is most likely that the pressure data read out from organic integrated circuits would be transferred to silicon chips. In this sense, some readers may want to claim that it is not necessary to build a decoder or selector with organic transistors. The denser and the larger-area integrated circuits, however, require the more complicated packaging and fine wiring, which cannot be easily achieved with silicon at reasonable cost. Thus our opinion is that it is very important to realize func-

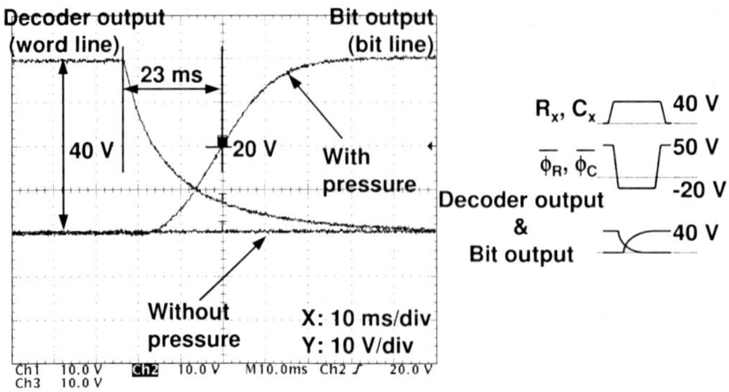

Fig. 16.4. A measured waveform of electronic artificial skin. When pressure is applied to the sensor matrix, the pressure-sensitive rubber becomes conductive and the bit line is pulled up to the supply voltage. (I) Input signals of column and row addresses. (II) Activation signal of the row decoder and column selector. (III) Measured waveform. The decoder output (word line) and bit output (bit line).

tionality of memories and processors and selectors and decoders with organic transistors.

16.4 Bending Experiments of Organic Transistors

To investigate the effects of strain on organic FETs on plastic films, electrical properties are measured under different compressive and tensile strain while changing the bending radius of the base plastic films systematically [18]. The saturation current increases regularly with increasing compressive strain, or reduction of the bending radius, whereas it decreases regularly with increasing tensile strain. The changes of saturation currents are +11% and −26%, respectively, under compressive and tensile strains of 1.5%. The changes are reversible and reproducible. Figures 16.5a and b show the normalized mobility as a function of strain or bending radius. The mobility increases regularly by 20% when we change the strain from 1.5% tension, passing through the flat state, to 1.5% compression. Such large changes cannot be explained solely by the deformation of device structures at 1.5% in strain. Transport properties of organic molecular systems are known to follow a hopping model rather than a band model. Under compressive strain, the energy barrier for hole hopping becomes lower, because of the shorter spacing between molecules, resulting in increased mobility. In contrast, tensile strains results in greater spacing, resulting in a decrease in mobility. Furthermore, the change of mobility does not depend on the directions of strains vs directions of current flows. Such an isotropic electrical property indicates that transport is also consistent with a hopping model in polycrystalline thin films – a conducting path from source to drain electrode is formed by coupled grain chains oriented along

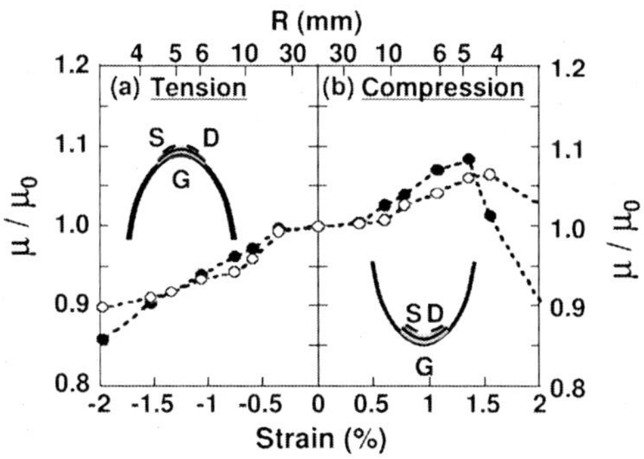

Fig. 16.5. Normalized mobility plotted as a function of strain or bending radius for (a) tension and (b) compression, while reducing the bending radius of the base film. (Filled and open circle are obtained from FETs with the direction of current flows parallel and perpendicular, respectively, to the direction of strain.)

random directions. The excellent mechanical flexibility of present organic FETs is attributed to excellent features of polyimide gate dielectric layers that have two-dimensional cross-linkage and excellent adhesion to gold. Irreversible deterioration can, on the other hand, be observed after removal of strains larger than 1.5%. Indeed, the critical strain for organic FETs is limited by buckling of gold electrodes, which is known to occur at a strain of approximately 1–2% [18].

16.5 High-temperature Operation of Organic Transistors

To investigate the high-temperature performance and the heat-resistance characteristics of organic FETs on plastic films, we measured transistor characteristics of the FETs at high temperatures [19]. The device was encapsulated by a parylene passivation layer. With increasing measurement temperatures up to 160 °C, saturation currents increase by 150%. Above 160 °C, significant change was observed, associated with deviation of the threshold voltage. Further increase of measurement temperatures above 190 °C degrades saturation current. The mobility of encapsulated FETs increases by 50% when the device is heated to 160 °C. Enhancement of mobility with increasing temperature can be described well by analysis of the carrier-hopping model. The on/off ratio is maintained above 10^4 even up to 200 °C. To investigate postannealing effects, transistor characteristics were measured at room temperature after removal of annealing temperatures as heat cycles. The device was heated from 30 to 50 °C, cooled to 30 °C, and the transfer curve was measured at 30 °C. Again, the same device was heated to 70 °C, then cooled, and the transfer characteristic was measured at 30 °C. The similar procedure was repeated with changing annealing temperatures up to 220 °C. After annealing temperatures below 160 °C no significant changes of transfer characteristics were observed. Figure

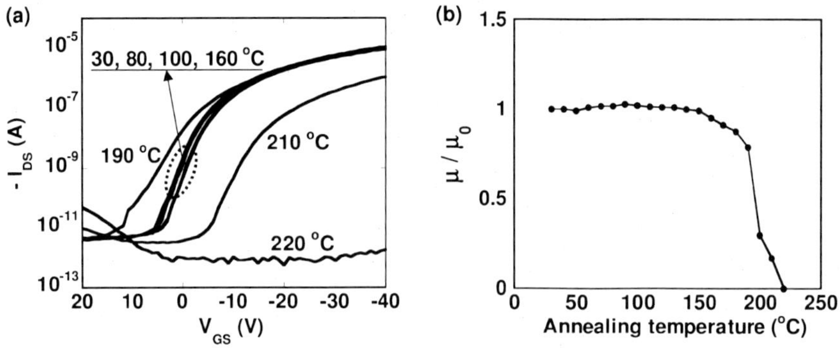

Fig. 16.6. (a) Transfer characteristics and (b) normalized mobility measured at room temperature after annealing.

16.6 shows that mobility did not change, demonstrating the reversible change of the transfer characteristics. Irreversible degradation of transistor performance was, on the other hand, observed after annealing temperatures above 170 °C. Such irreversible degradation of electronic performance is mainly because of decomposition of pentacene at high temperatures. Transistors with thermal stability can be produced by reducing the thermal stress at the interface between gate insulators and base films by using heat-resistant polyimide, for which matching of the coefficient of thermal expansion is perfect.

16.6 Sheet Image Scanners

In another new development of large-area electronics we have successfully demonstrated the first large-area, flexible, and lightweight *sheet image scanner*, fabricated on a plastic film and integrating organic field-effect transistors with organic photodiodes. The new sheet scanner does not require any mechanical or optical component such as focusing lenses. In conventional scanners, a linear array of photodetectors is moved from the top to the bottom of a page to capture images or characters. In the new design, a two-dimensional array of organic photodiodes coupled with organic transistors is used. Instead of a line-by-line mechanical scanning procedure, the signal of the photodiodes is read out by electrically probing the organic transistors, avoiding the need to use any movable part. As a result, the device is thin, lightweight, and mechanically flexible. The effective sensing area is four square inches and the current resolution of 36 dots per inch (dpi). The photodetectors can detect black and white tones by sensing the difference between reflected light from the dark and bright parts of an image.

The device structure is schematically illustrated in Fig. 16.7, with the chemical structure of each layer. Organic FET matrix and photodiode matrix have been man-

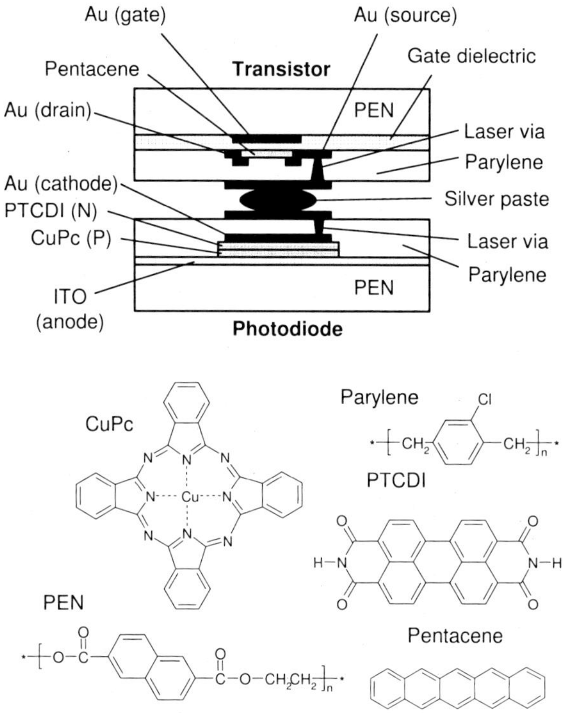

Fig. 16.7. The cross-sectional view of the device structure is shown schematically. Organic transistor matrix and organic photodiode matrix are separately manufactured on different PEN films and then laminated with each other with silver paste patterned by ultra fine printing technique. The chemical structure of each layer is also shown.

ufactured separately on different plastic films in a clean room (class 100–1000) and then laminated with each other with silver paste patterned by an ultrafine printing technique. Organic transistors are manufactured by the methods mentioned in Section 16.3. The thin-film pentacene transistors have an 18 μm channel length and a 0.7 $cm^2\ V^{-1}\ s^{-1}$ hole mobility. The base film of photodiodes is an ITO-coated PEN film with resistivity of 95 Ωm. A 30 nm thick p-type semiconductor of copper phthalocyanine (CuPc) and 50 nm thick 3,4,9,10-perylenetetracarboxylic acid diimide (PTCDI) are deposited in a vacuum sublimation system and 150 nm thick gold is deposited as cathode electrodes. The size of the cathode electrodes and the periodicity of photodiodes used to integrate with organic transistors are $450 \times 450\ \mu m^2$ and $700 \times 700\ \mu m^2$, respectively, but the smaller photodiodes are also fabricated for comparison.

Both films with organic FETs and photodiodes are transferred to the vacuum chamber without exposure to air after the manufacturing process and uniformly coated with a 2 μm poly(monochloro-*para*-xylylene) (parylene) passivation layer. Spots of parylene on electrodes are removed by a CO_2-laser drilling machine for

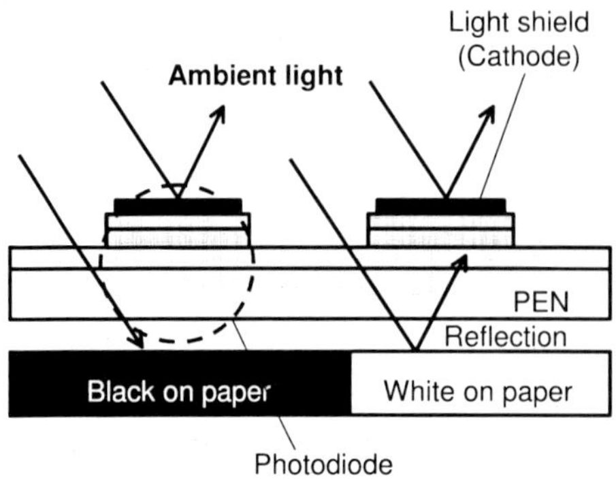

Fig. 16.8. Image capturing with a photodiode matrix sheet in reflection geometry. Organic photodiodes distinguish between black and white from the different reflectivity of the black and white parts on the paper. This image-capturing device requires no optics or mechanical scanning device.

electronic interconnections. The films are then laminated with each other with silver paste patterned by ultrafine printing technology.

We describe the principle of imaging with this sheet image scanner. As may be seen in Fig. 16.8, if all incident light reaches the active layers, photodetectors cannot distinguish black and white. We prepared light-shielding layers to prevent photodetectors from being exposed to directly incident light. Now direct light cannot reach the active layers. The incident light passing though transparent regions is reflected by the white part of the paper and reaches the active layers whereas the light incident on black does not go to the active layers. In this way the device distinguishes between black and white in the reflection geometry. Indeed, one of the organic photodetectors is positioned on a sheet of white paper that has a black region printed by a laser printer and we have found that a photocurrent ratio of 8:1 is obtained at voltage bias of -4 V.

We have prepared the 8×8 organic photodiode matrix without organic transistors. The effective sensing area of each sensor cell is $50 \times 50\ \mu m^2$ and periodicity is 100 μm, which corresponds to 250 dpi. The dispersion of photocurrent of photodiodes with illumination of light ($80\ mW\ cm^{-2}$) is shown in Fig. 16.9 under light illumination of black and white areas. We have positioned a sheet of paper with a white capital letter "T" prepared by a laser printer on the photodiode matrix and measured the photocurrent of each detector with light illumination ($80\ mW\ cm^{-2}$). The mapping of photocurrents is shown in Fig. 16.9b.

The new scanner is thin, ultra-light-weight and flexible, is suitable for mobile electronics and could be easily carried in a pocket. Because it can be bent to completely cover the bent page of an open book, it would be suitable for recording fragile, historically invaluable documents or other curved images, for example the labels of wine bottles (Fig. 16.10).

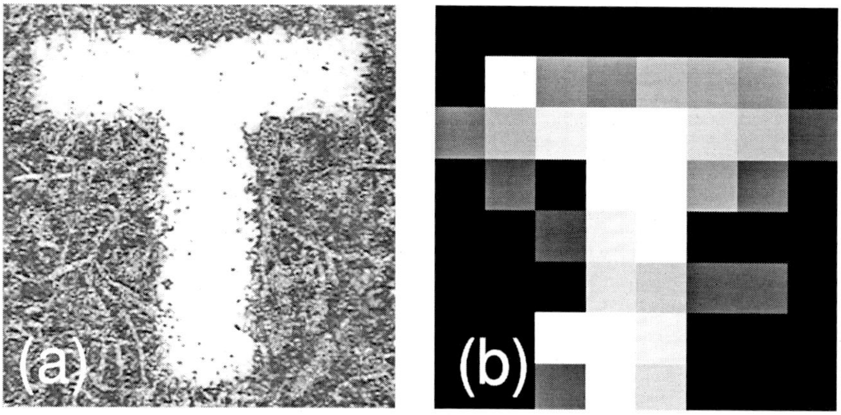

Fig. 16.9. A white capital letter "T" prepared by a laser printer (a) was placed on the 250-dpi organic photodetector matrix without organic transistors. The photocurrent of each detector is measured under light (80 mW cm^{-2}). The mapping of normalized photocurrents is shown in (b). Dark regions in (c) contains three sensor cells which were not functional. The size of each image is 0.8×0.8 mm^2.

Fig. 16.10. Because the new scanner is mechanically flexible, it can be bent to cover curved surfaces such as labels of wine bottles.

16.7 Three-dimensional Integrated Circuits

The drawback of organic transistor-based sheet image scanners so far, which has been a serious bottleneck in practical applications, has been the slow operating frequency. To overcome this problem we have used a new circuit concept, "double

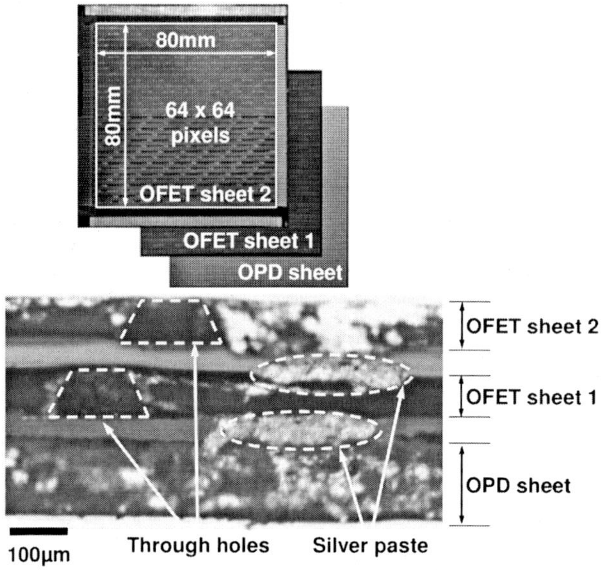

Fig. 16.11. Photograph of fabricated scanner system and cross-sectional view.

word-line and bit-line structure" which reduces the typical time delay of the circuit by a factor of five and the power consumption by a factor of seven. To realize the new structure, a new design is used in which two layers of organic transistors are stacked with the organic photodetectors using laser drilling, forming a three-dimensional circuit.

The physical structure of the double word-line/bit-line scanner is shown in Fig. 16.11. One organic photodetector sheet and two organic transistor sheets are separately fabricated and glued together with silver paste by means of a dispenser. Although the hierarchical word-line/bit-line structure is used for memory, the situation is different for sensor applications. In memory, the 1 word-line/1 bit-line selectors are laid out by shifting memory cells sideways. This is allowed, because memory is a logical device; for sensors, however, pixels cannot be shifted because the pixel density changes and uniform sensing becomes impossible. Moreover, because the organic transistor is large, only a single organic transistor is allowed per pixel, otherwise open-area ratio becomes almost zero and light cannot enter the organic photodetector. Thus, the 1 word-line/1 bit-line selector organic transistors are stacked on the top of the pixel selector organic transistors. This is why the development of 3D stacked organic transistor sheet is essential for this kind of circuit.

In the future it is supposed that the number of pixels per word-line and bit-line will be increased to more than 2048, and pixel size is reduced to less than 1/16. The delay for the conventional scheme will be of the order of 10^3 s whereas the proposed double word-line/bit-line scheme using stacked organic transistor sheets is estimated to reduce the delay by several order of magnitude. Power consumption can also be reduced by the same factor. The proposed approach will be applicable to other types of large-area organic transistor sensor, including artificial skin, and solves fundamental issues in large-area sensor electronics.

16.8
Future Prospects of Large-area Electronics

In this sheet image scanner, two-dimensional array of sensor cells cover the large area entirely and the data are read electrically, avoiding mechanical scans. We believe that the electronic scan method would be practical, because the manufacturing costs of organic transistors are expected low even for large-areas.

As mentioned in the introduction of this chapter, organic transistor-based integrated circuits play an important role in large-area electronics where the manufacturing cost per area must be very low. There is no doubt that one of the most important directions for future electronics is ambient intelligence or wireless sensor network. To realize such networks, one of the key technologies is a sensor to detect physical or chemical information distributed over a large area. We believe that large-area features of organic transistors would be suitable for realizing large-area sensors for this purpose.

We believe that large-area pressure sensors would find many applications beyond robotics, for example, security, homecare, entertainment, sports, and more. When pressure-sensitive carpet is spread on a floor in a house it will work as a good security system that distinguishes family members from a stranger merely from analysis of footprints. Furthermore, a tactile bed will be able to diagnose physical conditions instantly. A new class of application requiring large-area detection is increasing its importance in the ubiquitous electronics in the next generations. Although it is not trivial for organic transistors to compete with silicon from the standpoint of cost per function, organic transistors are really strong if the applications require low-cost features for a large area. Organic transistors should therefore fit well with large-area electronics.

16.9
Remaining Issues

We describe remaining issues of organic transistors from the standpoint of large-area electronics. First, stability and reliability are the central concerns for organic transistors. Pentacene transistors without encapsulation are functional after operation over a couple of days, but degradation of transistor characteristics, for example reduction of saturation current and/or threshold voltage shift, is observed. The degradation may be induced by oxygen and/or moisture, in the same way as for electroluminescent (EL) devices, and can therefore be suppressed substantially by appropriate encapsulation. Applications requiring mechanical flexibility, for example electronic skins and sheet image scanners also need flexible substrates with low gas permeability, although plastic films have usually high gas penetration. Thus one urgent problem is to develop flexible base films with low gas permeability.

The second issue is reduction of operating potential. The present device requires 40 V as power supply, but it is favorable for applications such as e-skin and scanner to have much lower voltage. Reduction of operating potential can be achieved by reducing device dimensions. Use of thinner gate dielectric layers and/or those with higher gate dielectric constants is a very effective means of reducing the operating potential. The thickness of the polyimide gate dielectric layer is currently 500–900 nm, which should be reduced to less than a few hundred nm. Making gate dielectric layers thinner usually increases the chance of gate leakage. To minimize the chance of this it is very important to use high-purity polyimide precursors. Although we have found that the filtering the source materials reduces the chance of gate leakage, further study is needed to achieve both low operation voltage and high yields.

As far as the sheet image scanner is concerned, another remaining issue is spatial resolution, which is currently 36 dpi. To improve the resolution we must reduce the size of both organic transistors and organic photo detectors. Reduction of device dimensions is not very difficult, but the bottleneck is the size of via interconnections. To reduce the diameter of via holes, we are currently working very hard to replace a CO_2 laser with other short wavelength lasers, for example excimer lasers and/or YAG lasers. We believe that 600 dpi would be feasible in the near future.

16.10 Conclusions

In this chapter, we have described large-area sensors as a promising application of organic transistors. First, we have demonstrated large-area pressure sensors produced by integrating an organic transistor matrix with pressure-sensitive rubber sheet. The new sensors will be useful for variety of applications beyond robot skins, for example regeneration medicine, new security systems, and intelligent transportation systems (ITS). Second, we have described sheet image scanners with integrated organic transistors and organic photodiodes.

This work makes full use of technical knowledge in the field of flexible printed circuits (FPC). Laser via processing, plating, and lamination approaches, whose reliability and throughput have been well realized in mass-production lines of FPC, are used as a manufacturing processes of organic transistor integrated circuits. It is very important for organic transistors to be manufactured by such low-cost and reliable processes.

Organic transistors are indispensable for realizing ambient electronics. We expect the multiplier effect to use organic transistors as a complementary function to silicon. We hope that remaining issues related to stability, operation voltage, speed, power consumption will be solved in the future and organic transistors will be pushed into practical use.

Acknowledgments

The authors acknowledge T. Sekitani, H. Kawaguchi, S. Iba, Y. Kato, Y. Noguchi, and Y. Murase for their invaluable contribution to the projects on large-area sensors. They also thank Professors Yasuhiko Arakawa, Hiroyuki Sakaki, and Makoto Kuwata-Gonokami, University of Tokyo, and Mr Tadashi Kobayashi, for fruitful discussions. The projects are partly supported by the IT Program, MEXT of Japan, NEDO, and MPHPT.

References

1 H. Klauk, D. J. Gundlach, and T. N. Jackson, IEEE Electron Device Lett. **20**, 289 (1999).

2 P. M. G. Kane, J. Campi, M. S. Hammond, F. P. Cuomo, B. Greening, C. D. Sheraw, J. A. Nichols, D. J. Gundlach, J. R. Huang, C. C. Kuo, L. Jia, H. Klauk, and T. N. Jackson, IEEE Electron Device Lett. **21**, 534 (2000).

3 H. Klauk, M. Halik, U. Zschieschang, G. Schmid, W. Radlik, and W. Weber, J. Appl. Phys. **92**, 5259 (2002).

4 C. D. Dimitrakopoulos and P. R. L. Malenfant, Adv. Mater. **14**, 99 (2002).

5 H. Sirringhaus, T. Kawase, R. H. Friend, T. Shimoda, M. Inbasekaran, W. Wu, and E. P. Woo, Science **290**, 2123 (2000).

6 T. Kawase, H. Sirringhaus, R. H. Friend, and T. Shimoda, Proceedings of SPIE, *Organic Field Effect Transistors*, D. Fichou, Z. Bao, Editors, **4466**, 80 (2001).

7 J. A. Rogers, Y.-L. Loo, T. Someya, K. W. Baldwin, Z. Bao, P. Ho, A. Dodabalapur, and H. E. Katz, Proceedings of National Academy of Science, USA, **99**, 10252 (2002).

8 F. Baude, D. A. Ender, M. A. Haase, T. W. Kelley, D. V. Muyres, and S. D. Theiss, Appl. Phys. Lett. **82**, 3964 (2003).

9 C. J. Drury, C. M. J. Mutsaers, C. M. Hart, M. Matters, and D. M. de Leeuw, Appl. Phys. Lett. **73**, 108 (1998).

10 H. Klauk, D. J. Gundlach, J. A. Nichols, and T. N. Jackson, IEEE Trans. Electron Devices **46**, 1258 (1999).

11 Takao Someya, Tsuyoshi Sekitani, Shingo Iba, Yusaku Kato, Hiroshi Kawaguchi, and Takayasu Sakurai, Proceedings of the National Academy of Sciences of the United States of America 101, 9966 (2004).

12 Hiroshi Kawaguchi, Takao Someya, Tsuyoshi Sekitani, and Takayasu Sakurai, IEEE Journal of Solid-State Circuits, Vol. 40, No. 1, pp. 177–185 (January 2005).

13 T. Someya and T. Sakurai, 2003 IEEE International Electron Devices Meeting (IEDM), December 8–10, 2003, No. 8.4, 203.

14 T. Someya, H. Kawaguchi, and T. Sakurai, Technical Digest of 2004 IEEE International Solid-State Circuits Conference (ISSCC 2004), February 14–19, 2004, No. 16.2, 288.

15 T. Someya, S. Iba, Y. Kato, T. Sekitani, Y. Noguchi, Y. Murase, H. Kawaguchi, and T. Sakurai, "A Large-Area, Flexible, and Lightweight Sheet Image Scanner", 2004 IEEE International Electron Devices Meeting (IEDM), 15.1, December 13–15, 2004.

16 H. Kawaguchi, S. Iba, Y. Kato, T. Sekitani, T. Someya, and T. Sakurai, "A Sheet-Type Scanner Based on a 3D-Stacked Organic-Transistor Circuit Using Double Word-Line and Bit-Line Structure", #32.3, 2005 IEEE International Solid-State Circuits Conference (ISSCC 2005) (February 6–10, 2005).

17 Yusaku Kato, Shingo Iba, Ryohei Teramoto, Tsuyoshi Sekitani, Takao Someya, Hiroshi Kawaguchi and Takayasu Sakurai, Applied Physics Letters, Vol. 84, No. 19, pp. 3789–3791 (10 May 2004).

18 Tsuyoshi Sekitani, Yusaku Kato, Shingo Iba, Hiroshi Shinaoka, Takao Someya, Takayasu Sakurai, and Shinichi Takagi, Applied Physics Letters vol. 86, 073511 (14 February 2005).

19 Tsuyoshi Sekitani, Shingo Iba, Yusaku Kato, and Takao Someya, Applied Physics Letters, Volume 85, Issue 17, pp. 3902–3904 (October 25, 2004).

17
Organic Semiconductor-based Chemical Sensors

Howard E. Katz and Jia Huang

17.1
Background

The field known as "organic" or "plastic" electronics is centered on field effect transistor (FET)-based circuits mounted on large-area and/or flexible substrates. When the semiconductor is organic, the device is referred to as an organic field-effect transistor (OFET). Work on OFET has been extensively reviewed, most notably and comprehensively in *Chemistry of Materials* and *Journal of Materials Research* special issues, to which the one of us has contributed three articles [1–3].

The attention paid to organic electronics is largely motivated by the desire to provide some silicon-like functionality at a small fraction of the cost, by use of inexpensive printing processes. An additional and perhaps greater opportunity is to use capabilities of organic semiconductors that are not as readily available in silicon. A compelling application of organic electronics is *chemical sensing*, because of the ability to covalently attach receptors for compounds of interest to the molecules that make up the semiconductor, in locations where analyte–receptor binding will strongly affect the current flowing across a transistor channel. Figure 17.1 shows a schematic diagram of an FET, with a suggested desirable mode of analyte vapor detection. Analyte molecules in a relatively noninteracting solvent could also bind selectively to the OFET surface.

The OFET consists of an organic semiconductor–dielectric bilayer addressed with source, gate, and drain electrodes, as illustrated in Fig. 17.1. The semiconductor, through which a lateral current is modulated, is a molecular solid film of pi-conjugated cores that can be as thin as 10 nm, and to which electronically, chemically, and morphologically influential functional groups may be conveniently attached. Also shown are typical plots of drain current against drain voltage for a set of gate voltages.

Several new approaches to designing semiconducting materials for enhanced chemical sensitivity have been reported very recently [4]. In some cases, these materials are intended to enhance the capability of an existing technique. In other cases, the materials are the basis of an entirely new sensing platform. The materials are employed as the current-carrying media in transistors or resistors whose

Organic Electronics, Materials, Manufacturing and Applications. Edited by Hagen Klauk

ISBN: 3-527-31264-1

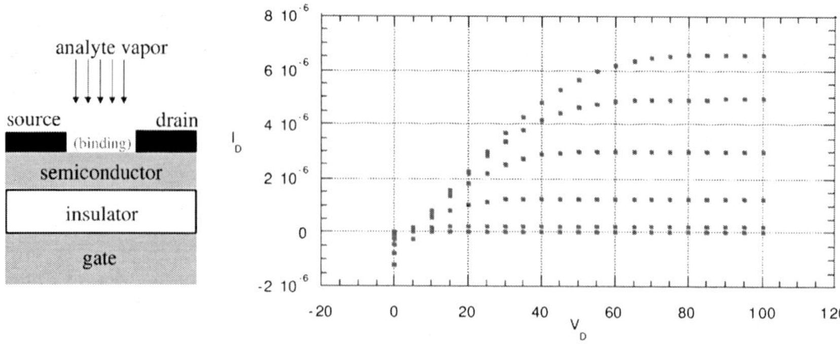

Fig. 17.1. OFET and current-drain voltage plot, current in amps, potential (negative) in volts, gate voltages 0–100 V in 20-V increments, for a diphenylbithiophene oligomer on Si/SiO_2.

current–voltage responses are altered in the presence of certain chemical compounds in the surrounding environment. Classes of materials considered include inorganic semiconductors [5], inorganic nanowire [6–11], polymer nanowire [12, 13], and carbon nanotube [14–19] species, electrochemically active films [20], and swellable polymer composites [21–23]. The devices can have a very high dynamic range for detection of low levels of specific analytes. Alternatively, an array of devices with smaller and less specific responses can provide information about the ambient environment via specificity in the pattern of the collected responses from the array [24].

The purpose of this chapter is twofold. First, some very recent approaches to semiconductor chemical sensors are discussed. The term semiconductor is taken broadly to include composites partially composed of conductive elements and nonmetallic charge-transport agents that may be inorganic, organic, or nanostructured. The second is to provide a brief electrostatic analysis of the binding of an analyte to a semiconductor surface and to begin to rationalize the observed electronic transduction. Such rationalization has been neglected in the overwhelming majority of descriptions of electrical response to chemical binding to semiconductors.

17.2
Inorganic and Nanostrctured Semiconductor Sensors

The conductivity of a generic semiconductor film is likely to be modulated by absorption of a polar species on the film surface. For example, a group at the Weizmann Institute has examined GaAs surfaces and found that porphyrin receptors linked to it will attract nitrogen oxide NO, and the binding of NO caused a change in resistance [5]. Other embodiments of GaAs sensors were found to be sensitive to ions in solution [25].

Organic semiconductors have been used as chemoresistive sensing elements in which the analyte penetrates into the bulk of the film, rather than simply binding to a surface. Versions of "conducting polymers" (polyaniline, polythiophenes, polypyrrole, and polyethylenedioxythiophene (PEDOT)) are subject to conductivity modulation via changes in dopant concentration and strength. Using a new interfacial polyaniline synthesis method, high surface area polyaniline nanowires have been produced and found to respond by a variety of chemical interaction mechanisms to small, volatile molecules [12, 13].

Composites containing solid particles respond to forces that change the connectivity of the particles. Sensing that uses both doping modulation and connectivity effects simultaneously has been reported by a Caltech group led by Nathan Lewis, who developed the original, commercially available "electronic nose" based on polymer–carbon black composites [21]. In this more recent work, carbon black dispersed in polyaniline doped with a variety of sulfonic acids responded powerfully and selectively to amines in the presence of other analyte vapors, and discriminated overwhelmingly between aniline and butylamine [22]. When the matrix polymer is the basic polyethyleneimine, sensitivity and selectivity for carboxylic acids are increased [26]. Most recently, the connectivity, and thus the resistance, of a nearly percolated film of gold nanoparticles networked with nonfunctional dithiols and then equilibrated with a thiolated single strand DNA was found to be sensitive to the presence of the strand complement in a sample solution [27]. The response could be calibrated to single base-pair mismatches.

Inorganic nanowire and carbon nanotube sensors are a major breakthrough from the last few years. Modulated conduction is observed in individual, chemically synthesized silicon nanowires in response to binding events and changes in ion distribution at their surfaces when immersed in solutions. This effect can be used to monitor external pH, ion concentrations, biopolymer–substrate binding, and nucleic acid hybridization [6]. In papers published a month apart two different groups reported sensitivity to DNA sequences of complements bound to Si nanowires. In one instance the nanowires were synthesized as free chemical entities and assembled on substrates [28]. In the other the nanowires were fabricated using wafer-processing methods to isolate and contact nanoscale silicon regions [29]. Other inorganic compounds have been formed in high aspect ratio, sensitive nanostructures, including In_2O_3 [10, 30], SnO_2, and ZnO [9].

Carbon nanotubes are an alternative nanostructured semiconductor with promise for chemical sensing. Orders of magnitude changes in conductance were obtained in response to exposure to ammonia (quenching) or NO_2 (doping) gas [14]. Predominantly semiconducting carbon nanotube bundles coated with filtering polymers have still greater sensitivity and selectivity as sensors for ammonia and NO_2, reaching parts per billion levels and reduced interference in gas mixtures [15]. Nitroaromatics, relevant to landmine detection and remediation, can be detected at parts per billion to parts per million levels by use of carbon nanotubes simply cast on interdigitated electrodes [16]. Another type of militarily relevant analyte that can be sensed with similar devices is volatile phosphonate ester, related to nerve agents [17]. Some selectivity was added to the system by using a prefilter composed of glass wool coated with an acidic polymer. Protein-absorption studies suggest that some carbon nanotube responses occur primarily at the contacts, rather than the length of the tube [31].

Though not strictly functioning as resistors/conductors, carbon nanotubes have just been reported in an extraordinarily vapor-sensitive capacitive device [19]. The electric field lines emanating from the nanotubes are responsible for a localized dielectric response that can be modulated by minute quantities of adsorbate on the nanotube surface. A layer of hydrogen-bonding polymer, or even a monolayer terminated in mildly acidic groups, increased sensitivity to parts per billion levels. Response strength was correlated with the dipole moment of the analytes.

A very recent paper details the biomolecular functionalization of a polypyrrole nanowire. ZnSe/CdSe quantum dots were functionalized with avidin and these quantum dots were incorporated into a polypyrrole nanowire during its electrochemical formation between two electrodes 100 nm apart [32]. Biotin with a DNA strand attached was shown to modulate the conductivity of the avidin nanowire, but not an unfunctionalized polypyrrole nanowire fabricated similarly. Presumably, the DNA–biotin–avidin complex would also be sensitive to changes in the hybridization state of the DNA.

Nanowires and nanobelts of inorganic oxides have been fashioned into chemically sensitive semiconductor devices. These include tin and zinc oxides [9], and indium oxide [30]. Once again, ammonia and NO_2 gases were used for initial demonstrations. Oxygen had very little effect on the sensing action. Because of the low concentrations detected and the speed of the response, it was suggested that single-molecule response could be within reach with these ultraminiaturized sensors.

17.3
Sensitive Organic Field-effect Transistors

A detailed manuscript by Guillaud and colleagues [33] introduces phthalocyanines as chemically sensitive semiconductors, and introduces the concept of sensitive OFET, with preliminary reports of phthalocyanine-based FET sensitive to oxygen, nitrogen dioxide, ammonia, carbon monoxide, and hydrogen sulfide. A second, independent investigation resulted in the selective detection of diisopropyl methylphosphonate by an interdigitated gate electrode FET using copper phthalocyanine as the chemically sensitive film. Selectivity was obtained from analysis of time and frequency-dependent responses in the milli to microsecond range, rather than from any property of the material [34].

Using naphthalenetetracarboxylic dianhydride as the semiconductor, changes in bulk conductivity, field-effect mobility, and threshold voltage were separately observed in response to exposure to water and oxygen [35, 36]. Another more elaborate kind of pattern was produced by a virtual array of eleven different semiconductor OFET monitoring on-current in response to polar and nonpolar organic vapors [37]. Responses (0.8–0.3-fold reductions and 1.5–2-fold increases) were dis-

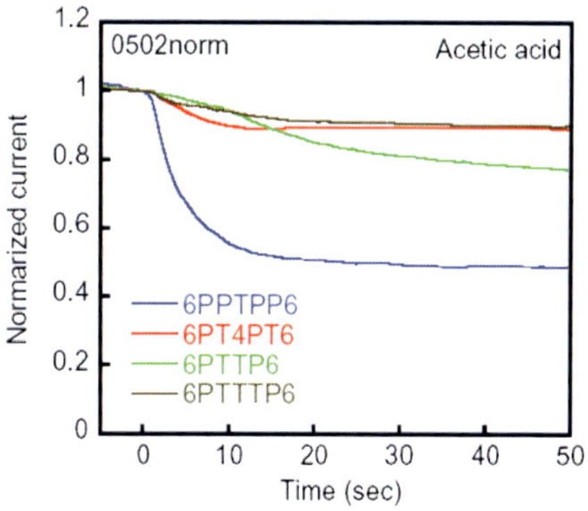

Fig. 17.2. Relative responses of oligomers to acetic acid vapor over time. 6 = hexyl, P = 1,4-phenylenediyl, T = 2,5-thienediyl.

tinguishable for different classes of analyte among semiconductors with different lengths of side chain, e.g. sexithiophene and dialkylsexithiophenes, and different carrier types, e.g. copper phthalocyanine and perfluorinated copper phthalocyanine, though in the absence of any particular receptor for the analytes. Hole currents are much further diminished, almost to zero, by more strongly electron donating compounds such as amines (H.E. Katz, unpublished work), and electron currents are correspondingly quenched by oxidizing agents [38]. Analyte concentrations of 10–100 ppm in the vapor phase have been detected, as, for example, shown in Fig. 17.2.

Differences between baseline and analyte-induced output are magnified if responses of more complex circuits are observed. For example, the frequency of a ring oscillator changes by a larger relative amount in response to analyte exposure than does the mobility of one of its constituent OFET [39]. Although responses to different analytes of a particular semiconductor and sensitivities of different semiconductors to a particular analyte were distinguishable, this selectivity was purely empirical, in no way designed into the active materials of the OFET.

There was, furthermore, no attempt to correlate the responses described above to physical attributes of the films. In later experiments, however, observations of grain structure were made on the same films used for sensing. The most specific experiment was to fabricate devices of different channel length on a polycrystalline film of dihexylquaterthiophene, in which grain structure is controlled by substrate temperature during deposition. Some devices were smaller than the crystallite size, and could be termed "single crystal devices", whereas others were larger and included multiple grain boundaries in the channels (Figure 17.3). The devices that included grain boundaries were much more sensitive to pentanol vapor than were the single-crystal devices, indicating that grain boundaries were a main site of chemical sensitivity of this class of film [40]. A subsequent study was published reporting distinctions between the nonspecific responses of pentacene OFETs to pentanol depending on whether the devices were grain boundary or single-grain

Fig. 17.3. Source–drain pairs traversing multiple (left) and single (right) grains of the same 6(T4)6 film.

dominated [41], consistent with the results of the preliminary study. Another study of oligothiophene films found correlations between vapor sensitivity and granularity revealed by transmission electron microscopy.

17.4
Mechanistic Rationale

Let us assume a simple electrostatic model for the electrical potential imposed by an adsorbed dipole having charges q separated by length L centered at point O, where the potential varies inversely with the square of the distance from the dipole [42] and is maximized for regions nearly collinear with the dipole (small angle θ). If a molecule with a dipole moment of 2 Debye (equivalent to ± 0.2 electronic charges separated by 2 Å, similar to the gas phase dipole moment of water) is physisorbed on to a nonpolar film, the perturbation in the electric potential is of the order of 0.05–0.1 V at a point X of the order of 10 Å from the molecule, and collinear with the dipole. As the point is moved laterally or vertically away from the dipole (towards the point r, θ), the potential falls rapidly, as shown in Fig. 17.4 below, where potential is plotted against lateral distance in Angstroms, with zero referring to starting points at 10, 20, and 30 Å collinear distance. Binding multiple molecules or larger dipoles increases the change in potential to approximately a quarter of a volt and increases the area over which the potential change is substantial. Figure 17.5 shows the potential perturbation as a function of lateral distance for odd numbers of assembled 2-Debye dipoles arrayed 5 Å apart, with zero lateral distance indicating a point 10 Å below the center of the array.

The maximum perturbation of this kind of dipole assembly converges to approximately 0.24 V, as seen from the plateau in Fig. 17.5, and from the plot of maximum potential against number of dipoles in Fig. 17.6.

The adsorption-induced perturbation of electrical potential in an organic semiconductor has been assessed experimentally by photoelectron spectroscopy [43]. Binding methanol to the nonpolar organic semiconductor film α-sexithiophene

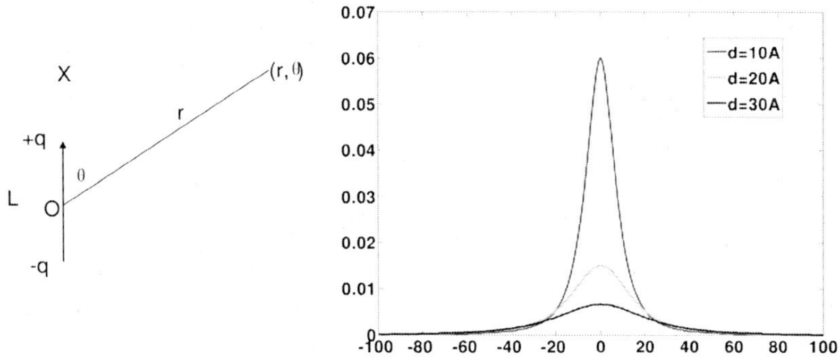

Fig. 17.4. Definition of position relative to dipole, and potential change (V) versus lateral distance from X (Angstrom), where d = XO and qL = 2 Debye.

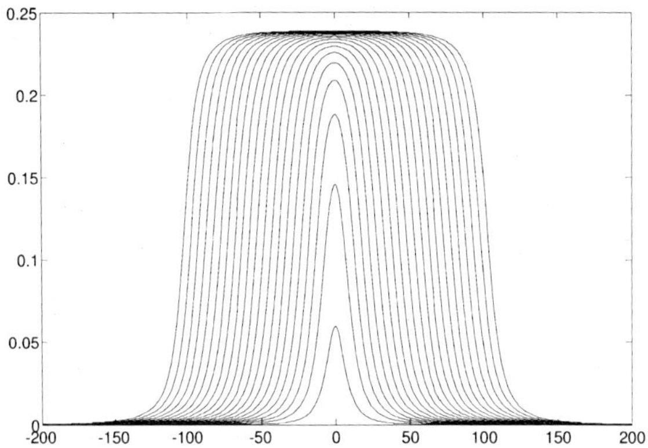

Fig. 17.5. Potential change versus lateral distance for arrays of 1,3,5, etc. dipoles spaced 5 Angstroms apart.

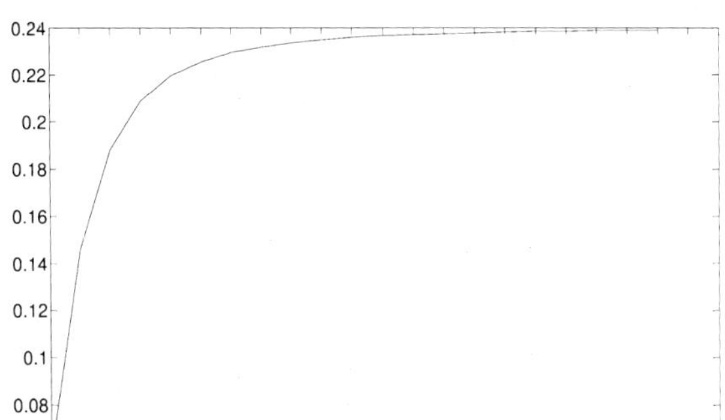

Fig. 17.6. Maximum potential change versus number of dipoles in the array.

results in binding energy shifts of approximately 0.2 eV. Larger effects are seen with NO_2, although the mechanism may be more dominated by doping effects. It has also been shown that water vapor significantly alters the performance of pentacene [44, 45] and poly(hexylthiophene) [46] OFET by up to an order of magnitude. The mobility and saturation current dropped similarly in two independent pentacene studies whereas off current increased and threshold voltage moved toward the depletion regime. Similar trends were observed for related molecular solids based on the fluorene subunit. Similar semiconductors are also sensitive to alcohol vapor [37]. For the polymer it was postulated that the dipole moment of water was sufficient to actually dope the film. Nitrogen and even dry oxygen did not have nearly as significant an effect.

The potential applied to most OFETs in printed electronics research is of the order of tens of volts. The potential that actually confines the gate-induced charge carriers to the channel is some fraction of that, and depends sensitively and in a complex way on the intrinsic conductivity and dielectric constant of the semiconductor [47–50]. Although plastic electronics prototypes often run on tens of volts, it is conceivable that a modest dipole moment associated with an assembly of simple functional groups, which imposes a change in potential of only 0.2 V, could have an impact on the mobile charge charier density equivalent to tens of volts of applied gate voltage, altering the conductivity by orders of magnitude. This is based on the observation that silane monolayers on oxide dielectric can have the effect of shifting the threshold voltage of an OFET by the equivalent of 50 V [51]. The absolute value of potential shifts from polar thiol monolayers on gold is of the order of 1 V, derived from studies on donor–acceptor functionalized diphenylethynes [52].

Another lever on channel conduction is trap formation. The activation energy for releasing carriers from traps in OFETs is often of the order of 10–100 meV [48, 53]. Bound analytes forming traps at similar energies to those already known to markedly affect OFET currents could provide a plausible means of transduction. For example, an analyte-induced trap with a depth of 100–200 meV (0.1–0.2 V) could have an activation energy 2–10 times that of the activation energy of traps otherwise in the material, altering the current by an order of magnitude at ambient temperature.

17.5
Conclusion

There are numerous examples of semiconducting materials with electrical conductivity altered by adsorption of analytes. Although a variety of mechanical mechanisms, for example swelling and changes in particle connectivity are sometimes operative, many other materials act because of gating effects of doping and dipole fields. Organic semiconductors are particularly suited to operate by this mechanism. A plausible connection has been made between an electrostatic model of dipole field gating and recent observations of organic semiconductors. The electrostatic principles, however, are applicable to inorganic and nanostructured semiconductors also.

References

1 H. E. Katz, "Recent advances in semiconductor performance and printing processes for organic transistor-based electronics", Chemistry of Materials **16**, 4748 (2004).

2 H. E. Katz, C. Kloc, Z. Bao, J. Zaumseil, and V. Sundar, "Field-effect transistors made from macroscopic single crystals of tetracene and related semiconductors on polymer dielectrics", Journal of Materials Research **19**, 1995 (2004).

3 A. Facchetti, J. Letizia, M. H. Yoon, M. Mushrush, H. E. Katz, and T. J. Marks, "Synthesis and characterization of diperfluorooctyl-substituted phenylene–thiophene oligomers as n-type semiconductors. Molecular structure–film microstructure–mobility relationships, organic field-effect transistors, and transistor non-volatile memory elements", Chemistry of Materials **16**, 4715 (2004).

4 H. E. Katz, "Chemically Sensitive Field-effect Transistors and Chemiresistors: New Materials and Device Structures", Electroanalysis **16**, 1837 (2004).

5 D. G. Wu, G. Ashkenasy, D. Shvarts, R. V. Ussyshkin, R. Naaman, A. Shanzer, and D. Cahen, "Novel NO biosensor based on the surface derivatization of GaAs by "hinged" iron porphyrins", Angewandte Chemie – International Edition **39**, 4496 (2000).

6 Y. Cui, Q. Q. Wei, H. K. Park, and C. M. Lieber, "Nanowire nanosensors for highly sensitive and selective detection of biological and chemical species", Science **293**, 1289 (2001).

7 X. T. Zhou, J. Q. Hu, C. P. Li, D. D. D. Ma, C. S. Lee, and S. T. Lee, "Silicon nanowires as chemical sensors", Chemical Physics Letters **369**, 220 (2003).

8 O. H. Elibol, D. Morisette, D. Akin, J. P. Denton, and R. Bashir, "Integrated nanoscale silicon sensors using top-down fabrication", Applied Physics Letters **83**, 4613 (2003).

9 M. S. Arnold, P. Avouris, Z. W. Pan, and Z. L. Wang, "Field-effect transistors based on single semiconducting oxide nanobelts", Journal of Physical Chemistry B **107**, 659 (2003).

10 C. Li, B. Lei, D. H. Zhang, X. L. Liu, S. Han, T. Tang, M. Rouhanizadeh, T. Hsiai, and C. W. Zhou, "Chemical gating of In_2O_3 nanowires by organic and biomolecules", Applied Physics Letters **83**, 4014 (2003).

11 C. Li, D. H. Zhang, B. Lei, S. Han, X. L. Liu, and C. W. Zhou, "Surface treatment and doping dependence of In_2O_3 nanowires as ammonia sensors", Journal of Physical Chemistry B **107**, 12451 (2003).

12 S. Virji, J. X. Huang, R. B. Kaner, and B. H. Weiller, "Polyaniline nanofiber gas sensors: Examination of response mechanisms", Nano Letters **4**, 491 (2004).

13 J. X. Huang, S. Virji, B. H. Weiller, and R. B. Kaner, "Polyaniline nanofibers: Facile synthesis and chemical sensors", Journal of The American Chemical Society **125**, 314 (2003).

14 J. Kong, N. R. Franklin, C. W. Zhou, M. G. Chapline, S. Peng, K. J. Cho, and H. J. Dai, "Nanotube molecular wires as chemical sensors", Science **287**, 622 (2000).

15 Q. F. Pengfei, O. Vermesh, M. Grecu, A. Javey, O. Wang, H. J. Dai, S. Peng, and K. J. Cho, "Toward large arrays of multiplex functionalized carbon nanotube sensors for highly sensitive and selective molecular detection", Nano Letters **3**, 347 (2003).

16 J. Li, Y. J. Lu, Q. Ye, M. Cinke, J. Han, and M. Meyyappan, "Carbon nanotube sensors for gas and organic vapor detection", Nano Letters **3**, 929 (2003).

17 J. P. Novak, E. S. Snow, E. J. Houser, D. Park, J. L. Stepnowski, and R. A. McGill, "Nerve agent detection using networks of single-walled carbon nanotubes", Applied Physics Letters **83**, 4026 (2003).

18 T. Someya, J. Small, P. Kim, C. Nuckolls, and J. T. Yardley, "Alcohol vapor sensors based on single-walled carbon nanotube field effect transistors", Nano Letters **3**, 877 (2003).

19 E. S. Snow, F. K. Perkins, E. J. Houser, S. C. Badescu, and T. L. Reinecke, "Chemical detection with a single-walled carbon nanotube capacitor", Science **307**, 1942 (2005).

20 P. N. Bartlett and Y. Astier, "Microelectrochemical enzyme transistors", Chemical Communications, 105 (2000).

21 M. C. Lonergan, E. J. Severin, B. J. Doleman, S. A. Beaber, R. H. Grubb, and N. S. Lewis, "Array-based vapor sensing using chemically sensitive, carbon black–polymer resistors", Chemistry of Materials **8**, 2298 (1996).

22 G. A. Sotzing, J. N. Phend, R. H. Grubbs, and N. S. Lewis, "Highly sensitive detection and discrimination of biogenic amines utilizing arrays of polyaniline/carbon black composite vapor detectors", Chemistry of Materials **12**, 593 (2000).

23 K. J. Albert, N. S. Lewis, C. L. Schauer, G. A. Sotzing, S. E. Stitzel, T. P. Vaid, and D. R. Walt, "Cross-reactive chemical sensor arrays", Chemical Reviews **100**, 2595 (2000).

24 J. J. Lavigne and E. V. Anslyn, "Sensing a paradigm shift in the field of molecular recognition: From selective to differential receptors", Angewandte Chemie – International Edition **40**, 3119 (2001).

25 R. Naaman, in *American Physical Society*, (2000), p. I14.

26 E. S. Tillman and N. S. Lewis, "Mechanism of enhanced sensitivity of linear poly(ethylenimine) – carbon black composite detectors to carboxylic acid vapors", Sensors And Actuators B – Chemical **96**, 329 (2003).

27 H. Shiigi, S. Tokonami, H. Yakabe, and T. Nagaoka, "Label-free electronic detection of DNA-hybridization on nanogapped gold particle film", Journal of The American Chemical Society **127**, 3280 (2005).

28 J. Hahm and C. M. Lieber, "Direct ultrasensitive electrical detection of DNA and DNA sequence variations using nanowire sensors", Nano Letters **4**, 51 (2004).

29 Z. Li, Y. Chen, X. Li, T. I. Kamins, K. Nauka, and R. S. Williams, "Sequence-specific label-free DNA sensors based on silicon nanowires", Nano Letters **4**, 245 (2004).

30 C. Li, D. H. Zhang, X. L. Liu, S. Han, T. Tang, J. Han, and C. W. Zhou, "In_2O_3 nanowires as chemical sensors", Applied Physics Letters **82**, 1613 (2003).

31 R. J. Chen, H. C. Choi, S. Bangsaruntip, E. Yenilmez, X. Tang, Q. Wang, Y.-L. Chang, and H. J. Dai, "An investigation of the mechanisms of electronic sensing of protein adsorption on carbon nanotube devices", Journal of The American Chemical Society **126**, 1563 (2004).

32 K. Ramanathan, M. A. Bangar, M. Yun, W. Chen, N. V. Myung, and A. Mulchandani, "Bioaffininty sensing using biologically functionalized conducting-polymer nanowire", Journal of The American Chemical Society **127**, 496 (2005).

33 G. Guillaud, J. Simon, and J. P. Germain, "Metallophthalocyanines – Gas sensors, resistors and field effect transistors", Coordination Chemistry Reviews **180**, 1433 (1998).

34 E. S. Kolesar, C. P. Brothers, C. P. Howe, T. J. Jenkins, A. T. Moosey, J. E. Shin, and J. M. Wiseman, "Integrated-Circuit Microsensor For Selectively Detecting Nitrogen Dioxide And Diisopropyl Methylphosphonate", Thin Solid Films **220**, 30 (1992).

35 L. Torsi, A. Dodabalapur, N. Cioffi, L. Sabbatini, and P. G. Zambonin, "NTCDA organic thin-film transistor as humidity sensor: weaknesses and strengths", Sensors And Actuators B – Chemical **77**, 7 (2001).

36 L. Torsi, A. Dodabalapur, L. Sabbatini, and P. G. Zambonin, "Multi-parameter gas sensors based on organic thin-film transistors", Sensors And Actuators B – Chemical **67**, 312 (2000).

37 B. Crone, A. Dodabalapur, A. Gelperin, L. Torsi, H. E. Katz, A. J. Lovinger, and Z. Bao, "Electronic sensing of vapors with organic transistors", Applied Physics Letters **78**, 2229 (2001).

38 H. E. Katz, J. Johnson, A. J. Lovinger, and W. Li, "Naphthalenetetracarboxylic Diimide-based n-Channel Transistor Semiconductors: Structural Variation and Thiol-enhanced Gold Contacts", Journal of The American Chemical Society **122**, 7787 (2000).

39 B. K. Crone, A. Dodabalapur, R. Sarpeshkar, A. Gelperin, H. E. Katz, and Z. Bao, "Organic oscillator and adaptive amplifier circuits for chemical vapor sensing", Journal of Applied Physics **91**, 10140 (2002).

40 T. Someya, H. E. Katz, A. Gelperin, A. J. Lovinger, and A. Dodabalapur, "Vapor sensing with alpha,omega-dihexylquarterthiophene field-effect transistors: The role of grain boundaries", Applied Physics Letters **81**, 3079 (2002).

41 L. Wang, D. Fine, and A. Dodabalapur, "Nanoscale Chemical Sensor Based on Organic Thin Film Transistors", Applied Physics Letters **85**, 6386 (2004).

42 D. Halliday and R. Resnick, *Physics (Part II)* (John Wiley & Sons, inc., New York, 1962).

43 C. Kendrick and S. Semancik, "Doping effects and reversibility studies on gas-exposed alpha-sexithiophene thin films", Journal of Vacuum Science & Technology A **16**, 3068 (1998).

44 D. Li, E.-J. Borkent, R. Nortrup, H. Moon, H. E. Katz, and Z. Bao, "Humidity effect on electrical performance of organic thin film transistors", Applied Physics Letters **86**, web042105 (2005).

45 Z.-T. Zhu, J. T. Mason, R. Dieckmann, and G. G. Malliaras, "Humidity sensors based on pentacene thin-film transistors", Applied Physics Letters **81**, 4643 (2002).

46 H. Hoshino, M. Yoshida, S. Uemura, T. Kodzasa, N. Takada, T. Kamata, and K. Yase, "Influence of moisture on device characteristics of polythiophene-based field-effect transistors", Journal of Applied Physics **95**, 5088 (2004).

47 G. Horowitz, "Organic thin film transistors: from theory to real devices", Journal of Materials Research **19**, 1946 (2004).

48 G. Horowitz, M. E. Hajlaoui, and R. Hajlaoui, "Temperature and gate voltage dependence of hole mobility in polycrystalline oligothiophene thin film transistors", Journal of Applied Physics **87**, 4456 (2000).

49 M. C. J. M. Vissenberg and M. Matters, "Theory of the field-effect mobility in amorphous organic transistors", Physical Review B **57**, 12964 (1998).

50 N. Tessler and Y. Roichman, "Two-dimensional simulation of polymer field-effect transistor", Applied Physics Letters **79**, 2987 (2001).

51 K. P. Pernstich, S. Haas, D. Oberhoff, C. Goldmann, D. J. Gundlach, B. Batlogg, A. N. Rashid, and G. Schitter, "Threshold voltage shift in organic field effect transistors by dipole monolayers on the gate insulator", Journal of Applied Physics **96**, 6431 (2004).

52 R. W. Zehner, B. F. Parsons, R. P. Hsung, and L. R. Sita, "Tuning the work function of gold with self-assembled monolayers derived from X-[C_6H_4-C C-](n)C_6H_4-SH ($n = 0, 1, 2$; X = H, F, CH_3, CF_3, and OCH_3)", Langmuir **15**, 1121 (1999).

53 R. J. Chesterfield, J. C. McKeen, C. R. Newman, P. C. Ewbank, D. Filho, J.-L. Bredas, L. L. Miller, K. Mann, and C. D. Frisbie, "Organic thin film transistors based on *N*-alkyl perylene diimides: Charge transport kinetics as a function of gate voltage and temperature", Journal of Physical Chemistry B **108**, 19281 (2004).

Index

Organic Electronics, Materials, Manufacturing and Applications. Edited by Hagen Klauk

ISBN: 3-527-31264-1

r